W9-AFU-197

HYDROLOGY AND THE MANAGEMENT OF WATERSHEDS

A lake scene in northern Minnesota.

Third Edition

A waterfall in Guizhou Province, China, illustrates both the beauty and power of flowing water.

Third Edition

HYDROLOGY AND THE MANAGEMENT OF WATERSHEDS

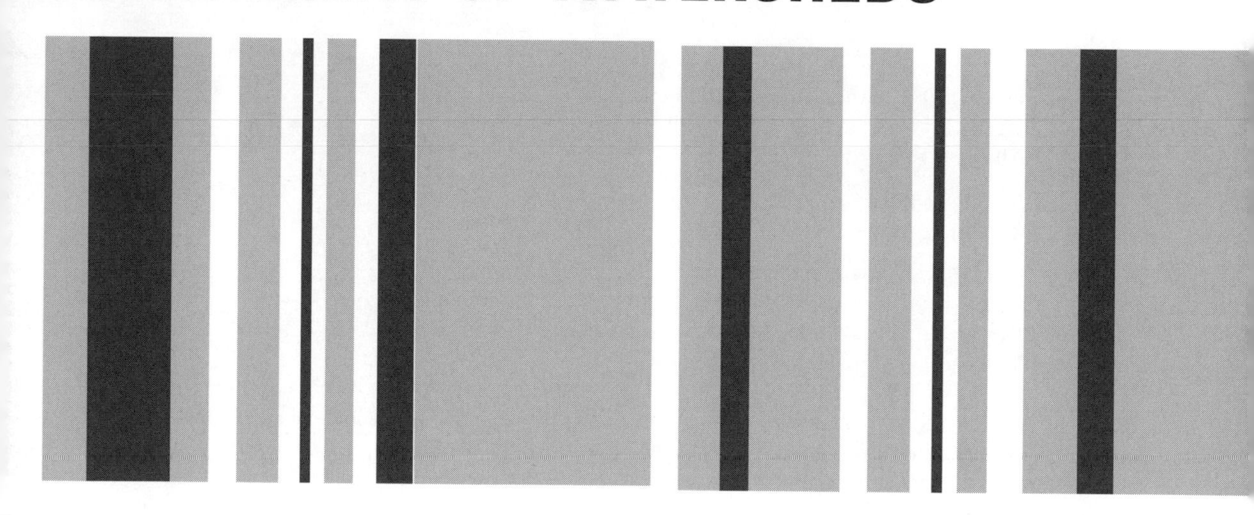

Kenneth N. Brooks
Peter F. Ffolliott
Hans M. Gregersen
Leonard F. DeBano

Iowa State University Press / Ames

To those who manage land and water resources

Kenneth N. Brooks is professor of forest hydrology, Department of Forest Resources, College of Natural Resources, University of Minnesota, St. Paul.

Peter F. Ffolliott is professor of watershed management, School of Renewable Natural Resources, College of Agriculture, University of Arizona, Tucson.

Hans M. Gregersen is professor emeritus with a joint appointment in the Department of Forest Resources and the Department of Agricultural and Applied Economics, University of Minnesota, St. Paul.

Leonard F. DeBano is professor of watershed management, School of Renewable Natural Resources, University of Arizona, Tucson.

Iowa State Press
2121 State Avenue, Ames, Iowa 50014

Orders: 1-800-862-6657
Office: 1-515-292-0140
Fax: 1-515-292-3348
Web site: www.iowastatepress.com

Printed on acid-free paper in the United States of America

First edition, 1991
Second edition, 1997
Third edition, 2003

Library of Congress Cataloging-in-Publication Data

Hydrology and the management of watersheds / Kenneth N. Brooks — [et al.] — 3rd ed.
p. cm.
Includes bibliographical references and index.
ISBN 0-8138-2985-2
1. Watershed management. 2. Watershed management — Economic aspects.
3. Watershed management — Social aspects. I. Brooks, Kenneth N.
TC409 .H93 2002
627 — dc21 2002008715

Last digit is the print number: 9 8 7 6 5 4 3 2

CONTENTS

Watersheds are managed for multiple purposes, including water, forage for livestock, and habitats for wildlife such as this chaparral shrub-covered watershed in the southwestern United States.

PREFACE

The third edition of this book is a major revision of the earlier two editions and is based largely on feedback we have received from those using the book in the classroom. The introduction and basic hydrology chapters in Part 1 have been updated, but the topics discussed are relatively unchanged from the second edition. Part 2 includes the previous chapters on erosion and sediment yield but has expanded the discussion of stream channel processes, morphology, and classification. A new Part 3 contains a chapter on basic water quality and a new chapter on water quality management. The topics of riparian and wetland hydrology and management have been expanded into two new chapters that comprise Part 4. Part 5, Special Topics, includes topics that may be of interest to managers and is intended to complement the earlier parts of the book. We recognize that snow hydrology and engineering applications may not be an integral part of many courses in hydrology and watershed management, but they are important topics to many. Chapter 17, Hydrologic Methods, remains as a highly technical chapter that can be used to quantify some of the hydrologic relationships discussed in earlier chapters. Similarly, Chapter 18 presents tools for analysis and research in watershed management. A new Part 6 is comprised of a single, comprehensive chapter that condenses and updates material from three chapters in the second edition that focus on socioeconomic considerations. Again, some courses in hydrology and watershed management may not explicitly deal with socioeconomic topics, but the authors feel that the topics of policy, planning methods, and economic assessments are essential components of any comprehensive book on watershed management.

The authors thank Clara M. Schreiber for her dedicated work and many contributions in the preparation of this third edition. The authors also want to acknowledge the excellent work of David L. Rosgen and Mark S. Riedel, who provided many of the ideas for and reorganization of the chapters on stream channel processes and morphology. We also thank E.S. (Sandy) Verry, whose contributions show up throughout this book, but are particularly evident in the chapters on wetlands and riparian systems. Thanks again are extended to Ben and Nancy Cole for the new figures developed for the third edition.

Stream measurements are essential for understanding hydrology and the effects of land use.

Student measuring suspended sediment with a DH-48 sampler.

Students measuring channel width.

DEFINITION OF TERMS

Hydrology is the science of water concerned with the origin, circulation, distribution, and properties of waters of the earth.

Forest hydrology/range hydrology/wildland hydrology refer to a branch of hydrology that deals with the effects of vegetation and land management, in respective settings, on water quantity and quality, erosion, and sedimentation.

Watershed, or catchment, is a topographically delineated area drained by a stream system; that is, the total land area above some point on a stream or river that drains past that point. The watershed is a hydrologic unit often used as a physical-biological unit and a socioeconomic-political unit for the planning and management of natural resources.

River basin is similarly defined but is of a larger scale. For example, the Mississippi River Basin, the Amazon River Basin, and the Congo River Basin include all lands that drain through those rivers and their tributaries into the ocean.

Watershed management is the process of organizing and guiding land and other resource use on a watershed to provide desired goods and services without adversely affecting soil and water resources. Embedded in the concept of watershed management is the recognition of the interrelationships among land use, soil, and water and the linkages between uplands and downstream areas.

Watershed management practices are those changes in land use, vegetative cover, and other nonstructural and structural actions taken on a watershed to achieve watershed management objectives.

HYDROLOGY AND THE MANAGEMENT OF WATERSHEDS

Water carves its way through the landscape in its journey to the ocean.

Third Edition

CHAPTER 1 *Introduction*

Our perspective of watershed management is that water and land resources must be managed in concert with one another. While hydrology and water quality are essential components of watershed management and are the subjects of much of this book, we also recognize the importance of land productivity, ecosystem functions, and stream channel morphology as integral parts of watershed management. That is, watershed management deals not only with the protection of water resources but also with the capability and suitability of land and vegetative resources to be managed for the production of goods and services in a sustainable manner. Few watersheds in the world are managed solely for the production of water. Some municipal and power company watersheds are the exception. Regardless of the management emphasis, watersheds serve as logical and practical units for analysis, planning, and management of multiple resources.

A basic understanding of hydrology is fundamental to the planning and management of renewable natural resources for sustainable use on a watershed. Hydrology enters explicitly and directly into the design of water resource projects, including reservoirs, flood control structures, navigation, irrigation, and water quality control. A knowledge of hydrology helps us in our attempt to balance demands for water with supplies, to avoid flood damages, and to protect the quality of streams and lakes. Therefore, basic hydrology and the hydrologic effects of land use are the subjects of Part 1.

One of our concerns, and an incentive for writing this book, is that hydrology is not always considered in the management of forests, rangelands, and agricultural croplands, even though it should be! Ignoring the effects of land management activities on soil and water resources is shortsighted and can lead to unwanted effects on a particular site and on downstream areas. Many of these effects are discussed in other parts of the book. For example, soil erosion can lead to losses of plant productivity and/or soil stability. Changes in vegetation and soils can alter streamflow quantity and quality.

Streamflow and sediment flow changes can consequently alter stream channel morphology, reducing the stability of rivers. Because of the growing importance of wetlands and riparian systems in watershed management, we have added a new Part 4 in this edition. Recognizing that the audience for this book is diverse, we have included Part 5 to address special topics that are of interest to particular regions and topical areas. Snow hydrology, watershed considerations for engineering applications, commonly applied hydrologic methods, tools for analysis and research, and socioeconomic considerations in watershed management are included. Also presented are methods to quantify hydrologic responses to land management practices. This part also serves as a supplement to parts 1 through 4.

In the past decade, there has been an emergence of locally led watershed management organizations and initiatives globally. Various names are used to describe these active management arrangements: watershed partnerships and associations, participatory rural watershed councils, and "co-management" schemes managed jointly by government and communities, for example. They all emphasize the growing awareness of the important hydrologic linkages between uplands and downstream areas. Furthermore, there is a growing recognition that institutional and policy changes are often needed to cope with these linkages and to develop sustainable solutions to many natural resource problems while, at the same time, avoiding many kinds of environmental degradation. Therefore, Chapter 19 focuses on the institutional, economic, and policy considerations involved in watershed management.

Watershed Management Strategies and Responses to Problems

Watershed management involves an array of nonstructural (vegetation management) and structural (engineering) practices. Soil conservation practices and land use planning activities can be tools employed in watershed management, as can building dams, establishing protected reserves, and developing regulations to guide road-building, timber-harvesting, agroforestry practices, and other types of activities. The unifying focus in all cases is on how these various activities affect the relationship between water and other natural resources on a watershed. The common denominator, or the integrating factor, is water. This focus on water and its interrelationship with other natural resources and their use is what distinguishes watershed management from other natural resource management strategies.

On the one hand, watershed management is an integrative way of thinking about human activities on a given area of land (the watershed) that have effects on, or are affected by, water. On the other hand, watershed management includes a set of tools or techniques—the physical, regulatory, or economic means for responding to problems or potential problems involving the relationship between water and land uses. What sometimes confuses people is the fact that these tools or techniques are not employed by a "watershed manager," as such, in most cases, but rather by foresters, farmers, soil conservation officers, engineers, and so forth.

This fact is both the dilemma and the strength of watershed management. In practice, activities using natural resources are decided upon and undertaken by individuals, local governments, and various groups that control land in a political framework that has little relationship to, and often ignores, the boundaries of a watershed. For example, in Figure 1.1, the forested uplands could be under the control of a federal or state forestry agency, but the middle elevations and lowlands could be a composite of private, city, and community ownerships. Activities are undertaken independently, often with little regard to how they affect other areas. Yet, despite this real world of disaggregated, independent political and economic actions, it remains a fact that water and its constituents flow downhill and ignore political boundaries. What one person or group does upstream can affect the welfare of those downstream. Somehow, the physical facts of watersheds and the political realities have to be brought together. That is the focus of *integrated watershed management*. Within this broad focus, there is concern with both how to prevent deterioration of an existing sustainable and productive relationship between the use of water and other natural resources and how to restore or create such a relationship where it has been damaged or destroyed in the past.

Watershed management actions and activities, therefore, are employed in *preventive strategies* aimed at preserving existing sustainable land use practices or in *restorative strategies* designed to overcome identified problems or restore conditions to a desirable level, where "desirable" is defined in both environmental and political terms. Both strategies respond to the same types of problems; however, in one case the objective is to prevent a problem from occurring, while in the other case, the objective is to restore conditions once the problem has occurred. In reality, we are dealing with a continuum, going from regulatory support and reinforcement of existing sustainable land

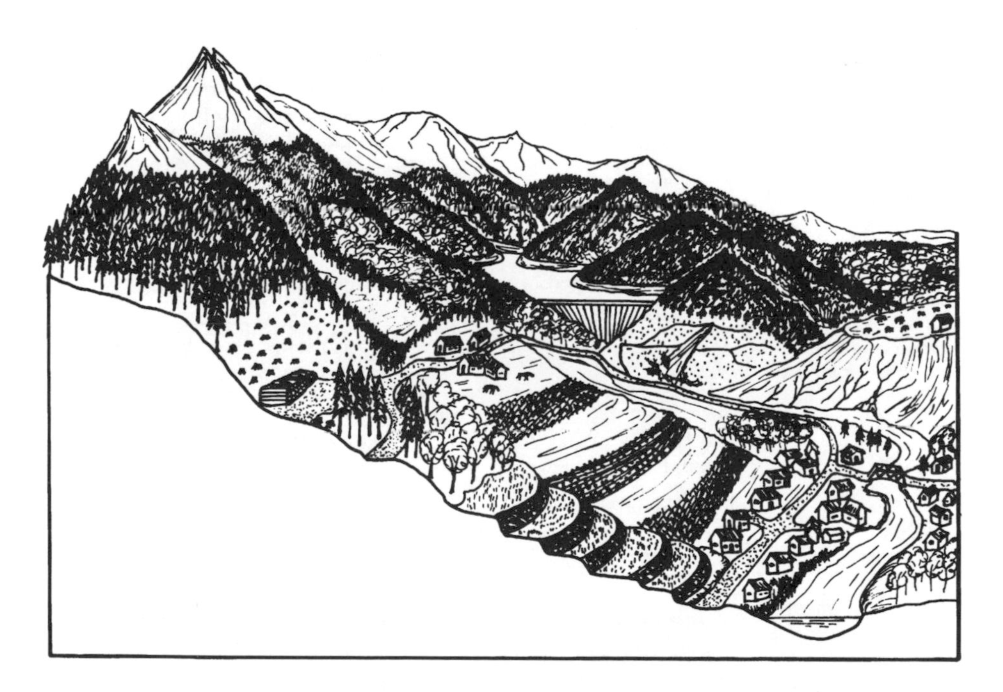

Figure 1.1. Many different types of land use can occur on a watershed; each affects a particular site but also has potential impact on downstream areas as well (redrawn from Food and Agriculture Organization of the United Nations, 1986).

use practices (preventive strategy) to emergency relief, building of temporary gully control structures, or restricting land use on fragile, eroded lands (restorative strategy). The point to stress is that routine preventive strategies and actions are fully as important as the more dramatic and visible restorative actions. Losses avoided (through preventive strategies) can be just as important to people as gains from solving a problem. In economic terms, the cost of preventing losses of productivity in the first place can be much lower than the cost of achieving the same benefit through more dramatic actions to restore productivity on problem lands that have already been degraded.

With this as background, Table 1.1 summarizes the most common situations encountered and alternative solutions (preventive or restorative) for overcoming them as they relate to watershed management objectives.

TABLE 1.1. The role of watershed management in developing solutions to natural resource problems

Problem	Possible Alternative Solutions	Associated Watershed Management Objectives
Deficient water supplies	Reservoir storage and water transport	Minimize sediment delivery to reservoir site; maintain watershed vegetative cover
	Water harvesting	Develop localized collection and storage facilities
	Vegetative manipulation; evapotranspiration reduction	Convert from deep-rooted to shallow-rooted species or from conifers to deciduous trees
	Cloud seeding	Maintain vegetative cover to minimize erosion
	Desalinization of ocean water	Not applicable
	Pumping of deep groundwater and irrigation	Management of recharge areas
Flooding	Reservoir storage	Minimize sediment delivery to reservoir site; maintain watershed vegetative cover
	Construct levees, channelization, etc.	Minimize sediment delivery to downstream channels
	Flood plain management	Zoning of lands to minimize human activities in flood-prone areas; minimize sedimentation of channels
	Revegetate disturbed and denuded areas	Plant and manage appropriate vegetative cover
Energy shortages	Utilize wood for fuel	Plant perpetual fast-growing tree species; maintain productivity of sites; minimize erosion
	Develop hydroelectric power project	Minimize sediment delivery to reservoirs and river channels; sustain water yield

Problem	Possible Alternative Solutions	Associated Watershed Management Objectives
Food shortages	Develop agroforestry	Maintain site productivity; minimize erosion; promote species compatible with soils and climate of area
	Increase cultivation	Restructure hill slopes and other areas susceptible to erosion; utilize contour plowing, terraces, etc.
	Increase livestock production	Develop herding/grazing systems for sustained yield and productivity
	Import food from outside watershed	Develop forest resources for pulp, wood and wildlife products, etc., to provide economic base
	Erosion control structures	Maintain life of structures by revegetation and management
	Contour terracing	Revegetate, mulch, stabilize slopes, and institute land use guidelines
	Revegetate	Establish, protect, and manage vegetative cover until site recovers
Poor quality drinking water	Develop alternative supplies from wells and springs	Protect groundwater from contamination
	Treat water supplies	Filter through wetlands or upland forests
Polluted streams/ reduced fishery production	Control pollutants entering streams	Develop buffer strips along stream channels; maintain vegetative cover on watersheds; develop guidelines for riparian zones
	Treat wastewater	Use forests and wetlands as secondary treatment systems for wastewater

Watershed Management: A Global Perspective

There are many examples of the need to manage watersheds better to meet the demands for water and other natural resources in a more sustainable manner. It must be recognized, however, that practices relating to resource use and management around the world do not depend solely on the physical and biological characteristics of watersheds. Institutional, economic, and social factors, such as the cultural background of rural populations and the nature of governments, need to be fully integrated into viable solutions that meet environmental, economic, and social objectives. How these factors are interrelated can best be illustrated by looking at specific issues and examples.

Land and Water Scarcity

Arable land and water resources are becoming more scarce with the earth's expanding human population. These scarcities and the human responses to these scarcities pose challenges to sustainable development and can have serious environmental consequences. Changing weather patterns and climatic conditions add uncertainty to future land and water resource management. It is unclear where and how much freshwater supplies will change with changes in climate, such as global warming. What is clear is that the increasing water demands caused by a changing population and economic development will pose far greater problems by the year 2025 (Vorosmarty et al. 2000).

Knowledge, information, and technologies will no doubt expand our capability to deal with shortages of resources. To harness these capabilities, however, will require an integrated and interdisciplinary approach to planning and managing natural resources. The lack of such an approach has been problematic, as emphasized by Falkenmark (1997), who stated, "In environmental politics, land and water issues are still seen as belonging to different worlds, taken care of by different professions with distinctly different education and professional cultures." As a consequence, land and water resources have traditionally been managed in isolation from one another. Yet, we recognize that land use impacts the quantity and quality of water flow in a watershed. Water development, conversely, affects land use. These relationships and linkages are the focus of much of this book.

Water scarcity has attracted attention globally and is considered the major environmental issue facing the 21st century. On March 22, 2001, the United Nations commemorated a World Day for Water, in which speakers concluded that demands for freshwater exceeded supplies by 17%, and over the next 25 yr, two-thirds of the world's population will experience severe water shortages. Although the 9000 to 14,000 km^3 of freshwater on earth should be sufficient to support expanding human populations for the foreseeable future, unequal distribution results in water scarcity (Rosegrant 1997). For example, per capita freshwater supplies in Canada are approximately 120,000 m^3, compared to Jordan's 300 m^3. In China, 400 of its 668 major cities have inadequate water supplies, even though the country encompasses the Yangtze River, the third largest river in the world. To enhance water supply in northern China, where more than one-third of the country's population lives, a 2400 km channel is planned to divert 16 billion m^3 of water annually from the Yangtze River Basin in the south (Environmental Data Interactive Exchange 2001). Other diversions are planned for western regions of the country. Such programs have dramatic effects on land and water, placing greater importance on improved watershed planning and management to protect the life of major reservoir and diversion investments. Extensive watershed management programs have already begun in the upper and middle Yangtze River Basin to complement and protect the $25 billion Three Gorges Reservoir and Dam.

Land scarcity has been exacerbated in many developing countries by the rural poor who clear forests, cultivate steep uplands, and overgraze the land to meet their food and natural resource needs. Watershed degradation often results, further reducing land productivity, which in turn causes more extensive and intensive land use. In other instances, irrigation practices intended to expand land productivity have been inappropriate and have caused land productivity to diminish through salinization. As watershed conditions deteriorate, those living in both upland and downstream areas become impacted. This cycle of deteriorating land productivity is a clear indicator of nonsustainable land use. Of the 8.7 billion ha of agricultural land, forest, woodland, and rangeland worldwide, over 22% has been degraded since the mid-1900s, with 3.5% being severely degraded (Scherr and Yadav 1996). Problems of desertification receive global attention as the productivity of some of the poorest countries and regions in the world continue to decline (Box 1.1). Ironically, we have the know-how to manage resources in a sustainable manner. What is most often lacking are the policies and institutions to promote and sustain sound resource use.

Box 1.1

Desertification: A Global Issue

Drylands cover more than one-third of the earth's land surface, and slightly over half that area is inhabited by over 900 million people. The remainder is climatically so arid and unproductive that it cannot support human life. The degradation of land and water resources by human activities, however, is turning potentially productive drylands into unproductive deserts in Asia, Africa, and America. This process is called *desertification*. It has been estimated that a total area larger than Brazil, with rainfall above the level classified as semiarid, has been degraded to desertlike conditions.

About 65 million people in the developing countries live on the dryland interface between deserts and more humid areas. Desert encroachment in the Sahelian region of West Africa has received the greatest international attention. More than 650,000 km^2 of land suitable for agriculture or grazing have undergone desertification in West Africa during the last 50 yr of the 20th century.

Source: United Nations Environment Programme, 1992.

Coping with Hydrometeorological Extremes: Role of Watershed Management

Floods, landslides, and debris torrents that result from excessive precipitation and the droughts that result from deficient precipitation represent both dimensions of hydrometeorological extremes. The disasters and famine that can result present some of the greatest challenges to land and water resource managers. Even though we call them extreme events, it should be recognized that floods and droughts occur naturally, will reoccur, and are not necessarily rare. Forecasting when and where they will occur, and at what magnitude, is at best uncertain. Whether such extremes end up being disasters, however, depends upon their impact on humans and on natural ecosystems. These aspects can be affected, to some extent, by our use of natural resources. What can be done to better cope with such extreme events? This is a question that natural resource planners and managers, as well as students of hydrology and watershed management, should be prepared to address.

Floods, landslides, and debris torrents result in billions of dollars being spent each year globally for flood prevention, flood forecasting, and hillslope stabilization. Yet the cost of lives and property damage due to floods, landslides, and debris flows is staggering. The impacts of these naturally occurring phenomena can be exacerbated by human encroachment on floodplains and other hazardous areas, which is often the result of land scarcity (Box 1.2).

Responses to disasters caused by too much water, as described above, or to too little water, resulting from droughts, are most often short term and limited to helping those who have been directly impacted rather than dealing with the causes. Interest in coping with extreme events flourishes immediately after they have taken their toll on human lives and property but becomes lower priority, with reduced funding, as memories fade with time. The term *crisis management* captures this approach. It is not unreasonable to

Box 1.2

Typhoons, Landslides, and Debris Flows: An Example in Taiwan

Between three and four typhoons strike the island of Taiwan each year, resulting in excessive rainfall that results in widespread landslides, debris flows, and flooding. The island is about 75% mountainous, with over 50% of the land area having slopes steeper than 20 degrees. Despite forest exploitation earlier in the 20th century, Taiwan now manages its almost 60% forest cover primarily for watershed protection with an emphasis on slope stabilization. Extensive engineering structures on hillslopes and in channels have been used throughout the country to further reduce landslides and debris flows. Even so, the heavy rainfall that accompanies typhoons saturates shallow soils on steep slopes and triggers landslides. Steep channels in the mountains become loaded with material from these landslides, causing frequent debris flows. Typhoon Herb, which occurred in 1996, was one of the most destructive typhoons on record, resulting in widespread death and destruction from landslides, debris flows, and flooding. Several rainfall gauges measured over 1300 mm of rainfall in less than two days. One of the high elevation gauges recorded 1987 mm of rainfall in 42 hr. Such events will trigger debris flows and floods no matter what the condition of hillslopes and watersheds. The disaster that resulted, however, must be attributed to human occupation on hillslopes and within steep drainage channels and flood plains as much as to the heavy rainfall. With Taiwan's population exceeding 600 people/km^2, land scarcity compounds the ability to cope with such disasters.

ask what, if anything, can be done to prevent such disasters. Different approaches may be needed for any given situation. To paraphrase Davies (1997), there are essentially three options to reduce or prevent disasters caused by excessive amounts of water (e.g., debris flows): (1) modifying the natural system, (2) modifying the human system of behavior, and (3) some combination of (1) and (2). It is important to recognize that *no matter how advanced our technologies*, the degree to which such events result in disasters is largely due to human behavior on the watershed.

Attempts to control events by modifying natural systems most often entail the use of engineering measures, for example, reservoirs, levees, and channelization aimed at "controlling" floods (see Table 1.1). Today there are increasing attempts to apply bioengineering and vegetative management measures along with structures to exert some measure of control over extreme events, although it is unrealistic to believe that we can mitigate the most extreme of hydrometeorological events. The development of any such measure requires a good understanding of natural systems and processes, how these systems function, and how mitigation actions impact other aspects of the environment.

Generally, we can view the options presented in Table 1.1 either as those that attempt to move or control water or those that attempt to move or manage people on the watershed. Furthermore, some of the measures listed tend to be the responsibility of water resource organizations with a strong engineering bias (governmental agencies, private utilities companies, irrigation projects, etc.), while others are in the domain of land management organizations with a strong land production or ecological viewpoint. Too often the separation of responsibilities and the lack of coordination lead to piecemeal programs that result in unwanted effects and furthermore, do not have the long-term perspective that is needed to develop sustainable solutions.

While it is true that land use might not affect the magnitude of the most extreme hydrometeorological events, human activities on the watershed and the cumulative effects of these activities determine the extent to which the events impact human life and welfare. The heavy reliance on large engineering structures to store or divert water with the intention of meeting water management objectives has for some time been met with objections from environmental groups. There is now concern by some professionals that an overreliance on engineering measures (dams, levees, channelization, etc.) has (1) led to unintended and unwanted effects, (2) reduced the hydrologic function and environmental values associated with natural rivers, floodplains, and estuaries, and (3) imparted a false sense of security to those living in downstream areas. Urban and agricultural expansion has further reduced natural wetlands, riparian systems, and floodplains. Postflood assessments of the 1993 Mississippi River flood emphasized some of these points (Box 1.3).

The aftermath of the record flood in the Upper Mississippi River in 1993, the 1997 rain-on-snow flood in the Upper Minnesota River and Red River of the North along the Minnesota–North Dakota border, and the widespread flooding along the Upper Mississippi River and its tributaries in 2001 led to questions concerning the extent to which the floods were the result of human modifications of watersheds. Since the beginning of the 20th century, efforts designed largely to expand farmland have resulted in extensive drainage of wetlands and the use of tile drains on farmlands. Flood control has involved levee construction and channelized rivers, with a resulting loss of riparian forests. The extent to which land use and channel modifications affected these floods was addressed in technical forums held at the University of Minnesota in 1997 and 2000. The forums concluded:

- Agricultural drainage was not an important contributor to the large magnitude floods of 1993 and 1997; the excessive precipitation that occurred would have caused floods no matter what the watershed condition.
- Channelization of natural streams has a more dramatic effect on peak flows for small watersheds and for average annual events than for large watersheds and extreme events.
- Drainage of wetlands can increase peak flows associated with the average annual flood.

These conclusions have subsequently been questioned. The importance of water storage by wetlands and by natural floodplains is discussed in greater detail in chapters 10 and 11.

Watersheds, Ecosystem Management, and Cumulative Effects

Taking a watershed perspective has utility in addressing many natural resource issues that cross different climatic and topographic regions and that occur over long periods. There is direct and obvious merit in taking a watershed perspective when dealing with issues of water supply and events such as floods and droughts. Not so obvious are the effects that become compounded spatially and temporally and that at first glance might

Box 1.3

The 1993 Flood on the Upper Mississippi River, USA

The Mississippi River flood of 1993 caused over $10 billion in damages (Tobin and Montz 1994), damaging or destroying more than 50,000 homes and causing some 54,000 people to be evacuated from flooded areas (NOAA 1994). It was the most devastating flood in the United States in the 20th century.

Following this flood, questions were raised by the public and by resource professionals concerning the extent to which changes on the land (urbanization, drainage of wetlands, etc.) contributed to the flooding. (Relationships between land use and flooding are discussed in Chapter 6.) Furthermore, given the extensive water resource management and engineering works in the basin, many were surprised at the magnitude and extent of damages in the basin. The extensive use of levees and river channelization measures for reducing flood damages might actually be offset by the loss of flood reduction benefits of natural meandering rivers, riparian systems, and floodplains (these relationships are discussed and explained in chapters 9 and 10).

Postflood evaluations pointed out that the flood was the result of abnormally heavy rainfall in the fall of 1992, which resumed in the spring of 1993 and persisted through mid-summer over large areas of the Upper Mississippi River Basin. More than 1000 mm of rainfall fell over extensive areas of the basin during this period. Record flood stages were measured at 44 locations in the Upper Mississippi River system, at 49 locations in the Missouri River system, and two locations in the Red River of the North system. Leopold (1994) found that levees in several locations caused river stages to be higher than they would otherwise be without levees being in place. As long as levees are not washed out or breached, the increased height of flood stages in the channel does not cause damage. However, when levees fail or are breached, as occurred in 1993, the increased stages result in greater devastation than would have occurred without levees. Another major flood in the spring of 2001 exceeded 1993 stages at some locations in the Upper Mississippi River. Damages were less at several of these locations, however, as many people did not rebuild on floodplains following the 1993 flood.

In reviewing this flood it can be concluded that the flood would have occurred no matter what types of land use had taken place in the basin. Hydrometeorological conditions caused this flood. A question that remains unanswered is the extent to which the cumulative effects of land use and channel changes have set the stage for more frequent flooding and downstream damages from events of a lesser magnitude. Subsequent chapters in this book provide some insight into this question.

not appear to be related to water and hydrologic processes. Examples are presented throughout this book that relate to these ideas.

During the past century in the United States, land use and vegetation changes on the landscape have dramatically altered the hydrologic characteristics of watersheds. Since the earliest European settlers, extensive areas of native grasslands and forests have been converted to agricultural crops. Urban areas and road systems have expanded. Wetlands have been drained, riparian areas altered, and natural river systems modified. These changes on the land are transformed into hydrologic changes through modifications of watersheds, their stream systems, and surface-groundwater linkages. As a result, water yields and the pattern of streamflow in many areas have been altered. We see adjustments in river channels and changes in water quality. The challenge to watershed management is that of mitigating the effects of all these changes that adversely impact human and natural ecosystems. Greater attention is now being paid to maintaining and restoring natural stream channel systems, riparian systems, wetlands, and floodplains as integral components needed to maintain watersheds in good hydrologic condition. Hey (2001) called for a major program in the upper Mississippi River basin that would maximize natural storage of wetlands and floodplains and minimize conveyance; such a program would, in effect, reverse the past 200 yr of engineering practice in the basin. The rationale for such a program is provided in chapters 9, 10, 13, and 14.

The role of watersheds as units for ecosystem management and analysis has gained international recognition in the past decades. *Ecosystem management*, a focus of natural resource management in the 1990s, is based on maintaining the integrity of ecosystems while sustaining benefits to human populations. The usefulness of a watershed approach in ecosystem management became apparent in the United States with the attempts to reconcile conflicts over the use and management of land and water in the Pacific Northwest. Environmental, economic, and political conflicts involving the use of old-growth forests for both spotted owl habitats and the production of timber were, and still are, heated. At the same time, concern over the effects of timber harvesting and hydropower dams on native salmon runs emerged. A watershed analysis approach facilitated the examination of the interrelationships between land and water use and many of the environmental effects. This approach was taken in the Pacific Northwest because watersheds "define basic, ecologically and geomorphologically relevant management units" and "watershed analysis provides a practical analytical framework for spatially explicit, process oriented scientific assessment that provides information relevant to guiding management decisions" (Montgomery et al. 1995, p. 371).

In Madagascar, efforts to establish a system of national parks and reserves to protect unique ecosystems were being threatened by the encroachment of shifting cultivators and poor land use practices that were depleting adjacent and surrounding forests, resulting in excessive soil erosion. One such area was the Andohahela Reserve, created in 1939. This reserve contains many endemic plants and animals unique to Madagascar and is the only reserve that includes two of the island's major vegetational formations: the eastern rain forest and the southern spiny bush. In addition to being an important center of biodiversity, the 76,000 ha reserve is also a key watershed for much of southeastern Madagascar. By taking a watershed perspective, government officials felt that a more interdisciplinary and comprehensive management plan could be developed for the reserve and the region overall, one that would account for the economic and environmental importance of the watershed to local people and downstream communities (Fenn 1993).

A characteristic effect that many types of land use can have on various environmental parameters is embedded in many of the above examples. We are referring to cumulative watershed effects, which are the combined environmental effects of activities in a watershed that can adversely impact beneficial uses of the land (Reid 1993, Sidle 2000). The view taken here is that there are interactions among different land use activities and that there can be incremental effects, which when added to past effects, can lead to more serious overall impacts. Individually, such environmental effects might not appear to be relevant, but collectively, over space and time, they can become significant. For example, conversion from forest to agricultural croplands in one part of a watershed can cause an increase in water and sediment flow. Road construction and drainage can have similar effects elsewhere in the watershed, as can the drainage of a wetland at another location. These activities can occur over a long period, and incrementally, each can have little obvious effect. At some point in time, however, the increased volumes of streamflow discharge, in combination with additional sediment loads, can lead to more frequent flooding in some channel reaches. River channels adjust to these changes in water flow and sediment, causing additional impacts downstream. These same changes can alter the aquatic habitat over time and can have unwanted effects. A watershed perspective forces one to look at multiple and cumulative effects in a unit and attempt to identify key linkages between terrestrial and aquatic ecosystems. Many such cumulative effects will be discussed throughout this book.

Reconciling Watershed and Political Boundaries: Emergence of Watershed Management Institutions

Historically, land and water have been managed not only by people with different technical backgrounds, but by organizations that have very focused and different missions that impact one another. There are often overlapping responsibilities. In contrast, it is rare to find national organizations that have the explicit mission of watershed management. As a result, there are few organizations with responsibilities that coincide with watershed boundaries. Yet there is a need to cope with issues that arise between upstream and downstream entities. The absence of watershed management organizations and institutions has resulted in an emergence of initiatives and organizations to plan and manage resources and to resolve upstream-downstream conflicts that arise.

By the late 1990s, more than 1500 locally led initiatives in watershed management had emerged in the United States in response to water resource problems that were not being addressed (Lant 1999). These initiatives were largely in response to the issues discussed above, including the effects of urbanization, intensification of agriculture, forest management activities, stream channelization, and wetland drainage on flooding, water yield, and water quality. Furthermore, we are seeing a shift in emphasis from federally funded engineering solutions and technological regulations to watershed management, water quality improvement, and restoration of river channels and wetlands. An example of an institutional response to New York City's water supply is presented in Box 1.4.

Institutional responses to international water issues are also emerging. Transboundary issues in the Nile Basin have focused attention on watersheds as a logical framework for analysis and understanding of the effects of human actions in one part of the basin on those living downstream. Here, concerns about water, human poverty, biodiversity, and wildlife are paramount. The disparity between watershed and political boundaries prompted the Nile Basin Initiative and an extensive study into transboundary issues of the Nile River Basin. The inequities in rainfall and water supply among the 10 riparian countries of the Nile River have threatened sustainable develop-

Box 1.4

New York City Watershed Management Strategy (National Research Council 2000)

On January 21, 1997, the New York City Watershed Memorandum of Agreement (MOA) was signed, providing a legal framework for protecting the drinking water supply of New York City's nine million people. Three main provisions in this MOA are (1) the purchase of watershed lands that are largely privately owned and supply 90% of the city's water supply, (2) key environmental regulations to protect water quality, and (3) payments to watershed inhabitants and communities for watershed protection and local economic development. Comprehensive land use planning is required that involves three watersheds that total 1970 mi^2, contain 600 billion gal of useable storage, and provide 2 billion gal/day of water.

A land acquisition program targeted 355,000 acres of "sensitive lands" on a willing buyer–willing seller basis. Conservation easements were an option for those not willing to sell their land. Components of the overall program included (1) a total maximum daily load (TMDL) program targeting phosphorus, (2) water treatment options, (3) a nonpoint pollution control program, and (4) rules and regulations for buffer zones and setback distances for separating waters from potential sources of pollution. This strategy encompasses the key elements of watershed management programs, the explicit consideration of sustaining production with environmental protection and dealing with the inequities that can arise when those on one part of watershed are asked to change land use for the benefit of downstream inhabitants.

ment in the region and have historically resulted in political conflict (Box 1.5). Developing solutions to water supply problems, although a necessary and daunting task, is not sufficient. The growth and distribution of human populations, widespread poverty, and inadequate policy responses to land and water resource issues compound problems in the basin. A major theme in the Nile Basin Initiative is that land and water must be managed in harmony so that environmental conditions and human welfare benefit. Integrated watershed management is one of the key themes in this effort.

Preventive Strategies: The Key to Watershed Management

The preceding discussion can leave the impression that watershed management most often involves the restoration of degraded lands. Problems of soil erosion, gully formation, localized flooding, or water shortages resulting from abusive land use call for action, usually in the form of engineering structures and widespread revegetation schemes. Much as in the popular press, the natural disasters and problems of human suffering associated with resource depletion and degradation get more attention than the

Box 1.5

The Nile River Basin: A Case for Watershed Management (Baecher et al. 2000)

The Nile River is the longest river of the world, traversing 6700 km from the rift valleys of East Africa, connecting with the Blue Nile from the Ethiopian highlands, and emptying into the Mediterranean Sea through the broad delta of Egypt. Annual rainfall amounts vary to the extreme within the Nile Basin. Areas in the Ethiopian highlands experience over 2000 mm of annual rainfall. These highland watersheds contribute from 60 to 80% of the water flow in the Nile River, yet represent less than 10% of the land area in the Nile Basin. Vast areas of the northern Basin, on the other hand, average less than 50 mm of annual rainfall. Consequently, the watersheds of Egypt and Sudan, which constitute over 75% of the Basin, contribute negligible flow to the Nile River.

Efforts to capture waters of the Nile Basin for economic development have led to environmental concerns and potential political conflict. Lake Nassar, created by the large Aswan Dam of the lower Nile River, is essential to Egypt's economy. However, the Lower Nile is vulnerable to water losses upstream. The controversial Jonglei Canal was proposed to enhance downstream water supplies by diverting water from the vast Sudd wetland (discussed in greater detail in Chapter 14) to the lower and drier north. In contrast, Ethiopia, as one of the poorest countries in the Basin, would like to expand its irrigation and hydropower production through construction of dams in the uplands. Any such projects would meet with objections from downstream riparian countries. Furthermore, such projects, whether in the uplands or lowlands, are threatened by intensive land use and watershed degradation that impairs the quantity and quality of streamflow. Forest cover in the Ethiopian Highlands decreased from more than 15% in the 1950s to less than 3% by 1990. Overgrazing and cultivation of hillslopes lead to soil erosion, which leads to both loss of productivity of the land and increased sediment delivery downstream. All these issues focus on the need for watershed management that extends beyond country boundaries.

quiet successes. Likewise, the mobilization of people and equipment to renovate devastated areas receives notice. It is sometimes an effective way to attract attention to serious problems; however, we do not want to leave the impression that watershed management comes into play only when problems arise.

A more proactive approach to watershed management involves establishing and sustaining preventive practices: this means instituting guidelines for land use and implementing land use practices on a day-to-day basis that result in long-term, sustainable resource development and productivity without causing soil and water problems. In fact, land and natural resource management agencies in the United States and many other countries have established policies that embody sound watershed management principles. These preventive measures, when listed and described, might not make for exciting reading, but in total they represent the ultimate goal of watershed management. To achieve this goal requires that managers recognize the implications of their actions and that they work effectively within the social and political setting in which they find themselves.

The role of watershed managers in any area should be to achieve the needed goods and services desired by society without adversely impacting long-term productivity and downstream communities and without causing unwanted environmental change. People who occupy watersheds must be an integral part of any watershed management solution. For example, creating new sources of water does not necessarily solve water shortages in arid areas; an increased use of new supplies or an influx of people from water-poor areas can quickly deplete new sources of water. Nor will reservoirs designed to store flood flows and attenuate flood peaks provide the solution to flooding. Floodplain occupancy is a continuing problem and one that cannot be solved only with hydrologic engineering practices.

The dilemma in watershed management is that land use changes needed to promote the survival of society over the long term can be at cross-purposes with what is essential to the survival of the individual over the short term. Requirements for food and natural resources today should not be met at the expense of future generations. Any discussion of sustainable natural resource development should consider watershed boundaries, the linkages between uplands and downstream areas, and the effects of land use practices on long-term productivity. Land use that is at cross-purposes with environmental capabilities cannot be sustained. However, sustainable productivity and environmental protection can be achieved with the integrated, holistic approach that explicitly considers hydrology and the management of watersheds.

References

Baecher, G.B., R. Anderson, B. Britton, K. Brooks, and J.J. Gaudet. 2000. *The Nile Basin, environmental transboundary opportunities and constraints analysis.* Report to the U.S. Agency for International Development. International Resources Group, Ltd., Washington, DC.

Davies, T.R.H. 1997. Using hydroscience and hydrotechnical engineering to reduce debris flow hazards. In *Debris-flow hazards mitigation: Mechanics, prediction, and assessment,* ed. C. Chen, 787–810. Proc. First International Conference, American Society of Civil Engineers, New York.

Environmental Data Interactive Exchange, www.edie.com. January 26, 2001.

Falkenmark, M. 1997. Society's interaction with the water cycle: A conceptual framework for a more holistic approach. *Hydrological Sciences-Journal-des Sciences Hydrologiques* 42(4): 451–466.

Fenn, M. 1993. Personal communication. September 16.

Food and Agriculture Organization of the United Nations. 1986. *Protect and Produce.* Rome: FAO.

Hey, D.L. 2001. Modern drainage design: the pros, the cons, and the future. Presentation in *Hydrologic Science: Challenges for the 21st Century.* Annual Meeting, American Institute of Hydrology, October 14–17, 2001, Bloomington, Minnesota.

Lant, C.L. 1999. Introduction, human dimensions of watershed management. *J. American Water Resour. Assoc.* 35:483–486.

Leopold, L.B. 1994. Flood hydrology and the floodplain. The Universities Council on Water Resources. *Water Resources Update* 94:11–14.

Montgomery, D.R., G.E. Grant, and K. Sullivan. 1995. Watershed analysis as a framework for implementing ecosystem management. *Water Resour. Bull.* 3:369–386.

National Oceanic and Atmospheric Administration (NOAA). 1994. *The great flood of 1993.* Natural Disaster Survey Report. U.S. Department of Commerce, Washington, DC.

National Research Council. 2000. *Watershed management for potable water supply, assessing the New York City strategy.* Washington, DC: National Academy Press.

Reid, L.M. 1993. *Research and cumulative watershed effects.* USDA For. Serv. Gen. Tech. Rep. PSW-GTR-141.

Rosegrant, M.W. 1997. Water resources in the twenty-first century: challenges and implications for action. *Food, Agriculture and the Environment Discussion Paper 20.* International Food Policy Research Institute, Washington, DC.

Scherr, S.J., and S. Yadav. 1996. Land degradation in the developing world: Implications for food, agriculture, and the environment in 2020. *Food, Agriculture, and the Environment Discussion Paper 14*, International Food Policy Institute, Washington, DC.

Sidle, R.C. 2000. Watershed challenges for the 21st century: A global perspective for mountainous terrain. In *Land stewardship in the 21st century: The contributions of watershed management*, tech. coords. P.F. Ffolliott, M.B. Baker, Jr., C.B. Edminster, M.C. Dillon, and K.L. Mora, 45–56. USDA For. Serv. Proceedings RMRS-13.

Tobin, G.A., and B.E. Montz. 1994. *The great midwestern floods of 1993.* New York: Saunders College Publ.

United Nations Environment Programme. 1992. *World atlas of desertification.* London: Edward Arnold.

Vorosmarty, C.J., P. Green, J. Salisbury and R.B. Lammers. 2000. Global water resources: Vulnerability from climate change and population growth. *Science* 289:284–288.

Rivers reflect the characteristics of their watersheds. The high sediment load in the Betsiboka River in Madagascar is indicative of soil erosion in the uplands.

PART 1

Hydrologic Processes and Land Use

The hydrologic cycle represents the processes and pathways involved in the circulation of water from land and water bodies to the atmosphere and back again. The cycle is complex and dynamic but can be simplified if we categorize components into input, output, and storage (see Fig. 2.1). Based on the principle of conservation of mass, inputs such as rainfall, snowmelt, and condensation must balance with changes in storage and outputs, which include streamflow, groundwater seepage, and evapotranspiration (inflow less outflow is equal to the change in storage). This hydrologic balance, or water budget, is an application of the conservation of mass law expressed by the equation of continuity:

$$I – O = \Delta S$$

where I = inflow; O = outflow; and ΔS = change in storage.

The water budget is both a fundamental concept of hydrology and a useful method for the study of the hydrologic cycle; this basic equation and its modifications allow the hydrologist to trace the pathways and changes in water storage in a watershed.

The quantities of water in the atmosphere, soils, groundwater, surface water, and other components are constantly changing because of the dynamic nature of the hydrologic cycle. At any one point in time, however, quantities of water in each component can be approximated. If we consider the total water resource on the earth, only about 2.6% is freshwater. About 77% of this freshwater is tied up in the polar ice caps and glaciers, and 11% is stored in deep groundwater aquifers, leaving about 12% for active circulation. Of this 12%, only 0.57% exists in the atmosphere and in the biosphere. The biosphere is from the top of trees to the lowest roots. The atmosphere redistributes evaporated water by precipitation and condensation. Components of the biosphere partition this water into runoff, soil and groundwater storage, groundwater seepage, and evapotranspiration back to the atmosphere.

The hydrologic processes of the biosphere and the effects of vegetation and soils on these processes are of particular interest in forest hydrology and watershed management.

Precipitation and the flow of water into, through, and out of a watershed all can be affected by land use and management activities. Likewise, human activities can alter the magnitude of various storage components including soil water, snowpacks, lakes, reservoirs, and rivers. With the water budget approach, we can examine existing watershed systems, quantify the effects of management impacts on the hydrologic cycle, and in some cases predict or estimate the hydrologic consequences of proposed activities.

Part 1 of this textbook (chapters 2 through 6) focuses on basic hydrology. Hydrologic processes are described, and the effects of land use and management activities on the respective processes are given special attention. The information contained herein is fundamental to understanding hydrology. Further, it provides the background necessary for a more complete understanding of land use impacts on soil and water resources.

CHAPTER 2 *Precipitation and Interception*

INTRODUCTION

Precipitation and interception (precipitation that is caught by vegetation) affect the amount, timing, spatial distribution, and quality of water added to a watershed from the atmosphere (Fig. 2.1). Hydrologists view precipitation as the major input to a watershed and a key to its water yield characteristics. Ecologists recognize the role of precipitation in determining the types of soils and vegetation that occur on a watershed. As a process, however, most people have only a cursory understanding of why precipitation occurs and why it occurs where it does.

Precipitation is the result of meteorological factors and is therefore largely outside human control. However, land use and associated vegetation alterations can affect the deposition of precipitation by changing interception, at least to some extent. This chapter examines the process of precipitation and its deposition and occurrence in time and space. Basic methods of analyzing precipitation and of estimating interception are presented.

MOISTURE IN THE ATMOSPHERE

Air masses take on the temperature and moisture characteristics of underlying surfaces, particularly when they are stationary or move slowly over large water or land surfaces. Air masses moving from an ocean to land bring to that land surface a source of moisture. Air masses from polar regions will be dry and cold. The movement of air masses modifies the temperature and moisture characteristics of the atmosphere over a watershed and determines the climatic and, more specifically, the precipitation conditions that occur. To understand the precipitation process, the relationship between atmospheric moisture and temperature must be understood.

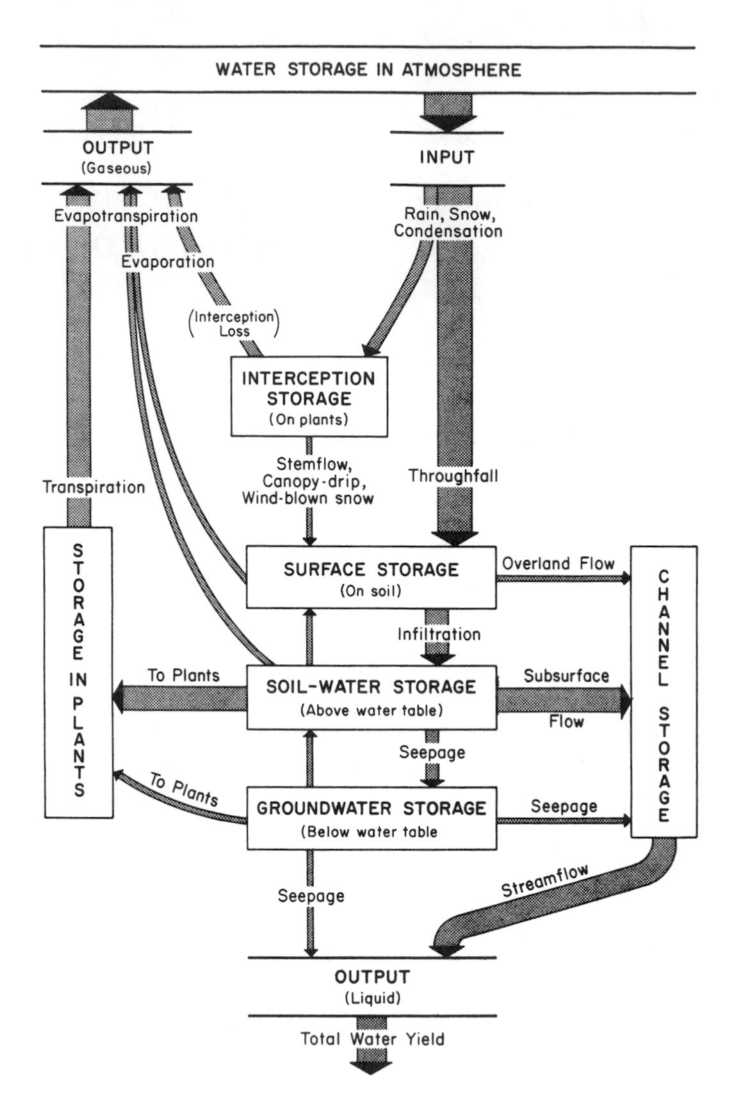

Figure 2.1. The hydrologic cycle consists of a system of water storage compartments and the solid, liquid, or gaseous flows of water within and between the storage points (from Anderson et al. 1976).

Moisture is added to the atmosphere by the process of *evaporation*, the change in state from liquid water to water vapor, resulting in a net loss of liquid water from the underlying surface (a detailed discussion follows in Chapter 3). The pressure resulting from evaporation at the water-air interface is called the vapor pressure of water. Once in air, the water vapor exerts its own partial pressure, called simply *vapor pressure*.

The relationships among moisture content in the atmosphere, temperature, and vapor pressure determine the occurrence and amounts of evaporation and precipitation (Fig. 2.2).

The *saturation vapor pressure* (e_s) is determined by air temperature alone and is the partial pressure of water vapor in a saturated atmosphere as described by Lee (1978):

$$\ln e_s = 21.382 - \frac{5347.5}{T} \quad \textbf{(2.1)}$$

where T = absolute temperature in kelvins (K).

A parcel of unsaturated air (point A in Figure 2.2) can become saturated by either cooling (A to C) or the addition of moisture to the air mass (A to B). Unsaturated air can be characterized by its *relative humidity*, the ratio of vapor pressure of the air to saturation vapor pressure at a given temperature, which is expressed as a percentage. For example, at 25°C in Figure 2.2, the relative humidity of air parcel A is 100(17 mb/31 mb) = 55%. The difference between the saturation vapor pressure (e_s) and the actual vapor pressure of an unsaturated air parcel (e_a) is the *vapor pressure deficit* (for A it is 31 mb –17 mb = 14 mb). The temperature at which a parcel of unsaturated air reaches its saturation vapor pressure is the *dew point temperature* (15.56°C in Figure 2.2 for the air parcel A).

Atmospheric pressure decreases with height, causing a rising parcel of air to expand. Therefore, the temperature of the air will decrease, while that of a descending parcel of air will become compressed and warmed. The process is called *adiabatic* if no heat is gained or lost by mixing with surrounding air. The rate at which unsaturated air cools by adiabatic lifting is approximately 1°C/100 m, called the *dry adiabatic lapse* rate. Movements of large air masses can change the lapse rate, and conditions can even reverse the lapse rate when large, cold air masses underlie less-dense, warm air, a condition called an inversion (temperature increases with height).

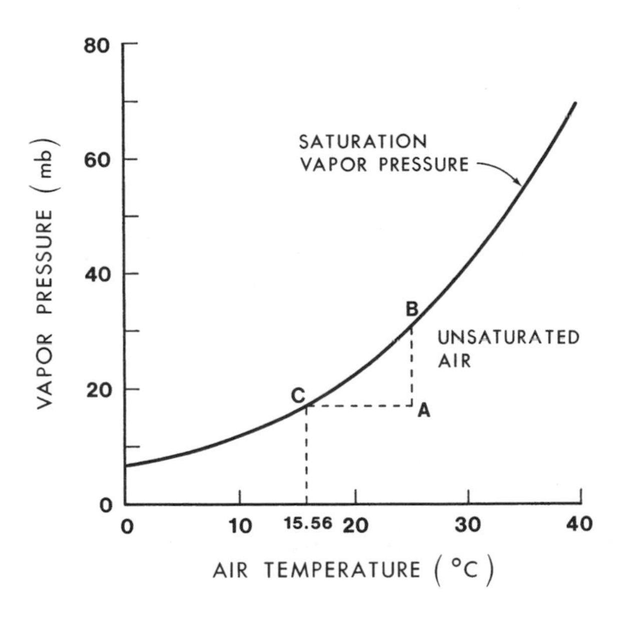

Figure 2.2. Saturation vapor pressure–air temperature relationship: a parcel of unsaturated air (A) must be cooled (A to C) or moisture must be added (A to B) before saturation occurs.

PRECIPITATION

Precipitation occurs when three conditions are met:

1. The atmosphere becomes saturated.
2. Small particles or nuclei are present in the atmosphere (e.g., dust and ocean salt) upon which condensation or sublimation can take place.
3. Water or ice particles coalesce and grow large enough to reach the earth against updrafts.

Saturation results when either the air mass is cooled until the saturation vapor pressure is reached or moisture is added to the air mass (Fig. 2.2). Rarely does the direct introduction of moist air cause precipitation. More commonly, precipitation occurs when an air mass is lifted, becomes cooled, and reaches its saturation vapor pressure. Air masses are lifted as a result of (1) frontal systems, (2) orographic effects, and (3) convection. Different storm and precipitation characteristics result from each of these lifting processes (Fig. 2.3).

Frontal precipitation occurs when general circulation brings two air masses of different temperature and moisture content together and air becomes lifted at the frontal surface. A cold front results from a cold air mass replacing and lifting a warm air mass, which already has a tendency to rise. Cold fronts are characterized by high-intensity rainfall of relatively short duration and usually cover a zone narrower than that of warm fronts. Conversely, a warm front results when warm air rides up and over a cold air mass; it is generally characterized by widespread, gentle rainfall.

Orographic precipitation occurs when general circulation forces an air mass up and over a mountain range. As the air mass becomes lifted, a greater volume of the air mass reaches saturation vapor pressure, resulting in more precipitation with increasing elevation. A striking example of orographic precipitation is on the windward side of Mount Rainier, Washington, where annual precipitation varies from about 1346 mm at 305 m to 2921 mm at 1690 m. Once the air mass passes over mountains, a lowering and warming of the air occurs, creating a dry, rain shadow effect on the leeward side.

Convective precipitation, characterized by summer thunderstorms, is the result of excessive heating of the earth's surface and of the adjacent layer of moist air. When the air adjacent to the surface becomes warmer than the air mass above it, lifting occurs. As the air mass rises, condensation takes place, the latent heat of vaporization is released, more energy is added to the air mass, and consequently, more lifting occurs. Rapidly uplifted air can reach high altitudes where water droplets become frozen and hail forms or becomes intermixed with rainfall. Such rainstorms or hailstorms are some of the most severe precipitation events anywhere and are characterized by high-intensity, short-duration rainfall over rather limited areas. When thunderstorms occur over a large enough area, flash flooding can result.

Measurement of Precipitation

Precipitation characteristics of interest include the total amount (or depth) over some time period (daily, monthly, seasonally, or annually), the intensity (depth per unit time), and the distribution over time and space. Although precipitation is often measured routinely at major towns and airports, precipitation is not routinely measured in many rural parts of a watershed. In addition, many records of precipitation are short term or discontinuous, which hampers our ability to understand better the likelihood of having droughts or floods.

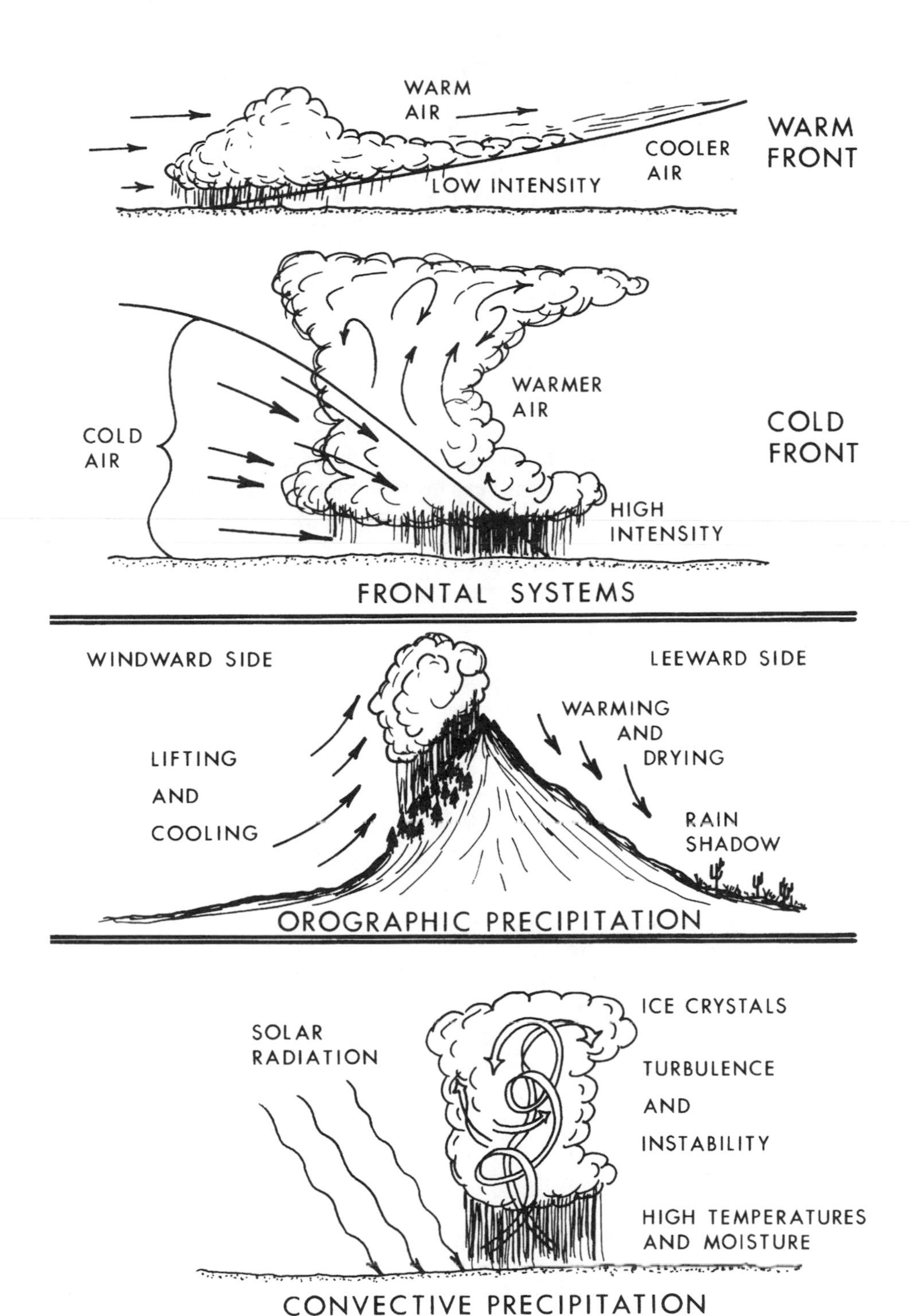

Figure 2.3. Examples of different mechanisms causing air masses to lift and cool, resulting in precipitation.

Types of Precipitation

Drizzle—water drops less than 0.5 mm in diameter; intensity less than 1 mm/hr.

Rain—water drops greater than 0.5 mm in diameter; upper limit is 6 mm in diameter.

Sleet—small frozen raindrops.

Snow—ice crystals formed in the atmosphere by the process of sublimation.

Hail—ice particles greater than 0.5 mm in diameter formed by alternate freezing and thawing in turbulent air currents; usually associated with intense convective cells.

Fog, dew, and frost—not actually precipitation; the result of interception, condensation, or sublimation; can be important sources of moisture to watersheds in coastal areas and other areas subjected to persistent fog and/or clouds.

Determining the depth of precipitation over watersheds requires precipitation to be measured at selected points within the watershed (or in adjacent areas), and then these measurements must be extrapolated to estimate the depth of precipitation over the ungauged parts of the watershed. The purpose of such measurements is to estimate the total amount of precipitation that is representative of the area.

In higher latitudes and at high elevations of mountainous regions, both rainfall and snow may need to be measured. The following discussion will concentrate on rainfall. Snow measurements are discussed in Chapter 15.

Methods of Measurement

The most common method of measuring rainfall is with a series of gauges, typically cylindrical containers 20.3 cm (8 in.) in diameter (Fig. 2.4A). Three types of precipitation gauges in general use are (1) the standard gauge, (2) the storage gauge, and (3) the recording gauge. Standard (or nonrecording) gauges are often used because of economy. Such gauges must be read periodically, normally every 24 hr at the same time each day. The standard gauge magnifies rainfall depth 10-fold because it funnels rainfall into an internal cylinder of 10-fold smaller cross-sectional area. Storage gauges have the same size opening but have a greater storage capacity, usually 1525–2540 mm of rainfall. They may be read periodically, for example, once a week, once a month, or seasonally. To suppress evaporation, a small amount of oil is usually added to gauges read less frequently than every 24 hr.

The use of recording gauges, which allow for continuous measurement of rainfall, is more limited because of their higher cost. Examples of recording rain gauges are the weighing-type (Fig. 2.4B) and the tipping-bucket gauge. The weighing-type gauge records the weight of water with time by means of a calibrated pen on a clock-driven drum; the chart on the drum indicates the accumulated rainfall with time. Rainfall inten-

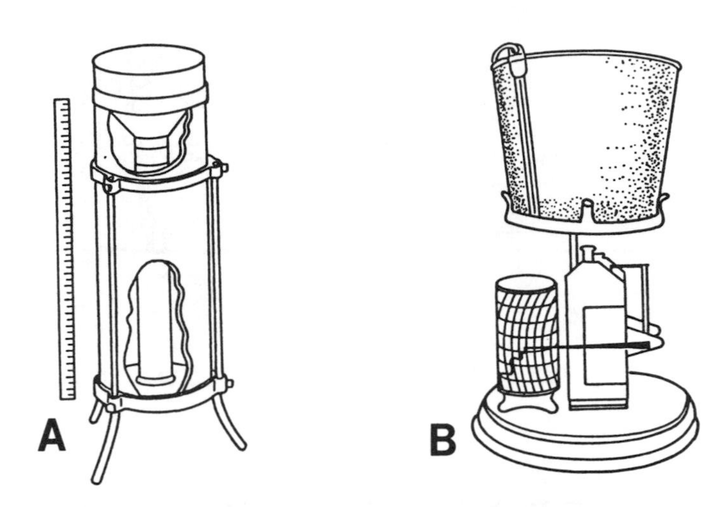

FIGURE 2.4. Rain gauges. A: cutaway of the cylindrical standard Weather Service rain gauge; B: a weighing-type recording gauge with its cover removed to show the spring housing, recording pen, and storage bucket (from Hewlett 1982, © University of Georgia Press, by permission).

sity is obtained by determining incremental increases in the amount per unit of time (typically 1 hr). A tipping-bucket gauge records intensity, making a recording each time a small cup (usually 1 mm deep) fills with water and then empties as it tips back and forth. Because about 0.2 sec is required for the bucket to tip, high-intensity rainfall may not be accurately measured.

The accuracy of rainfall measurements is affected both by gauge site characteristics and by the relationship of the location of gauges to the watershed. As a rule, a rain gauge should be located in a relatively flat area with the funnel opening in a horizontal plane. In the United States, the standard is to situate the gauge so that the funnel orifice is 1 m above the ground surface. The gauge should be far enough away from surrounding objects that the rainfall catch is not affected. A clearing defined by a 30–45° angle from the top of the gauge to the closest object is usually sufficient (Fig. 2.5). If gauges are located too close to trees or structures, the wind patterns around the gauge can result in gauge catches far different from the rainfall that actually occurred. Ideally, one should select small openings in a forest or other area that are sheltered from the full force of the wind but that meet the above criteria. Sometimes, gauges are located in a large enough opening when they are initially installed but become affected by forest growth in adjacent areas over time; this is particularly important in areas where vegetation grows quickly. Also, high wind speeds diminish the efficiency of gauge catch (Table 2.1). Wind shields, such as Nipher or Alter shields, should be used to reduce eddy effects in areas of high wind speeds. Errors caused by catch deficiencies are called instrumentation errors.

Gauges should be located throughout watersheds so that spatial and elevational differences in precipitation can be measured. Such factors as topographic barriers, elevational differences, and storm track patterns, or tracking, should be considered in developing a rain gauge network. Practical considerations such as economics and accessibility usually limit the number, type, and location of gauges.

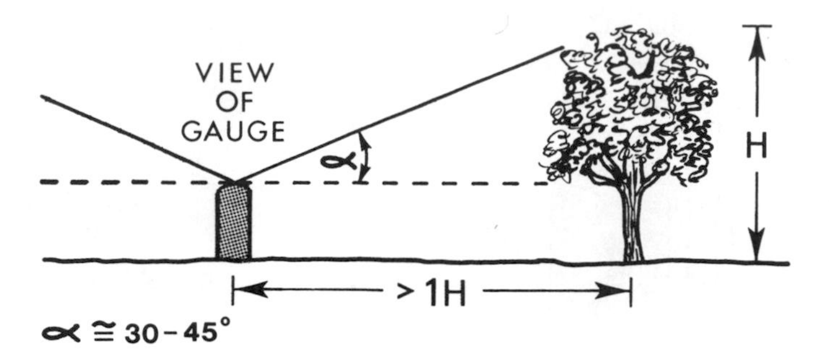

FIGURE 2.5. Proper siting of a rain gauge with respect to the nearest object.

TABLE 2.1. **Effect of wind velocity on the catch of precipitation by standard precipitation gauges**

Wind Velocity (mi/hr)	Catch Deficiency (% of true amount)	
	Rainfall	Snowfall
0	0	0
5	6	20
15	26	47
25	41	60
50	50	73

Source: Gray 1973, by permission.

Precipitation over an area can also be estimated indirectly by radar sensing if the radar coverage is sufficient. Radar senses the backscatter of radio waves caused by water droplets and ice crystals in the atmosphere. The area and relative intensity of precipitation can be estimated up to a distance of 250 km. Radar echoes can be correlated with measured precipitation, but such a calibration is hampered by ground barriers, drop size, distribution of rainfall, and other storm factors. As a consequence, instantaneous radar data on a local scale can contain large errors (Sauvageot 1994). Radar is most useful for tracking storm systems and identifying areas of relative rainfall amounts and intense or violent storm activity.

Number of Gauges Required

The number of rain gauges required to measure precipitation should generally increase with the size of the watershed and with the variability of precipitation. Sampling requirements can be determined with standard statistical methods (see Chapter 18). Use of random sampling as a means of excluding bias in the selection of gauge sites and for estimating the number of gauges needed is suggested. However, in areas of dense brush or forest, this type of rainfall sampling might not be practical owing to the difficulty of obtaining adequate sampling sites. Accessibility also limits the ideal siting of gauges in remote watersheds. As a result, insufficient sampling is more often the norm than the exception.

Using a regular network that is read after each storm event can estimate rainfall variability on a watershed for monthly, seasonal, or annual periods. By reading storage gauges monthly or seasonally, the effects of different storm types on variability may be lost, but systematic differences in precipitation between parts of the watershed for these longer periods can be estimated.

Methods of Calculating Mean Watershed Precipitation

The mean depth of precipitation over a watershed is required in many hydrologic investigations. Several methods are used in deriving this value. The three most common are the arithmetic mean, the Thiessen polygon, and the isohyetal methods (Fig. 2.6).

Arithmetic Mean Method. A straight arithmetic average is the simplest of all methods for estimating the mean rainfall on a watershed (Fig. 2.6). This method yields good estimates if the gauges are numerous and uniformly distributed. Even in mountainous country, averaging the catch of a dense rain gauge network will yield good estimates if the orographic influence on precipitation is considered in the selection of gauge sites. However, if gauges are relatively few, irregularly spaced, or precipitation over the area varies considerably, more sophisticated methods are warranted.

Thiessen Polygon Method. When gauges are nonuniformly distributed over a watershed, the Thiessen polygon method can improve estimates of precipitation amounts over the entire area. Polygons are formed from the perpendicular bisectors of lines joining nearby gauges (Fig. 2.6). The watershed area within each polygon is determined and is used to apportion the rainfall amount of the gauge in the center of the polygon. It is assumed that the depth of water recorded by the rain gauge located within the polygon represents the depth of rain over the entire area of the polygon. The results are usually more accurate than the arithmetic average when the number of gauges on a watershed is limited and when one or more gauges are located outside the watershed boundary.

The Thiessen method allows for nonuniform distribution of gauges but assumes linear variation of precipitation between gauges and makes no attempt to allow for orographic influences. Once the area-weighing coefficients are determined for each station, they become fixed, and the method is as simple to apply as the arithmetic method.

Isohyetal Method. With the isohyetal method, gauge location and amounts are plotted on a suitable map, and contours of equal precipitation (isohyets) are drawn (Fig. 2.6). Rainfall measured within and outside the watershed can be used to estimate the pattern of rainfall, and isohyets are drawn according to gauge catches. The average depth is then determined by computing and dividing by the total area. Many investigators indicate this as theoretically the most accurate method of determining mean watershed precipitation. But it is also by far the most laborious.

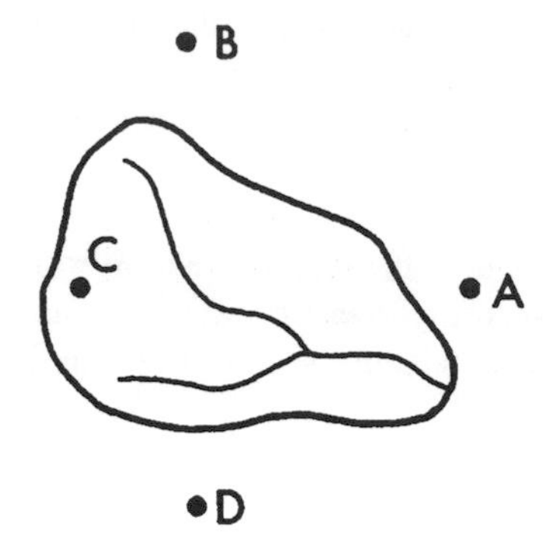

Arithmetic Mean Method

Source Data: Daily rainfall measured at each gauge in centimeters

A	B	C	D
4	8	10	6

$$\text{Arithmetic mean} = \frac{4 + 8 + 10 + 6}{4}$$

$$= 7 \text{ cm}$$

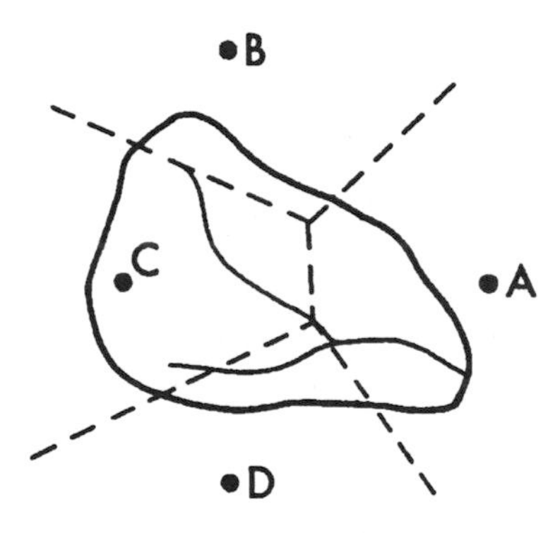

Thiessen Polygon Method

Thiessen Polygon:

Station	Depth (cm)		Area in Polygon*		Volume (cm)
A	4	×	0.28	=	1.12
B	8	×	0.09	=	0.72
C	10	×	0.49	=	4.90
D	6	×	0.14	=	0.84
			sum	=	7.58

*As a fraction of total area.

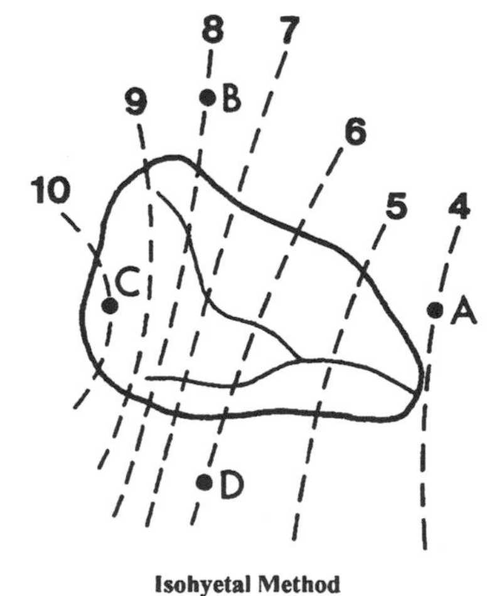

Isohyetal Method

Isohyetal:

Mean Depth (cm)		Area between Isohyets*		Volume (cm)
4.5	×	0.12	=	0.54
5.5	×	0.25	=	1.38
6.5	×	0.14	=	0.91
7.5	×	0.13	=	0.98
8.5	×	0.18	=	1.53
9.5	×	0.14	=	1.33
10.5	×	0.04	=	0.42
		sum		7.09

*As a fraction of total area.

FIGURE 2.6. Methods of estimating the average rainfall for a watershed: arithmetic mean; Thiessen polygon; isohyetal.

The isohyetal method is particularly useful when investigating the influence of storm patterns on streamflow and for areas where orographic precipitation occurs. In some instances, relationships between precipitation and elevation can be used advantageously and only a few gauges need to be set out (typically located in the lower elevations). Where orographic precipitation occurs, contour intervals can sometimes be used to help estimate (locate) the lines of equal precipitation. Precipitation amounts then are determined for each elevation zone or band and the respective areas weighed to obtain estimates for the entire watershed.

The accuracy of the isohyetal method depends upon the skill of the analyst. An improper analysis can lead to serious error. If linear interpolation between stations is used, the results will be essentially the same as those obtained with the Thiessen method.

Errors Associated with Precipitation Measurement

Hydrologic studies of watersheds are often constrained by an inadequate number of precipitation gauges, by the absence of long-term precipitation records, or both. Two types of error must be considered when determining precipitation measurement: *instrumentation error* is related to the accuracy with which gauges catch the true precipitation amount at a point; *sampling error* is associated with how well the gauges in a watershed represent the precipitation over the entire watershed area.

Care in siting a gauge correctly and proper maintenance can minimize instrumentation error. For standard 8 in. (324 cm^2) precipitation gauges in the United States, there are biases in gauge catch from the effects of wind and from wetting of the gauge orifice (Legates and DeLiberty 1993). Legates and DeLiberty report up to 18 mm/mo of systematic undercatch bias for winter precipitation in the northeastern and northwestern United States. A 28% undercatch bias was reported for winter precipitation in the northern Rockies and Upper Midwest. The biases can be corrected by using:

$$P_c = k_r\,(P_{gr} + dP_{wr}) + k_s\,(P_{gs} + dP_{ws}) \tag{2.2}$$

where P_c is the gauge-corrected precipitation, P_g is measured precipitation, k is the wind correction coefficient (usually > 1), dP_w is the correction for wetting loss (usually 0.15 for rain events and half that for snowfall), and subscripts r and s denote rainfall and snowfall, respectively. Wetting loss refers to the amount of rain or snow that is initially stored on the interior surface of the rain gauge at the beginning of the storm.

The wind correction coefficient for rain (Sevruk and Hamon 1984) is:

$$k_r = \frac{100}{100 - 2.12w_{hp}} \tag{2.3}$$

and for snowfall (Goodison 1978) it is:

$$k_s = e^{0.1338 w_{hp}} \tag{2.4}$$

where w_{hp} is the wind speed at the height of the gauge orifice.

Properly designing a network with an adequate number of gauges, on the other hand, minimizes sampling error. Accessibility and economic considerations usually determine the extent to which sampling errors can be reduced.

Analysis of Precipitation

Once we have precipitation records, a number of analyses can be performed to enhance our knowledge of hydrology and climate. This section briefly describes some of the more common types of analysis.

Estimating Missing Data

All too often, one or more gauges in a precipitation network become nonfunctioning for some period. One way to estimate the missing record for such a station is to use existing relationships with adjacent gauges. For example, if the precipitation for a storm is missing for station C, and if precipitation at C (seasonal or annual) is correlated with that at stations A and B, the normal ratio method can estimate the storm precipitation for station C:

$$P_C = \frac{1}{2}\left(\frac{N_C}{N_A}P_A + \frac{N_C}{N_B}P_B\right) \quad \textbf{(2.5)}$$

where P_C = estimated storm precipitation for station C (mm); N_A, N_B, N_C = normal annual (or seasonal) precipitation for stations A, B, and C (mm), respectively; and P_A and P_B = storm precipitation for stations A and B (mm).

The generalized equation for missing data is:

$$P_x = \frac{1}{n}\left(\frac{N_x}{N_1}P_1 + \cdots + \frac{N_x}{N_n}P_n\right) \quad \textbf{(2.6)}$$

Equation 2.6 is recommended only if there is a high correlation with other stations.

Double Mass Analysis

Double mass analysis is a convenient method of checking the consistency of a precipitation station against that of one or more nearby stations. An example best explains the application of this method. Consider that station E has been collecting rainfall data for 45 yr. Originally, the station was located in a large opening in a conifer forest, but over the years the surrounding forest has grown up to the point where you suspect that the catch of this gauge is now affected. Based on your knowledge of the precipitation patterns in the region, you recognize that the same storm patterns influence stations H and I, although their elevational differences and other factors cause annual rainfall to differ. There was a consistent correlation between the average of stations H and I and that of E in the early years of station E. By plotting the accumulated annual rainfall of E against the accumulated average annual rainfall of H and I, you find that the relationship clearly changed after 1970 (Fig. 2.7). The relationship then can be used to correct existing rainfall catch at E so that it better represents the true catch at the location without the interference of nearby trees.

Frequency Analysis

Water resource systems such as small reservoirs, waterways, irrigation networks, and drainage systems for roads should be planned and designed for future events, the magnitude of which cannot be accurately predicted. Weather systems vary from 1 yr to the next, and no one can accurately predict what the next year, season, or even month will bring. Therefore, we rely on statistical analyses of rainfall amounts over certain periods. From these analyses, the frequency distributions of past events are determined; the

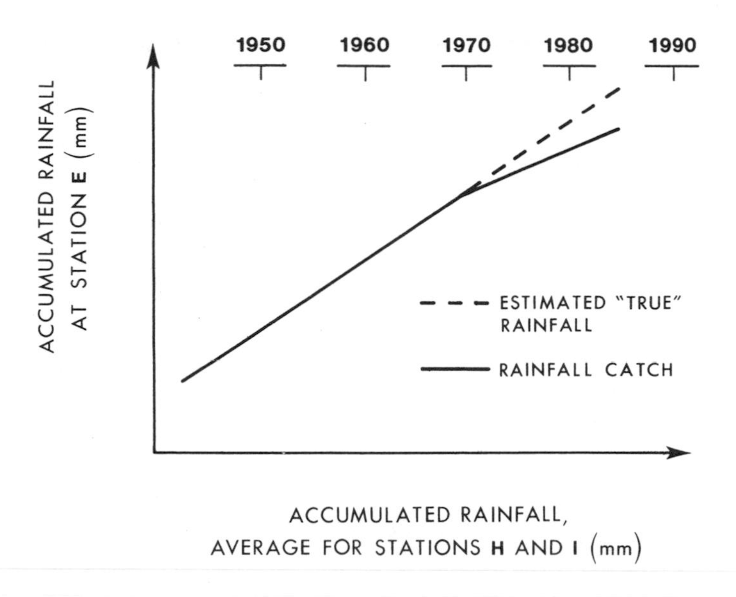

FIGURE 2.7. Double mass plot of annual precipitation at station E vs. average annual precipitation at stations H and I.

probability or likelihood of having certain events occur over a specified period then can be estimated. This approach, called frequency analysis, will be discussed for precipitation events.

The objective of frequency analysis is to develop a frequency curve, which is a relationship between the magnitude of events and either the associated probability or the recurrence interval (Fig. 2.8). The details of developing a frequency distribution function are outlined in Chapter 18 (see also Haan 1977). The recurrence interval (T_r) can be approximated by:

$$T_r = \frac{n + 1}{m} \tag{2.7}$$

where n = number of years of record and m = rank of the event.

The recurrence interval, or return period, that corresponds to a given probability (p) is determined as the reciprocal of the probability:

$$T_r = \frac{1}{p} \tag{2.8}$$

For example, in Figure 2.8, the recurrence interval associated with a 0.05 event is 20 yr. The chance or risk of having certain events occur in any given year can also be estimated from such a curve. In Figure 2.8, the probability of having a 24 hr rainfall of 100 mm or more in any given year is about 0.02 (or a 50 yr recurrence interval). Precipitation frequency curves can be developed to evaluate maximum annual events (Box 2.1) or to evaluate rainfall during dry periods or droughts (e.g., a frequency curve can be developed for the minimum 12 mo precipitation amounts).

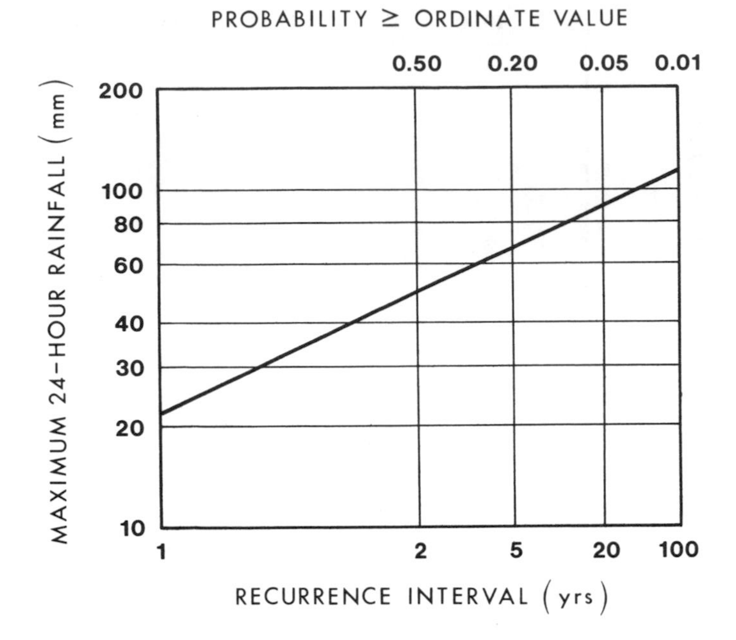

FIGURE 2.8. Frequency curve of daily rainfall for a single station.

Box 2.1

Using Rainfall Frequency Information to Determine the Correct Size of a Culvert for a Road System

A road is to be constructed in a forested area and should be designed with sufficient culverts (numbers and sizes) to minimize washouts. It has been decided that the acceptable risk for this road is 5%; that is, we only want to take a 5% chance in any given year that it will be washed out. From a frequency curve of maximum 1 hr rainfall amounts, the rainfall corresponding to the 0.05 probability was determined to be 20 mm. The 20 mm of rainfall over a 1 hr period then would be used to estimate the corresponding runoff (using a method such as the rational method discussed in Chapter 17). As watershed size and complexity increase, it becomes less likely that the 0.05 rainfall event would produce a runoff event with the same probability. Therefore, this approach should only be used for small, relatively homogeneous watersheds.

Once a frequency curve is developed, the probability of exceeding certain rainfall amounts over some specified period can be determined. The probability that an event with probability p will be equaled or exceeded x times in N years is determined by:

$$\text{Prob}(x) = \frac{N!}{x!(N-x)!}(p)^x(1-p)^{N-x} \quad \textbf{(2.9)}$$

This relationship can be simplified by considering the probability of at least one event with probability p being equaled or exceeded in N years as follows:

$$\text{Prob (no occurrences in } N \text{ yr)} = (1-p)^N \quad \textbf{(2.10)}$$

Therefore:

$$\text{Prob (at least one occurrence in } N \text{ yr)} = 1-(1-p)^N \quad \textbf{(2.11)}$$

For example, the probability of having a 24 hr rainfall event of 100 mm or greater (Fig. 2.8) over a 20 yr period is determined by:

$$\text{Prob }(x > 100) = 0.025$$

or:

$$\text{Prob }(x > 100 \text{ mm over 20 yr}) = 1 - (1 - 0.025)^{20} = 0.40, \text{ or } 40\%$$

Depth-Area-Duration Analysis

The above frequency analysis is based upon rainfall characteristics at a point or a specific location. The design of storage reservoirs and other water resource–related activities require that the watershed area be taken into account. Because rainfall does not occur uniformly, it is expected that as larger watersheds are considered, the depth of rainfall associated with any given probability will decrease (Fig. 2.9). Furthermore, on a given area, the greater the depth of rainfall for a given duration, the lower the probability of equaling or exceeding that amount.

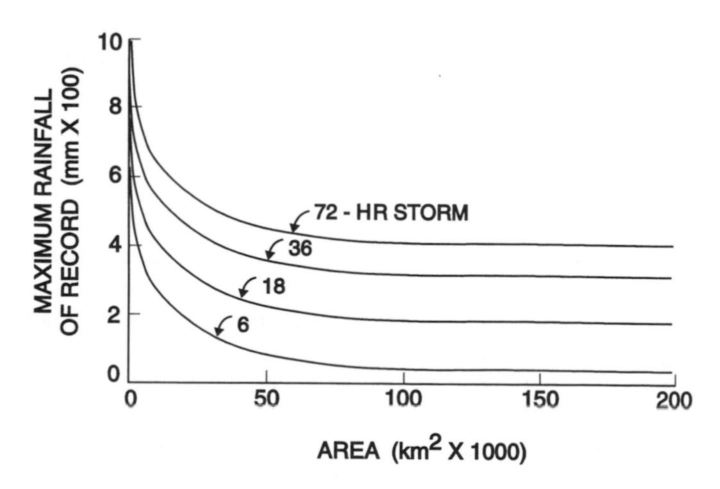

FIGURE 2.9. Relationship between maximum rainfall amounts for specified durations and area (from Hewlett 1982, © University of Georgia Press, by permission).

INTERCEPTION AND NET PRECIPITATION

Once rainfall or snowfall occurs, the type, extent, and condition of vegetation influence the pattern of deposition and amount of precipitation reaching the soil surface. Dense coniferous forests in northern latitudes and the multistoried canopies of the Tropics catch and store large quantities of precipitation, which directly return to the atmosphere by evaporation and thus become a loss of water from the watershed. Interception losses are less in arid and semiarid environments, which have sparser vegetation.

Although we generally consider forests to have the highest interception losses, shrublands and prairie vegetation can intercept 10–20% of gross precipitation during periods when maximum growth has been attained. Forest floor litter can also store large quantities of precipitation, part of which evaporates directly to the atmosphere.

Not all the precipitation caught by a forest canopy is lost to the atmosphere. Much can drip off the foliage or run down the stems (stemflow), ultimately reaching the soil surface. Similarly, drainage from forest floor litter will reach the soil surface.

Components of Interception

The components of the interception process, methods of measurement, and resulting deposition of precipitation are illustrated in Figure 2.10. Interception by the forest canopy is defined as:

$$I_c = P_g - T_h - S_f \qquad \textbf{(2.12)}$$

where I_c = canopy interception loss (mm); P_g = gross precipitation (mm); T_h = throughfall, precipitation that passes through the vegetative canopy or as drip from vegetation (mm); and S_f = stemflow, water that flows down the stems to the ground surface. Collars are fixed to the stems of trees and divert stemflow to containers for measurement (mm).

The partitioning of a given quantity of rainfall into the above pathways depends upon vegetative cover characteristics such as leaf type, leaf and branch surface area, branch attitude, shape of the canopy, and roughness of the bark. The interception components of a growing forest, from seedling stage to mature forest, change as follows: (1) T_h diminishes over time as the canopy cover increases; (2) S_f increases over time but is always a small quantity; and (3) the storage capacity of vegetation and litter, as related primarily to leaf surface area, increases substantially.

Throughfall

Canopy coverage, total leaf area, the number of layers of vegetation, and rainfall intensity determine how much of the gross rainfall reaches the forest floor. The size and shape of canopy openings affect the amount, intensity, and spatial distribution of throughfall. The shape of the overstory, particularly the branch and leaf angles, can concentrate throughfall in drip points, which can result in greater amounts and intensities of rainfall at these points, with a corresponding greater kinetic energy than in the open. Other locations under dense forest canopy can receive little throughfall. Throughfall relationships have been determined for forest types in many parts of the world. Relationships between throughfall and precipitation (Box. 2.2) have been developed for mature forest stands in particular areas. Because throughfall depends on tree canopy surface area and cover, descriptors of forest stands that relate to these characteristics should be included in prediction equations to allow for wider application.

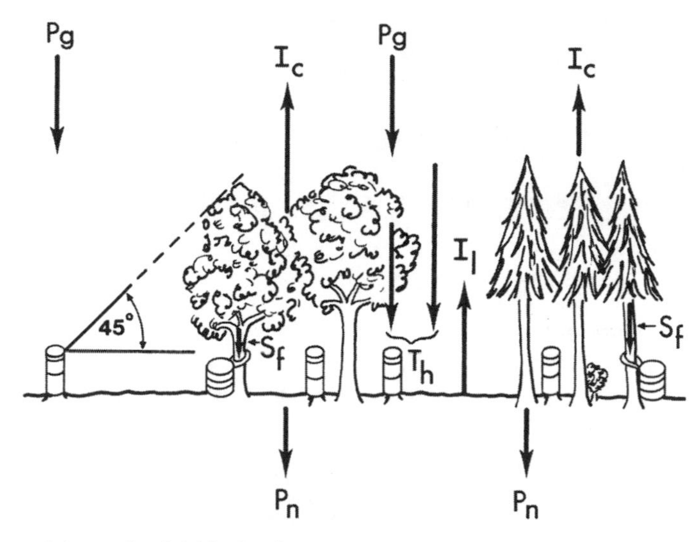

Interception total $I = I_c + I_l$
The amount reaching the forest floor $= T_h + S_f$
Interception by canopy (overstory + understory) $I_c = P_g - T_h - S_f$
Net precipitation $P_n = T_h + S_f - I_l$

FIGURE 2.10. Components of interception (from Hewlett 1982, © University of Georgia Press, by permission): I_c = canopy interception loss; I_l = litter interception; P_g = gross precipitation; P_n = net precipitation; S_f = stemflow; T_h = throughfall.

Stemflow

Stemflow, usually < 2% of gross annual precipitation, is affected by branch attitude, shape of tree crowns, and roughness of bark. Generally, tree species with rough bark retain more water and exhibit less stemflow than those with smooth bark.

Higher stemflows have been reported, however. In Australia, stemflow amounted to 5 and 9% of precipitation for dry sclerophyll eucalyptus and radiata pine forests, respectively (Crockford and Richardson 1990). Stemflow varies considerably among eucalyptus species; some have smooth, nonabsorptive bark and exhibit high stemflow amounts, whereas others have thick, fibrous, absorptive bark and exhibit little stemflow.

Although stemflow is not a large quantity in terms of an annual water budget, the process can be an important mechanism of replenishing soil moisture. Stemflow concentrates water in a small area near the base of the tree stem; this water can flow quickly and penetrate deeply into the soil through root channels.

Interception Process

Observations of interception losses for individual storms indicate that the precipitation and associated storm characteristics as well as vegetation characteristics influence the interception process. The type of precipitation, whether rain or snow, the intensity and duration of rainfall, wind velocity, and evaporative demand affect interception losses. The interception of snow, although clearly visible for a conifer forest immediately after snowfall, is not a significant loss in some cases. Much of the snow caught by foliage of

Box 2.2

Throughfall Relationships for Different Vegetative Canopies

1. Eastern U.S. Hardwood Forests (Helvey and Patric 1965):

Growing Season

$T_h = 0.901P - 0.031n$

Dormant Season

$T_h = 0.914P - 0.015n$

where T_h = throughfall (in.); P = total precipitation (in.); and n = number of storms.

2. Southern U.S. Pine Forests (Roth and Chang 1981):

Longleaf Pine

$T_h = 1.002P - 0.0008P^2 - 1.397$

Loblolly Pine

$T_h = 0.930P - 0.0011P^2 - 0.610$

where T_h = throughfall (mm); and P = total rainfall (mm).

3. New Zealand Vegetation Communities (Blake 1975):

Kauri Forest

$T_h = 0.60P - 3.71$

Manuka Shrub

$T_h = 0.44P - 0.10$

Mountain Beech Forest

$T_h = 0.69P - 1.90$

individual trees often reaches the soil surface by the mechanical action of wind or by melt and drip (Box 2.3).

The process of interception during a rainstorm usually results in greater losses than with snowfall. The total interception loss is the sum of (1) water stored on vegetative surfaces (including forest litter) at the end of a storm and (2) the evaporation from these surfaces during the storm. If a storm were to last over a long period under windy conditions, the interception loss would be expected to exceed that from a storm of equal duration with calm conditions. Conversely, a high-intensity, short-duration thunderstorm with high wind velocities can have the least amount of interception loss; this can be explained by the action of wind, which can mechanically remove water from the canopy and, therefore, not allow the storage capacity of the canopy to be reached. The effects of wind on evaporative loss would be minimal for a storm of short duration.

Potential interception loss (I) for a storm can be expressed as (Horton 1919):

Box 2.3

Deposition of Intercepted Snowfall

Photograph records in Colorado (Hoover and Leaf 1967) and Arizona (Tennyson et al. 1974) indicated that intercepted snowfall remains in coniferous trees for a short time period. Much of the intercepted snow eventually reaches the ground because of wind erosion and snowmelt with subsequent dripping and freezing in the snowpack. While snowfall interception by a tree might not be a significant loss to the small-scale water budget surrounding an individual tree, the deposition of intercepted snow on the ground results in a redistribution of the snowpack. On a much larger scale, it appears that both small patch cuts and thinning in subalpine forests in Fraser Experimental Forest of central Colorado increases snow accumulations on the ground of treated areas (Meiman 1987). While the responsible processes are not completely known, data on the effects of patch cuts and thinning treatments suggest that a reduction in snow interception loss is a major factor. Still inseparable, however, are the processes of interception and differential deposition of snow on a larger scale during snowfall events.

$$I = S + RtE \tag{2.13}$$

where S = water storage capacity of vegetative surfaces, expressed as depth over the projection area of canopy (mm); R = ratio of evaporating surface to the projected area (decimal fraction); E = evaporation rate (mm/hr) during the storm; and t = time duration of storm (hr).

Equation 2.13 assumes that rainfall is sufficient to satisfy the total storage capacity (S). Equation 2.14 has been used to account for rainfall amounts (Merriam 1960, as presented by Gray 1973):

$$I = S(1 - e^{-P/S}) + RtE \tag{2.14}$$

where P = rainfall (mm); and e = the base of natural logarithms.

The storage capacity of mature conifers is generally greater than that of mature hardwoods (deciduous trees). Comparisons among conifer stands have yielded a wide range of values, in contrast to less variability among mature hardwood stands in North America. Interception losses of deciduous hardwoods vary with season as a result of leaf fall. In many instances, the total interception loss from a natural forest is attributed to both understory and overstory species and to both conifers and hardwoods.

Hydrologic Importance of Interception

The hydrologic importance of interception is dependent upon several climatic, physical, and vegetative characteristics. For water budget studies covering a period greater than a few days, interception can be an important storage term that should be subtracted from gross precipitation. The result is net precipitation, or that amount of precipitation available either to replenish soil water deficits or to become surface, subsurface, or groundwater flow. Net precipitation can be determined from:

$$P_n = P_g - I \tag{2.15}$$

where P_n = net precipitation (mm); P_g = gross precipitation measured by rain gauges in openings (mm); and I = interception loss (mm).

The above terms can be measured using individual plots, as shown in Figure 2.10. Determining net precipitation over a watershed is more difficult. The spatial variability of canopy cover type and extent, canopy stratification (layering), and the storage capacity of plant litter all affect the total interception loss for a watershed. Under certain climatic conditions, the interception storage differences among species result in water yield differences. In regions where annual precipitation exceeds potential evapotranspiration (see Chapter 3) and soil water rarely limits transpiration, differences in interception between conifers and hardwoods also can result in differences in water yield. Converting from hardwoods to conifers in the humid southeastern United States, for example, has increased interception losses and reduced annual streamflow volume. Such differences likely would not be observed in semiarid regions because of the higher ratio of annual potential evapotranspiration to annual precipitation. The reasoning is that the difference in net precipitation reaching the soil surface, due to differences in interception, will satisfy soil water deficits rather than contribute to streamflow. The increase in net precipitation will simply be transpired at some other time and will not necessarily result in an increase in water yield.

Forest canopies affect the deposition pattern of precipitation. In the case of snow, such effects have management implications for water yield improvement. Snow has a high surface to mass ratio and, consequently, is strongly affected by wind patterns. Small openings within a conifer stand, for example, experience eddies, which deposit more snow than is deposited in the adjacent forest. Also, wind mechanically deposits much of the snow intercepted by surrounding trees into these openings. By clearcutting strips in conifer stands and orienting them perpendicular to the wind, the deposition of snow can be increased within the strips and thus increase the water yield in some cases (see Chapter 15).

Up to this point in our discussion, interception has been considered a loss from a watershed. However, in some coastal areas and high elevations in the humid tropics that experience many days with low clouds or fog, interception can add moisture to the soil. Foliage intercepts fog which condenses, coalesces, and drips off, adding moisture that would otherwise remain in the air. In such cases, the greater the foliage surface area, the greater the interception input to the water budget.

Interception has been studied widely and the literature cites interception values for many different vegetative types and tree species. Unfortunately, few have related interception storage values or total interception loss to forest stand (or other vegetative) characteristics so that the values can be estimated from field data. Examples of stand characteristics and associated interception storage values for red pine stands in Minnesota are presented in Table 2.2. These storage relationships were used with measured precipitation data to calculate (Eq. 2.14) the growing-season interception losses for the respective stands (Table 2.3). Interception storage represents both canopy and

TABLE 2.2. Interception storage for red pine stands at the Cloquet Forestry Center, Minnesota

	Stand Characteristics			Canopy Storage		
Stand	Age (yr)	Basal Area (ft^2/acre)	Stems (no./acre)	P_g<1 in.(in.)	P_g>1 in.(in.)	Litter Storage Capacity (in.)
A	21	85	1030	.06	.14	.07
B	20	165	1512	.14	.28	.12
C	29	234	1150	.10	.22	.16
D	71	174	427	.09	.15	.16

Source: Fox 1985.

TABLE 2.3. Simulated interception components for the growing season (June, July, and August) for four red pine stands at Cloquet Forestry Center, Minnesota

Year	Gross Rainfall (P_g)(in.)	Stand	Simulated Net Rainfall (in.)	Canopy Interception (in.)	Litter Interception (in.)	Stemflow (in.)
1953	21.55	A	18.19	3.50	0.42	0.57
		B	16.49	5.07	0.62	0.64
		C	17.46	4.02	0.62	0.45
		D	17.43	3.46	0.74	0.08
1970	6.13	A	4.68	1.16	0.41	0.12
		B	3.48	2.10	0.67	0.13
		C	3 87	1.66	0.69	0.09
		D	4.06	1.30	0.78	0.02
1976	11.74	A	9.79	1.77	0.46	0.28
		B	8.48	2.90	0.67	0.31
		C	9.02	2.23	0.71	0.22
		D	9.12	1.82	0.85	0.04

Source: Fox 1985.

associated understory vegetation. Based on these tables, the interception loss for dense conifer plantations can be > 30% of the precipitation occurring during the growing season.

The amount, duration, intensity, and pattern of rainfall all influence the amount of interception. For example, in the Pacific Northwest, the amount of interception in coniferous forests ranges from 100% for storms dropping < 1.5 mm of precipitation, to 15% for storms dropping > 75 mm (Rothacher 1963).

The highest annual losses of water due to interception have been reported in coniferous forests in humid temperate regions and in tropical regions. Depending on density, annual interception losses of conifer forests in North America vary from 15 to 40% of annual precipitation.

In the humid Tropics, interception losses of forests are more variable than in temperate regions. Conifer and broadleaf plantations in India have been reported to intercept 20–25% and 20–40% of annual rainfall, respectively (Ghosh et al. 1982; Chandra 1985). Interception by secondary lowland tropical forests in West Java, Indonesia, has amounted to 21% of rainfall (Calder et al. 1986). In montane rain forests of the Colombian Andes, interception ranged from 12.4 to 18.3% of annual rainfall (Veneklaas and Van Ek 1990).

Many tree species in the humid Tropics have large, waxy leaves. This type of leaf and high-intensity rainfall tend to favor smaller percentages of interception. Low-intensity, long-duration rainfall in temperate climates in conifer forests, with their high leaf surface area, tend to favor higher interception losses.

In dryland forests, annual interception losses are generally lower than those in either humid temperate or humid tropical forests because of lower canopy densities. However, in such climates, even small amounts of water loss can be important. Between 5 and 10% of annual rainfall is intercepted in conifer woodlands of the semiarid southwestern United States (Skau 1964). Up to 70% of the late-summer rainfall is intercepted by trees in oak woodland communities of the southwestern United States and northern Mexico (Haworth and McPherson 1991).

The hydrologic role of litter interception is twofold: (1) the storage of part or all of the throughfall and (2) the protection that litter provides for the mineral soil surface against the energy of rainfall. The storage capacity of forest litter depends on the type, thickness, and level of decomposition of the litter. Generally, the storage capacity of conifer litter exceeds that of hardwood litter. Litter storage capacities of conifer plantations in Minnesota were similar to canopy storage capacities (see Table 2.2). However, the moisture content of litter generally remains high because the forest floor is usually protected from wind and direct solar radiation. As a result, litter might not be able to absorb much additional water, a point emphasized by contrasting litter storage capacities in Table 2.2 with seasonal litter interception losses in Table 2.3. Forest stands that are more open, such as ponderosa pine in the southwestern United States, can experience significant litter interception; ponderosa pine litter storage capacities of > 200% by weight have been reported.

The protection against rainfall that litter provides for the soil surface influences the surface soil conditions directly and, therefore, infiltration, surface runoff, and surface soil erosion (see chapters 4 and 7).

Summary

You should now have a general understanding of the important factors that influence the occurrence of precipitation over a watershed and how vegetation affects the amount and spatial deposition of that precipitation. Specifically, you should be able to:

- Describe the conditions necessary for precipitation to occur.
- Explain the different precipitation and storm characteristics associated with frontal storm systems, orographic influences, and convective storms.
- Understand how precipitation is measured at a point and how such measurements can be used to estimate the average depth of precipitation over a watershed area.
- Estimate values of precipitation that are missing for a particular storm.
- Explain the purpose of performing double mass analysis and frequency analysis.
- Understand the ways in which vegetation influences the deposition of precipitation.
- Explain and be able to calculate stemflow, throughfall, and interception storage when given appropriate data.
- Calculate net precipitation, given values of gross precipitation and interception values.
- Explain and discuss the hydrologic importance of interception under different vegetative cover and climatic regimes.
- Discuss the relative significance of intercepted snowfall to a water budget.

References

Anderson, H.W., M.D. Hoover, and K.G. Reinhart. 1976. *Forests and water: Effects of forest management on floods, sedimentation, and water supply.* USDA For. Serv. Gen. Tech. Rep. PSW-18.

Blake, G.J. 1975. The interception process. In *Prediction in catchment hydrology, national symposium on hydrology,* ed. T.G. Chapman and F.X. Dunin, 59–81. Melbourne: Aust. Acad. Sci.

Calder, I.R., I.R. Wright, and D. Murdiyarso. 1986. A study of evaporation from a tropical rain forest-West Java. *J. Hydrol.* 86:13–31.

Chandra, S. 1985. Small watersheds and representative basin studies and research needs. In *Proceedings of the national seminar on watershed management,* 63–71. Dehradun, India: National Institute of Hydrology.

Crockford, R.H., and D.P. Richardson. 1990. Partitioning of rainfall in a eucalypt forest and pine plantation in southeastern Australia, II: Stemflow and factors affecting stemflow in a dry sclerophyll eucalypt forest and a Pinus radiata plantation. *Hydrological Processes* 4:145–155.

Fox, S.J. 1985. *Interception-net rainfall relationships of red pine stands in northern Minnesota.* Plan B Paper. College of Forestry, Univ. of Minnesota.

Ghosh, R.C., O.N. Kaul, and B.K. Subba Rao. 1982. Environmental effects of forests in India. *Indian For. Bull.* 275.

Goodison, B.E. 1978. Accuracy of Canadian snow gage measurements. *J. Applied Meteorology* 17 (10):1542–1548.

Gray, D.M., ed. 1973. *Handbook on the principles of hydrology.* Reprint. Port Washington, NY: Water Information Center, Inc.

Haan, C.T. 1977. *Statistical methods in hydrology.* Ames: Iowa State Univ. Press.

Haworth, K., and G.R. McPherson. 1991. Effect of *Quercus emoryi* on precipitation distribution. *J. Arizona Nevada Acad. Sci. Proc. Suppl.* 1991:21.

Helvey, J.D. 1971. A summary of rainfall interception by certain conifers of North America. In *Biological effects of the hydrologic cycle,* 103–113. Proc. Third Int. Sem. Hydrol. Professors. West Lafayette, IN: Purdue Univ.

Helvey, J.D., and J.H. Patric. 1965. Canopy and litter interception by hardwoods of eastern United States. *Water Resour. Res.* 1:193–206.

Hewlett, J.D. 1982. Forests and floods in the light of recent investigations. In *Proceedings of the Canadian Hydrology Symposium '82 on hydrological processes of forested areas,* 543–559. Ottawa: National Research Council, Canada.

Hoover, M.D., and C.F. Leaf. 1967. Process and significance of interception in Colorado subalpine forests. In *Forest hydrology,* ed. W.E. Sopper and H. W. Lull, 213–224. New York: Pergamon.

Horton, R.E. 1919. Rainfall interception. *Mon. Weather Rev.* 47:603–623.

Jetten, V.G. 1996. Interception of tropical rain forest: Performance of a canopy water balance. *Hydrological Processes* 10:671–685.

Lee, R. 1978. *Forest climatology.* New York: Columbia Univ. Press.

Legates, D.R., and T.L. DeLiberty. 1993. Precipitation measurement biases in the United States. *Water Res. Bull.* 29(5): 855–861.

Leonard, R.E. 1967. Mathematical theory of interception. In: *Forest Hydrology,* ed. W.E. Sopper and H. W. Lull, 131–136. New York: Pergamon Press.

Meiman, J.R. 1987. Influence of forests on snowpacks. In *Management of subalpine forests: Building on 50 years of research*, tech. coord. C.A. Troendle, M.R. Kaufmann, R.H. Hamre, and R.P. Winokur, 61–67. USDA For. Serv. Gen. Tech. Rep. RM-149.

Merriam, R.A. 1960. A note on the interception loss equation. *J. Geophys. Res.* 65:3850–3851.

Roth, F.A., II, and M. Chang. 1981. Throughfall in planted stands of four southern pine species in east Texas. *Water Resour. Bull.* 17:880–885.

Rothacher, J. 1963. Net precipitation under a Douglas-fir forest. *For. Sci.* 9:423–429.

Sauvageot, H. 1994. Rainfall measurement by radar: A review. *Atmos. Res.* 35:27–54.

Sevruk, B., and W.R. Hamon. 1984. *International comparison of national precipitation gauges with a reference pit gauge.* Instruments and Observing Methods, Report 17. Geneva, Switzerland: World Meteorological Organization.

Skau, C.M. 1964. Interception, throughfall, and stemflow in Utah and alligator juniper cover types of northern Arizona. *For. Sci.* 10:283–287.

Tennyson, L.C., P.F. Ffolliott, and D.B. Thorud. 1974. Use of time-lapse photography to assess potential interception in Arizona ponderosa pine. *Water Resour. Bull.* 10:1246–1254.

Veneklaas, E.J., and R. Van Ek. 1990. Rainfall interception in two tropical montane rain forests, Colombia. *Hydrological Processes* 4:311–326.

CHAPTER 3 *Evapotranspiration and Soil Water Storage*

INTRODUCTION

Evaporation from soils, plant surfaces, and water bodies and water losses through plant leaves are considered collectively as *evapotranspiration (ET).* Evapotranspiration affects water yield, largely determines what proportion of precipitation input to a watershed becomes streamflow, and is influenced by forest, range, and agricultural management practices that alter vegetation. The water budget equation can be used to estimate *ET* over a period of time as follows:

$$ET = P - Q - \Delta S - \Delta l \quad \textbf{(3.1)}$$

where ET = evapotranspiration (mm); P = precipitation (mm); Q = streamflow (mm); ΔS = change in the amount of storage in the watershed = $S_2 - S_1$ (mm), where S_2 = storage at the end of a period, and S_1 = storation at the beginning of a period; and Δl = change in deep seepage = $l_o - l_i$ (mm), where l_o = seepage out of the watershed, and l_i = seepage into the watershed.

The *ET* component of the water budget is > 95% of the 300 mm of annual precipitation in Arizona; it is > 70% of the annual precipitation for the entire United States (Gay 1993). Generally, for dry climates the ratio *ET/P* is close to unity, meaning that the ratio *Q/P* is small. For humid climates, *ET/P* is less, and the magnitude of *ET* is governed by available energy rather than the availability of water. Vegetative cover affects the magnitude of *ET* and thus, *Q*. For dry climates, vegetative effects on *ET* are limited by plant water availability. Changes in vegetation that reduce annual *ET* will increase streamflow and/or groundwater recharge; increases in annual *ET* have the opposite effect.

Rates of *ET* influence water yield by affecting the antecedent water status of a watershed: high rates deplete water in the soil and in surface water impoundments;

more storage space is then available for precipitation. Low rates leave less storage space in the soil and surface water impoundments. The amount of storage space in the watershed determines the amount and, to some extent, the timing of streamflow resulting from precipitation.

THE PROCESS

Evaporation is the net loss of water from a surface resulting from a change in the state of water from liquid to vapor and the net transfer of this vapor to the atmosphere. *Transpiration* is the net loss of water from plant leaves by evaporation through leaf stomata. Before evaporation or transpiration can occur, there must be (1) a flow of energy to the evaporating or transpiring surfaces, (2) a flow of liquid water to these surfaces, and (3) a flow of vapor away from these surfaces. If one or more of these flows are changed, there is a corresponding change in the total *ET* loss from a surface.

Energy Flow

Solar energy drives the hydrologic cycle. Conditions that control the net flow of energy determine the amount of energy available for the latent heat of vaporization. The flow of energy to evaporating and transpiring surfaces is usually described with an energy budget, components of which can be partitioned and related to parts of the water budget. The linkage between water and energy budgets is direct; the net energy available at the earth's surface is apportioned largely in response to the presence or absence of water. Reasons for studying the energy budget and the relation to the water budget are to develop a better understanding of the hydrologic cycle and be able to quantify or estimate evaporation from bodies of water, potential evapotranspiration for terrestrial systems, and snowmelt.

The earth's surface neither gains nor loses significant quantities of energy over long periods of time, but there can be a net gain or loss for any given interval. The following discussion emphasizes the general concepts of the energy budget.

Radiation

All substances with a temperature above absolute zero (0°K) emit electromagnetic radiation, as determined by:

$$W = \varepsilon \sigma T^4 \quad \textbf{(3.2)}$$

where W = emission rate of radiation in cal/cm^2/min (langleys/min); ϵ = emissivity, the radiation emitted from a substance divided by the radiation emitted from a perfect blackbody (solid terrestrial objects have emissivities of 0.95–0.98, often approximated as 1.0); σ = Stefan-Boltzmann constant (8.132×10^{-11} cal/cm^2/°K^4/min); and T = absolute temperature in °K (°C + 273).

The amount of radiation at a particular wavelength is temperature dependent, just as the emission rate of radiation is. A perfect blackbody absorbs and emits radiation in all wavelengths. By convention, radiation is separated into (1) *shortwave*, or solar, radiation (sometimes called *insolation*), which includes wavelengths up to 4.0 μm, and (2) *longwave*, or terrestrial, radiation, which is radiation above 4.0 μm. As temperature increases, the greatest magnitude of emitted radiation occurs at shorter wavelengths; that is, the hotter the substance, the shorter the wavelength. Therefore, the sun emits radiation at shorter wavelengths than do terrestrial objects. A doubling of the absolute temperature increases the emission rate of radiation 16-fold. The sun has a temperature of about 6000°K and emits about 10^5 cal/cm^2/min, whereas a soil surface with a temperature of 300°K (27°C) emits about 0.66 cal/cm^2/min. The radiant environment of soil, plant, water, and snow surfaces is determined by both shortwave and longwave processes of radiation.

Shortwave radiation comprises direct solar radiation (W_s) and diffuse radiation (w_s). The latter includes scattered and reflected solar radiation. Scattering is caused mainly by air molecules; reflection is from clouds, dust, and other atmospheric particles. Diffuse skylight averages about 15% of the total downward stream of solar radiation.

The total amount of shortwave radiation that terrestrial surfaces absorb depends on the *albedo,* or shortwave reflectivity, of terrestrial objects. *Albedo* (α) is the proportion of the total shortwave radiation ($W_s + w_s$) reflected by an object. Light-colored surfaces have a higher albedo than dark-colored surfaces (Table 3.1). The net shortwave radiation at a surface is then determined as $(W_s + w_s) - \alpha(W_s + w_s)$, or $(1 - \alpha)(W_s + w_s)$.

The atmosphere and all terrestrial objects emit *longwave radiation.* The primary longwave-emitting constituents in the atmosphere are CO_2, O_3, and liquid and vapor forms of H_2O. Soil and plant surfaces reflect only a small portion of the total downward longwave radiation (I_a); therefore, terrestrial objects are usually considered blackbodies in terms of longwave radiation. The net longwave radiation at a surface is the difference between incoming (I_a) and emitted (I_g) longwave radiation: $I_a - I_g$.

Net radiation, or the net all-wave radiation (R_n) is the resulting radiant energy available at a surface:

$$R_n = (W_s + w_s)(1 - \alpha) + I_a - I_g \tag{3.3}$$

By measuring incoming and outgoing shortwave and longwave radiation over a surface, net radiation is the residual term.

TABLE 3.1. Albedos of natural terrestrial surfaces

Terrestrial Surface	Albedo (α) (%)
New snow	80–95
Old snow	40
Dry, light sand	35–60
Dry grass	20–32
Cereal crops	25
Eucalyptus	20
Mixed hardwood forests (in leaf)	18
Rain forests	15
Pine forests	10 –14
Bare wet soil	11

Source: Lee 1980, Reifsnyder and Lull 1965, U.S. Army Corps of Engineers 1956.

ENERGY RELATIONSHIPS OF WATER

Latent heat of fusion: 80cal/g required to change water from solid ice (at 0°C) to liquid without changing the temperature.

Specific heat of ice: 0.5 cal/g/°C.

Latent heat of vaporization (L): energy required to change from liquid to vapor state without changing temperature; varies with temperature as follows:

Temperature (°C)	L (cal/g)
0	597.3
5	594.5
10	591.7
15	588.9
20	586.0
30	580.4

Specific heat of liquid water: 1 cal/g/°C.

Energy Budget

Net radiation can be either positive or negative for a particular interval. A positive R_n represents excess radiant energy for some time interval, which according to the conservation of energy principle must be converted into other nonradiant forms of energy. When positive, R_n can be allocated at a surface as follows (for a snow-free condition):

$$R_n = (L)(E) + H + G + P_s \tag{3.4}$$

where L = latent heat of vaporization (cal/g); E = evaporation (g/cm^2 or cm^3/cm^2); H = energy flux that heats the air, or sensible heat (cal/cm^2); G = heat of conduction to ground, or rate of energy storage in terrestrial system (cal/cm^2); and P_s = energy of photosynthesis (cal/cm^2). The latent heat of vaporization (L) and evaporation (E) are usually expressed as a product (LE), which represents the energy available for evaporating water.

Net radiation is important from a hydrologic standpoint because it is the primary source of energy for evaporation, transpiration, and snowmelt. The allocation of net radiation in snow-free systems is dependent largely upon the presence of liquid water. If water is abundant and readily available at the evaporating surface, as much as 80–90% of the net radiation can be consumed in the evaporative process (LE). Little energy is left to heat the air (H) or ground (G). If water is limited, a greater amount of net radiation energy is available to heat the air, the ground surface, and other terrestrial objects. Losses (or gains) of energy to the interior earth do not change rapidly with time and are usually small in relation to the net radiation. Similarly, energy consumed in photosynthesis (although of immeasurable importance to life on earth) is a small portion of the net radiation and is usually not considered. When snow is present, the majority of net radiation can be apportioned to snowmelt (see Chapter 15).

TABLE 3.2. Energy budget measurements at Akron, Colorado, USA, for three different conditions

	R_n	G	H	LE
Site Description		(cal/cm²/day)		
Dry bare soil	284	18	220	46
Wet bare soil	226	58	–52	220
Oasis condition	388	18	–180	550

Source: Adapted from Hanks et al. 1968, as reported by Hanks and Ashcroft 1980.
Note: R_n = net radiation; G = heat of conduction to ground; H = sensible heat; LE = energy used in evaporation.

Energy budget applications to watersheds are concerned mainly with net radiation (R_n), latent heat (LE), and sensible heat (H). When the energy available for evaporating water from a watershed, latent heat, is considered, the contribution of sensible heat from adjacent areas cannot be neglected. The best example of lateral sensible-heat contributions is that of an oasis, where a well-watered plant community can receive large amounts of sensible heat from the surrounding dry, hot desert.

Energy budgets for an oasis and two different surface conditions at the same site are compared in Table 3.2. Note that for the dry bare soil condition, much of the net radiation is used to heat the air (H = 220 cal/cm²/day). In contrast, for the oasis condition, 180 cal/cm²/day of sensible heat are added into the oasis. This energy is in addition to net radiation. Box 3.1 illustrates the energy budget calculations for an oasis condition. Similarly, an island of tall forest vegetation presents more surface area than low-growing vegetation does (e.g., agricultural crops). The greater, laterally exposed surface area intercepts more solar radiation and more sensible heat from moving air masses if the vegetation is cooler than the air. The total latent heat flux then is determined by:

$$LE = R_n + H \tag{3.5}$$

Such lateral movement of warm air to cooler plant-soil-water surfaces is called *advection*. *Convection*, in contrast, describes the vertical component of sensible-heat transfer. The combination of advection and convection over and within irregular tree canopies can contribute significant quantities of sensible heat for evapotranspiration.

Water Flow

The flow of water through the soil-plant-atmosphere system is analogous to the flow of electrical current in an electrical circuit (Fig. 3.1A). The soil can be represented as a variable resistor that changes with soil water content and soil-root interfaces. As the soil dries, more resistance is offered to flow. However, the resistance to flow becomes offset as roots grow into moist soil.

Evaporation and transpiration from the soil-plant system require that liquid water must flow either to the *soil-atmosphere* interface or to plant roots, from where it then moves to the *leaf-atmosphere* interface (Fig. 3.1B). Therefore, we must also understand how liquid water flows in soils and plants.

Water Flow in Soil

Processes of evaporation and transpiration in most watersheds are controlled by the water flow through unsaturated soils. The driving force in the system is the difference in

Box 3.1

Energy Budget Components for Oasis Conditions at Aspendale, Australia (adapted from Penman et al. 1967)

LE = energy used in evaporation
R_n = net radiation = 433 cal/cm²/day
G = heat of conduction to ground = 21 cal/cm²/day
H = sensible heat = –183 cal/cm²/day

The energy consumed in evapotranspiration was:

$LE = R_n - G - H = 433 - 21 - (-183) = 595$ cal/cm²/day

Assuming the latent heat of vaporization to be 585 cal/cm³, the actual evapotranspiration (ET) would be:

$$E = \frac{595 \text{ cal/cm}^2\text{/day}}{585 \text{ cal/cm}^3} = 1.02 \text{ cm/day}$$

water potential (ψ) between two points; the concept being that water flows from a region of higher free energy (higher ψ) to a region of lower free energy (lower ψ). *Water potential* is the amount of work that a unit volume of water is capable of doing in reference to an equal unit of pure, free water at the same location in space. It can also be considered as the minimum work needed to move a unit of water from the soil that is in excess of the work needed to move an equal unit of pure, free water from the same location in space.

The water potential concept is derived from the second law of thermodynamics and relates to the free energy of water. A system that physically or chemically restricts the free energy of water results in negative values of water potential ($-\psi$). Gradients of negative water potentials are most common in soil-plant-atmosphere systems.

Soil water potential (ψ_s) is determined by several potentials:

$$\psi_s = \psi_g + \psi_p + \psi_o + \psi_t + \psi_m \tag{3.6}$$

where ψ_g = gravitational potential; ψ_p = pressure potential; ψ_o = osmotic, or solute, potential; ψ_t = thermal potential; and ψ_m = matric potential.

The gravitational potential exerts a downward pressure as a function of the weight of water as determined by the height of the water column, gravity, and density. It is the difference in elevation of a point in the system with respect to a stable reference datum, often mean sea level.

For a point in saturated soil or groundwater situations, the pressure potential is positive. The sum of gravitational and pressure potential is the total hydraulic potential. In the example in the top frame of Figure 3.2, the pressure potential (ψ_p at C) corresponds to the depth below the free-water surface, a depth of 15 cm. The pressure potential is

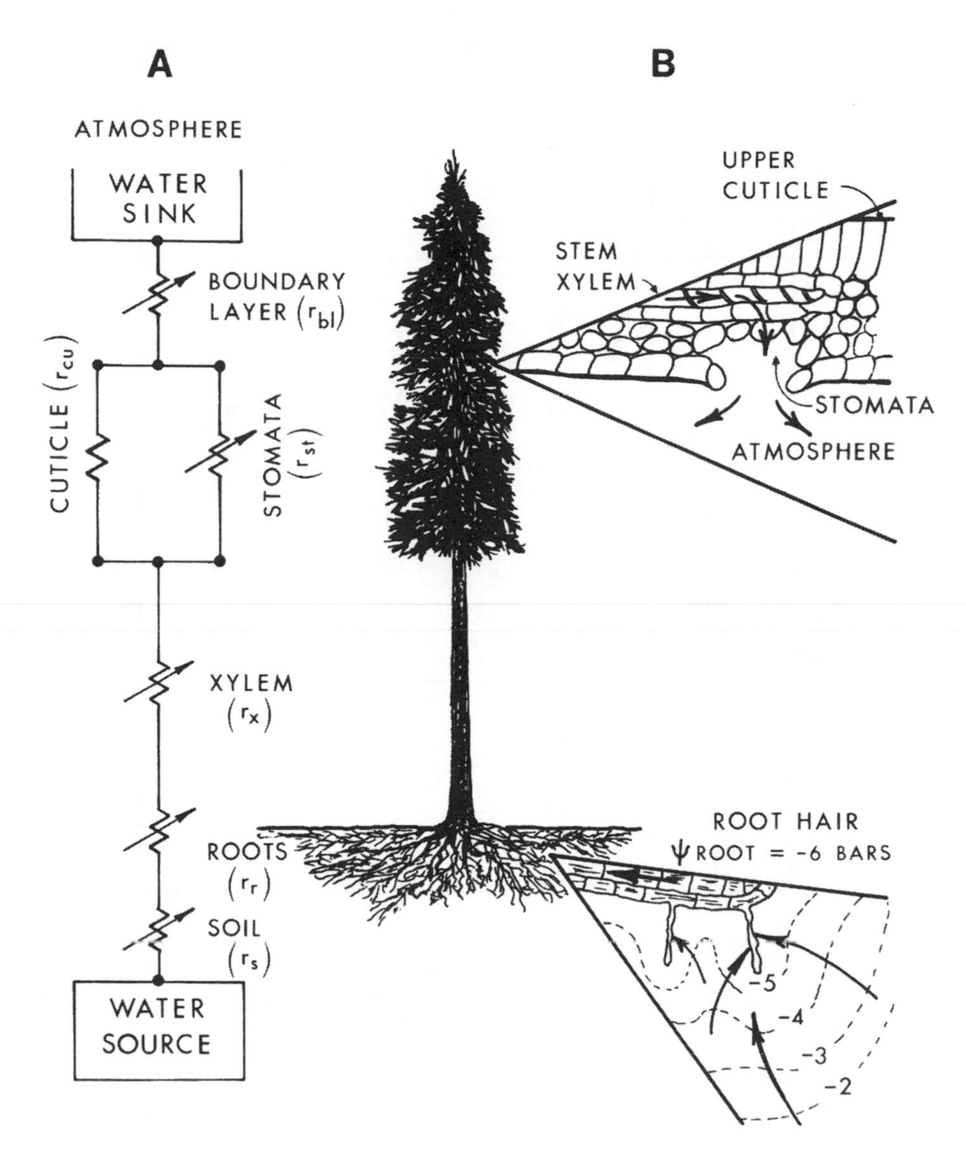

FIGURE 3.1. A representation of soil-plant-atmosphere resistances to water flow (A) and corresponding flow through roots and leaves (B) (adapted from Rose 1966 © Pergamon Books Ltd., by permission). Flow of water can be described by $v = \Delta\psi/r$, analogous to Ohm's law.

zero at the water table level (ψ_p at W.T. in Figure 3.2), and soil water potential becomes negative above the water table because of matric potential. *Matric potential* (ψ_m) describes the physical attraction of water to soil particles by both capillary and adsorptive forces. As the soil dries, matric forces in the soil increase and tend to oppose water flow. Therefore, more energy must be exerted to move a quantity of water in a drier soil

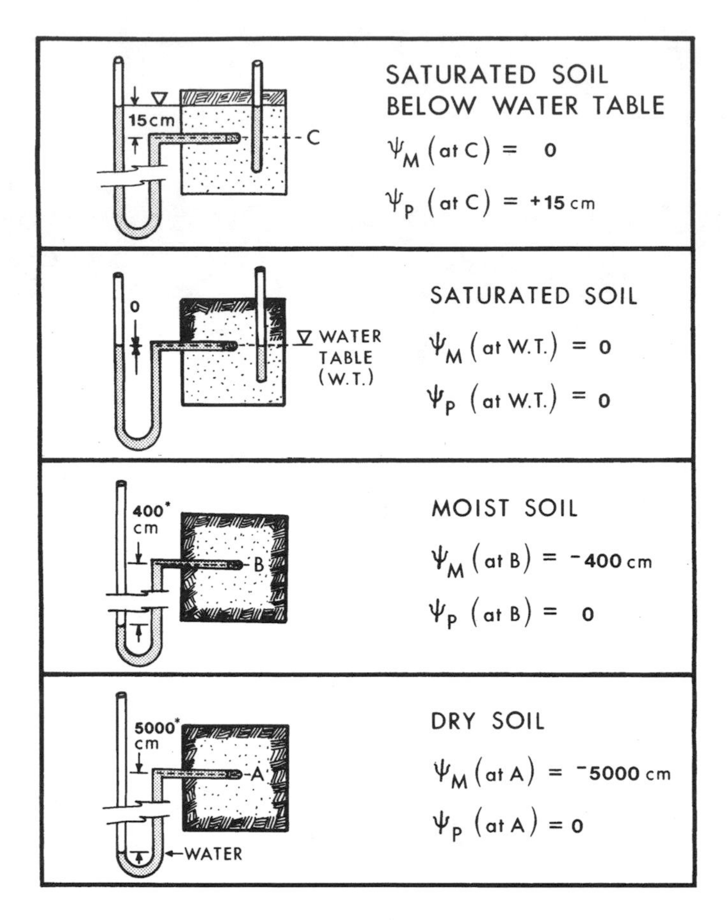

FIGURE 3.2. Matric potential (ψ_m) and pressure potential (ψ_p) of a dry soil (A), moist soil (B), at the water table (W.T.), and in a saturated soil (C) measured with a water manometer or tensiometer (adapted from Hanks and Ashcroft 1980). *Vertical distance not to scale; porous ceramic cup at A,B,W.T.,C.

UNITS COMMONLY USED TO EXPRESS WATER POTENTIAL

1 kilopascal (kPa) = 10 millibars (mb)
= 10.2 cm H_2O
= 0.01 atmosphere (atm)
= 0.75 cm Hg

than in a more moist soil. If the moist and dry soils in Figure 3.2 were connected, water would flow from B to A.

Solutes in the soil water solution cause osmotic potential (ψ_o). Solutes lower soil water potential (ψ_s) because they attract water molecules in the form of hydrated shells. This component normally has little effect on liquid-water flow in soils except at the soil-root interface. Here, a semipermeable membrane (permeable only for water molecules) must be crossed for water to enter the root vascular system.

Thermal potential (ψ_t) is usually neglected. It would appear, however, that soil temperature gradients (over time and space) would affect (ψ_s) under certain conditions. Higher thermal energy would increase (ψ_s) and, thereby, change the flow of water in soil.

Water flow in unsaturated soils is primarily a function of matric potential gradients ($d\psi_m/d_x$), which are directly related to gradients in soil water content. As the roots take up water, the soil water content immediately adjacent to the roots is depleted. The lower moisture content results in a greater attraction between water and the soil particles next to the root. A gradient in soil water content and, hence, water potential is then established. In all cases, water flow is from a region of higher (ψ_s) to a region of lower (ψ_s) (Fig. 3.1).

The driving forces operating in unsaturated flow have been described, but the velocity of soil water flow (v) is determined by:

$$v = \frac{\Delta\psi}{r_s} = -k_v\frac{d\psi}{dx} \qquad \textbf{(3.7)}$$

where v = velocity (cm/sec); $\Delta\psi$ = water potential difference (kPa); r_s = resistance of any component (kPa • sec/cm); k_v = hydraulic conductivity (cm/kPa • sec); x = distance over which gradient is present (cm); and $d\psi/dx$ = total water potential gradient.

This relationship (Eq. 3.7) is analogous to Ohm's Law for electrical current. The velocity of water flow is proportional to $d\psi/dx$ and k_v. The pore size, pore geometry of the soil, and the soil water content affect the value of k_v for unsaturated conditions. Soil texture and structure affect soil water flow in unsaturated conditions and, as discussed in Chapter 5, in saturated conditions as well. Total flow through a cross-sectional area becomes:

$$Q = Av \qquad \textbf{(3.8)}$$

where Q = flow (cm^3/sec); A = cross-sectional area (cm^2); and v = velocity (cm/sec).

Water Flow in Plants

Water flow through soils is relatively passive until intercepted by the roots of plants. Once the root absorbs soil water, different forces become operative as the major constituents of water potential. The primary components of *plant water potential* are:

$$\psi_{pl} = \psi_o + \psi_p + \psi_t + \psi_g + \psi_m \qquad \textbf{(3.9)}$$

where ψ_{pl} = plant water potential; and ψ_o, ψ_p, ψ_t, ψ_g, and ψ_m are as defined in Equation 3.6.

Isothermal conditions are usually assumed within the plant system, thereby eliminating ψ_t. Matric potential, an important part of soil water potential, is a minor constituent of ψ_{pl} and is usually ignored. Gravitational potential is not generally considered for herbaceous plants but can be important in tall trees. For example, about 0.3 bar/m of tree height must be overcome for water to move to the top of a tree. Water potential gradients between cells of plants, therefore, are due to the interaction of the osmotic potential (ψ_o) with the pressure potential (ψ_p).

In metabolizing plant cells, fluctuations of solute concentrations affect the energy status of cellular water. When solutes are added to cellular water, the ψ_{pl} of the cell is lowered. This steepens the water potential gradient between surrounding cells, causing water to move through differentially permeable membranes into the cell. Water can enter from intercellular regions as well. The increased water content in the cell causes an increased turgidity (like blowing up a balloon), which (in turn) opposes water entry. The final water potential of the cell is determined by these opposing forces, which can be expressed in terms of pressure as follows:

$$\psi_{pl} = P_t - P_o \tag{3.10}$$

where P_t = turgor pressure; and P_o = osmotic pressure. Whereas the addition of solutes lowers ψ_{pl}, increased turgor pressure increases ψ_{pl}.

The analogy to Ohm's law can again be used in describing water flow through the plant:

$$q = -k_w A(\Delta\psi_{pl}) = -A\frac{\Delta\psi_{pl}}{r_{pl}} \tag{3.11}$$

where q = water flow (cm^3/sec); k_w = water permeability (cm/sec • kPa); A = membranous area (cm^2); $\Delta\psi_{pl}$ = water potential difference (kPa); and r_{pl} = resistance of plant components (kPa • sec/cm).

The leaf of a plant, which is the primary food-manufacturing center of the plant, maintains the water potential gradient since solute concentrations are increased by photosynthesis. This osmotic potential gradient alone is probably sufficient to cause some water flow up the stem. Transpiration reduces the pressure potential in the leaf and, thereby, steepens the total potential gradient from root to leaf. Liquid water moves through the cells in the leaf and eventually reaches substomatal cavities of the leaf, from where it is evaporated. Water vapor then diffuses out and through the leaf boundary layer to be dissipated by turbulent mass transport into the atmosphere.

Vapor Flow

Evapotranspiration requires both energy and conditions that permit water vapor to flow away from evaporating or transpiring surfaces. Water molecules migrate from the liquid surface as a result of their kinetic energy. This transfer involves a change of state from liquid to vapor, during which the energy inputs to the vaporization (or transpiration) process occur. Vapor flow is initially a *diffusion* process in which water molecules diffuse from a region of higher concentration (evaporating surface or source) toward a region of lower concentration (sink) in the atmosphere. Water molecules at the soil-atmosphere or leaf-atmosphere interface must first diffuse through the *boundary layer.* This is also the layer through which sensible heat is transferred by molecular conduction only. The thin layer of air adjacent to evaporating and transpiring surfaces, which is at maximum thickness under still-air conditions, can be as thin as only 1 mm or less. Wind and air turbulence reduce the boundary layer thickness, but there is no turbulent flow in the boundary layer itself.

Once water molecules exit the boundary layer, they move into a turbulent zone of the atmosphere where further movement is primarily by *mass transport* (turbulent eddy movement). In mass transport, whole parcels of air or eddies with water vapor and sensible-heat flow in response to atmospheric pressure gradients, which cause the air parcels to flow both vertically and horizontally.

Evaporation describes the *net* flow of water away from a surface. Therefore, water molecules also return to the evaporating surface by mass transport and diffusion processes. If the amount of vapor arriving equals the amount leaving, a steady state exists and no evaporation occurs. If more molecules arrive than leave, a net gain results, called *condensation*.

The vapor pressure of water molecules at the evaporating surface must exceed the vapor pressure in the atmosphere for evaporation to occur. Under natural circumstances, the vapor pressure of liquid water is mainly a function of its temperature, although solute content, atmospheric pressure, and water surface curvature in capillaries can also be important. The vapor pressure of water molecules in the atmosphere is primarily a function of air temperature and the humidity of the air (Fig. 2.2). The vapor pressure gradient between evaporating surfaces and the atmosphere is the driving force that causes a net movement of water molecules. For this reason, the vapor pressure deficit between an evaporating surface and the atmosphere (the difference between points A and B in Figure 2.2) is often a component of empirical equations like the following (Dunne and Leopold 1978):

$$E_o = N(e_s - e_a)f(u) \qquad \textbf{(3.12)}$$

where E_o = evaporation from a water body (mm); N = mass transfer coefficient, determined empirically; e_s = vapor pressure of water surface (mb); e_a = vapor pressure of the air (mb); and $f(u)$ = function of wind speed (km/day).

Formulas like Equation 3.12 are useful for estimating vapor flow away from free-water surfaces, such as a lake, but they cannot be directly applied to *nonsaturated* conditions that prevail in watersheds. The vapor pressure deficit (vpd) or gradient ($e_s - e_a$) above a free-water surface can be determined from measurements of surface water temperature, air temperature, and the relative humidity of the air.

Conceptual relationships of the evaporative process that are applicable to complex surfaces, including plants and soil, have been developed. Such models assume that the vapor flow away from evaporating (E) or transpiring (T) surfaces is directly proportional to the vapor pressure deficit and inversely proportional to the resistance (R_v) of air to the molecular diffusion and mass transport of water vapor. That is:

$$E \text{ or } T = \frac{vpd}{R_v} \qquad \textbf{(3.13)}$$

The R_v term includes a resistance for the turbulent layer of the atmosphere (r_e) and the boundary layer (r_{bl}) and internal resistance characterizing air-filled soil pores (r_s) or plant pores (r_{st}), called *stomates*. For convenience, the two atmospheric resistances are sometimes combined as r_a and the two internal resistances as r_n. The internal resistance is necessary because the liquid-air interface is often beneath the external surface of soil or plants. Water vapor diffusing outward encounters resistance from the air in soil and plant pores before reaching the boundary layer. For a wet external surface, the diffusive path length consists only of the boundary and turbulent layers of the atmosphere. The extra path length for *dry* soil or plants increases the total resistance to flow.

EVAPORATION

Previous sections of this chapter described the flow of energy and liquid water to evaporating or transpiring surfaces, and the flow of water vapor away from these surfaces into the atmosphere. In this section, these three flows are discussed for the simplest watershed situation—evaporation from a water body (pond, lake, or reservoir) or from a bare soil surface.

Evaporation from Water Bodies

Evaporation from ponds, lakes, or reservoirs is determined only by energy and vapor flows. There are no direct methods of measuring lake evaporation. Evaporation from a lake can be estimated using a water budget approach in which all components of the water budget except evaporation are either measured or estimated over a period of time (t):

$$E = Q_i + GW_i + P - Q_o - GW_o - \Delta S \quad \textbf{(3.14)}$$

where E = lake evaporation (m³/t); P is the precipitation on the lake surface (m³/t); Q_i and Q_o are surface water flows into and out of the lake (m³/t), respectively; GW_i and GW_o are subsurface or groundwater flows into and out of the lake (m³/t) respectively; and ΔS is the change in storage (m³/t).

Measuring of all the inflow and outflow components of a lake is difficult. Although it is straightforward to measure surface streams that enter and leave a lake, surface runoff directly into the lake from the surrounding area cannot be directly measured. Neither can the subsurface or groundwater flows be directly measured. In some circumstances, these components can be estimated with the use of either hydrologic models or methods of proportioning surface and groundwater components (discussed in chapters 5 and 17).

Because of the difficulty and costs of conducting detailed water budget analyses, lake evaporation is often estimated using methods such as Equation 3.12 or the more standard *pan evaporation* method. With the latter approach, evaporation is measured from a pan and a coefficient is applied as follows:

$$E = C_e E_p \quad \textbf{(3.15)}$$

where C_e = pan coefficient; and E_p = pan evaporation (mm/day).

The standard pan in the United States, a National Weather Service Class A pan, is a metal cylinder 122 cm in diameter and 25 cm deep. Water depth is maintained at 18–20 cm and measured daily with a hook gauge in a stilling well.

The *pan evaporation method* has been used extensively to estimate evaporation from lakes and reservoirs, for which C_e usually ranges from 0.5 to 0.8. Average annual pan coefficients of 0.70–0.75 are often used for lakes where they have not been derived experimentally. Seasonal relationships of C_e for lakes tend to be less than one for the warm season and can be greater than one for deep lakes during periods of cold air temperature. Pan evaporation data have also been used as estimates of potential evapotranspiration as described later in this chapter. In some instances, values of C_e have been derived to simulate changes in transpiration for certain plant species.

Evaporation from Soil

Evaporation from a soil is a more complex phenomenon than that from a water body. Given a bare, flat, wet soil surface, the water supply initially is unlimited and the amount of evaporation depends on the *energy supply* and the *vapor pressure gradient.* Under open atmospheric conditions, sufficient vapor pressure gradients usually exist and are maintained, causing only the energy supply to limit evaporation from the wet, exposed soil. If a piece of transparent plastic sheeting were laid over this soil, evaporation would cease because vapor flow would be blocked, even though energy and water flows would still converge at the active surface. In natural environments, however, energy inputs to the active surface increase the vapor pressure of water, steepening the vapor pressure gradient. In this case, evaporation proceeds at rates similar to those of free-water surfaces, assuming equal energy input. As evaporation occurs, lost water is replaced by water moving up from below the active surface in connecting water films around soil particles and through capillary pores. As the soil surface dries, a gradient in total water potential ($d\psi/dx$) is established; water is forced to move from the zone of higher potential (lower layer, wet zone) to the region of lower potential (upper layer, drier evaporating surface). As the soil dries, hydraulic conductivity also decreases. After a period of drying, the rate of water flow through the soil limits the rate of evaporation at the soil surface.

Water flow in moist soil is primarily liquid, but as soils dry, vapor diffusion through pores becomes more dominant. At about –1500 kPa water potential, the flow must be mainly as vapor or some combination of vapor and liquid. Liquid flow is reduced greatly at –1500 kPa because the continuity of capillary water and water films becomes disrupted; this is the point at which most plants become wilted and is referred to as the permanent wilting point. Water cannot move as rapidly through soils in vapor form as in liquid form; consequently, as the soil dries, a deficient water supply limits evaporation at the active surface, regardless of energy input. This condition is reached sooner in sandy soils than in clay soils, which have smaller pores that permit water films to remain intact for a longer period. Fine-textured soils retain pore water continuity at lower water contents than coarse-textured soils.

Evaporation from wet soil will occur initially at a rate limited by energy flow to the active surface. With time, evaporation rates decrease because the water flow to the active surface is too slow to keep pace with the energy input. How much water will be evaporated from a soil under these conditions depends largely upon soil texture.

TRANSPIRATION

Transpiration is a biological modification of the evaporation process, because it is a function of the plant system and the environment. This modification is more efficient because of the large evaporating surface presented by plant foliage exposed directly to turbulent airflows above the soil boundary layer. Plants affect the amount of water transpired by stomatal regulation (the variable stomatal resistor in Figure 3.1), structural and physiological adaptations, and rooting characteristics. In essence, plants provide a variable conduit for water to flow from the soil water reservoir to the active evaporating surface at the leaf-atmosphere interface. This conduit bypasses the higher resistance offered by dry surface soils. To understand the importance of transpiration and associated land management implications, the basic process is examined below.

Once liquid water reaches cell surfaces within the leaf, 585–590 cal/g (for temperatures of most terrestrial systems) are required for vaporization. After vaporization, the water vapor flows through intercellular spaces to the substomatal cavity, between guard cells of stomata, and into the atmosphere in response to the *vapor pressure gradient* at the leaf surface. Water vapor can also take a parallel path through leaf cuticles, but this pathway usually offers more resistance (r_{cu}) to flow than that through stomata, except when stomata are closed tightly. Consequently, r_{cu} is considered large, making the variable resistor of stomata (r_{st}) the primary regulator of transpiration. The magnitude of r_{st} is proportional to the degree of opening of the stomatal pore, or stomatal aperture, while the magnitude of r_{cu} is a function of cuticular integrity and thickness.

The total resistance offered to vapor flow by the leaf (r_l) is:

$$\frac{1}{r_l} = \frac{1}{r_{st}} + \frac{1}{r_{cu}} \qquad \textbf{(3.16)}$$

or

$$r_l = \frac{r_{st} r_{cu}}{r_{st} + r_{cu}} \qquad \textbf{(3.17)}$$

The effect that the stomatal opening has on transpiration depends on the thickness of the boundary layer (r_{bl}) surrounding leaf surfaces; this is evident in Figure 3.1, because r_{bl} is in series with r_{st}. The total diffusive resistance (r) is described by:

$$r = r_{bl} + r_l \qquad \textbf{(3.18)}$$

Because boundary layer resistance is related inversely to wind speed, r_{bl} will be large under still-air conditions, causing r_{st} to have less effect on transpiration. Under windy conditions, changes in stomatal aperture strongly affect rates of transpiration (Fig. 3.3).

The vapor pressure gradient, which causes vapor to flow from the substomatal cavity, is usually created and maintained by energy inputs to the leaf. This energy causes the vapor pressure of leaf water to be greater than the partial pressure of water vapor in the surrounding atmosphere. Consequently, more water molecules exit the liquid-air interface than enter, and the gradient in liquid-water potential, $d\psi/dx$, is steepened from the leaf down to the root surface. Water flows into the plant until some permanently limiting level of soil water content is reached. This critical level varies for each plant

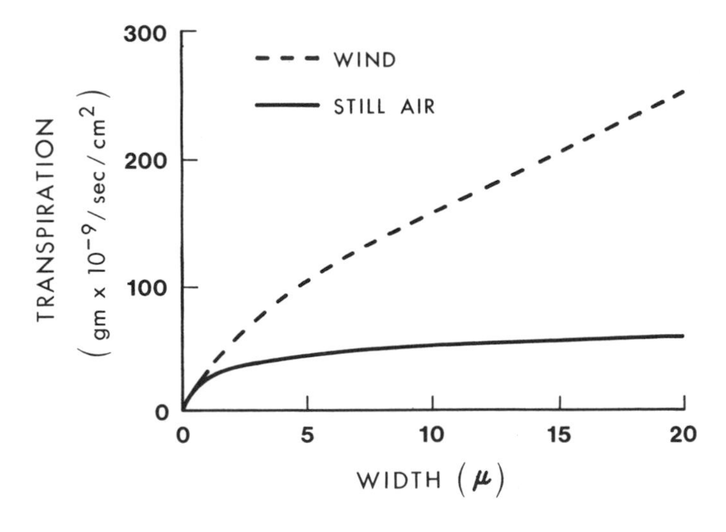

FIGURE 3.3. Relation between stomatal width and transpiration in still air and in wind (from Slatyer 1967, after Bange 1953, by permission).

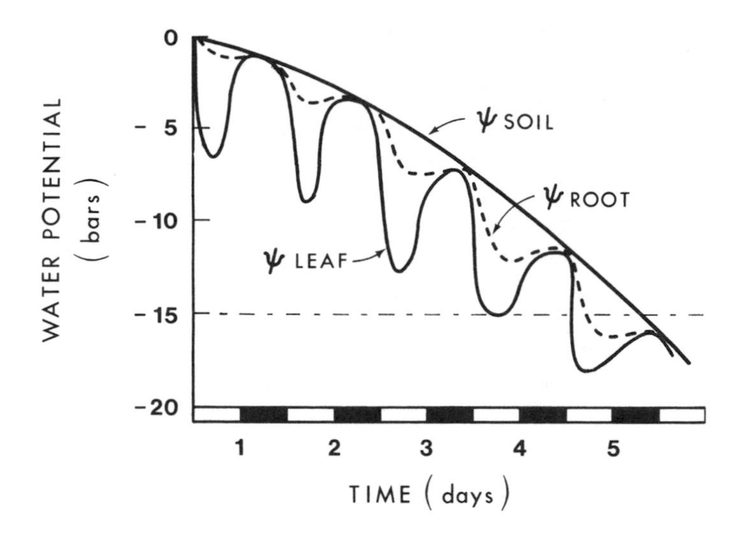

FIGURE 3.4. Changes in water potential (ψ) of soil, root, and leaf as transpiration occurs, being with a soil near field capacity and proceeding until the permanent wilting point (–15 bars) is reached (from Slatyer 1967, by permission).

species, although $\psi_s = -15$ bars (–1500 kPa) is often considered the limit for most plants (Fig. 3.4).

Measurement of Transpiration

Most methods of measuring transpiration have been developed for agricultural crop plants. Therefore, few are applicable for field measurements of larger trees and shrubs that are of interest in forest hydrology. For the most part, transpiration cannot be measured directly in the field without some type of major disturbance to the plant. This section of the chapter provides a brief overview of some of the most common methods used in the field.

Potted Plants and Lysimeters

Soil-plant systems contained in small covered pots or larger tanks, called *lysimeters,* can be used to measure transpiration. Water budget analyses, in which every component is measured directly except transpiration, are performed on these systems.

The potted-plant method is suited for small, individual plants. The bottom of the pot is perforated to allow water to drain freely. Typically, the soil is wetted thoroughly, the soil surface is covered (usually with plastic sheeting) to prevent soil evaporation, and the soil is allowed to drain. After all free water drains from the soil, the potted plant is weighed; the soil is assumed to be near field capacity (see below) at this point. At some determined time, the potted plant is reweighed, and the difference in weights is equated to transpiration loss over the period. This method is not suited for large plants, but small trees and shrubs can be measured and transpiration rates compared for different environmental conditions or treatments.

Lysimeters are tanks designed to hold a larger mass of soil and usually more than one plant and can be either a weighing type (similar to the potted-plant method) or a drainage type (see Chapter 18). With the drainage type, any surface runoff or drainage from the bottom is collected and measured. As with the potted-plant method, the only unmeasured part of the water budget is transpiration or total *ET*. To obtain estimates of transpiration, the soil surface must be covered.

Because of the greater volume of soil, lysimeters allow plant roots to develop more naturally, and boundary conditions are not as severe as with potted plants. Although usually designed for smaller plants, a lysimeter was developed by Fritschen et al. (1977) to measure transpiration of a 28 m high Douglas-fir tree. The lysimeter weighed 28,900 kg and could detect changes in weight of 6.3 g. The difficulty of construction and costs associated with such lysimeters make them impractical for most situations.

Tent Method

In the tent method, a plant is enclosed with plastic sheeting, and the rate and moisture content of air entering and leaving the tent are monitored (Fig. 3.5). When the amount of moisture in the air leaving the tent exceeds that entering the tent, the difference is due to transpiration if the soil surface is covered, or *ET* if soil evaporation is permitted. The method excludes rainfall interception from the *ET* process.

This method facilitates measurement of the transpiration of rather large shrubs and small trees without transplanting or disturbing the soil. One disadvantage is the buildup of heat in the tent caused by the trapping of radiation, that is, the greenhouse effect. This problem can be partly overcome by maintaining adequate air circulation in the tent. The same principle is used for small enclosures for individual leaves. Nevertheless, the environment surrounding the plant is artificial, and the measured transpiration rates might not coincide with rates outside a tent. The method can be used to indicate relative transpiration differences between plant species in adjacent plots or the same species under different treatments.

Measurement of Sap Flow

Transpiration rates for woody plants can be determined by measuring the net rate of the ascent of xylem sap through stems. The basis for this method is that most of the water that moves up the stem of a plant is largely in response to transpiration. Other causes of sap flow can be ignored, since there is a negligible amount of change in stem storage over time and only 2–5% of sap flow is used in photosynthesis. Two basic sap flow methods can be used: (1) the heat pulse or tracing method and (2) the steady state or stem heat balance method.

The heat pulse or tracing method measures the velocity of a short pulse of heat, or a tracer such as deuterium, moving up the xylem sap of woody stems. Sensors are located in the xylem both upstream and downstream of the source of heat (or tracer), and measurements are taken over time to determine net upstream flow of sap (Fig. 3.6). Sap flow velocity is useful in detecting when and at what velocity sap flows in response to transpiration losses. To estimate the volume of water loss per plant, the sap flow velocity needs to be related to the area of the stem through which sap flows. Hinckley et al. (1978) reported maximum sap velocities ranging from 1550–6000 cm/hr for ring porous species such as *Quercus*, and 75–200 cm/hr for nonporous coniferous species.

FIGURE 3.5. Triple-inlet evaporation tent (from Mace and Thompson 1969): 1 = inlet; 2 = squirrel cage blower; 3 = inlet humidity thermometer; 4 = perforated polyvinyl curtain; 5 = outlet; 6 = outlet humidity thermometer.

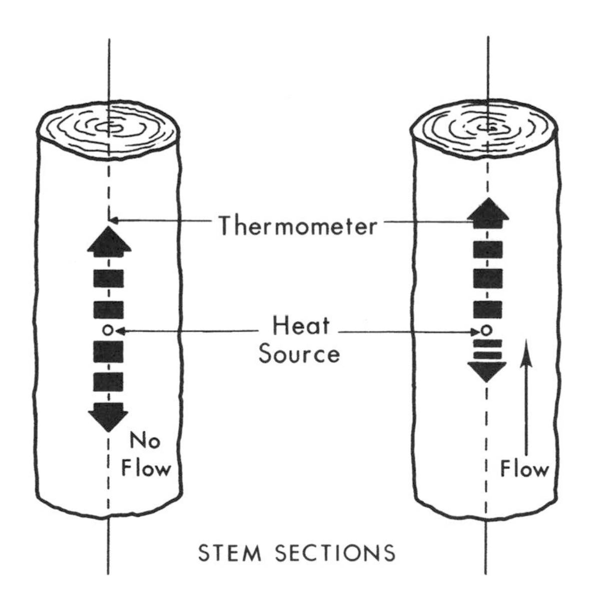

FIGURE 3.6. Sapflow velocity method using heat as a tracer; heat dissipation in stem on the left indicates no transpiration, while the stem on the right indicates transpiration (from Swanson and Lee 1966).

The steady state or stem heat balance method directs a known amount of heat into the stem and uses sensors that are external to the stem. The method is a heat balance approach that relates heated stem tissue to sap flow rates. When water flow through the stem is relatively constant, the temperature of the sapwood will reach a stable value (Swanson 1994). If there is minimal heat loss, the heat that is transported out of a section of stem is related to the amount of heat added to the section.

Other Methods

There are several indirect methods of estimating transpiration activity that do not indicate rates or volumes over time. One such method, called *quick-weighing*, has been used to overcome the difficulty of weighing large plants in the field. A leaf or small branch is cut off, weighed immediately, then reweighed after some short period. The change in weight is related to transpiration. The severity of plant disturbance makes this a questionable method. However, one can get comparative data in the field that can be used to indicate relative transpiration activity.

Another indirect method uses small chambers, or *porometers*, to indicate stomatal openings. Usually, the degree of penetration of some solvent or dye is measured and related to stomatal aperture. This method indicates relative transpiration activity but does not measure volume over time.

Methods of estimating evapotranspiration (discussed later in this chapter) can sometimes be modified to estimate transpiration rates of plants. Soil water depletion measurements in the field, for example, can be indicators of transpiration rates when soil surface evaporation is prevented.

Interception and Transpiration Relationships

When a vegetative surface intercepts rainfall, part of the energy normally allocated for transpiration is used in evaporation at the leaf surface. Some compensation occurs in that transpiration rates are often reduced when the foliage is wet. The net effect, however, is usually a greater total loss of water by vaporization than would have occurred via transpiration alone. Evaporation rates of wet canopies have been reported to be two to three times greater than transpiration rates for forest stands, largely because evaporation of a wet surface is not affected by stomatal resistance. Also, forest canopies are rough surfaces projected into the more turbulent upper air, where a greater exchange of advective energy results in high evaporation rates. Wet forest canopies generally exhibit higher evaporation rates than wet low-growing crops or grasses, and the evaporation rates can exceed potential evapotranspiration rates as estimated by traditional methods discussed later in this chapter. Furthermore, evaporation rates at night can far exceed transpiration rates because the stomata of most plants close at night.

Effects of Vegetative Cover

The type, density, and coverage of plants on a watershed influence transpiration losses over time. Differences in transpiration rates among individual plants and plant communities can be attributed largely to differences in rooting characteristics, stomatal response, and albedo of plant surfaces. Annual transpiration losses are affected by the length of a plant's growing season. Grasses, other herbaceous vegetation, and agricultural crops generally have shorter growing seasons, and hence shorter active transpiration seasons, than forest vegetation. Likewise, deciduous forests normally transpire over a shorter season than do conifers. Transpiration rates of trees are usually considered to be the greatest of all plants. Wullschleger et al. (1998) reviewed 52 studies conducted since 1970 and reported whole tree water use for 67 species in over 35 genera. Some of these maximum daily rates of transpiration are presented in Table 3.3.

Comparing a bare soil, a herbaceous grass cover, and a mature forest can illustrate effects of changing vegetative cover on transpiration and total *ET* (Fig. 3.7). If soil water is abundant in all three sites, evaporation and transpiration will occur at rates primarily dependent on available net energy, vapor pressure gradients, and wind conditions. Differences in overall vapor loss will be largely the result of differences in advective energy. In such instances, transpiration by plants with a large leaf area and a canopy extending higher above ground can surpass that of smaller plants. With an extensive, dense forest canopy, advection might only affect transpiration at the edges of stands or stand openings. Plant canopies can also increase the rate of vapor flow by creating more turbulent airflow around transpiring surfaces. This effect would be particularly pronounced with some conifers, whose needlelike leaves create numerous small eddies.

TABLE 3.3. Maximum daily transpiration rates for selected tree species (examples from Wullschleger et al. 1998)

	Method	Height (m)	Diameter (cm)	Leaf Area (m^2)	Water Use (kg day^{-1})
Conifers					
Abies amabilis	TM	18	40	151	98
Larix gmelinii	TM	20	25	...	67
Picea abies	TM	25	36	447	175
	TM	17	15	...	66
Pinus caribea	R/SI	7	13	...	100
Pinus contorta	L/P	...	20–26	...	44
Pinus radiata	TM	25	42	300	349
Pseudotsuga menziesii	L/P	28	38	...	64
	R/SI	76	134	...	530
Hardwoods					
Eucalyptus grandis	L/P	6	41	219	174
	TM	34	30	71	141
Eucalyptus regnans	TM	58	89	330	285
Populus x euramericana	L/P	5	...	26	109
Populus trichocarpa x P. deltoides	TM	15	15	...	51
Quercus patraea	TM	15	9	...	10
Salix matsudana	L/P	5	...	28	106
Spondias mombin	TM	23	44	...	80

Note: TM = thermal balance or heat dissipation (sap velocity); R/SI = radioactive or stable isotopes (sap velocity); L/P = lysimeters or large tree potometers.

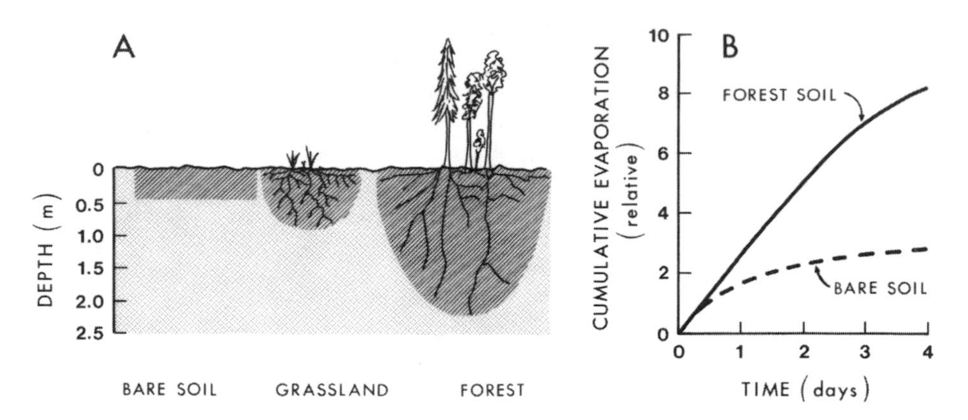

FIGURE 3.7. Effects of changing vegetative cover on transpiration and total evapotranspiration. A: soil water depletion (cross-hatched areas) of bare soil, grass cover, and a mature forest (vertical scale exaggerated); B: the associated accumulated evaporation/evapotranspiration for the bare soil and forest conditions (from Lee 1980, © 1980 Columbia University Press, by permission).

Therefore, larger-canopied trees can potentially transpire larger amounts of water than would otherwise evaporate from a bare soil or transpire from communities dominated by herbaceous plants of smaller stature, such as grasses and forbs.

Once the soil begins to dry, the film of water around soil particles becomes thinner, the pathways of flow become more tortuous, and the hydraulic conductivity decreases. Eventually, the slower rate of water movement through the drier soil limits the rate of evaporation at the soil surface.

Soil water depletion can occur only to a depth of 0.4 m after a given period. Except for very coarse soils, evaporation seldom depletes soil water below a depth of 0.6 m. The flow of water to the evaporating (transpiring) surface of herbaceous vegetation can continue for a longer time because plant roots grow and extend into greater depths (1 m in grassland, Figure 3.7A) and extract water that would otherwise not evaporate from a bare soil in the given time. Deep-rooted forest vegetation can extract water to depths > 2 m and, thereby, has greater access to the soil water reservoir. Over time the differences in evaporation from a bare soil versus *ET* from a forest can be substantial (Fig. 3.7B). Such differences in soil water depletion result in differences in water yield. For a given rainfall or snowmelt event, more water is required to recharge soils under forest vegetation than soils with herbaceous cover. The least amount of water would be needed to recharge bare soil areas. Consequently, the proportion of rainfall or snowmelt that will be yielded as streamflow will be greatest for the bare soil and least for the forested area.

By reasoning alone, it is sometimes possible to estimate relative differences in *ET* among different soil-plant systems. One obstacle to many hydrologic investigations, however, is that of quantifying *ET*. Approaches that can be used to approximate *ET* losses are discussed below.

POTENTIAL EVAPOTRANSPIRATION

The concept of *potential evapotranspiration (PET)* has its origin in evapotranspiration studies of irrigated crops. Potential evapotranspiration was defined as the amount of water transpired in unit time by a short green crop, completely shading the ground, of uniform height, and never short of water (Penman 1948). This definition was supposedly an expression of the maximum *ET* that could occur and was limited only by available energy. This led to thinking that all well-watered soil-plant systems and open bodies of water will lose equal amounts of water, amounts that are controlled by available energy. Of course, available energy (discussed earlier) can differ appreciably among vegetative types, different soils, and bodies of water. Empirical models of energy availability have been developed as workable definitions or indices of *PET*.

Several methods of estimating *PET* are described in the literature, but only a few will be reviewed here. In all cases, equations and relationships used to approximate *PET* should be considered only as indices of *PET*. One should review its origin and understand its limitations and range of applications before applying any *PET* equation.

There are several empirical relationships that can be used to approximate potential evapotranspiration. The *PET* indices that require only air temperature data are attractive to hydrologists because they only require one simple variable to be measured. Thornthwaite's equation (Thornthwaite and Mather 1955) estimates *PET* for 12 hr days and a 30-day month with the equation:

$$PET = 1.6\left(\frac{10T_a}{I}\right)^a \tag{3.19}$$

where *PET* = annual potential evapotranspiration (cm); T_a = mean monthly air temperature (°C); I = annual heat index =

$$\sum_{i=1}^{12}\left(\frac{T_{ai}}{5}\right)^{1.5} \tag{3.20}$$

and $a = 0.49 + 0.0179I - 0.0000771I^2 + 0.000000675I^3$.

Values of *PET* determined thus must be adjusted for the number of days per month and day length (latitudinal adjustment).

Hamon's equation (1961) estimates *PET* by:

$$PET = 5.0\rho_s \tag{3.21}$$

where *PET* = daytime potential evapotranspiration (mm/30-day mo); and ρ_s = saturation vapor density at mean air temperature (g/cm^3).

Penman (1948) combined a simplified energy budget with aerodynamic considerations to estimate evaporation. The Penman equation, perhaps the most widely known method of estimating daily *PET*, is defined as:

$$PET = \frac{\Delta R_n + \gamma E_a}{(\Delta + \gamma)L} \tag{3.22}$$

where *PET* = evaporation or potential evapotranspiration (g/cm^2/day); Δ = slope of saturation vapor curve (mb/°C); γ = psychometric constant (0.66 mb/°C); R_n = net radiation (cal/cm^2/day); $E_a = (e_s - e_a)f(u)$, $e_s - e_a$ = vapor pressure deficit at 2 m height (mb), $f(u)$ = wind function (km/day), approximating atmospheric diffusivity near evaporating surface (g/cm^2/day); and L = latent heat of vaporization (585 cal/g at 21.8°C).

Penman's equation was originally developed to predict evaporation from an open water surface rather than *PET* from a vegetated surface. Modifications to Equation 3.22 have been numerous and include the addition of plant coefficients that express physiological and aerodynamic resistance of vegetation. Although such equations represent more physically based approaches than earlier empirical models, they have limited application in hydrology because of their extensive data requirements.

Estimating Actual Evapotranspiration

Evapotranspiration losses from a watershed are affected by net radiation, rainfall interception, advection, turbulent transport, leaf area, stomatal resistance, canopy resistance, and plant water availability (Zhang et al. 2001). As moisture conditions change, the factors that have major control over evapotranspiration differ. For wet conditions, net radiation, advection, leaf area, and turbulent transport dominate the process. Under dry conditions, soil water content (plant water availability) and stomatal resistance dominate. Under intermediate moisture conditions, these factors vary with climate, soil, and vegetation. As a result of the factors that affect actual evapotranspiration, actual evapotranspiration from watersheds cannot be measured directly by any practical field method.

The best estimates of vegetative effects on actual evapotranspiration come from water budget analyses and paired-watershed experiments (Box 3.2). An example of the latter was an experiment conducted at Hubbard Brook, New Hampshire, in which two similar watersheds with mixed hardwood forests were calibrated and one was subsequently cleared of all living vegetation (Hornbeck et al. 1970). By suppressing vegetative regrowth with herbicides on the cleared watershed, water yield was increased an average of 290 mm/yr over a 3 yr period; these changes were largely due to reduced transpiration. An accurate determination of total transpiration was not possible because evaporation from soil and litter was not suppressed.

The absence of practical methods of measuring *ET* requires that hydrologists apply their knowledge of *ET* processes of plant and soil systems with which they are working. A commonly used approach is to employ some index of *PET* and relate *PET* to available water in the watershed. Such an approach requires knowledge of soil water characteristics and plant response to soil water changes.

A review of evaporation and transpiration experiments of forest vegetation exposed the weaknesses of *PET* methods (Calder 1982, Morton 1990). Early work implied that all wet vegetative surfaces experienced similar *ET* (differences being attributed to albedo). This assumption was attractive to practitioners because maximum possible rates of *ET* could be estimated from physical and meteorological measurements. Experimental evidence indicates, however, that:

- *ET* losses vary significantly among vegetative types (e.g., forest vs. grasses) even if both systems have abundant available water.
- Forest interception losses can cause *PET* (as predicted by Penman's equation) to be exceeded.
- Some forest stands, well supplied with water, will periodically transpire significantly less water than would be predicted from net radiation methods.

The departures from earlier *PET* concepts addressed above can be explained using the Penman-Monteith equation (Monteith 1965):

Box 3.2

The Paired Watershed Method of Estimating Vegetative Effects on Actual Evapotranspiration

This method is often used for evaluating effects of timber harvesting on streamflow. Two watersheds first must be selected on the basis of similar soils, vegetative cover, geology, and topography. They should be close to one another and of similar size. Streamflow is measured at the outlet of both watersheds (as discussed in Chapter 4) over a sufficient time so that streamflow from one (watershed B in the figure below) can be predicted from streamflow measurements at the other (watershed A in the figure below). Depending on the similarity of the two watersheds, usually from 5 to 10 yr is needed to achieve an acceptable regression relationship for annual streamflow (see Chapter 18 for discussion on regression analysis); this time period is called the *calibration period*. After calibration, the vegetation on watershed B can be cleared; however, streamflow measurements continue for both watersheds. Observed streamflows Q_B after treatment then are compared to streamflow values from watershed B that were predicted from the regression relationships based on streamflow measurements at Q_A. The treatment period following vegetation clearing shows an increase in streamflow at watershed B. The change in streamflow is usually attributed primarily to changes in evapotranspiration.

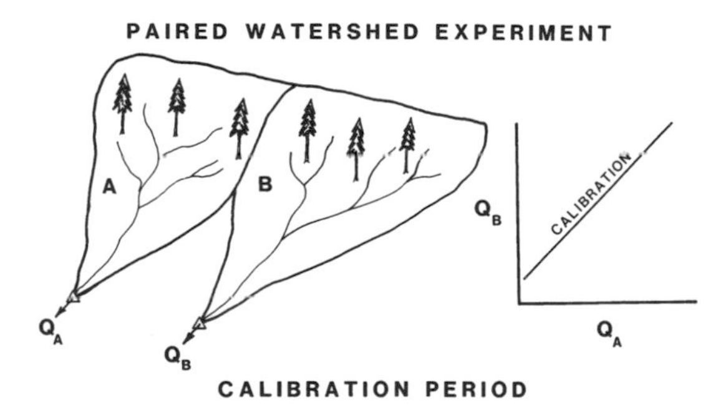

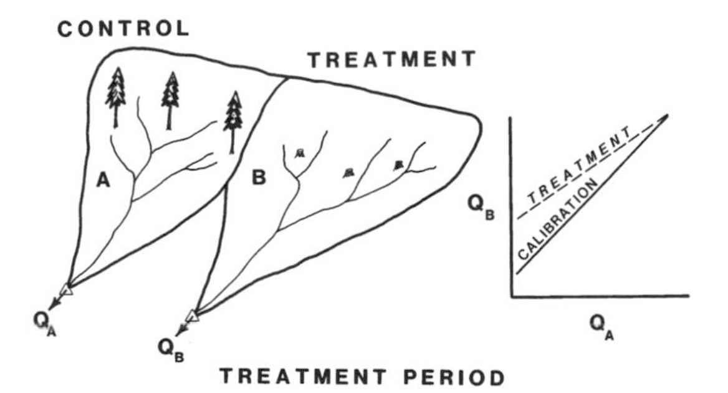

The paired watershed method as a means of estimating the effects of timber harvesting on streamflow.

$$PET = \frac{\Delta R_n + \rho C_p(e_s - e_a)/r_a}{\Delta + \gamma(1 + r_{st}/r_a)} \tag{3.23}$$

where ρ = air density; C_p = specific heat of air; r_a = aerodynamic resistance; and r_{st} = plant stomatal resistance.

When a film of water covers leaves (interception), r_{st} becomes negligible. In addition, r_a for forest vegetation is smaller than that for grasses and other low-growing herbaceous vegetation. Both factors result in high interception losses for forests, particularly conifers. As much as 80% of the total energy input to wet forest canopies can be derived from advection; even in humid climates such as that in England, latent heat flux can exceed net radiation by 12%.

Equation 3.23 also helps to explain differences in transpiration losses among species that cannot be explained based on energy exchange alone. Plants control transpiration through stomatal response. Stomata respond to changes in light intensity, soil moisture, temperature, and vapor pressure deficit. Most plant species close their stomata at night. Some close their stomata in response to wind. Many species close stomata when soils become dry.

Evapotranspiration/Potential Evapotranspiration Approach

Definitive relationships between transpiration or *ET* and soil moisture deficits (soil water content below field capacity) have been developed for only a few wildland species. Relationships of the form below have been used to relate actual *ET* to *PET*:

$$ET = (PET)f\left(\frac{AW}{AWC}\right) \tag{3.24}$$

where f = functional relationship; AW = available soil water (mm) = (soil moisture content – permanent wilting point) × rooting depth of mature vegetation; and AWC = available water capacity of the soil (mm) = (field capacity – permanent wilting point) × rooting depth of mature vegetation.

Field capacity (FC) refers to the maximum amount of water that a given soil can retain against the force of gravity. If a soil were saturated and allowed to drain freely, the amount of water remaining in the soil after all drainage ceased would be its field capacity. As a soil dries, the permanent wilting point eventually can be reached. Relationships of *FC* and the permanent wilting point for different soil textures are illustrated in Figure 3.8.

Actual *ET* would normally be expected to be at or near *PET* when soil moisture conditions are near field capacity. However, quantifying the relationship $f(AW/AWC)$ requires experimental evidence of plant or plant community response to soil moisture deficits. Tan and Black (1976) found that transpiration rates of Douglas-fir were halved when $\psi_s = -1000$ kPa. They also indicated that high vapor pressure deficits accentuated the effects of soil moisture deficits. Leaf and Brink (1975) developed relationships for forest types and stand conditions in the Rocky Mountain region of the western United States. The effects of clearcutting and regrowth on $f(AW/AWC)$ are illustrated in Figure 3.9. In all such approaches, remember that *PET* values are only an index.

The point at which soil water deficit begins to limit *ET* varies both with stand condition (from clearcut to old growth) and tree species (Fig. 3.10) and can be estimated from:

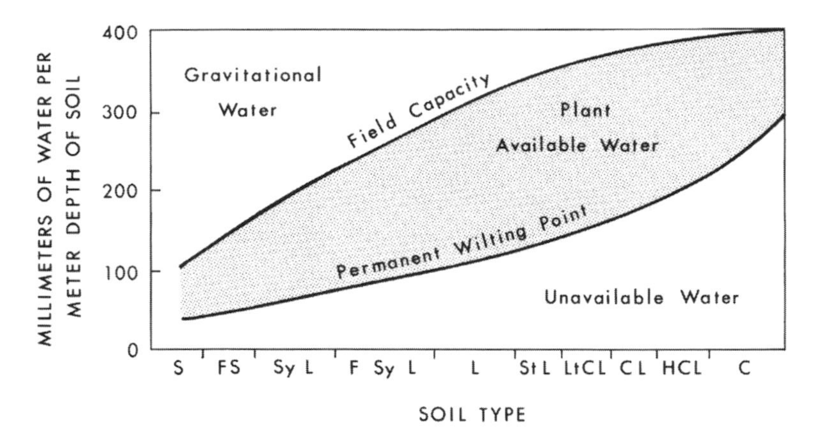

FIGURE 3.8. Typical water-holding characteristics of soil of different textures (redrawn from USDA Forest Service 1961): C = clay; F = fine; H = heavy; L = loam; Lt = light; S = sand; St = silt; Sy = sandy.

$$\tau = (FC)e^{-k(t - t_c)} \quad \textbf{(3.25)}$$

for conditions $\tau = FC$, $t < t_c$, and $\tau = FC/2$, $t > t_r$, where τ = soil water deficit value at which *ET* becomes limited; k = index of rate of decline of τ; t_c = time (yr) when soil water begins to limit *ET*; and t_r = time (yr) when the hydrologic effect of clearcutting becomes negligible.

The approach used in the above example requires information that might not be available for many types of vegetation. Simpler approaches can be used to approximate *ET* on a watershed basis.

Water Budget Approach

Application of a water budget as a hydrologic tool is relatively simple: if all but one component of a system can be either measured or estimated, then the unknown component can be solved directly.

A simplified water budget can be used to estimate annual *ET* of a watershed if changes in storage over a 1 yr period are normally small. Computations for the water budget could be made beginning and ending with wet months (AA') or dry months (BB'), as illustrated in Figure 3.11. In either case, the difference in soil water storage between the beginning and ending of the period should be small.

The above assumes that all outflow of liquid water from the watershed has been measured, there was no loss of water by deep seepage to underground strata, and all groundwater flow from the watershed was measured at the gauging site. If geologic strata such as limestone underlie a watershed, the surface watershed boundaries might not coincide with boundaries governing groundwater flow. In such cases, there are two unknowns in the water budget, *ET* and groundwater seepage (l), which result in:

$$ET + l = P - Q \quad \textbf{(3.26)}$$

If groundwater seepage is suspected, it can sometimes be estimated by specialists in hydrogeology, who have knowledge of geologic strata and their water-conducting properties.

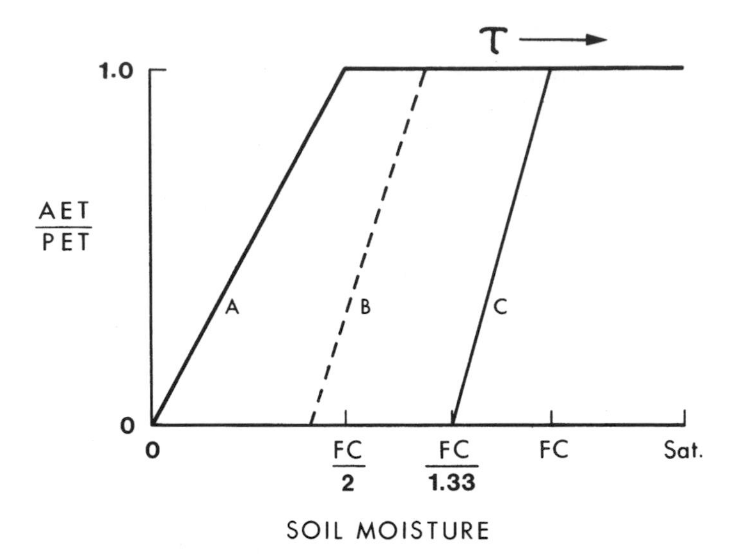

Figure 3.9. Actual evapotranspiration (*AET*) to potential evapotranspiration (*PET*) ratio as a function of soil water conditions for an old-growth forest (A); an intermediate forest cover condition (B); and open or clearcut condition (C) (adapted from Leaf and Brink 1975). *FC* = field capacity.

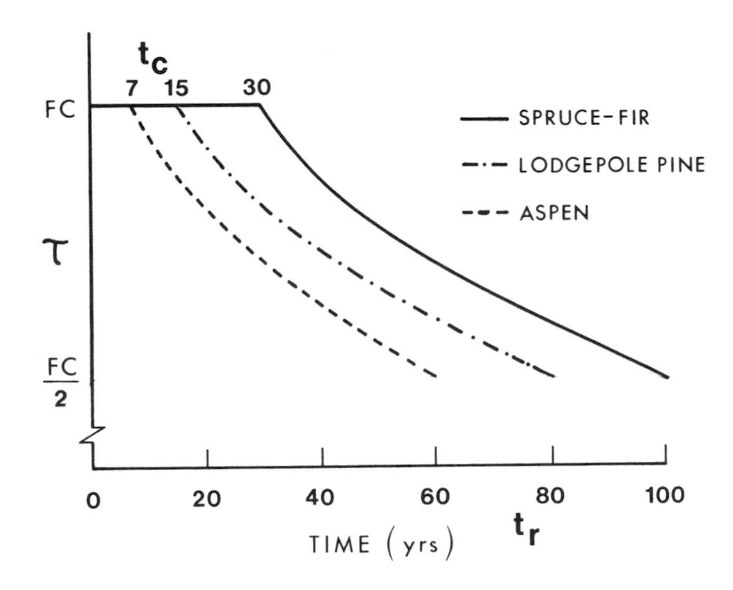

Figure 3.10. Effect of time and forest tree species on soil water deficit (adapted from Leaf and Brink 1975): τ = limiting soil water deficit with time for forest tree species; t_c = time (yr) when soil water begins to limit *ET*; t_r = time (yr) when the hydrologic effect of clearcutting becomes negligible.

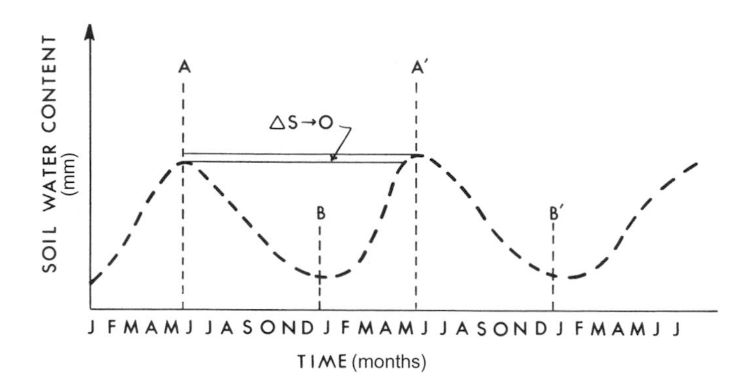

FIGURE 3.11. Hypothetical fluctuation of soil moisture on an annual basis.

When annual changes in storage are considered significant, they must be determined. Estimates of change in storage become more difficult as the computational interval diminishes and as the size of the area increases. The change of storage for a small vegetated plot can involve only periodic measurements of soil water content. Soil water content can be estimated gravimetrically (weighing a known volume of soil, drying the soil in an oven, and reweighing), with neutron attenuation probes, or by other methods. As the size of the watershed area increases, storage changes of surface reservoirs, lakes, and groundwater must also be considered. Reservoir elevation outflow data are needed to evaluate changes in lake or reservoir storage. These data are easier to analyze than storage changes in surface soils and geologic strata.

The water budget approach can be used for purposes other than estimating actual *ET*. For the example in Box 3.3, actual *ET* is assumed to equal *PET* if soil moisture content is sufficient. Once available soil moisture is less than *PET*, the actual *ET* equals the available soil moisture content. This approach likely overestimates *ET*, but it is useful for providing conservative estimates of water yield (runoff) from a watershed. If *ET/PET* relationships (such as those illustrated in Figure 3.9) are known, they should be used in the analysis.

SUMMARY

The importance of evapotranspiration and its influence on the water budget of a watershed should be recognized after reading this chapter. Evapotranspiration is one of the most significant hydrologic processes affected by human activities that alter the type and extent of vegetative cover on a watershed. Managers who manipulate soil-plant systems on a watershed should have a good fundamental understanding of the process of evapotranspiration and the factors that influence its magnitude. After reading this chapter, you should be able to:

- Explain and differentiate among the processes of evaporation from a water body, evaporation from soil, and transpiration from a plant.

Box 3.3

Water Budget Exercise for a Forest-Covered Watershed, near Chiang Mai, Thailand

	Year 1				
	Oct	Nov	Dec	Jan	Feb
			(mm)		
1. Average rainfall[a]	130	46	10	5	10
2. Initial soil moisture[b]	192	192	131	39	0
3. Total available moisture	322	238	141	44	10
4. Potential *ET*[c]	114	107	102	99	85
5. Actual *ET*[d]	114	107	102	44	10
6. Remaining available moisture	208	131	39	0	0
7. Final soil moisture[e]	192	131	39	0	0
8. Runoff[f]	16	0	0	0	0

[a]Average over the watershed for each month of record.
[b]At start of each month. Same as "final soil moisture" of previous month.
[c]Average evapotranspiration (*ET*) annual values for the month, as estimated by Thornthwaite's method.
[d]Total available moisture, or potential *ET*, whichever is smaller.
[e]At end of month. Same as "initial soil moisture" for next month. This value cannot be larger than the soil-water holding capacity determined for the watershed, for this watershed 192 mm; effective rooting depth = 1.2 m x 160 mm/m of available water (field capacity – permanent wilting point).
[f]Runoff occurs when the remaining available moisture exceeds the water holding capacity for the watershed (192 mm).

- Understand and be able to solve for evapotranspiration using a water budget and an energy budget method.
- Explain potential evapotranspiration and actual evapotranspiration relationships in the field. Under what conditions are they similar? Under what conditions are they different?
- Understand and explain how changes in vegetative cover affect evapotranspiration.
- Describe methods used in estimating potential and actual evapotranspiration.

References

Amiro, B.D., and E.E. Wuschke. 1987. Evapotranspiration from a boreal forest drainage basin using an energy balance/eddy correlation technique. *Bound. Layer Meteorol.* 38:125–139.

Bange, G.G.J. 1953. On the quantitative explanation of stomatal transpiration. *Acta Bot. Neerl.* 2:255–297.

Barrett, D.J., T.J. Hatton, J.E. Ash and M.C. Ball. 1996. Transpiration by trees from contrasting forest types. *Aust. J. Bot.* 44:249–263.

Calder, I.R. 1982. Forest evaporation. In *Hydrological processes of forested areas*, 173–193. Proc. Can. Hydrol. Symp. 1982, National Research Council, Canada. Fredericton, NB.

Dunne, T., and L.B. Leopold. 1978. *Water in environmental planning.* San Francisco: W. H. Freeman and Co.

Federer, C.A. 1970. *Measuring forest evapotranspiration: Theory and problems.* USDA For. Serv. Res. Pap. NE–165.

Fritschen L.J., J. Hsia, and P. Doraiswamy. 1977. Evapotranspiration of a Douglas-fir determined with a weighing lysimeter. *Water Resour. Res.* 13:145–148.

Gay, L. W. 1993. Evaporation measurements for catchment scale water balances. In *Proceedings of the first international seminar of watershed management*, ed. J. Castillo Gurrola, M. Tiscareno Lopez, and I. Sanchez Cohen, 68–86. Hermosia, Sonora, Mexico: Universidad de Sonora.

Hamon, W.R. 1961. Estimating potential evapotranspiration. *J. Hydrol. Div., Proc. Am. Soc. Civil Eng.* 87(HY3): 107–120.

Hanks, R.J., and G.L. Ashcroft. 1980. *Applied soil physics.* New York: Springer-Verlag.

Hanks, R.J., H.R. Gardner, and R.L. Florian. 1968. Evapotranspiration-climate relations for several crops in the central Great Plains. *Agron. J.* 60:538–542.

Hinckley, T.M., J.P. Lassoie, and S.W. Running. 1978. Temporal and spatial variation in the water status of forest trees. *For. Sci. Monogr.* 20:72.

Hornbeck J.W., R.S. Pierce, and C.A. Federer. 1970. Streamflow changes after forest clearing in New England. *Water Resour. Res.* 6:1124–1132.

Leaf, C.F., and G.E. Brink. 1975. *Land use simulation model of the subalpine coniferous forest zone.* USDA For. Serv. Res. Pap. RM–135.

Lee, R. 1980. *Forest hydrology.* New York: Columbia Univ. Press.

Mace, A.C., Jr., and J.R. Thompson. 1969. *Modifications and evaluation of the evapotranspiration tent.* USDA For. Serv. Res. Pap. RM–50.

Monteith, J.L. 1965. Evaporation and the environment. *Symp. Soc. Exp. Biol.* 19:205–234.

Morton, F.I. 1990. Studies in evaporation and their lessons for the environmental sciences. *Canadian Water Resour. J.* 15:261–286.

Penman, H.L. 1948. Natural evaporation from open water, bare soil, and grass. *Proc. R. Soc. London, Ser. A* 193:120–145.

Penman, H.L., D.E. Angus, and C.H.M. van Bavel. 1967. Microclimate factors affecting evaporation and transpiration. In *Irrigation of agricultural lands.* Vol. 2, *Agromony,* eds. R.M. Hagen, H.R. Haise, and T.W. Edminster, 483–505. Madison, WI: Am. Soc. Agron.

Reifsnyder, W.E., and H.W. Lull. 1965. *Radiant energy in relation to forests.* USDA For. Serv. Tech. Bull. 1344.

Rose, C.W. 1966. *Agricultural physics.* New York: Pergamon Press.

Shuttleworth, W.J. 1993. Evaporation. In *Handbook of hydrology*, ed. D.R. Maidment, 4.1–4.53. New York: McGraw Hill.

Slatyer, R.O. 1967. *Plant-water relationships.* New York: Academic Press.

Smith, D.M., and S.J. Allen. 1996. Measurement of sap flow in plant stems. *J. Exp. Bot.* 47:1833–1844.

Swanson, R.H. 1994. Significant historical development in thermal methods for measuring sap flow in trees. *Agric. For. Meteorol.* 72:113–132.

Swanson, R.H., and R. Lee. 1966. Measurement of water movement from and through shrubs and trees. *J. For.* 64:187–190.

Tan, C.S., and T.A. Black. 1976. Factors affecting the canopy resistance of a Douglas-fir forest. *Boundary-Layer Meteorol.* 10:475–488.

Thom, A.S. 1975. Momentum, mass, and heat exchange of plant communities. In *Vegetation and the atmosphere*, ed. J.L. Monteith, 1:57–109. London: Academic Press.

Thornthwaite, C.W., and J.R. Mather. 1955. *The water balance*. Laboratory of Climatology, Publ. 8. Centerton, NJ.

U.S. Army Corps of Engineers. 1956. *Snow hydrology*. Portland, OR: U.S. Army Corps of Engineers, North Pacific Division. (Available from U.S. Dept. Commerce, Clearinghouse for Federal Scientific and Technical Information, as PB 151660.)

USDA Forest Service. 1961. *Handbook on soils*. No. 2212.5.

Vogt, R., and L. Jaeger. 1990. Evaporation from a pine forest—Using the aerodynamic method and Bowen ratio method. *Agr. Forest Meteorol*. 50:39–54.

Wullschleger, S.D., F.C. Meinzer and R.A. Vertessy. 1998. A review of whole-plant water use studies in trees. *Tree Physiol*. 18:499–512.

Zhang, L., W.R. Dawes and G.R. Walker. 2001. Response of mean annual evapotranspiration to vegetation changes at catchment scale. *Water Resour. Res*. 37:701–708.

CHAPTER 4 *Infiltration, Runoff, and Streamflow*

INTRODUCTION

Once net precipitation reaches the ground, it either moves into the soil, forms puddles on the soil surface, or flows over the soil surface. Precipitation that enters the soil, and is not retained by the soil, moves either downward to groundwater or downward and laterally to a stream channel. This is called either *effective precipitation* or *excess precipitation*. Water flowing over the soil surface reaches the stream channel in a shorter time than that flowing through the soil. The allocation of excess precipitation at the soil surface into either surface or subsurface flow determines, to a large extent, the timing and amount of streamflow that occurs.

The discussion in this chapter shifts emphasis from soil moisture movement governed by matric potential to soil moisture movement governed by gravity.

INFILTRATION

The process by which water enters the soil surface is called *infiltration*, which results from the combined forces of capillarity and gravity. If water is applied to a dry, unfrozen soil, a rapid initial infiltration rate will normally be observed (Fig. 4.1). This high initial rate is due to the physical attraction of soil particles to water (the matric potential gradient). As water fills the micropores (pores that are associated with soil texture), the rate of infiltration typically drops and eventually becomes constant. At this time, infiltration is only as rapid as the rate at which water moves through the soil macropores or drains under the influence of gravity; this downward movement of water through the soil is *percolation*. Land use can affect infiltration relationships, as is illustrated in Figure 4.1 for a medium-textured soil subjected to different grazing practices.

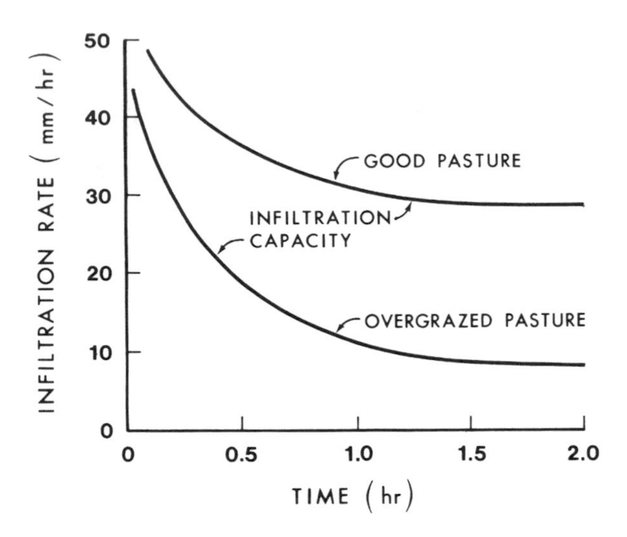

FIGURE 4.1. Infiltration capacity curves for soils subjected to different levels of grazing.

The rate at which net precipitation enters the soil surface depends upon several soil surface conditions and the physical characteristics of the soil itself. Plant material or litter on or near the soil surface influences infiltration and can be viewed as two hydrologically distinct layers: (1) an upper horizon composed of leaves, stems, and other undecomposed plant material and (2) a lower horizon of decomposed plant material that behaves much like mineral soil. The upper layer protects the soil surface from the energy of raindrop impact, which can displace smaller soil particles into pores and effectively seal the soil surface. Plant debris also slows or detains surface runoff, allowing water to enter the soil. The lower layer can have a substantial water storage capacity, > 200% by weight in some instances. Plant litter, therefore, is important as both a storage component and a protective cover that maintains an open soil surface condition favorable for high rates of infiltration. Conditions that reduce vegetative cover and compact the soil surface, such as the overgrazed pasture condition in Figure 4.1, cause infiltration to diminish.

Infiltration Capacity

The maximum rate at which water can enter the soil surface is called *infiltration capacity*. Infiltration capacity diminishes over time in response to several factors that affect the downward movement of the wetting front. The size of individual pores and the total amount of pore space in a soil generally decrease with increasing soil depth. Air entrapment within the pores and the swelling of soil colloids can also reduce infiltration rates.

The actual infiltration rate equals the infiltration capacity only when the rainfall (or snowmelt) rate equals or exceeds the infiltration capacity. When rainfall rates exceed infiltration capacity, surface runoff or ponding of water on the soil surface occurs (Fig. 4.2). When surface ponding reaches a sufficient depth, the positive pressure of water (head) can cause infiltration to exceed the infiltration capacity. Conversely, when rainfall intensity is

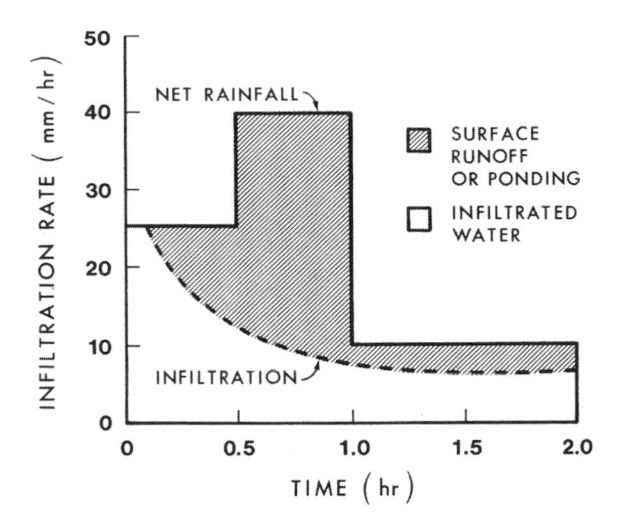

FIGURE 4.2. Relationship between rainfall rate and infiltration rate resulting in surface runoff or ponding.

less than the infiltration capacity, the rate of infiltration equals rainfall intensity. In such instances, water enters the soil and either is held within the soil (when soil moisture content is less than or equal to field capacity) or percolates downward under the influence of gravity (when soil moisture content is greater than field capacity).

The infiltration capacity of a soil depends on several factors, including texture, structure, surface conditions, the nature of soil colloids, organic matter content, soil depth or presence of impermeable layers, and the presence of *macropores* within the soil. Macropores function as small channels within a soil and are nonuniformly distributed pores created by processes such as earthworm activity, decaying plant roots, and the burrowing of small animals. Soil water content, soil frost, and the temperature of soil and water all influence infiltration characteristics of a soil at any point in time. Land use and vegetation management practices also influence many of the above factors.

The size and interconnections of pores within a soil affect infiltration and the subsequent movement of a wetting front through the soil. As pore size increases, all other factors being equal, the maximum amount of water moving through a pore within the soil is proportional to the fourth power of the radius of the pore:

$$Q = \frac{\pi R^4 \Delta p}{8 \eta L} \tag{4.1}$$

where Q = volume of water moving through a pore of length L; R = radius of pore; Δp = pressure drop over length L; and η = viscosity.

This equation, called Poiseuille's law, is a fundamental relationship for the laminar flow of water in saturated soils or groundwater systems and explains why water moves at a faster rate through coarse-textured soils than fine-textured soils. Even though matric forces within fine-textured soils exceed those of coarse-textured soils at a given water content, pore size and the interconnected nature of pores ultimately govern infiltration

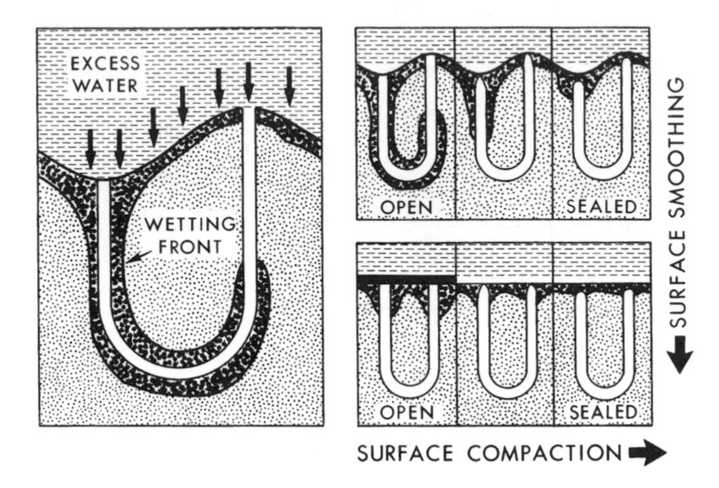

FIGURE 4.3. Effects of surface roughness and surface sealing on infiltration of a soil with a macropore and micropore system (from Dixon and Peterson 1971, as reported by Dixon 1975, © Am. Soc. Civil Eng., by permission).

capacity. Similarly, soil structure and the presence of soil fauna and old root channels create a macropore system that increases water conveyance through a soil.

The condition of the soil surface influences infiltration capacities. Surface roughness and the nature of pore openings at the surface largely govern the flow of water into (and air out of) the soil (Fig. 4.3). Air entrapment reduces infiltration capacity and storage. Soils with rough surfaces have a greater amount of depression storage; water in depressions is under a positive pressure (due to the depth of the water) that is greater than atmospheric pressure. Large pores that are open to the atmosphere likewise promote rapid infiltration; water moves freely into the soil and displaced air freely escapes. Surface compaction or sealing diminishes the effectiveness of large pores, which then become barriers to water entry because of the back pressure of air trapped in the sealed pores. Different activities on the soil surface can affect surface and macropore relationships to either enhance or diminish infiltration capacity.

Measurement of Infiltration

The infiltration capacity of a soil can be estimated in the field with *infiltrometers*, the two most common of which are the flooding type and the rainfall-runoff plot type. With either instrument, the entry of water into the soil surface is measured on a small plot of soil.

The flooding type infiltrometer uses a cylinder driven into the soil. Water is added and maintained at a specified depth (usually 10 cm) in the cylinder, and the amount of water needed to maintain the constant depth is recorded at specific times. Most often, a double-ring infiltrometer, in which one cylinder is placed inside another, is used. Typically, the inner ring is about 30 cm in diameter and the outer ring is 46–50 cm in diameter. Water is added to both cylinders, or rings, but measurements are made only in the inner ring. The outer ring provides a buffer that reduces boundary effects caused by the cylinder and by lateral flow at the bottom of the ring. This method is easy and relatively inexpensive to apply, but the positive head of water is usually thought to cause higher infiltration rates than might occur from rainfall. Double-ring infiltrometers are useful for obtaining comparisons of infiltration rates for different soils, sites, vegetation types, and treatments.

In the rainfall-runoff plot method, either water is applied to the soil surface in a way that simulates rainfall (sprinklers) or natural rainfall events are evaluated. The runoff plot has a boundary strip that forces any surface runoff to flow through a measuring device. Rainfall simulators can be adjusted to represent different drop sizes and rainfall intensities. Rainfall intensities can be increased until surface ponding or surface runoff occurs, at which time the infiltration capacity has been reached. The rainfall simulator approach is more costly and difficult to apply in remote areas, particularly where thick brush or dense trees interfere with sprinkler rainfall simulation. Fewer replications are possible over a given period unless more than one rainfall simulator is available. Infiltration capacities determined by this approach should be more representative of actual infiltration capacities than those determined by flooding type infiltrometers. However, studies have shown a consistent relationship between infiltration capacities determined by the two methods. Once the relationship is defined, the double-ring approach can be used and the values adjusted with a coefficient to represent more accurate estimates of infiltration.

Infiltration Equations

Infiltration rates have been determined for many types of soils and plant cover conditions. Typically, the initial phase of infiltration (dry soils) is high and decreases to a relatively constant value as the soil becomes thoroughly wetted (Fig. 4.1). In the case of agricultural soils and under pasture or rangeland conditions, the curves can usually be approximated with several equations of the type described in Box 4.1.

Infiltration measurements on wildland soils, particularly forested soils, indicate that infiltration rates change erratically and often do not conform to the smooth curves characterized by the equations in Box 4.1. Forested soils are usually porous and open at the soil surface, with an extensive macropore system caused by old root cavities, burrowing animals, and earthworms. Initial and final infiltration rates of such soils can be orders of magnitude higher than those of agricultural soils of similar texture. Deep, forested soils in humid climates often have infiltration capacities far in excess of any expected rainfall intensity. Under such conditions, surface runoff rarely occurs.

A simplified approach for quantifying infiltration has been frequently used in which infiltration (I_t) is estimated by:

$$I_t = A_i + f_c t \tag{4.5}$$

where A_i = initial loss or storage by the soil (mm); f_c = transmission rate or net infiltration rate (mm/hr); and t = time period after A_i is satisfied to the end of the rainfall event.

This simplified approach can be used to estimate precipitation excess for stormflow and flood analyses. Precipitation excess (P_e) at the surface can be determined by:

$$P_e = P_n - A_i - f_c t \tag{4.6}$$

where P_n = net precipitation available at the soil surface (mm).

Box 4.1

Infiltration Equations

$$\text{Horton (1940): } I_t = f_c t + de^{-kt} \quad \textbf{(4.2)}$$

where I_t = cumulative infiltration (cm^3/cm^2) in time t, f_c = constant rate of infiltration after prolonged wetting of the soil (cm/hr), e = base of natural logarithms, and d,k = constants.

$$\text{Philip (1957): } I_t = S_p t^{1/2} + at \quad \textbf{(4.3)}$$

where I_t = cumulative infiltration (cm^3/cm^2) at time t, S_p = "sorptivity" parameter that relates to capillarity or soil matrix forces, and a = soil parameter relating to transmission of water through the soil or gravity forces.

$$\text{Holtan (1971): } f_m = ci\, S_a^n + f_c \quad \textbf{(4.4)}$$

where f_m = infiltration capacity (cm/hr); c = 0.69 for cm (1.0 for in.); i = infiltration capacity per unit of available storage (cm/hr); S_a = available storage, which is the difference between the potential soil moisture storage and the cumulative infiltration (cm); n = coefficient that relates to soil texture; and f_c = constant rate of infiltration after prolonged wetting of soil (cm/hr), as in Equation 4.2.

Values of f_c (net infiltration rates) have been estimated for a variety of soils, vegetation, and land uses (Table 4.1). Shallow bedrock or impervious layers within a soil can reduce net infiltration rates because of the limited available storage in the soil. Such influences have been reported even for soils under undisturbed forests. For example, in a tropical forest in Australia an impervious layer at 0.2 m depth caused the upper soil layer to become quickly saturated during rainstorms (Bonell et al. 1982). This resulted in surface runoff even though the pores of the undisturbed soils were apparently quite open at the surface.

Net infiltration rate refers to the relatively constant rate of infiltration that occurs after infiltration has taken place for some time (often 2 hr) and is usually governed by the saturated hydraulic conductivity of a soil unless shallow bedrock or impervious layers are present. Hydraulic conductivity is a physical characteristic of a soil that is constant under saturated conditions. It is best illustrated with Darcy's law (Fig. 4.4):

$$Q = k_v A \frac{\Delta H}{L} \quad \textbf{(4.7)}$$

where Q = rate of flow (cm^3/sec); k_v = hydraulic conductivity (cm/sec); A = cross-sectional area (cm^2); ΔH = change in head (cm); and L = length of soil column (cm).

If the soil-pore system remains unchanged following saturation, Darcy's law approximates the rate of flow. Again, wildland soils with a heterogeneous pore system may not respond as do more homogeneous media, such as agricultural soils or groundwater aquifers, for which the equation is more useful.

TABLE 4.1. Net infiltration rates for unfrozen soils

Soil Category	Bare Soil	Row Crops	Poor Pasture	Small Grains	Good Pasture	Forested
			(mm/hr)			
I	8	13	15	18	25	76
II	2	5	8	10	13	15
III	1	2	2	4	5	6
IV	1	1	1	1	1	1

Source: Gray 1973, by permission.
Note: I = coarse- to medium-textured soils over sand or gravel outwash; II = medium-textured soils over medium-textured till; III = medium- and fine-textured soils over fine-textured till; IV = soil over shallow bedrock.

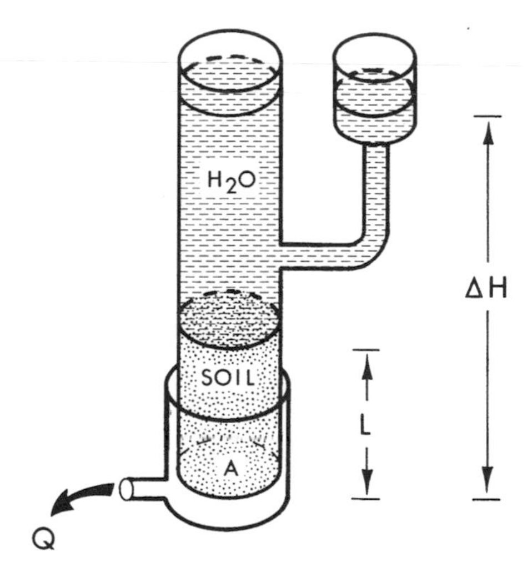

FIGURE 4.4. Darcy's laboratory method of determining the hydraulic conductivity (k_v) of a soil for a given head of water (ΔH) and length of soil column (L) and measuring the quantity of flow per unit time (Q).

Land Use Impacts on Infiltration

Activities that compact or alter the soil surface, soil porosity, or the vegetative cover can reduce the infiltration capacity of a soil. Driving vehicles or pulling logs over a soil surface, intensive grazing, and intensive recreational use can compact the surface and reduce infiltration. Exposing a soil to direct raindrop impact also will diminish the openness of the surface soil and reduce infiltration capacities.

Logging of a 110 yr old Douglas-fir stand in Oregon with a low ground pressure, torsion-suspension skidder resulted in 25–45% increases in soil bulk density to a depth of 15 cm (Sidle and Drlica 1981). The area affected by skidding amounted to 13.6% of

Table 4.2. **Comparisons of soils and infiltration relationships (double-ring infiltrometers) for three land use conditions in northern Morocco**

	Heavily Grazed, Doum Palm Vegetation	Moderately Grazed, Brushland	Ungrazed, Afforested (Aleppo Pine)
Soil texture	***Coarse***	***Medium***	***Fine***
Soil organic matter content (%)	1.47	1.77	2.7
Soil bulk density (g/cm^3)	1.44	1.42	1.22
Vegetative cover (%)	12.5	41.3	99
Slope (%)	0–10	5–25	5–40
Initial infiltration rate (mm/hr)	179	194	439
Infiltration rate after 2 hr (mm/hr)	43	65	226

Source: Adapted from Berglund et al. 1981.

the total area logged plus 1.5% of the area used as a landing. The greatest compaction resulted from frequent travel over wet soils.

The compaction of surface soils by yarding and skidding of logs reduces infiltration capacities and can result in surface runoff and erosion. If soils are allowed to recover, the effects of such operations are usually negligible after 3–6 yr except where soils were heavily disturbed (Johnson and Beschta 1980).

Livestock grazing can reduce infiltration capacities by removing plant material, exposing mineral soil to raindrop impact, and compacting the surface. Surfaces compacted by intensive grazing can reduce infiltration capacities over a wider area than can activities such as skidding logs. Infiltration capacities for different grazing and vegetative conditions in Morocco are compared in Table 4.2. Differences in soils and vegetative cover often confound comparisons of infiltration capacities among different grazing conditions. As a rule, rangelands in good to excellent condition with light grazing exhibit infiltration capacities at least twice those of rangelands in poor condition with heavy grazing.

Land use also can affect infiltration capacities indirectly by altering soil moisture content and other soil characteristics.

Water-Repellent Soils

Dry soils normally have an affinity for absorbing liquid and vapor water because of the strong attraction between the mineral soil particles and water. Water droplets that are placed on a wettable soil surface are absorbed immediately. However, when water droplets are placed on the surface of a dry water-repellent soil, they tend to "bead up" and not penetrate the soil because the mineral particles are coated with hydrophobic substances that repel water. The presence of water-repellent soils impacts the infiltration process in a manner that is similar to that of any dense or hardpan layer that restricts water movement through the soil. Infiltration of water into a hydrophobic soil is inhibited and often completely impeded, in which case the net precipitation reaching

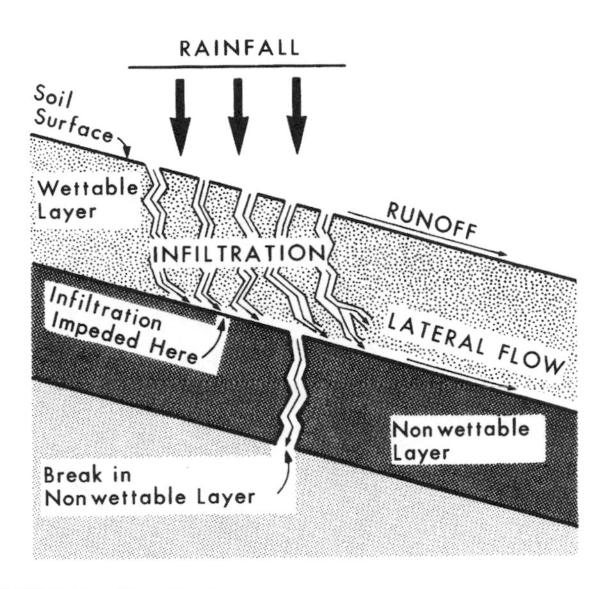

FIGURE 4.5. Effect of water-repellent layer on infiltration and surface runoff (from DeBano 1981).

the ground quickly becomes surface runoff or shallow lateral flow (Fig. 4.5). The change in the infiltration rate because of a water-repellent layer is illustrated in Figure 4.6. When a water-repellent soil becomes wetted up, it tends to transmit water as rapidly as the wettable soil.

Water repellency appears to be caused by several mechanisms involving the presence of organic matter consisting of long-chain hydrocarbon substances coating the mineral soil particles. These mechanisms include:

- Irreversible drying of the organic matter, e.g., the difficulty encountered when rewetting dried peat materials.
- Plant leachates that coat mineral soil particles, e.g., coarse-grained materials easily made water repellent by plant leachates.
- Coating of soil particles with hydrophobic microbial by-products, e.g., fungal mycelium.
- The intermixing of dry mineral soil particles and dry organic matter.
- The vaporization of organic matter and the condensation of hydrophobic substances on mineral soil particles during fire, e.g., heat-induced water repellency.

Water-repellent soils are found throughout the world on both wildlands and agricultural lands and have been a concern to watershed managers since the early 1900s. The nature of the water repellency of soils under chaparral vegetation in southern California has received widespread attention since the 1960s (DeBano 2000a). Organic matter that accumulates in the litter layer under chaparral shrubs is leached to the soil. Water repellency then results as the organic substances accumulate and mix in the upper soil profile. The naturally occurring water repellency in soils is intensified by the frequent wildfires in the region, which volatilize the organic substances, driving the

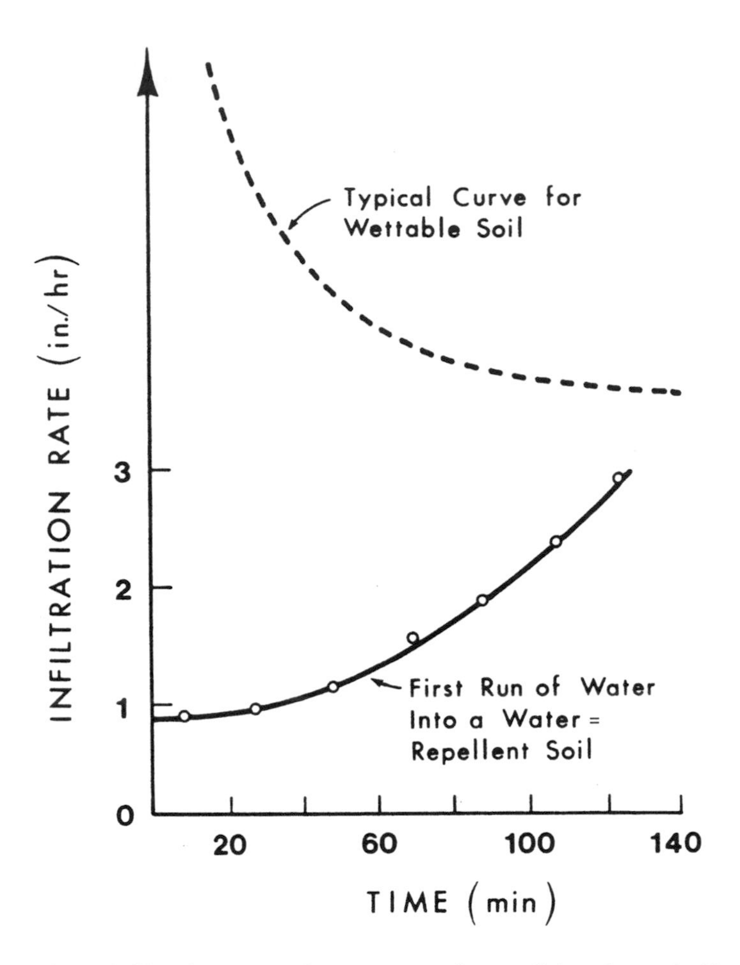

FIGURE 4.6. Infiltration rates of a water-repellent soil (as shown in Fig. 4.5) and a wettable soil (from DeBano 1981).

water-repellent layer deeper into the soil (Box 4.2).

Wetting agents or *surfactants* have been used in treating fire-induced water repellency to minimize its effects on infiltration. However, while non-ion surfactants have been somewhat effective in reducing runoff and erosion from small plots, they are often ineffective on larger watersheds. Because the magnitude of fire-induced water repellency is related to fire severity, a remedial approach on a watershed basis is to prescribe low severity fire to reduce fuel accumulations that lead to uncontrolled wildfire conditions and the resulting severe water repellency.

There is also a widespread concern about the effects of natural water repellency on the productivity of agricultural lands. About 5 million ha of agricultural and pasture lands in Australia and New Zealand have been adversely affected by water repellency, resulting in large financial losses for the land owners (Blackwell et al. 1993). Remedial treatments to mitigate this problem include applications of wetting agents, direct

Box 4.2

The Role of Fire and Soil Heating in Intensifying Water Repellency

Field and laboratory studies have confirmed that water repellency is intensified by soil heating during a fire, because the combustion and heat transfer during a fire produces steep temperature gradients in the surface layers of the mineral soil (DeBano 2000b). Temperatures in the canopy of burning chaparral shrubs in southern California can be over 1100°C, while temperatures reach about 850°C at the soil-litter interface. However, temperatures at 5 cm deep in the mineral soil probably do not exceed 150°C because dry soil is a good insulator. Heat produced by combustion of the litter layer on the soil surface vaporizes organic substances, which are then moved downward in the soil along the steep temperature gradients until they reach the cooler underlying soil layers where they condense. Incipient water repellency at different soil depths is also intensified in place by heating, because organic particles are heated to the extent that they coat and are chemically bonded to mineral soil particles. Movement of hydrophobic substances downward in the soil occurs mainly during the fire.

After a fire has passed a site, the continued heat movement downward through the soil can revolatilize some of the hydrophobic substances, resulting in thickening the water-repellent soil layer or fixing the hydrophobic substances in situ. The final result is a water-repellent layer below and parallel to the soil surface on the burned area, which restricts water movement into and through the soils.

drilling, wide furrow sowing, and the use of microorganisms and fertilizers to stimulate microbial breakdown of water repellency. Mixing large amounts of clay in the upper water-repellent layer has also been used.

Undisturbed peat soils are porous and exhibit high infiltration capacities, but a hydrophobic condition has been observed by the authors for these soils when they have been mincd or allowed to become dry. Mining of peat for horticulture or fuel exposes large areas of peat soil (see Chapter 14). Once the upper layer of sphagnum is removed and the darker, more decomposed peat becomes exposed to direct solar radiation and wind, the soil surface can dry to a point where it becomes hydrophobic. After this occurs, surface runoff can increase. The cause of this hydrophobic condition is unknown, but it may be similar to that induced by fire. The easiest way to alleviate this condition is to prevent the water table from dropping and allowing the surface of the peat deposit to become air dry.

Soil Frost

Soil frost is common during winter and spring in cold continental climates. It can also occur periodically in milder climates and can lead to serious flooding, particularly when

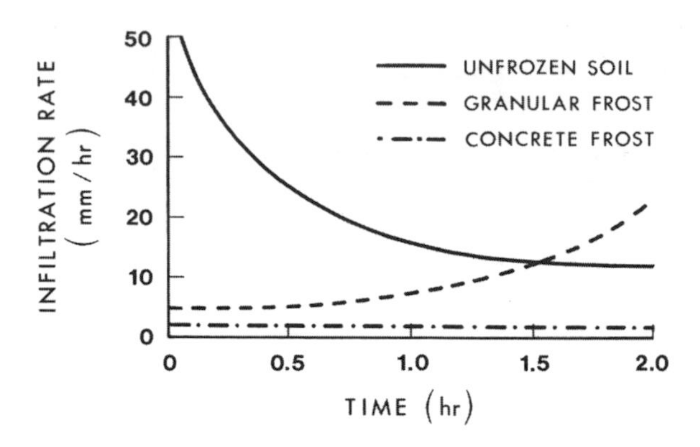

FIGURE 4.7. Effects of soil frost on infiltration rates (adapted from Gray 1973).

high-intensity rainfall occurs on frozen soils. The influence of soil frost on infiltration capacity is determined largely by the moisture content of the soil when it freezes. A saturated soil upon freezing can act as a pavement with little infiltration, a condition referred to as concrete frost (Fig. 4.7). If the soil is not saturated when freezing occurs, a granular, more porous frost develops, and the infiltration capacity is affected less. Under conditions of granular frost, some melting of soil frost occurs as infiltration continues; the soil pores then become, in effect, larger and able to transmit water at a faster rate. This explains the increase in infiltration over time.

The occurrence of soil frost is affected by vegetative cover, soil texture, depth of litter, and depth of snow. Snow and litter act as insulators; the deeper the snow before freezing, the less likelihood of soil frost. By altering vegetative cover, particularly forest cover, frost can be affected indirectly because of the influence of the forest canopy on snow accumulation and distribution. In general, removal of forest cover results in more frequent and deeper occurrences of soil frost. Compaction of soil surface horizons also can increase the depth of frost penetration. The effects of different forest cover conditions on soil frost in northern Minnesota are illustrated in Figure 4.8. The balsam fir stand intercepts more snow than hardwood stands, resulting in less snow depth and deeper frost than under the hardwood stands. Unforested areas experience the deepest frost penetration.

RUNOFF AND STREAMFLOW

Runoff and streamflow result from excess precipitation occurring on a watershed and its stream channels. In discussing watersheds and their stream channels, it is useful to refer to an established nomenclature of stream orders (Horton 1945, Strahler 1952). The commonly used method of stream orders classifies all unbranching stream channels as *first order streams* (Fig. 4.9). A *second order stream* is one with two or more first order stream channels; a *third order stream* is one with two or more second order stream channels, and so forth. Any single lower stream juncture above a larger order stream does not change the order of the larger order stream. That is, a third order stream that has a juncture with a second order stream remains a third order stream below the juncture. The drainage basin or watershed that feeds the stream takes on the same order as the stream; the watershed of a second order stream is a second order watershed, etc.

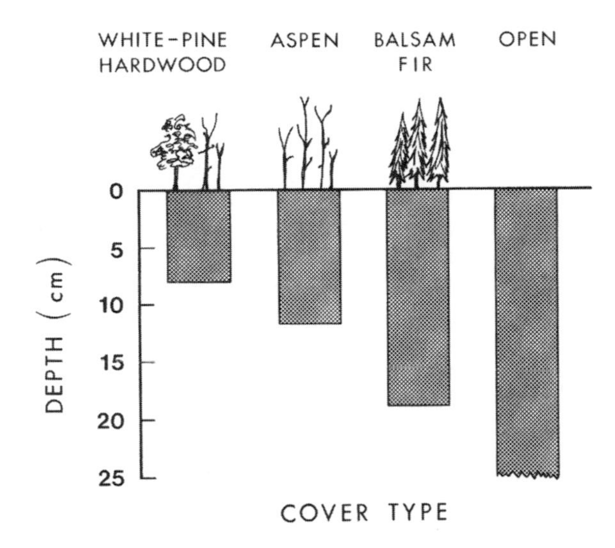

FIGURE 4.8. Depth of concrete frost during midwinter (January to mid-March) on loamy soils in northern Minnesota under different cover types; the vertical scale is exaggerated (from Weitzman and Bay 1963).

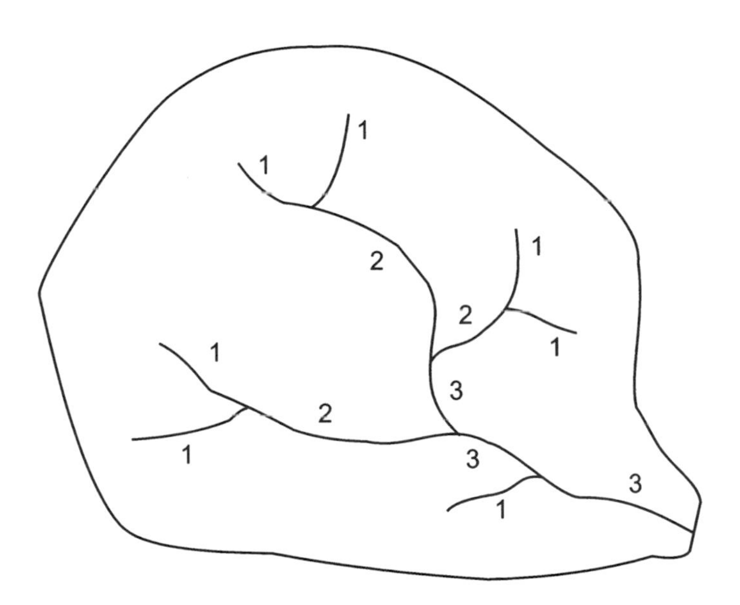

FIGURE 4.9. Stream order system by Horton (1945) as modified by Strahler (1964).

Although there is little evidence that streamflow characteristics are related to stream order, the use of this terminology helps one place a stream channel or watershed in the context of the overall drainage network of a river basin.

Various processes and pathways determine how much and how fast precipitation becomes streamflow. When trying to understand how watershed and stream channel conditions affect streamflow, it is helpful to think in terms of *storage* and *conveyance*. As discussed in Chapter 3, a considerable amount of precipitation can be stored in the soil, or in surface water bodies, and evapotranspired into the atmosphere. When rainfall or snowmelt rates exceed infiltration capacity, surface runoff occurs. Infiltrated water that is not stored in the soil drains through the soil and eventually reaches a stream channel, another surface water body, or a groundwater aquifer. Some water flows directly into a channel and quickly produces streamflow. Other pathways have a detention storage time, and weeks or months can pass before excess precipitation enters a stream channel. Therefore, the magnitude of water flowing through the various pathways determines the ultimate shape and size of a *streamflow hydrograph*. A streamflow hydrograph is the graphical relationship of streamflow discharge (m^3/sec) plotted against time (Fig. 4.10).

Groundwater is most likely feeding a *perennial stream* (one that flows continuously throughout the year), pathway D in Figure 4.10. Groundwater plus any long-term

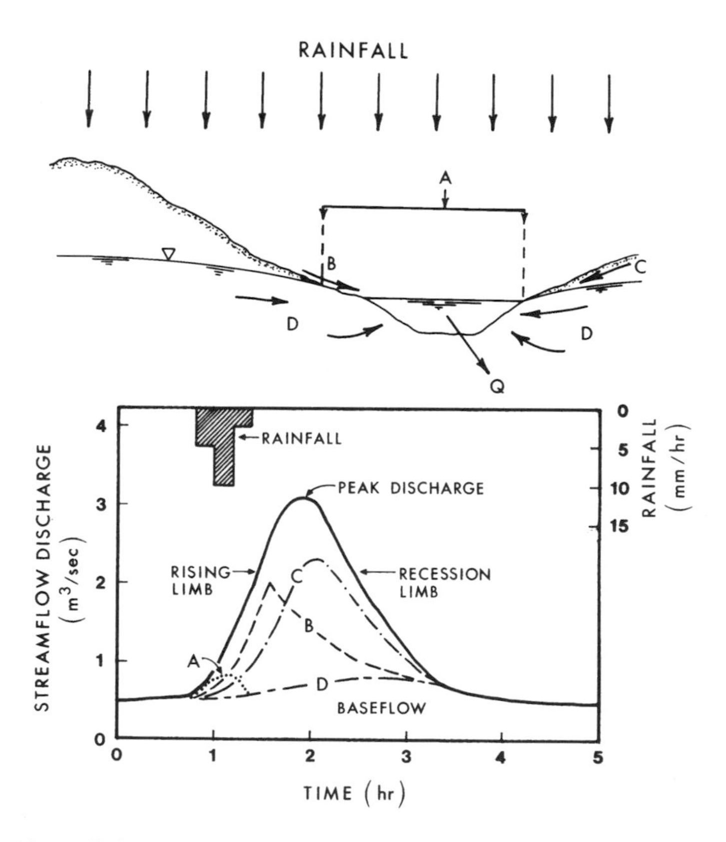

FIGURE 4.10. Relationship between pathways of flow from a watershed and the resultant streamflow hydrograph: A = channel interception; B = surface runoff, or overland flow; C = subsurface flow, or interflow; D = groundwater or baseflow; Q = streamflow discharge.

subsurface drainage from uplands sustains streamflow between periods of rainfall or snowmelt. Because of the long and tortuous pathways involved, this graphically level part of the hydrograph, called *baseflow*, does not respond quickly to excess precipitation.

Once rainfall or snowmelt occurs, several additional pathways of flow contribute to streamflow. The most direct pathway is precipitation that falls directly on the stream channel and adjacent saturated areas, called *channel interception* (A in Figure 4.10), causing the initial rise in the streamflow hydrograph; it ceases after precipitation stops. *Surface runoff*, or *overland flow*, is water that flows over the soil surface and occurs from areas that are impervious or locally saturated or from areas where the rainfall rate exceeds the infiltration capacity of the soil (B in Figure 4.10). Some surface runoff is detained by the roughness of the surface and can be slowed. Overall, surface runoff represents a quickflow response that reaches the outlet of a watershed second only to channel interception. Surface runoff is a relatively large component of the hydrograph for impervious urban areas, but in contrast, it is insignificant for forested areas with well-drained deep soils.

Overland flow in one part of a watershed can infiltrate at some downslope location before reaching a stream channel. This pathway results in flow reaching the channel later than surface runoff but quicker than groundwater.

Subsurface flow, or *interflow*, is that part of excess precipitation that infiltrates but arrives at the stream channel over a short enough period to be considered part of the storm hydrograph (C in Figure 4.10). This is considered the major flow pathway in most well-drained forested watersheds.

The sum of channel interception, surface, and subsurface flow is called *direct runoff*, *stormflow*, or *quickflow*. This is the part of the hydrograph of interest when floods and flood-producing characteristics of watersheds are analyzed. (The term *stormflow* will be used when describing this part of the hydrograph in the remaining chapters.)

Although the four major pathways of flow can be conceptually visualized, measuring each pathway and separating one from the others is impossible physically. The actual pathway from rainfall to streamflow usually involves a combination of surface and subsurface flows. Water can infiltrate in one area and exfiltrate (return to the surface) downslope and run over the land surface for some distance. Conversely, some surface runoff can collect in depressions to be evaporated or infiltrated later. The total streamflow hydrograph depicts the integrated response of a watershed to a given quantity of moisture input with a given set of watershed conditions.

Most hydrograph studies do not attempt to separate the various pathways of flow as illustrated in Figure 4.10. Rather, the streamflow response is evaluated by separating the stormflow component from the slow-responding baseflow. Because the hydrograph represents the integrated response to a precipitation event, the separation of a hydrograph in terms of time response rather than flow pathway is more realistic and useful for flood analysis (see chapters 16 and 17).

Hillslope Hydrology

At times, hillslopes are studied rather than entire watersheds because they help us better understand processes of streamflow generation, and they represent the smallest unit on which most of the processes occur that one would expect to see in a watershed

(Troendle 1985). The fate of precipitation with respect to interception, infiltration, soil moisture storage, surface runoff, and percolation can be observed within a hillslope. The pathways through which water flows are largely determined by infiltration capacities and the characteristics of soil strata on hillslopes. In many areas, particularly those with deep soils and forest vegetation, the runoff at the toe of a slope can be largely the product of subsurface flow through the soil profile rather than overland flow. As infiltrated water moves through the soil profile, constricting layers are encountered and saturated zones develop. Along hillslopes, the saturated zone develops a head of pressure that can force water to move along soil layers or through the soil profile, eventually reaching the toe of the slope. When the zone of saturation is continuous in the profile, any addition of water at the top of the zone results in an increased rate of flow at the toe of the slope. This type of flow is called *translatory flow*. When large continuous pores (macropores) are present, water can move through the soil rapidly, as through a pipe (recall Equation 4.1); this is called *pipeflow*.

Hillslope studies, particularly in forested regions, indicate that stormflow production can be largely the product of subsurface flow (Box 4.3).

Variable Source Area Concept

Wildland watersheds typically are heterogeneous mixtures of soils, vegetative cover, and land use with various hillslope and channel configurations. As such, there can be a wide range of rainfall (or snowmelt) runoff responses. At one extreme, forested watersheds with deep, permeable soils can have high infiltration capacities and exhibit predominantly subsurface flow. On the other extreme, rangelands with shallow soils can have low infiltration capacities and exhibit a quick, or flashy, streamflow response dominated by surface runoff. Perhaps the most common situation is a watershed in which some areas produce surface runoff for any rainfall event (rock outcrops, roads, etc.) and other areas seldom, if ever, produce surface runoff.

The *variable source area* concept (VSAC) explains the mechanisms of stormflow generation from watersheds that exhibit little surface runoff (Hewlett and Troendle 1975). Early concepts of stormflow runoff suggested that the only mechanism capable of producing quick-responding peak discharge was surface runoff. The VSAC suggests two mechanisms that are primarily responsible for the quickflow response: (1) an expanding source (saturated) area that contributes flow directly to a channel and (2) a rapid subsurface flow response from upland to lowland areas (Fig. 4.11).

A stream channel and the wet areas immediately adjacent to the channel respond most quickly to a rainfall event. As it rains, the wetter areas and shallow soil areas become saturated; this saturated zone expands upstream and upslope. Therefore, the area contributing directly to the channel becomes larger with the duration of the storm. This source area slowly shrinks again after the rain stops.

Water stored in the soil upslope from the channel system contributes to flow downstream by displacement and by direct flow (pipeflow) out of saturated zones near the channel. Midslope and low areas can respond quickly due to the displacement of upslope water into the saturated zone. Ridgetop areas may contribute little to stormflow; much of the infiltrated water can be stored or the pathway can be long enough to delay the flow until long after the storm event has ended.

Box 4.3

Stormflow Response of Hillslopes and Small Watersheds (from Beasley 1976)

Runoff from plots (less than 0.1 ha) on forested hillslopes in Mississippi produced negligible overland flow and negligible shallow subsurface flow (above the B-horizon) from rainfall events. Streamflow peaks from 1.86 and 1.62 ha watersheds, averaged over 36 storms, responded within 5 min of peak subsurface flow from the plots (see hydrograph below). It was concluded that flow through macropores (pipeflow), not translatory flow, could be the only explanation for such a quick response.

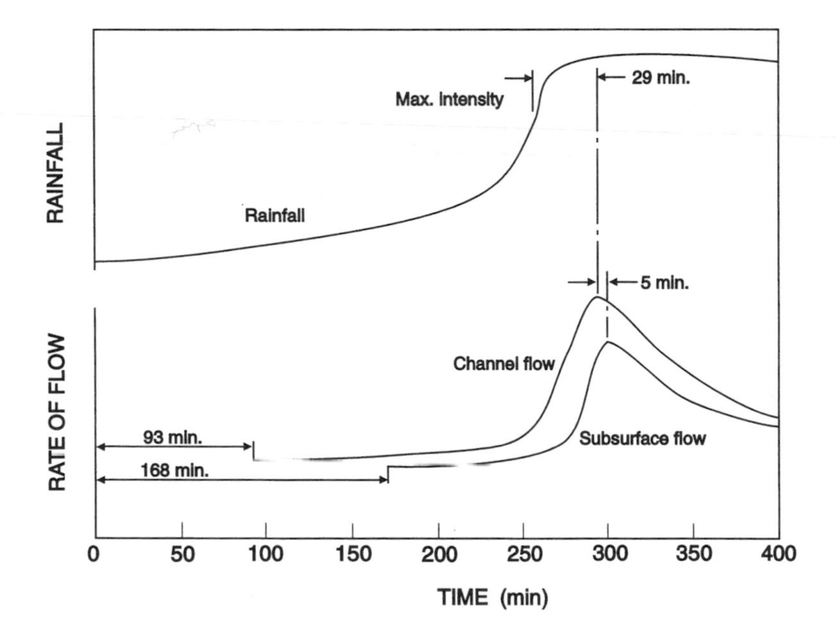

Relationship between timing of rainfall and corresponding subsurface flow and channel flow responses averaged for 36 storms for hillslopes in Mississippi (Beasley 1976, as presented by Troendle 1985).

Stormflow Response

The stormflow response of a watershed is often characterized by separating stormflow from baseflow. The *unit hydrograph* (UHG), a widely used method of characterizing the stormflow response of a watershed, is defined as the stormflow (direct runoff) response of a given watershed to one unit (1 mm) of precipitation excess that occurs uniformly over the area and over a given time increment. The method is described in Chapter 17.

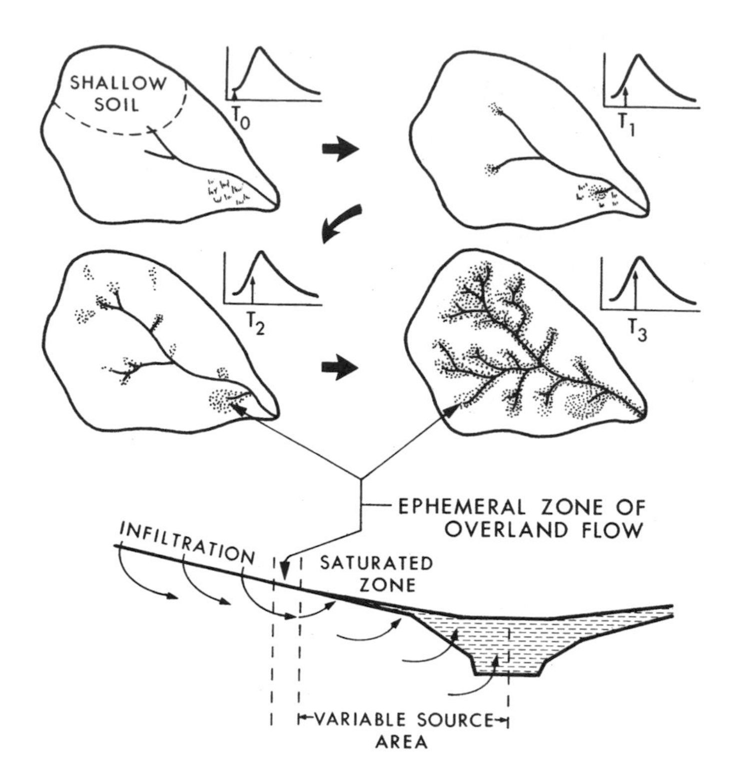

FIGURE 4.11. Schematic of the "variable source area" of stormflow and the relationship between overland flow and the zone of no infiltration. The small arrows on the hydrographs indicate streamflow response changes as the variable source area expands (modified from Hewlett and Troendle 1975 and Hewlett 1982, © University of Georgia Press, by permission).

Simpler approaches can be used, such as the hydrologic response (Woodruff and Hewlett 1970):

$$R_s = \left(\frac{\text{annual stormflow}}{\text{annual precipitation}}\right)100 \quad \textbf{(4.8)}$$

where R_s = hydrologic response (%).

The hydrologic response gives some indication of the stormflow response, or flashiness, of a particular watershed to rainstorms. For individual storms the hydrologic response can vary considerably (<1% to >75%), depending largely upon antecedent moisture conditions.

Factors Affecting Stormflow Response

Many factors determine the magnitude of stormflow volume and peak flow; some are fixed and some vary in time for a given watershed. From earlier discussions, some of these factors should be apparent. Here, they are discussed in qualitative terms. Methods of estimating stormflow characteristics for a given watershed are discussed in Chapter 17.

Watershed characteristics that are fixed and have a pronounced influence on stormflow response include size and shape of the watershed, channel and watershed slopes, drainage density, and presence of wetlands or lakes. The larger the watershed, the greater the volume and peak of streamflow for rainfall or snowmelt events. Watershed shape affects how quickly surface and subsurface flows reach the outlet of a watershed. For example, a round-shaped watershed concentrates runoff more quickly at the outlet than an elongated watershed and, all other factors being equal, will tend to have higher peak flows. Likewise, the steeper the hillslopes and channel gradients, the quicker the response and the higher the peak flows. *Drainage density*, defined as the sum of all stream channel lengths divided by the watershed area, also affects the rapidity with which water can flow to the outlet. The higher the drainage density, the quicker the flow response and the higher the peak. As the percentage of area in wetlands, lakes, or reservoirs increases, a greater attenuation or flattening of the stormflow hydrograph occurs. Again, the effect is due to the impact on travel time through the watershed to the outlet; wetlands and lakes detain (slow down) and retain (store) water flowing into them.

Factors affecting stormflow response that vary with time can be separated into precipitation and watershed factors. The magnitude of rainfall or snowmelt affects the magnitude of streamflow response. For rainstorms, the intensity and duration are important. As a rule, the higher the intensity and the longer the duration, the higher the magnitude of peak flow. The areal distribution of rainfall or snowmelt and the movement, or tracking, of a storm affect the peak and volume of stormflow. For example, a storm that moves from the upper reaches of a watershed to the outlet will tend to concentrate flow at the outlet; one that tracks upland will tend to spread out the flow response over time. Such a response is due to the timing of drainage from upland and from downstream parts of a watershed. The length of time between rainfall or snowmelt events affects the *antecedent conditions* of a watershed, which refers to the relative moisture storage status of a watershed at some point in time. If a watershed has experienced rainfall or snowmelt of any magnitude recently, it is primed to respond quicker and with a greater volume of streamflow (because there is less available storage) than one that has not had precipitation for weeks.

Watershed conditions that can vary and influence stormflow response include vegetation type and extent, soil surface conditions, and a variety of human-caused changes, such as roads, reservoirs, drainage systems, waterways, and stream channel alterations. These watershed factors are the ones that can be manipulated to achieve desired hydrologic objectives or become altered as parts of other management activities or projects.

All the above factors exert some influence on stormflow response. It is difficult to separate and quantify the contributions of individual components or factors. Hydrologic methods (discussed in Chapter 17), including computer simulation models, have been developed to study and quantify the various watershed and meteorological factors affecting stormflow.

Streamflow Measurements

Streamflow (discharge) data are perhaps the most important information needed by the engineer and the water resource manager. Peak flow data are needed in planning for flood control or engineering structures (e.g., bridges and culverts). Streamflow data during low-flow periods are required to estimate the dependability of water supplies. Total runoff and its variation must be known for design purposes (e.g., reservoir storage discussed in Chapter 16).

The *stage*, or height, of water in a stream is measured readily at some point on a stream reach with a staff gauge or water level recorder. The problem is to convert a record of the stage of a stream to discharge (quantity of flow per unit of time); this is done either by *stream gauging* or with precalibrated structures such as *flumes* or *weirs* constructed in the stream.

Measuring Discharge

One of the simplest ways of measuring discharge is to observe the time it takes a floating object tossed into the stream to travel a given distance. A measurement of the cross section of the stream should be made simultaneously, and the two are then multiplied together:

$$Q = VA \tag{4.9}$$

where Q = discharge (m^3/sec); V = velocity (m/sec); and A = cross section (m^2).

This simple method is not always accurate, particularly for a large stream, because velocity varies from point to point with depth and width over the cross section of the stream. The velocity at the surface is greater than the mean velocity of the stream. Generally, actual velocity is assumed to be about 80–85% of surface velocity.

If the cross section of a stream is divided into finite vertical sections, the velocity profile can be estimated by individually measuring the mean velocity of each section (Fig. 4.12). The area of each section can be determined, and the average discharge of the entire stream is then computed as the sum of the product of area and velocity of each section, as follows:

$$Q = \sum_{1}^{n} A_i V_i \tag{4.10}$$

where n = the number of sections.

The greater the number of sections, the closer the approximation. However, for practical purposes, between 10 and 20 sections commonly are used. The actual number of sections depends upon the channel configuration and the rate of change in the stage with discharge. Depth and velocity should not vary greatly between points of measurement. Also, all measurements should be completed before the stage changes too much.

The velocity and depth of vertical sections can be measured by wading into the stream or from cable car, boat, or bridge. Velocity is usually measured with a current meter, using the following rules:

- For depths > 0.5 m, two measurements are made for each section at 20 and 80% of the total depth and then averaged.
- For depths < 0.5 m, one measurement is made at 60% of the depth.
- For shallow streams less than about 0.5 m deep, a pygmy meter or similar instrument is used instead of a standard current meter.

The most critical aspect of stream gauging is the selection of a control section, that is, a section of the stream for which a rating curve (Fig. 4.13) is to be developed. Such a section of stream should be stable and have a sufficient depth for velocity measurements at the lowest of streamflows. The control section should be in a straight reach without turbulent flow.

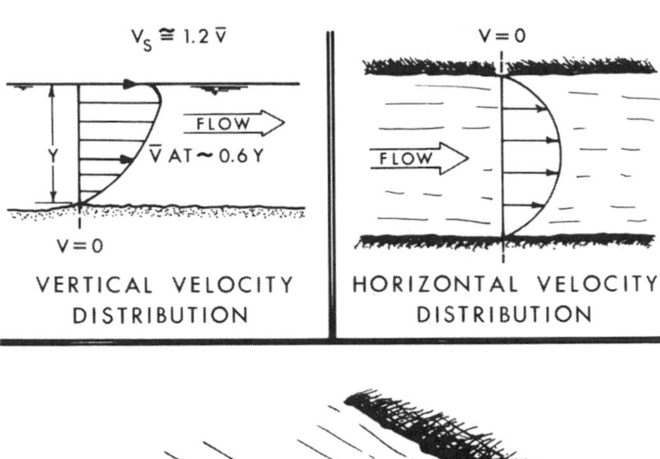

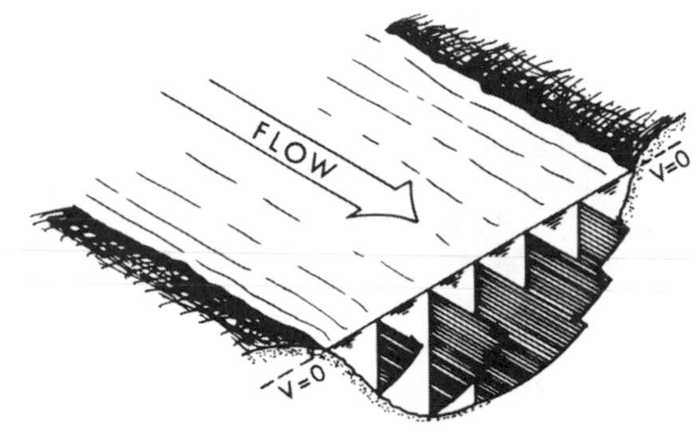

FIGURE 4.12. Measurements of channel cross section and velocity needed to determine streamflow discharge.

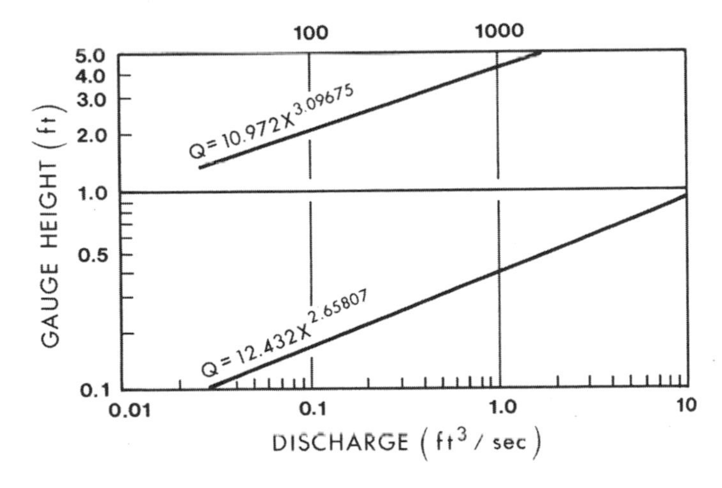

FIGURE 4.13. Example of a rating curve of streamflow discharge vs. water surface evaluation (from Brown 1969).

Precalibrated Structures for Streamflow Measurement

On small watersheds, usually < 800 ha in size, and particularly on experimental watersheds, precalibrated structures are often used because of their convenience and accuracy. The most common types of precalibrated structures are *weirs* and *flumes*. Because of their greater accuracy, weirs generally are preferred for gauging small watersheds, particularly those in which flows can become quite low. Where sediment-laden flows are common, flumes are preferred.

Weirs and flumes can be constructed of concrete, concrete blocks, treated wood, metal, fiberglass, and other materials. The notch of a weir is often a steel blade set into concrete, and flumes are frequently lined with steel for permanence.

Weirs

As used here, a weir includes all components of a stream-gauging station that incorporates a notch control (Fig. 4.14). The notch can be V-shaped, rectangular, or trapezoidal. An impoundment of water—the stilling basin—is formed upstream from the wall (dam) containing the notch, and a stilling well with a water level recorder is connected to the stilling basin. A gaugehouse or some other type of shelter is provided to protect the recorder. The cutoff wall or dam, used to divert water through the notch, is seated into bedrock or other impermeable material where possible so that no water can flow under or around it. Where leakage is apt to occur, the stilling basin is sometimes constructed as a watertight box.

The edge or surface over which the water flows is called the crest. Weirs can be either sharp crested or broad crested. A sharp-crested weir has a blade with a sharp upstream edge so that the passing water touches only a thin edge and clears the rest of

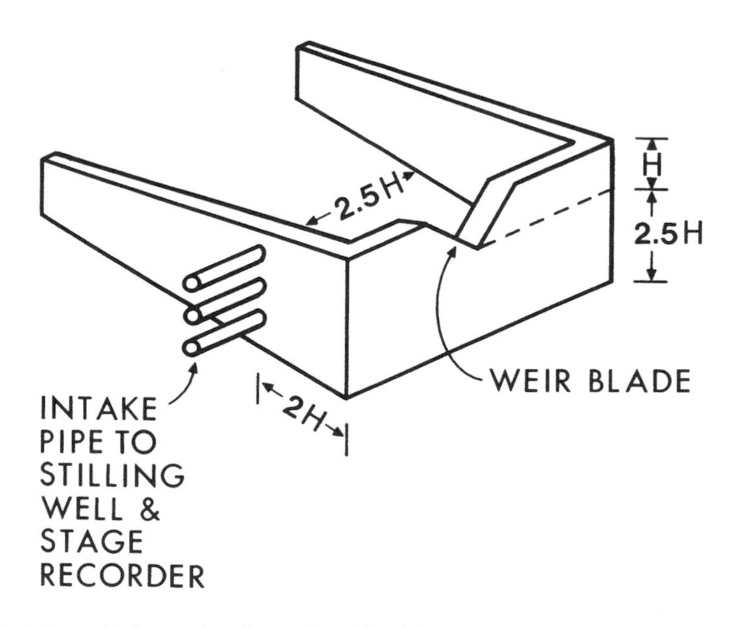

FIGURE 4.14. Schematic of a weir with a V-shaped notch.

the crest. A broad-crested weir has a flat or broad surface over which the discharge flows. Broad-crested weirs are generally used where sensitivity to low flows is not critical and where sharp crests would be dulled or damaged by sediment or debris.

The rectangular weir has vertical sides and horizontal crests. Its major advantage is its capacity to handle high flows. However, the rectangular weir does not provide precise measurement of low flows.

The trapezoidal weir is similar to the rectangular weir, but it has a smaller capacity for the same crest length; the discharge is approximately the sum of discharges from the rectangular and triangular sections.

Sharp-crested V-notch or triangular weirs are often used where accurate measurements of low flows are important. The V-notch weirs can have a high rectangular section to accommodate infrequent high flows.

Flumes

A flume is an artificial open channel built to contain flow within a designed cross section and length (Fig. 4.15). There is no impoundment, but the height of water in the flume is measured with a stilling well. The types of flumes that have been used on small watersheds are described here.

HS-, H-, and HL-type flumes developed and rated by the USDA Natural Resources Conservation Service (formerly the USDA Soil Conservation Service) have converging vertical sidewalls cut back on a slope at the outlet to give them a trapezoidal projection. These have been used largely to measure intermittent runoff.

The Venturi flume is rectangular, trapezoidal, triangular, or any other regular shape with a gradually contracting section leading to a constricted throat and an expanding section immediately downstream. The floor of the Venturi flume is the same grade as the stream channel. Stilling wells for measuring the head are at the entrance and at the throat; the difference in heads at the two wells is related to discharge. This type of flume is used widely in measuring irrigation water.

The Parshall flume (a modification of the Venturi flume) measures water in open conduits and is also used frequently for measuring irrigation water. It consists of a contracting inlet, a parallel-sided throat with a depressed floor, and an expanding outlet, all

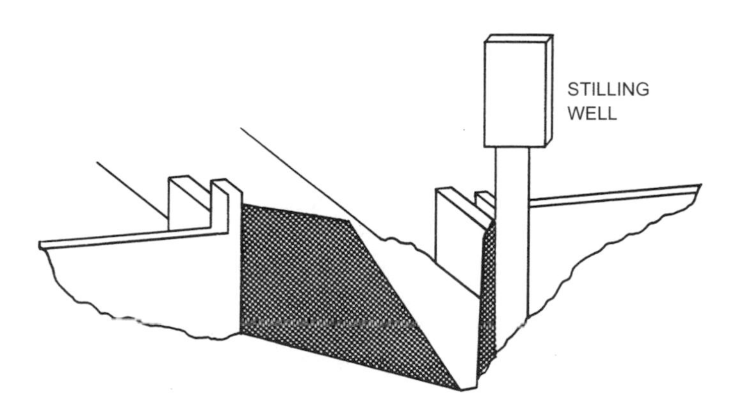

FIGURE 4.15. Schematic illustration of an H-flume.

of which have vertical sidewalls. It can measure flows under submerged conditions. Two water level recorders are used when measuring submerged flow, one in the sidewall of the contracting inlet and the other slightly upstream from the lowest point of the flow in the throat. When measuring free flow, only the upper measuring point is used.

The San Dimas flume measures debris-laden flows in mountain streams. It is rectangular, has a sloping floor (3% gradient), and functions as a broad-crested weir except that the contraction is from the sides rather than the bottom; therefore, there is no barrier to cause sediment deposition. Depth measurements are made in the parallel-walled section at about the midpoint. Rapid flow keeps the flume scoured clean.

The need to measure both high- and low-volume sediment-laden flows accurately prompted the development of an improved design of a supercritical flume by the USDA Agricultural Research Service (Smith et al. 1981). The flume was first used at the Santa Rita Experimental Range and later tested on the nearby Walnut Gulch experimental watersheds in southeastern Arizona. The flume was intended for use in small channels with flows of generally < 4 m^3/sec; however, the flume size is not limited as long as certain proportions are maintained. This supercritical flume differs from earlier models in that the slope of the floor breaks at the entrance to the throat defined by the walls rather than at the entrance to the curved approach. Also, curvature of the approach wall has been reduced to decrease the tendency of waves to develop in the throat when flow direction changes too rapidly at the entrance walls. Rating relationships have been developed by both experimental and theoretical means.

Considerations for Using Precalibrated Structures

The type of flume or weir to be used depends upon several factors: magnitude of maximum and minimum flows, accuracy needed in determining total discharge for high flows and low flows, amount and type of sediment or debris expected, channel gradient, channel cross section, underlying material, accessibility of site, and the length of the project and associated funding (costs) for gauging. In general, weirs are more accurate than flumes at low flows, but flumes are preferred when streams transport high volumes of sediment or debris.

Maximum and minimum flows must be estimated before construction. Such estimates can be made from observation of high and low flows and of high watermarks and from information given by local residents. Flow estimates also might be based on the area of the watershed and records from other gauging stations in the region. Maximum expected flood peaks can also be estimated from rainfall, soil, and cover data, using a method developed by the U.S. Natural Resources Conservation Service (see Chapter 17). The maximum and minimum flows to be measured at any degree of precision depend upon the objectives of the project and the extremes that might occur.

Flumes and weirs have been used together in tandem. Under high flows, the discharge flows through the flume and over the weir that is immediately downstream. Under low flows, there is not a sufficient jump to clear the downstream weir. In such cases, the water trickles into the impoundment above the weir, and accurate measurements of low flow can be obtained. Such configurations are costly and can be justified usually only for experimental purposes.

Empirical Estimations of Streamflow

In practice, streamflow data are often needed where there is no gauge. In many rural areas (particularly in developing countries), stream gauges are few and data are completely lacking for large areas. An estimation of streamflow, no matter how rough, is often essential for appraising the condition of catchments or for design purposes. Most frequently, a main interest is in flood flows. Sometimes, reasonable estimates can be made by extrapolating information from a similar basin, but this must be done cautiously and by an experienced analyst. It must be emphasized that estimates, no matter how sophisticated, are never as good as direct measurements.

Several empirical methods are available to estimate streamflow where no gauge exists. Two commonly used methods for estimating stream discharge at known stages (depths) of flow are the Manning and the Chezy equations.

The *Manning equation* is:

$$V = \frac{1.49}{n} R_h{}^{2/3} s^{1/2} \quad \textbf{(4.11)}$$

where V = the average velocity in the stream cross section (ft/sec); R_h = the hydraulic radius (ft) = A/WP (Fig. 4.16), where A = cross-sectional area of flow (ft^2) and WP = wetted perimeter (ft); s = energy slope as approximated by the water surface slope (ft/ft); and n = a roughness coefficient.

The *Chezy equation* is:

$$V = C\sqrt{R_h S} \quad \textbf{(4.12)}$$

where C = Chezy roughness coefficient.

Equations 4.11 and 4.12 are similar. The relationship between the roughness coefficients is:

$$C = \frac{1.49}{n} R_h{}^{1/6} \quad \textbf{(4.13)}$$

The equations are used in similar fashion: the hydraulic radius and water surface slope are obtained from cross-sectional and bed slope data in the field (Fig. 4.16). The roughness coefficient is estimated (Table 4.3), and the average discharge, Q, is calculated by multiplying the velocity by the cross-sectional area (A in Fig. 4.16).

In practice, the above equations are most often used to estimate some previous peak flow. High watermarks can be located after the stormflow event and used to estimate the depth of flow. Sometimes, this estimate can be obtained by measuring the height of debris caught along the stream channel (Fig. 4.17) or by watermarks on structures. This should be obtained for a reach of channel where the cross section of the flow of interest can be measured with reasonable accuracy. The slope of the water surface can be approximated by the slope of the stream channel along the reach. The wetted perimeter can be measured by laying a tape on the channel bottom and sides between the high watermarks. The cross-sectional area should be measured by summing several segments.

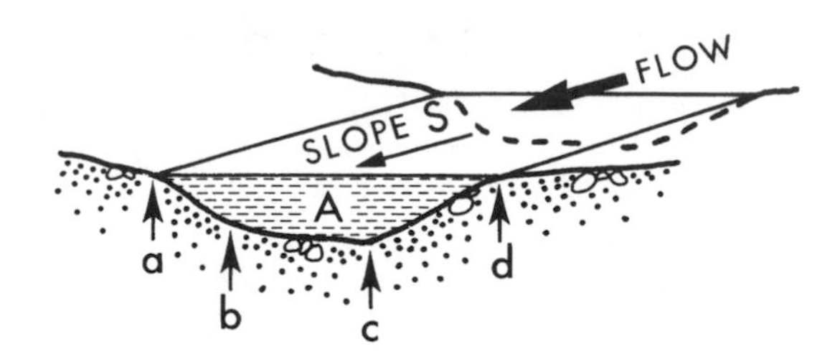

FIGURE 4.16. Stream channel section showing slope or gradient of streambed, wetted perimeter WP (line a-b-c-d), and cross-sectional area A.

TABLE 4.3. Examples of Manning's roughness coefficient *n*

Type of Channel	Minimum	Average	Maximum
A. Excavated or dredged			
Earth, straight, clean	0.016	0.018	0.020
Gravel, uniform, clean	0.022	0.027	0.033
Earth, winding, sluggish grass, some weeds	0.025	0.030	0.033
Dragline-excavated, light brush on banks	0.035	0.050	0.060
Channels not maintained, weeds/brush not cut	0.050	0.080	0.120
B. Natural streams			
1. Minor streams with width at flood stage < 30 m			
A. Streams on plains			
Clean, straight, full stage, no pools	0.025	0.030	0.033
Clean, winding, some pools and bars	0.033	0.040	0.045
Same as above but some weeds and stones	0.035	0.045	0.050
Sluggish reaches, weedy, deep pools	0.050	0.070	0.080
Very weedy, deep pools or floodways with heavy stand of timber and underbrush	0.075	0.100	0.150
B. Mountain streams, no vegetation in channel, banks usually steep, brush along banks submerged at high stages			
Bottom consists of gravels, cobbles, and few boulders	0.030	0.040	0.050
Bottom consists of large boulders and some large organic debris, sinuous flow	0.050	0.070	0.100
2. Floodplains			
Pasture, no brush, short grass	0.025	0.030	0.035
Pasture, no brush, tall grass	0.030	0.035	0.050
Cultivated areas, no crop	0.020	0.030	0.040
Cultivated areas, mature row crops	0.025	0.035	0.045

Type of Channel	Minimum	Average	Maximum
Scattered brush, heavy weeds	0.035	0.050	0.070
Medium to dense brush in winter	0.045	0.070	0.110
Medium to dense brush in summer	0.070	0.100	0.160
Dense willows, summer, straight channel	0.110	0.150	0.200
Heavy stand of timber, few downed trees, little undergrowth, flood stage below branches	0.080	0.100	0.120
Same as above but flood stage above branches	0.100	0.120	0.160
3. Major streams with width at flood stage > 30 m			
A. Streams on plains			
Sand channels	0.025	0.035	0.045
Boulder channels	0.028	0.040	0.045
Vegetation-lined channels at flood stage	0.045	—	0.120
B. Mountain streams			
Cobbly bottoms, no debris dams	0.028	0.035	0.040
Cobbly bottoms with debris dams	0.032	—	0.06
Large boulders, debris dams in channel	0.050	—	0.100

Source: Adapted from Gray 1973 and Van Haveren 1986.

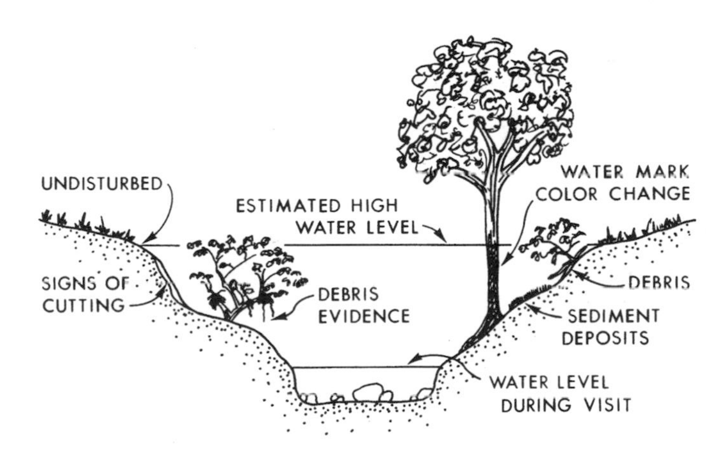

FIGURE 4.17. Looking for evidence of high water level in the field.

SUMMARY

The streamflow response from a watershed due to rainfall or snowmelt events is the integrated effect of many factors. Some of these factors are affected directly by human activities on the watershed, while others are not. To this point in this book, many of the precipitation and watershed characteristics that affect the amount and pattern of streamflow have been discussed. By now, you should be able to:

- Explain how the following affect infiltration rates:
  - soil moisture content,
  - hydraulic conductivity of the soil,
  - soil surface conditions, and
  - presence of impeding layers in the soil profile.
- Explain how land use activities affect infiltration capacities of a soil through each of the above.
- Discuss how changes in infiltration capacities can result in different flow pathways through the watershed.
- Illustrate and discuss the different pathways and mechanisms of flow that result in a stormflow hydrograph and baseflow for a forested watershed with deep soils versus an urban or agricultural watershed; explain how the major pathways of flow differ in each case, and how these differences change the total streamflow hydrograph.
- Determine streamflow discharge, given velocity and cross-sectional area data.
- Describe different ways in which streamflow can be measured; discuss the advantages and disadvantages of each.
- Define the terms in the Manning and Chezy equations and give an example of how they can be used to estimate peak discharge of a streamflow event that was not measured directly.

References

Beasley, R.S. 1976. Contribution of subsurface flow from the upper slopes of forested watersheds to channel flow. *Soil Sci. Soc. Am. Proc.* 40:955–957.

Berglund, E.R., A. Ahyoud, and M. Tayaa. 1981. Comparison of soil and infiltration properties of range and afforested sites in northern Morocco. *For. Ecol. Manage.* 3:295–306.

Blackwell, P., W. Crabtree, D. McGhie, and E. Spadek. 1993. Improving sustainable production from water repellent sands. *West. Aust. J. Agri.* 34:160–167.

Bonell, M., D.S. Cassells, and D.A. Gilmour. 1982. Vertical and lateral soil water movement in a tropical rainforest catchment. In *First national symposium on forest hydrology,* eds. E.M. O'Loughlin and L.J. Bren, 30–38. National Committee on Hydrology and Water Resources of the Institution of Engineers. Melbourne, Aust.

Brown, H.E. 1969. A combined control-metering section for gaging large streams. *Water Resour. Res.* 5:888–894.

DeBano, L.F. 1981. *Water repellent soils: A state-of-the-art.* USDA For. Serv. Gen. Tech. Rep. PSW-46.

DeBano, L.F. 2000a. Water repellency in soils: A historical overview. *J. Hydrol.* 231–232:4–32.

DeBano, L.F. 2000b. The role of fire and soil heating on water repellency in wildland environments. A review. *J. Hydrol.* 231–232:195–206.

Dixon, R.M. 1975. Infiltration control through soil surface management. In *Proceedings of a symposium on watershed management,* 543–567. New York: American Society of Civil Engineers.

Dixon, R.M. and A.E. Peterson. 1971. Water infiltration control: A channel system concept. *Soil Sci. Soc. Am. Proc.* 35:968–973.

Gray, D.M., ed. 1973. *Handbook on the principles of hydrology.* Reprint. Port Washington, NY: Water Information Center, Inc.

Hewlett, J.D. 1982. *Principles of forest hydrology.* Athens: Univ. Of Georgia Press.

Hewlett, J.D., and C.A. Troendle. 1975. Nonpoint and diffused water sources: A variable source area problem. In *Proceedings of a symposium on watershed management*, 21–46. New York: American Society of Civil Engineers.

Holtan, H.N. 1971. A formulation for quantifying the influence of soil porosity and vegetation on infiltration. In *Biological effects in the hydrological cycle*, 228–239. Proc. Third Int. Sem. Hydrol. Professors. West Lafayette, IN: Purdue Univ.

Horton, R.E. 1940. An approach to the physical interpretation of infiltration capacity. *Proc. Soil Sci. Soc. Am.* 5:399–417.

Horton, R.E. 1945. Erosional development of streams and their drainage basins: Hydrophysical approach to quantitative morphology. *Geol. Soc. Am. Bull.* 56:275–370.

Johnson, M.G., and R.L. Beschta. 1980. Logging, infiltration capacity, and surface erodibility in western Oregon. *J. For.* 78:334–337.

Maidment, D.R. (Ed.) 1993. *Handbook of hydrology*. New York: McGraw-Hill.

Philip, J.R. 1957. The theory of infiltration: The infiltration equation and its solution. *Soil Sci.* 83:345–357.

Sidle, R.C., and D.M. Drlica. 1981. Soil compaction from logging with a low-ground pressure skidder in the Oregon coast ranges. *Soil Soc. Am. J.* 45:1219–1224.

Smith, R.E., D.I. Cherry, K.G. Renard, and W.R. Gwinn. 1981. *Supercritical flow flumes for measuring sediment-laden flow.* USDA Tech. Bull. 1655.

Strahler, A.N. 1952. Dynamic basis of geomorphology. *Geol. Soc. Am.*

Strahler, A.N. 1964. Quantitative geomorphology of drainage basins and channel networks. In *Handbook of applied hydrology*, ed. V.T. Chow, Section 4.2. New York: McGraw-Hill.

Troendle, C.A. 1985. Variable source area models. In *Hydrologic forecasting*, ed. M.G. Anderson and T.P. Burt, 347–403. New York: John Wiley & Sons.

Van Haveren, B.P. 1986. *Water resources measurements: A handbook for hydrologists and engineers.* Denver: American Water Works Association.

Weitzman, S., and R.R. Bay. 1963. *Forest soil freezing and the influence of management practices, northern Minnesota.* USDA For. Serv. Res. Pap. LS–2.

Woodruff, J.F., and J.D. Hewlett. 1970. Predicting and mapping the average hydrologic response for the eastern United States. *Water Resour. Res.* 6:1312–1326.

CHAPTER 5 *Groundwater*

Introduction

Water that occurs in saturated zones beneath the soil surface is groundwater. In contrast with the more visible surface water in streams, rivers, ponds, lakes, and reservoirs, groundwater comprises more than 97% of all liquid freshwater on the earth. Groundwater contributes about 30% of all streamflow in the United States. Furthermore, nearly 50% of drinking water in the United States comes from wells. Globally, more than one-half of the world's population depends on groundwater. Although groundwater is an important source of liquid freshwater, it does not always occur where it is most needed and is sometimes difficult to extract. Without proper management, large quantities of this valuable resource can become unusable because of contamination or due to deep pumping to the point that further extraction is not feasible economically.

The purpose of this chapter is to describe the linkages between watershed and groundwater management. A comprehensive discussion of groundwater is beyond the scope of this book, but we intend to familiarize the reader with some basic terminology and concepts. The emphasis is on land use impacts on groundwater, particularly those associated with upland watersheds and riparian-wetland systems that are discussed in Chapters 13 and 14.

Basic Concepts

Some perceive groundwater to occur as vast underground lakes and rivers, but (for the most part) groundwater occurs in voids between soil and rock particles in the *zone of saturation*. To understand how a zone of saturation is formed, we first discuss the forces that govern the downward movement of water in the soil.

The process of infiltration and the subsequent movement of water through the soil are both the result of matric, or capillary, forces and gravity. Capillary forces represent the physical attraction of soil and rock particles to water; water flows from wetted particles (high energy potential) to drier particles (low energy potential). As the openings between particles become filled with water, gravity becomes more dominant. Once the field capacity of a well-drained soil is exceeded, a flow of water begins in a downward direction. This downward movement of water continues through pores in the soil, parent material, and underlying rock. In general, pores become fewer and smaller with increasing depth, generating more resistance to flow. Furthermore, as water moves into deeper zones, it is no longer subjected to evaporation or transpiration. Although impervious layers (or strata) can exist at various depths below the soil surface, porosity in general becomes negligible at depths below 600 m. These conditions result in the formation of a zone of saturation.

The *zone of aeration* is that part of the profile that occurs between the soil surface and the top of the zone of saturation (Fig. 5.1). The zone of aeration consists of the soil water zone, which extends through the rooting zone; the *vadose zone*, which extends

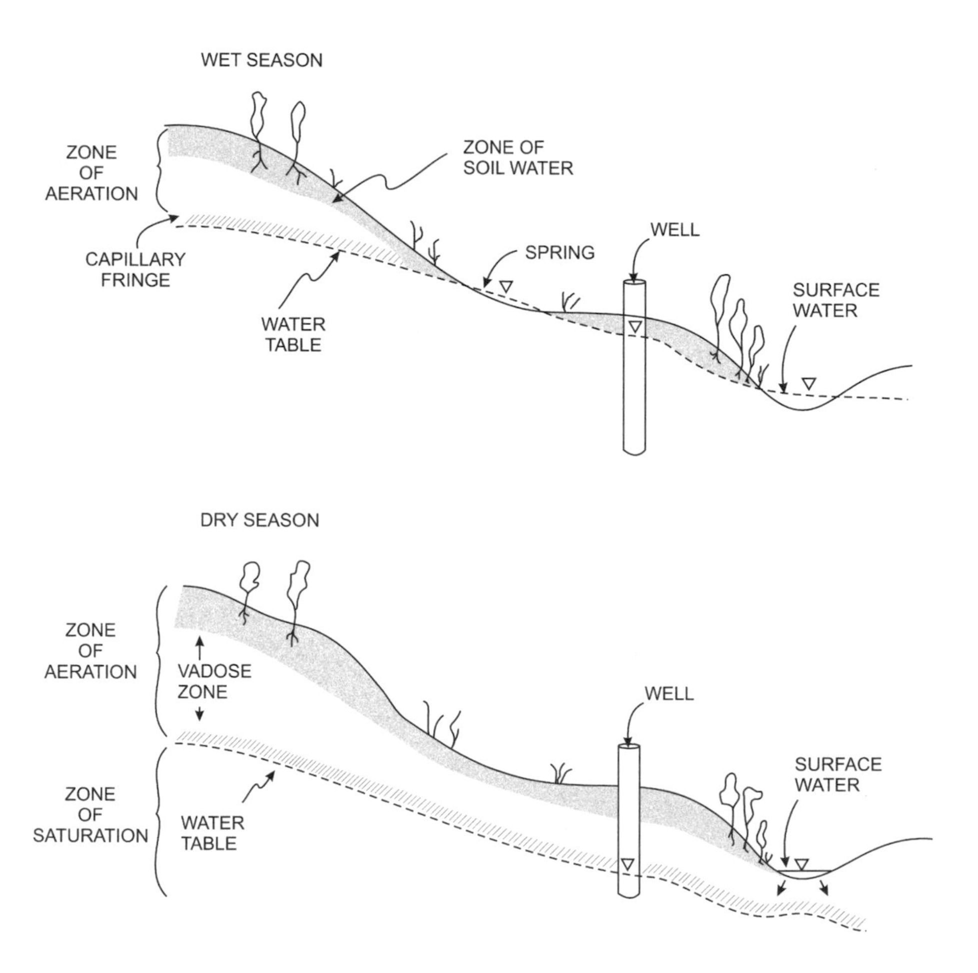

FIGURE 5.1. Groundwater characteristics and water table changes from a wet to a dry season.

from the soil water zone to the *capillary fringe*; and the capillary fringe. The top of the zone of saturation, where the water potential is zero, is called the *water table*. It is measured by the elevation of water surfaces in wells that penetrate into the zone of saturation. Immediately above the water table is the capillary fringe, a zone in which water from the zone of saturation is "pulled up" by capillary forces into the zone of aeration. This capillary fringe has a negative water potential, and its irregular position varies with changes in water table elevation. The height of the capillary fringe above the water table is determined by the type of matrix; it is insignificant in coarse-grained sediments but can be several centimeters high in silts and clays.

The above terminology should not suggest that the zone of aeration cannot become saturated. In fact, this zone can frequently become saturated in some areas when rainfall or snowmelt becomes excessive. The distinguishing feature of the aeration zone is that saturated conditions are only temporary. It must also be emphasized that the water table is not a static surface. The elevation of a water table moves up and down in response to changing precipitation and evapotranspiration patterns, as illustrated in Figure 5.1. During the wet season, springs occur where the water table comes in contact with the soil surface and the groundwater system can discharge water into streams. Streams that are groundwater fed are called *effluent streams*; if they are effluent year-round, they also are *perennial streams*. During dry periods, the water table can drop and create situations where springs no longer flow and where streams are no longer fed by groundwater. Streams that lose runoff to the groundwater are *influent streams* (Fig. 5.1).

Storage and Movement of Groundwater

Groundwater occurs in many different types of soil and rock strata. It can occur between individual soil or rock particles, in rock fracture openings, and in solution openings (formed when water dissolves constituents in the rock strata, leaving a void). The amount of groundwater stored and released from water-bearing strata depends on the porosity, the size of pore spaces, and the continuity of pores. Water-bearing porous soil or rock strata that yield significant amounts of water to wells are called *aquifers*. Any water-bearing soil or rock strata that are effectively impermeable, such as shales, slates, or thick clay lenses, are referred to as *aquicludes*. Geologic strata that are slowly permeable and retard groundwater, such as silts and mudstone, are called *aquitards*.

An aquifer can be an underground lens of sand or gravel, a layer of sandstone, a zone of highly fractured rock (even granite), or a layer of cavernous limestone. An aquifer can be from a few meters to hundreds of meters thick and can underlie a few hectares or thousands of square kilometers. The Ogallala aquifer underlies several states in the midwestern United States.

Porosity, the total void space between the grains and in the cracks and solution cavities that can fill with water, is defined in terms of percent pore space as:

$$\text{porosity} = \frac{100V_v}{V_t} \qquad \textbf{(5.1)}$$

where V_v = the volume of void space in a unit volume of rock or soil; and V_t = the total volume of earth material, including void space.

Porosity ranges from 10 to 20% for glacial till, from 25 to 50% for well-sorted sands or gravels, and from 33 to 60% for clay. The effective porosity is the ratio of the void space through which water can flow to the total volume.

If all the grains in a consolidated or unconsolidated material are about the same size and are well sorted, the spaces between them account for a large part of the total volume. If grains are sorted poorly, the larger pores can fill with smaller particles rather than water. Well-sorted materials tend to hold more water than materials that are sorted poorly.

Pores must be connected to each other if water is to move through a soil or rock. If pores are interconnected and of sufficient size to allow water to move freely, the soil or rock is *permeable*. Aquifers that contain small pores or pores that are connected poorly can yield only small amounts of water, even if their total porosity is high.

Water flows through a soil or rock material in response to hydraulic head gradients and follows the pathway of least resistance. It will move through permeable materials and around impermeable ones. As the complexity of the geology of an area increases (i.e., the amount of folding, uplifting, and fracturing of strata increases), the pathways of water flow likewise become complex. In some instances, groundwater can flow slowly for hundreds of kilometers before emerging as a natural spring, seeping into a stream, being tapped by a well, or emerging into the ocean. Just as discharge areas can be varied, recharge for an aquifer also can occur in many areas dispersed spatially over a large area. Recharge areas can be far distant from discharge areas of an aquifer, which sometimes makes it difficult to identify all important recharge areas. Because water moves through aquifers under the influence of gravity, one thing is certain: recharge zones are higher in elevation than areas of discharge.

Recharge usually takes place in areas where permeable soil and rock materials are relatively close to the land surface and where there is an excess of water from precipitation. The rate of recharge and the area over which recharge takes place are important considerations when groundwater pumping is being contemplated. If pumping removes more groundwater than is being recharged, the aquifer is being *mined*.

Unconfined and Confined Aquifers

Aquifers that contain water that is in direct contact with the atmosphere through porous material are called *unconfined aquifers*. The groundwater system illustrated in Figure 5.1 is unconfined; the soil system immediately above the water table readily allows the exchange of gases and water. In contrast, a *confined aquifer* is separated from the atmosphere by an impermeable layer, or *aquiclude* (Fig. 5.2). A confining stratum often forms a perched water table. An unconfined aquifer can become a confined aquifer at some distance from the recharge area.

Confined aquifers, also called *artesian aquifers*, contain water under pressure, in some cases sufficient to produce freely flowing wells. Water pressure (P), or pressure potential, is a function of the height of the water column at a point (h_p), the density of water (ρ), and the force of gravity (g). For a system without energy loss due to flow friction, the pressure can be approximated by:

$$P = \rho g h_p \quad \textbf{(5.2)}$$

Pressure is directly proportional to the height of the water column above some point in the system. The total hydraulic head (h_t) includes the water pressure from that point down to an arbitrary but stable reference datum:

$$h_t = z + h_p \quad \textbf{(5.3)}$$

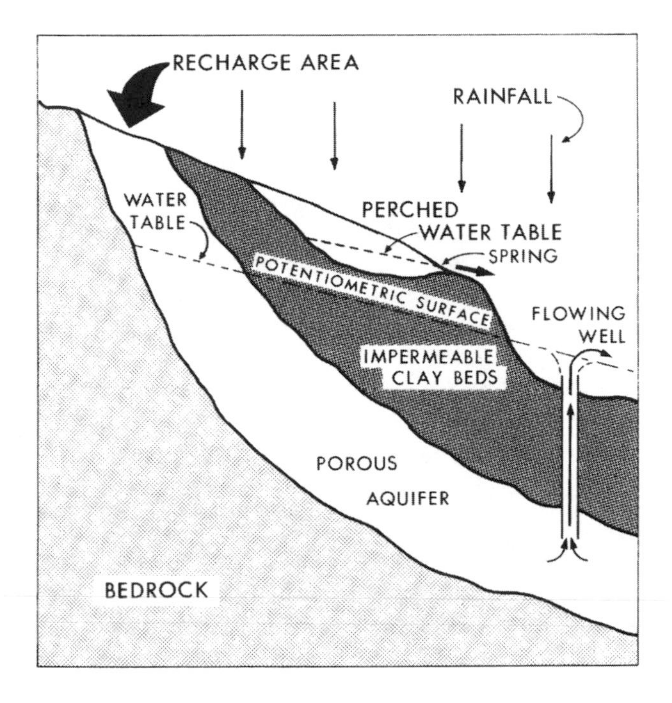

FIGURE 5.2. Artesian (confined) aquifer and recharge area with a perched water table above an impermeable layer (adapted from Baldwin and McGuinness 1963).

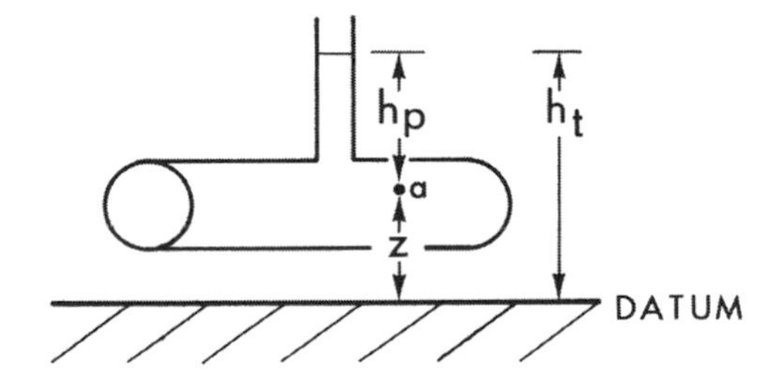

FIGURE 5.3. Pressure head (h_p), elevation head (z), and total hydraulic head (h_t) of water. Elevation head (z) is the distance from an arbitrary but stable reference datum to a point (a) where pressure head (h_p) is measured.

These components are illustrated in Figure 5.3. The difference in total hydraulic head from one point to another creates the hydraulic gradient dh_t/dx, where x is the distance between points. In unconfined aquifers, the elevation of the water surface measured by wells can be used to construct a water table contour map, which is similar to a surface contour map for surface runoff. The direction of groundwater flow, or *flow lines*, can be determined by constructing lines perpendicular to the water table contours from higher to lower elevation contours.

The piezometric, or *potentiometric*, surface of an artesian aquifer describes the imaginary level of hydraulic head to which water will rise in wells drilled into the confined aquifer (Fig. 5.2). The potentiometric surface declines because of friction losses between points, but when the land surface falls below the potentiometric surface, water will flow from the well without pumping (artesian, or flowing, well). Therefore, artesian pressure is the result of the actual water table in the downstream discharge area being at a much lower level than in the upstream recharge area and its suppression by confining layers above the aquifer.

As with unconfined aquifers, points of equal potentiometric head can be connected, forming contours that are used to construct a potentiometric surface map. Such a map represents the slopes of the potentiometric surface and indicates the direction of groundwater flow in artesian aquifers.

Aquifer Characteristics

When considering the development of groundwater for pumping, certain characteristics of the aquifer(s) from which the groundwater is to be extracted need to be understood. An important characteristic is the *transmissivity* of an aquifer, which is the amount of water that can flow horizontally through the entire saturated thickness of the aquifer under a hydraulic gradient of 1 m/m. It is defined as:

$$T_r = bk_v \quad \textbf{(5.4)}$$

where T_r = transmissivity (m²/unit time); b = saturated thickness (m); and k_v = hydraulic conductivity of the aquifer (m/unit time).

Any time the hydraulic head in a saturated aquifer changes, water will be either stored or discharged. *Storativity* is the volume of water that is either stored or discharged from a saturated aquifer per unit surface area per unit change in head. The storativity characteristic of an aquifer is related to the specific yield of the soil or rock material that constitutes the aquifer. *Specific yield* (S_y) is the ratio of the volume of water that can drain freely from saturated earth material due to the force of gravity to the total volume of the earth material.

The amount of water discharged from an aquifer can be approximated with Darcy's law (see Chapter 4):

$$Q = k_v A \frac{dh_t}{dx} \quad \textbf{(5.5)}$$

The discharge from an aquifer (Box 5.1) is dependent, therefore, on the cross-sectional area through which flow occurs (A), the hydraulic conductivity of the material constituting the aquifer (k_v), and the hydraulic gradient (dh_t/dx). The value of k_v is dependent on the properties of the porous medium and of the fluid passing through it; the more viscous the fluid, the lower the k_v. Examples of hydraulic conductivities of earth material for pure water at a temperature of 15.6°C are listed in Table 5.1.

Groundwater Development

Assessing the potential for groundwater development requires knowledge of the local geology and aquifers. Surface features ordinarily do not allow one to determine the

Box 5.1

Application of Darcy's law

The problem presented is to determine the discharge of flow through a well-sorted gravel aquifer, given that k_v = 0.01 cm/sec, the change in head is 1 m over a distance of 1000 m, and the cross-sectional area of the aquifer is 500 m^2.

$$Q = k_v A \frac{dh_t}{dx}$$

$$= (0.01 \text{ cm/sec})\,(500 \text{ m}^2)\left(\frac{10{,}000 \text{ cm}^2}{1 \text{ m}^2}\right)(0.001 \text{ m/m})$$

$$= 50 \text{ cm}^3/\text{sec}$$

$$= 4.32 \text{ m}^3/\text{day}$$

TABLE 5.1. Examples of hydraulic conductivities for unconsolidated sediments (pure water, 15.6°C)

Material	Hydraulic Conductivity (cm/sec)
Well-sorted gravel	10^{-2}–1
Well-sorted sands, glacial outwash	10^{-3}–10^{-2}
Silty sands, fine sands	10^{-5}–10^{-3}
Silt, sandy silts	10^{-6}–10^{-4}
Clay	10^{-9}–10^{-6}

Source: Adapted from Fetter 2001.

location, depth, and extent of water-bearing material or strata. Geologic maps can be used to help identify potentially productive water-bearing strata by examining the direction and degree of dipping strata, locating faults and fracture zones, and determining the stratigraphy of rocks with different water-bearing and hydraulic characteristics. Information from geologic maps can be used to determine whether special techniques, such as horizontal wells, may be appropriate. For example, areas that have old lava flows often exhibit considerable vertical development of secondary openings, such as lava tubes and fissures caused by escaping gases. Horizontal wells increase the chances of intercepting these larger water-bearing pores that generally are not widespread and are difficult to locate by vertical drilling.

As a rule, opportunities for groundwater development increase as one moves from upland watersheds to lower basins and floodplains. Extensive and high-yielding aquifers occur in most major river valleys and alluvial plains. On a smaller scale, the same features can be important sources of groundwater in upland areas. Small valleys in uplands and associated stream channel systems can have locally high water-yielding deposits of alluvium. Although usually not extensive, such deposits can provide water

TABLE 5.2. Water-bearing and yield characteristics of some common aquifers

Type of Material	Special Yield (%)	Well Yields	
		GPM	l/sec
Metamorphic/plutonic igneous	0–25	10–25	0.6–1.6
Volcanic	variable	< 1500	< 95
Granite	< 1	neg.	neg.
Sedimentary rocks			
Shales/claystones	0–5	< 5	< 0.3
Sandstones	8	5–250	0.3–16
Limestone, solid	2	neg.–5	neg.–0.3
Limestone with solution cavities	variable	> 2000	> 126
Unconsolidated deposits			
Clay	0–5	neg.	neg.
Sand/gravel	10–35	10–>3000	0.6–>189

Source: Adapted from Davis and DeWiest 1966 and Fetter 2001.
Note: GPM = gallons per minute; neg. = negligible.

supplies during critically dry periods for local consumption or as a backup for other water supply systems.

Given sufficient aquifers and proper well location, groundwater can supply most of the water needs of many communities. Upland wells are often dug for local drinking water for humans and livestock. Some upland wells can also supply water for larger communities and limited irrigation. The amount of water that can be supplied from upland wells depends on the types of rocks underlying the area, the degree of weathering, the presence of faults or fracture zones, and the extent of unconsolidated sands and gravels that occur as alluvium or below stream channels. Well yields from consolidated and unconsolidated materials can vary considerably (Table 5.2).

Wells

There are many types of wells, ranging from those that are hand dug to wells that are driven into an aquifer (using well-points) to those drilled with a cable-tool drilling rig. Dug wells are generally not deep, but even a well dug by hand must be lined to keep the sides from falling in. Driven wells consist of pipes that are pushed into shallow gravel and sand aquifers, usually < 20 m below the surface. Such wells are simple and cheap to install.

A drilled well differs from a dug well in that the hole is made with drilling rigs, enabling much deeper wells to be developed. Cable-bucket and rotary drill rigs commonly are used. A cable-bucket rig churns a heavy bit up and down, pounding it through the soil or rock. A rotary rig drills its way through. In either case, the hole is *cased* with

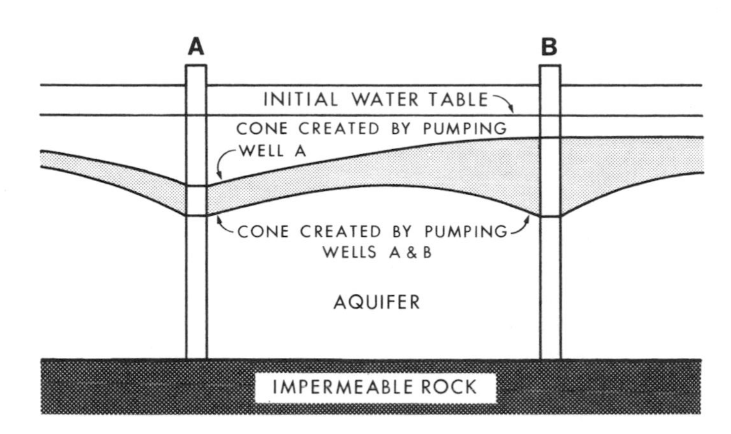

FIGURE 5.4. Cones of depression (from Baldwin and McGuinness 1963).

a pipe to prevent a cave-in. When a hole has been drilled some distance below the water table, the drilling is stopped and a water pipe is lowered inside the casing. The *well-point* is the lower end of the pipe to which a screen is attached; this screen consists of a length of pipe with many fine perforations that allow water to enter the pipe but exclude soil material. Water is forced out of the well by a motor-driven submersible pump or a pump driven by a windmill (unless it is an artesian well).

To test a well, one measures the water level, then pumps the well at a steady rate. The water level will drop quickly at first and then more slowly as the rate at which water is flowing into the well approaches the pumping rate. The difference between the original water level and the water level after a period of pumping is called the *drawdown*. The discharge rate is determined by a flowmeter attached to the discharge pipe. The ratio between the discharge rate and the drawdown characterizes the well's specific capacity (m^3/sec). Hydrologists can use the result of controlled pumping tests to predict effects of future pumping on water levels.

Pumping water from a well lowers the water table around the well and creates a *cone of depression* (Fig. 5.4). Around small-yield wells in productive aquifers, the cone of depression is small and shallow. A well pumped for irrigation or industrial use can withdraw so much water that the cone of depression extends for many kilometers.

Locating wells too close together causes more lowering of a water table than spacing them far apart. This process is called *interference*. Interference can draw water levels so low that pumping costs will be greatly increased.

Management of Groundwater Resources

The continued use of large quantities of groundwater can create water problems. Under natural conditions, the hydrologic cycle tends to be in balance, but people's use of the water can upset this balance. Use of groundwater resources without knowledge of the effects of use or in disregard of them is unwise. In contrast, *good management* of groundwater is use with knowledge of the probable effects and with plans to minimize adverse effects.

Good management of groundwater depends upon knowledge of basic water facts. Detailed studies of groundwater in local areas are needed, and basic research on recharge and movement of groundwater is required. Also, information on groundwater quality is necessary, and methods of storing surplus water in underground reservoirs must continue to improve.

Groundwater can be managed using the concept of *safe yield*, which refers to the annual draft of groundwater at levels that do not produce undesirable effects. For example, groundwater should not be withdrawn at rates that result in excessive lowering of the water table and, hence, high pumping costs, or that result in saltwater intrusion. A water budget analysis can be performed on aquifers to study the quantitative aspects of safe yield; various inputs can be compared with outputs as follows:

$$I - O = \Delta S \quad \textbf{(5.6)}$$

where I = inputs to groundwater, including groundwater recharge by percolation of rainwater and snowmelt, artificial recharge through wells, and seepage from lakes and streams; O = outputs from groundwater, including pumping, seepage to lakes and streams, springs, and evapotranspiration; and ΔS = change in storage, determined as the product of change in water table elevation and specific yield for an unconfined aquifer, or the product of change in potentiometric head and storativity for a confined aquifer.

Ideally, groundwater should be managed over long periods of time so that there is zero change in storage. Artificial recharge can sometimes offset pumping so that $\Delta S = O$. Depletion of groundwater storage not only affects groundwater use but also can affect surface water supplies by reducing groundwater contributions to lakes and streams.

The sustained withdrawal of groundwater can have unanticipated effects, many of which are detrimental. In karst systems, sinkholes can form where groundwater withdrawal reduces water pressure to the point where limestone collapses. Sinkholes can become large and seriously damage property. Similarly, subsidence can occur where thick, compressible clays and silts overlie aquifers that are mined. Again, the loss of groundwater pressure can lead to compaction and subsidence over large areas. This loss has been observed in many parts of the world, including Mexico City; Bangkok, Thailand; Venice, Italy; and southern Arizona, where groundwater has been pumped excessively for decades.

Excessive groundwater pumping in coastal areas can lead to saltwater intrusion. Reduced water pressure in freshwater aquifers creates hydraulic gradients favoring saltwater movement into the aquifer. Once an aquifer is contaminated by saltwater, the use of groundwater in an area becomes severely restricted. Costly methods of artificial recharge (adding freshwater back into aquifers through wells) can sometimes be used to reduce saltwater intrusion.

Groundwater that seeps into streams provides the baseflow for those streams. Therefore, if water levels decline because of heavy pumping, the baseflow of the streams will also be reduced (Box 5.2). Surface water and groundwater are linked inextricably and, therefore, should be evaluated together. River basin development can affect groundwater reservoirs and vice versa. However, plans for river basin development commonly neglect groundwater. The rate of natural replenishment need not limit the use of groundwater if floodwaters can be used to increase recharge. River basin development should include a coordinated program of flood control and artificial

Box 5.2

Sources of Baseflow to a Flowing River—One Example (Wirt and Hjalmarson 2000)

Multiple lines of evidence have been used to identify source aquifers, quantify their respective contributions, and trace groundwater flow paths that supply baseflow to the uppermost reach of the Verde River in north central Arizona. Groundwater discharge through springs provides the baseflow for approximately a 40 km long reach. This flowing reach is important to the increasing number of downstream water users, maintains critical habitats necessary for the recovery of native fish species, and has been designated a Wild and Scenic River because of its unique characteristics. Sources of baseflow in the reach were deduced from baseline geologic information, historical groundwater levels, long-term precipitation and streamflow records, downstream changes in baseflow measurements, analysis of water budget components, and stable-isotope geochemistry of groundwater, surface water, and springs. Taken together, this information clearly indicates that the interconnected aquifers in the headwater area are the primary sources of springs presently supplying at least 80% of the upper Verde River's baseflow. It is possible, therefore, that increased pumping of these aquifers to furnish water to the rapidly increasing human population downstream could contribute to a dewatering of this section of the river.

recharge and, importantly, must recognize the linkages between surface water and groundwater systems.

Special Considerations—Upland Areas

Several problems can result from extensive pumping of groundwater in both upland areas and the valleys below. Land subsidence, a problem in many lowland areas, is usually not of major concern in upland areas. Perhaps of greatest concern is when upland watershed inhabitants become overly dependent upon groundwater resources that are not sufficient to support long-term, sustained demands. Increasing human and livestock populations in remote watersheds can deplete local groundwater supplies quickly. Prolonged dry spells or droughts can then cause loss of life and serious economic losses. A point can be reached easily where digging another well and deepening existing wells are no longer viable solutions to water needs.

In some water-scarce areas, the development of wells for livestock watering can have detrimental effects on the watershed vegetation. For example, areas where water availability has previously limited livestock numbers can become subjected to overgrazing after the development of wells. In such instances, the design and development of wells should be accomplished in a manner that is compatible with a sound range management program.

To sustain water yields from wells, the rate of extraction or pumping cannot exceed the rate of recharge over long periods. The rate of recharge is governed by the availability of water and the infiltration and hydraulic characteristics of the soil and rock strata in the recharge area. The time required for water to move from recharge zones to well sites can be days, months, or years. For coarse sands and fine gravels, water can travel at rates of 20–60 m/day. Rates of travel in finer clays and dense-rock aquifers can be < 0.001 m/day. Altering the hydraulic properties of the soil system in recharge zones can affect rates of infiltration and recharge, but these impacts might not be observed at downstream well sites for long periods.

Naturally occurring springs can provide local sources of water for upland inhabitants and their livestock and are useful indicators of the location and extent of aquifers. The permeability of the aquifer and its recharge determine the discharge of a spring. The areal extent of the recharge area and its hydraulic characteristics govern the amount of recharge that takes place. As discussed earlier, springs occur where the water table intercepts the ground surface and when discharge is sufficient to flow in a small rivulet most of the time. If such flow is not evident, the resulting wet areas are called *seeps*. Springs can normally be found at the toes of hillslopes, along depressions such as stream channels, and where the ground surface intercepts an aquifer.

Wells and springs can enhance water resource development in upland watersheds if they provide dependable, high-quality water. Dependability is a function of recharge, the extent of the aquifer, and its yield characteristics. Many perched, or temporary, zones of saturation occur in upland watersheds and can be identified by seeps or springs that flow only during the wet season. Even when flow occurs year-round, extreme variability of flow can indicate an unreliable or temporary groundwater system.

The quality of groundwater sometimes can indicate whether the aquifer is perched or is part of a regional groundwater system. For example, the specific conductance (see Chapter 11) of groundwater in northern Minnesota is a good indicator of the type of groundwater present. Specific conductance readings of > 120 μmhos/cm indicate significant contributions of water from regional groundwater sources (Hawkinson and Verry 1975). Readings of < 50 μmhos/cm indicate a short residence time underground and low concentrations of minerals and salts in the water. In this illustrative case, specific conductance can be used to predict the dependability of the groundwater source. Areas that have calcareous soils would not show the same distinction, because any underground water would tend to have high specific conductance readings.

Groundwater Recharge Zones

Upland forested watersheds commonly are viewed as being important recharge zones for aquifers, because forests occur in areas with high annual precipitation and are associated with soils that have high infiltration capacities. Given this to be true, then what are the effects of land management activities, including forest cutting and regeneration, on groundwater recharge? In considering this question, we will first examine the processes affected and then the implications of such changes for groundwater supplies. To provide a focal point for this discussion, we will examine forest management implications.

The removal of forest cover would normally increase the amount of water in soil storage and the amount available for groundwater recharge. If extensive road and skid trail development accompany timber harvesting, however, total infiltration capacity can become reduced. If such disturbance is widespread and in proximity to stream channels, surface runoff can be increased at the expense of subsurface and groundwater flow. The conversion of cut-over areas to agricultural crops or pasturelands could result in a more widespread and permanent impact on infiltration capacity and recharge. The net effect of such activities depends on whether reductions in evapotranspiration or reductions in infiltration have the greatest impact on recharge.

Although not well documented through controlled catchment experiments, it is possible that widespread soil disturbance in a recharge zone could cause groundwater-fed, perennial streams to become dry during seasonal low-flow periods. Such occurrences would be rare, however, and would be significant only where small catchments feed a localized groundwater aquifer. Otherwise, the opportunities for water to recharge a groundwater aquifer are too great, especially when large distances and vast regional aquifers are involved. As stated in Chapter 6, most controlled watershed experiments have shown increases in recession flow and baseflow following timber-harvesting and logging activities.

A realistic appraisal of land use impacts on the recharge of large, regional groundwater aquifers indicates that little impact would be expected under most conditions. For example, if forest management in the United States is considered, the effects on recharge of major aquifers would be slight because (1) at any point in time only a small portion of any recharge area is under clearcut or logging conditions; (2) recharge usually occurs over vast areas, and aquifers store large amounts of water and do not respond quickly or noticeably to small changes in recharge; and (3) changes in evapotranspiration, infiltration, and permeability normally are not severe, especially when compared to changes in precipitation and energy associated with natural fluctuations in climate.

Effects of Vegetation on Groundwater

Much of the focus of groundwater management centers on the geological aspects of location, volumetric extent, and hydraulic characteristics of aquifers that relate to dependability and performance. However, there are situations where vegetative cover can also affect groundwater directly. The effects of vegetation on groundwater are most often manifested in riparian-phreatophyte communities and other wetland communities. These effects are considered briefly below, with the reader referred to Chapters 13 and 14 for more in-depth discussions of these vegetation communities and their linkages with groundwater.

Riparian communities consist of plants that grow adjacent to streams or lakes and often have root systems in close proximity to the water table. These stream-side plant communities are found in both wet and dry climates. Many riparian plant species have adapted themselves to conditions of shallow water tables or wet sites near to streams and lakes and, therefore, consume large amounts of water, including groundwater. Since water is available throughout their growing season, transpiration by these plants can occur at rates near potential evapotranspiration. Large quantities of groundwater can be extracted annually from underlying aquifers as a result.

In dry climates, extensive riparian plant communities can sometimes be found along ephemeral stream channels and on expansive floodplains, both of which have shallow water tables. Many of the plants in these communities, called *phreatophytes*, have extensive rooting systems that allow them to extract water from the water table or from the capillary fringe. Extensive stands of saltcedar occur on floodplains throughout the southwestern United States. Ash, willow, cottonwood, and alder are often indicators of shallow water tables and potable water throughout North America. However, the presence of such tree species does not mean that the shallow groundwater is necessarily recoverable, because these plants can extract water from fine silts and clays that might not yield water to a well. Under conditions of high potential evapotranspiration, phreatophytes can transpire large quantities of groundwater (see Table 6.3).

Wetlands are usually low-lying areas that are connected with the groundwater system. Wetlands occur where there is an excess of water, either as a result of drainage to a depression on the landscape or where annual precipitation amounts exceed potential evapotranspiration over a large area with little topographic relief. On these sites, the water table is at or near to the ground surface throughout the year; therefore, wetlands exhibit high rates of evapotranspiration. Vegetation on wetlands can be forests, shrubs, mosses, grasses, and sedges. Although wetlands normally occur in lowlands or coastal areas, they frequently form the headwater areas for streams or lakes.

The hydrologic behavior of any wetland is dependent largely on whether regional groundwater feeds it. The water table of some wetlands is an expression of regional groundwater. Such wetlands are characterized by a relatively stable water table, and those that have a surface stream outlet exhibit an even pattern of streamflow throughout wet and dry seasons. Wetlands that have perched water tables, or are otherwise separated from the regional groundwater, have greater seasonal fluctuations in both the water table and streamflow discharge. In either case, the linkage between wetlands and the associated groundwater requires that a water budget of wetlands must explicitly account for groundwater inputs, outputs, and changes in storage.

SUMMARY

A basic understanding of groundwater storage and flow characteristics, linkages between surface water and groundwater, and a knowledge of land use impacts on groundwater are important to watershed management. At this point in the text, you should be able to:

- Define and illustrate a regional water table, perched groundwater, a potentiometric (piezometric) surface, a water table well, an artesian and an unconfined aquifer, a capillary fringe, a spring, and a cone of depression.
- Explain the components of a water budget for a groundwater aquifer; contrast a groundwater budget with a water budget for a watershed.
- Explain the important factors that govern groundwater flow, using Darcy's equation as a point of reference.
- Describe characteristics of an aquifer that would yield high quantities of groundwater on a sustainable basis.
- Understand how surface water and groundwater can be linked together in a hydrologic system.
- Describe and explain how land use activities, including changes in vegetative cover and soil characteristics, can affect groundwater storage.

References

Baldwin, H.I., and C.L. McGuinness. 1963. *A primer on groundwater.* U.S. Geol. Surv.

Davis, S.N., and R.J.M. DeWiest. 1966. *Hydrogeology*. New York: John Wiley & Sons.

Fetter, C.W., Jr. 2001. *Applied hydrogeology*. 4th ed. New Jersey: Prentice-Hall.

Freeze, R.A. and J.A. Cherry. 1979. *Groundwater.* Englewood Cliffs, NJ: Prentice-Hall.

Hawkinson, C.F., and E.S. Verry. 1975. *Specific conductance identifies perched and groundwater lakes.* USDA For. Serv. Res. Pap. NC–120.

Heath, R.C. 1983. *Basic groundwater hydrology.* U.S. Geological Survey Water Supply Paper 2220.

Moore, J.E., A. Zaporozec, and J.W. Mercer. 1995. *Groundwater—A primer*. Alexandra, VA: American Geological Institute.

Pierce, M. 1995. *Introducing groundwater*. 2nd ed. London: Chapman & Hall.

Todd, D.K. 1980. *Ground water hydrology*. 2nd ed. New York: John Wiley & Sons.

Wirt, L., and H.W. Hjalmarson. 2000. *Sources of springs supplying base flow to the Verde River headwaters, Yavapai Country, Arizona*. U.S. Geol. Surv. Open-File Report 99–0378.

CHAPTER 6

Vegetation Management, Water Yield, and Streamflow Pattern

INTRODUCTION

Land use activities that alter the type or extent of vegetative cover on a watershed will frequently change water yields and, in some cases, maximum and minimum streamflows. Changes in vegetative cover occur as a normal part of natural resource management and rural development. Timber harvesting, shifting cultivation, and conversions of forests or brushlands to agricultural croplands or pastures are examples of changes that can alter streamflow response. The hydrologic implications of extensive and long-term changes in vegetative cover are controversial. Extensive deforestation in the Tropics, for example, has fostered widespread debate about possible changes in regional and global climate and precipitation.

The purpose of this chapter is to examine vegetation–water yield relationships and to point out the implications for water resource development and management. This chapter addresses such questions as:

- What are the real or anticipated effects of vegetative removal on the amount and distribution of precipitation?
- To what degree can water yields be manipulated by altering vegetative cover?
- Can vegetation be manipulated to complement water resource management objectives?
- To what extent are seasonal streamflow patterns, including flood flows, altered by changing vegetative cover?

VEGETATION MANAGEMENT FOR WATER YIELD

Studies conducted throughout the world have demonstrated that annual water yields change when vegetation type or extent is substantially altered on a watershed. In general,

TABLE 6.1. Increases in water yield associated with reductions in vegetative cover, for noncloud forest or coastal forest conditions

	Increase in Water Yield per 10% Reduction in Cover		
Vegetative Cover Type	**Average (mm)**	**Maximum (mm)**	**Minimum (mm)**
Conifer and eucalypt[a]	40	65	20
Deciduous hardwood	25	40	6
Shrub	10	20	1

Source: Adapted from Bosch and Hewlett 1982, as reported by Gregersen et al. 1987.

[a]Pilgrim et al. (1982) indicated that in Australia, pine used more water than eucalypts. Dunin and Mackay (1982) also indicated that interception losses of *Pinus radiata* were 10% more than eucalypt forests on an annual basis; in Australia annual differences between pine and eucalypt forests were estimated at 35–100 mm/yr.

changes that reduce evapotranspiration (ET) increase water yields. ET can be reduced by changing the structure and/or composition of vegetation on the watershed.

Evaporative processes generally account for most of the annual precipitation on watersheds; consequently, the potential to increase water yield by decreasing ET is attractive. For example, 85–95% of the annual precipitation is evaporated or consumptively used by plants on many watersheds in arid and semiarid regions, leaving only 5–15% available to either recharge groundwater aquifers or produce streamflow. High-elevation mountain watersheds in the snow zones of the world yield as much as 50% of annual precipitation, but ET still remains significant and potentially subject to reduction through vegetative management.

Water yield usually increases when (1) forests are clearcut or thinned, (2) vegetation on a watershed is converted from deep-rooted species to shallow-rooted species, (3) vegetative cover is changed from plant species with high interception capacities to species with lower interception capacities, or (4) species with high annual transpiration losses are replaced by species with low annual transpiration losses.

The amount of water yield change depends largely upon the soil and climatic conditions and the percentage of the watershed affected. The largest increases in water yield often result from clearcutting forests. The length of time that water yields continue to exceed precutting levels is influenced by the type of vegetation that regrows on the site and the rate of regrowth. Higher water yield responses would be expected in regions with deep soils and high annual precipitation, whereas responses would be lower in magnitude in dry climates. Nevertheless, improving water yields has been emphasized in many drylands, where small increases in water yield can be important for human and livestock needs.

The general relationships indicated in Table 6.1 can be used to determine approximate changes in water yield. Regionalized relationships and exceptions to the rule will be examined in the following paragraphs.

Humid Temperate Regions

Water yield response to vegetation management in humid temperate climates would seem to be of questionable interest at first glance. There are situations, however, where either decreasing or increasing water yield can be of benefit. On one hand, areas with excessive water may call for management aimed at decreasing water yield. On the other hand, increasing demands for water by expanding urban areas may warrant management of municipal watersheds for increased water yield, even in areas with relatively high annual precipitation (say 1000 mm). The following discusses relationships between vegetative cover and water yield in this climatic regime.

Forested Uplands

Paired watershed experiments in the eastern United States indicate a consistent relationship between water yield and forest cover. For example, Douglas (1983) has summarized water yield responses to the cutting of eastern hardwood forests as follows:

$$Y_H = 0.00224\left(\frac{BA}{PI}\right)^{1.4462} \quad \textbf{(6.1)}$$

$$D_H = 1.57Y_{H_I} \quad \textbf{(6.2)}$$

$$Y_{Hi} = Y_H + b\log(i) \quad \textbf{(6.3)}$$

where Y_H = first-year increase in water yield (in.) after cutting hardwoods (H); BA = percent basal area cut; PI = annual potential solar radiation in cal/cm$^2 \times 10^{-6}$ for the watershed; D_H = duration of the increase in water yield (yr); Y_{Hi} = increase in water yield for the ith year after cutting (in.); and b = coefficient derived by solving Equation 6.3 when $i = D_H$ and $Y_{Hi} = 0$.

For conifers, the relationships were:

$$Y_C = Y_H + (I_C - I_H) \quad \textbf{(6.4)}$$

$$D_C = 12 \quad \textbf{(6.5)}$$

$$Y_{Ci} = Y_C + b\log(i) \quad \textbf{(6.6)}$$

where $I_C - I_H$ is the difference in interception between conifers and hardwoods (in.).

The above relationships apply for conditions where annual precipitation exceeds 1015 mm and is uniformly distributed and where solar radiation indices (defined by Lee 1964) are 0.20–0.34.

Humid forest areas of the Pacific Northwest of the United States also show potential for augmenting water yield by manipulating forest cover. The climatic regime is different from that of the humid East. Annual precipitation can exceed 4000 mm at the higher elevations on the windward side of the Cascade Mountains. Much of this precipitation falls during the winter months and occurs as snow in the higher elevations. Distinctive dry periods usually occur during July through September.

Clearcutting Douglas-fir forests has increased annual water yield 360–540 mm/yr, compared to 100–200 mm/yr increases from partial clearcuts. Increases in yield diminish as forest vegetation grows back on the site. The water yield increase expected for any year following clearcutting on one Douglas-fir watershed in Oregon was calculated as (Harr 1983):

$$Y = 308.4 - 18.1\,(X_1) + 0.087\,(X_2) \qquad \textbf{(6.7)}$$

where Y = annual water yield increase (mm); X_I = number of years after clearcutting; and X_2 = annual precipitation (mm).

The above relationship would not apply for Douglas-fir forests in areas with persistent fog and long periods of low clouds; such situations exist in the coastal areas of the same region and show quite opposite responses to clearcutting. These "cloud forests" exhibit a reduction in water yield following clearcutting and will be discussed later.

In another high-rainfall area, in New Zealand, the effects of forest harvesting and various land preparation techniques for conversion from native forests to plantations indicated substantial water yield increases (Rowe and Pearce 1994). For native evergreen forested watersheds, Rowe and Pearce determined the annual water budget expressed as $P = Q + I + T + L$ to be:

$$2370 \text{ mm} = 1290 \text{ mm} + 620 \text{ mm} + 360 \text{ mm} + 100 \text{ mm}$$

When trees were clearcut on one watershed, a 550 mm increase was observed the first year. Other watershed experiments noted a 200–250 mm increase in harvesting the first year. These cleared watersheds were planted with *Pinus radiata*, but the rapid growth of bracken and Himalayan honeysuckle caused streamflow increases to decline to preharvest levels within 5 yr.

Wetlands

Wetlands cover vast areas in the humid temperate regions of North America and Europe (see Chapter 14). Watershed management on wetlands poses different hydrologic questions from those related to mineral-soil watersheds previously discussed. Because wetlands occur as a result of excess water, their management to enhance water yield is a moot point. However, the water yield implications of widespread commercial peat extraction and forest harvesting on wetlands are of interest, as they pertain to flooding and low-streamflow regimes.

Clearcutting black spruce in a northern Minnesota peatland, for example, resulted in little change in annual water yield (Verry 1986). However, water tables in the clearcut peatland rose as much as 10 cm higher during wet periods and dropped during dry periods to a level 19 cm lower than those of a mature forested peatland (control). Differences in interception explain the wet-period response, whereas differences in transpiration explain the dry-period response. The sedge understory apparently responded to overstory removal with equal or higher transpiration rates than the original forest stand. This response likely would not have been predicted from models that ignore the physiological response of plant species on the watershed.

Drylands

Opportunities for increasing water yield by manipulating vegetative cover are limited on arid and semiarid watersheds. Unfortunately, these are the areas where water is usually scarce. An analysis by Hibbert (1979, 1983) showed that vegetative manipulations could increase water yields only on watersheds receiving > 450 mm of annual precipitation. He reasoned that precipitation below this amount is effectively consumed by residual overstory vegetation and subsequent increases in herbaceous cover on the watersheds.

Forested Uplands in Dryland Regions

Watersheds in mountainous regions exhibit a variety of soils, vegetation, and climate, which are accentuated by differences in elevation, slope, and aspect. As a rule, precipitation and water yield increase with elevation; therefore, the greatest potential for increasing water yield usually lies in the mid- to upper-elevation watersheds. The proximity of many such watersheds to agricultural and urban centers in drier valleys downstream makes water yield enhancement opportunities attractive to water resource managers. As a result, numerous watershed experiments have been conducted in the mountainous western United States to develop vegetative management schemes that increase water yield.

Opportunities for increasing water yield in the mountainous western United States depend largely upon snow management as well as reducing *ET*. Much of the watershed research in this region has concentrated on timber-harvesting alternatives that redistribute the snowpack to achieve more runoff from snowmelt (see Chapter 15). The reduced transpiration associated with timber harvesting increases runoff efficiency by leaving more water in the soil. Higher soil water content during the fall and winter months results in a greater percentage of snowmelt ending up as streamflow, rather than being stored in the soil.

Streamflow from high-elevation watersheds in the Rocky Mountains exceeds 1000 mm/yr. The region as a whole, however, yields < 30 mm streamflow/yr, with < 15% of the land area contributing the majority of streamflow. Water yield from these watersheds can be increased 20–60 mm/yr by harvesting coniferous forests in small patches. Although increases diminish with the regrowth of vegetation, water yield in excess of precutting conditions can persist for 60–80 yr.

In the Sierra Nevada range of the western United States, annual water yields vary from 350 to 1000 mm/yr, with the higher-elevation watersheds exhibiting the highest yields. If large forested watersheds were managed exclusively for water yield improvement, annual harvesting schedules could increase water yields from 2 to 6% (Kattelmann et al. 1983). Under multiple-use and sustained-yield harvesting schedules, annual water yield increases of < 20 mm would be expected.

The potential for increasing water yield from ponderosa pine forests in the southwestern United States has been of interest because of the scarcity of water and a rapidly expanding population. Depending upon the percentage of forest cover removed and annual precipitation, water yield increases of 25–165 mm/yr have been reported (Hibbert 1983, Baker 1986, 1999). The effects normally persist for only 3–7 yr because of vegetative regrowth.

Rangelands

Most rangelands in the western United States are arid or semiarid ecosystems with little potential for water yield improvement. However, as pointed out by Hibbert (1983), water yield can be increased from watersheds that receive > 450 mm precipitation/yr where deep-rooted shrubs can be replaced with shallow-rooted species such as grasses. Based on studies in Arizona and California, annual water yield increases (Q) from such conversions can be estimated by:

$$Q = -100 + 0.26P \tag{6.8}$$

where P = mean annual precipitation (mm). Other vegetative management opportunities for increasing water yield are summarized in Table 6.2.

TABLE 6.2. Potential water yield increases by vegetative management on forested uplands and rangelands in the western United States

Vegetative Type	Annual Water Yield Increases (mm)
Forest	
Aspen (*Populus tremuloides*)	100–150
Ponderosa pine (*Pinus ponderosa*)	25–165
Pinyon juniper (*Pinus,* spp. *Juniperus* spp.)	0–10
Rangeland	
Sagebrush (*Artemisia* spp.)	0–12
Semidesert shrublands	Negligible

Source: Adapted from Hibbert 1983 and Baker 1986.

TABLE 6.3. Annual estimates of evapotranspiration from phreatophytes in the southwestern United States

Species	Water Table Depth (m)	Annual Evapotranspiration (m)	Reference
Saltcedar (*Tamarix* spp.)	1.5	2.2	Van Hylckama (1970)
	2.1	1.5	
	2.7	1.0	
Mesquite (*Prosopis* spp.)		0.3–0.5	Horton and Campbell (1974)
Cottonwood (*Populus*, spp.)		1.1	Horton and Campbell (1974)

Livestock graze on many rangelands in the western United States and generally are considered a component to be managed in these ecosystems. Most studies indicate that grazing has little effect on water yield, however.

Riparian-Phreatophyte Communities

Riparian-phreatophyte communities consume large quantities of water. Annual consumption by such plant communities represents significant losses of subsurface water supplies, including groundwater (Table 6.3). Generally, the higher *ET* losses occur when the water table is shallow.

The removal of riparian and phreatophyte vegetation in water-short areas such as the southwestern United States can result in groundwater savings. The removal of cottonwood along a stream channel in northwestern Arizona salvaged about 0.5 m of water from an 8.9 ha area (Bowie and Kam 1968, as reported by Horton and Campbell 1974). One of the most troublesome phreatophytes in the southwestern United States, saltcedar

(*Tamarix* spp.) has been the object of eradication efforts aimed at salvaging groundwater supplies. By significantly reducing or eradicating saltcedar transpiration over an area of 200 ha, for example, between 2 million and 4 million m^3 of water could be saved per year. If this groundwater could be retrieved, there would be enough water to supply municipal needs for a small town. Mechanical and chemical measures to eradicate saltcedar communities have met with opposition, even in locales where water is in short supply, because of the wildlife habitat and aesthetic values of saltcedar communities. This opposition is an example of a multiple-use conflict that can arise when vegetation is being manipulated for some specific purpose, in this case, increasing water supplies (see Chapters 13 and 14 for further discussions on the hydrology and management of riparian-phreatophyte and wetland ecosystems, respectively).

Humid Tropical Regions

Relatively little is known about forest management–water yield relationships in the humid tropics. One reason is that only a few controlled watershed experiments have been conducted in tropical ecosystems. This lack of research is surprising because of the claims by some people that humid tropical forests exert a strong influence on regional precipitation, climate, and global weather systems. Large-scale and complex studies would be needed to understand and quantify such influences. As a result, questions dealing with regional or global implications of vegetative changes in humid tropical forest ecosystems have been addressed more with conjecture than with fact (Box 6.1).

The few controlled watershed experiments in the humid Tropics show that water yield responses to changes in forest cover are similar to those in temperate climates except for cloud forest conditions (discussed below). For example, rain forests logged and cleared for pasture in North Queensland, Australia, resulted in a 10%, or 293 mm, increase in water yield for > 2 yr (Gilmour et al. 1982). Weekly comparisons indicated that minimum discharges increased from 14 to 60% following land clearing and pasture development. These hydrologic responses were similar to those reported for cleared Temperate Zone forests during the wet season.

The replacement of rain forests with tea plantations in Kenya, East Africa, showed little effect on annual water yields, surface runoff, or sediment loss (Edwards and Blackie 1981). Clearing a bamboo forest in the same area, but followed by the establishment of a *Pinus patula* plantation, resulted in an initial increase in water yield. However, once the pine canopy closed, there was no difference in water yield from the original bamboo forest. Cultivation of small areas in evergreen forests increased water yield (expressed as a percentage of annual rainfall) by 12%.

The effects of forest removal on water yield would normally be of shorter duration in the humid Tropics than in temperate climates because of rapid regrowth of vegetation. Most forest management practices and logging activities would not be expected to influence water yield for more than a few years. When forests are converted to agricultural croplands or pastures, however, long-term and substantial increases in water yield can result. The larger the percentage of a watershed affected, the greater the increase in yield.

Cloud Forests—A Special Case

Forests along coastal areas or situated on high mountainous islands near coastal regions sometimes produce more water for a watershed than they consume by transpiration. In

Box 6.1

Deforestation in the Amazon Basin: Evapotranspiration and Rainfall Implications

Although convincing evidence has been presented by Lee (1980) and others that removing forest cover has little effect on precipitation in temperate regions, the issue has been controversial concerning tropical forests. Salati and Vose (1984) suggest that in the humid tropics, and particularly the Amazon Basin, deforestation effects on atmospheric moisture can change the climate drastically and reduce rainfall. The following examines the possible changes in *ET* and rainfall that might result from deforestation in the Amazon Basin and considers whether such changes might be expected in other tropical forests. The Amazon Basin, an area of about 5.8 million km^2, is one of the largest river basins in the world. The Amazon River yields between 15 and 20% of the total freshwater flow in the world, which amounts to about 950 mm when expressed as a depth over the basin (Salati and Vose 1984). Assuming annual rainfall averages 2000 mm over the basin, *ET* is 1050 mm/yr (2000 mm–950 mm). Marques et al. (1977) estimated that 48% of annual rainfall, or 960 mm, in the east central part of the basin is derived from *ET* within the basin itself. If we assume that this holds for the basin as a whole, then 91% (960/1050) of the basin *ET* is recycled back to the basin as rainfall. The amount of rainfall that originates from moisture outside the basin, therefore, would be 1040 mm.

Maximum reductions in *ET* due to clearcutting forest vegetation could vary from 40 to 65 mm/yr for each 10% of the area clearcut (Bosch and Hewlett 1982, Gilmour et al. 1982). If we assume a maximum reduction of 65 mm per 10% cleared and that crops or shrub-type vegetation occupy the cleared sites, the net reduction in *ET* would be offset by about 25 mm/yr per 10% of the area affected. Deforestation followed by conversion to crops of 30% of the Amazon Basin would result in:

$$ET \text{ reduction} = [(65 - 25) \text{ mm}/10\%] \times 30\% = 120 \text{ mm/yr}$$

$$\text{New basin } ET = 1050 \text{ mm} - 120 \text{ mm} = 930 \text{ mm/yr}$$

$$\text{Amount of } ET \text{ that contributes to } P = 0.91 \times 930 \text{ mm/yr} = 846 \text{ mm/yr}$$

$$\text{Atmospheric moisture for } P = 846 \text{ mm} + 1040 \text{ mm} = 1886 \text{ mm/yr}$$

$$\text{Annual water yield} = 1886 \text{ mm} - 930 \text{ mm} = 956 \text{ mm/yr}$$

Therefore, a 30% conversion of the Amazon would result in a 6% reduction of annual rainfall. Water yield would be increased by about 1%.

The above exercise could be repeated for a variety of different changes in vegetative cover, but the likelihood of major changes in rainfall over the Amazon Basin due to realistic projections in land use change is minimal (see table and figure). The clearcut conditions under A in the table represent the first year or maximum changes in *ET* following clearing. Conditions under B would approximate those of conversion to crops or shrublands. Salati and Vose (1984) suggest that changes in annual rainfall of 10–20% would be detrimental to the ecosystem; to achieve such changes, more than 35% of the basin would have to be denuded and converted to crops or shrublands.

The possible effects of deforestation in other tropical areas would be expected to be smaller than the above because of the proximity of most tropical forests to large bodies of water. Atmospheric moisture likely would not be limiting to most of the humid tropics.

Estimated effects of clearcutting forest vegetation in the Amazon Basin on evapotranspiration, precipitation and streamflow

Forest Vegetation Cleared (%)	Basin Evapotranspiration (mm)	Average Annual Rainfall (mm)	Average Annual Streamflow (mm)	Percent Reducation in Average Annual Rainfall (mm)
A. No regrowth				
0	1050	2000	950	0
10	985	1936	951	3
20	920	1877	957	6
40	790	1759	969	12
60	660	1641	981	18
B. Followed by conversion to crops/shrubs				
10	1010	1959	949	2
20	970	1923	953	4
40	890	1850	960	6
60	810	1777	967	11
80	730	1704	974	15

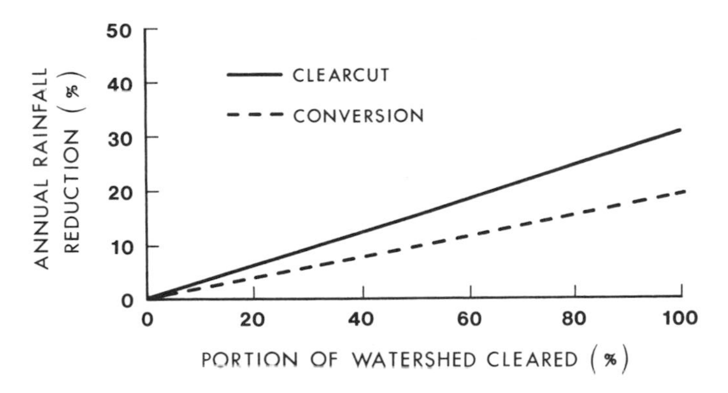

Estimated changes in annual rainfall (average) due to clearing forest cover in the Amazon Basin.

these areas of frequent and persistent low clouds or wind-blown fog, water droplets are deposited on foliage and stems of trees in these *cloud forests*, where they coalesce into larger droplets. This deposition is part of the interception process. Once on the vegetation, fog drip behaves like rain or snow in its interception storage (Satterlund and Adams 1992). Some evaporates into the atmosphere, some drips from the foliage, and some flows down stems. While not a precipitation input to a watershed in the more traditional sense, fog drip adds water to the system that would not be added if the area were devoid of forests or contained only low-growing vegetation.

Cloud forests are usually localized and do not cover large watersheds or river basins. Their contributions to the annual water budget of a watershed depend largely on the density of trees, total surface areas of the foliage, and the exposure of trees to low clouds or fog. As the frequency and duration of clouds and fog engulfing these forests increase with increasing elevation, the likelihood of gains in net interception increases with elevation in most mountainous regions.

The importance of fog drip has been suggested by many investigators, but it has been quantified in only a relatively few studies. Fog drip from mature Douglas-fir forests near Portland, Oregon, has been observed to add nearly 880 mm of water each year (Harr 1983). Fog drip represented 35% of the total input of water to the coastal redwood forests of northern California over a 3 yr study (Dawson 1998). Using an isotopic mixing model to determine how much of this fog input was used by the dominant plants, it was also found that almost 20% of the water contained within the redwood trees and 65% of the water in understory plants came from fog after it had dripped from the foliage of trees. It was concluded that there was a significant reliance by plants on fog as a water source, especially in the summer months when rainfall was lowest. Isotopic studies in California and northern Kenya indicate that fog drip from trees of cloud forests contributes significantly to the recharge of groundwater in fog-laden areas (Ingraham and Mathews 1988, 1995). Varying amounts of the fog drip were used year-round by the arboreal vegetation in California.

Cloud forests in Central America, which occur mostly on old volcanoes or high mountains, have been reported to add significant amounts of water to local watersheds (Zadroga 1981). Cloud forests that occur in narrow bands along the Pacific Ocean in northern Chile, though limited in extent, can contribute significantly to the annual water budget in this arid region. For example, in a region where the annual input of water is approximately 200 mm, measurements indicate that two-thirds of this amount is due to the interception of fog that flows inward from the Pacific Ocean and coalesces on the foliage of the trees. To increase this input further, artificial barriers have been constructed to intercept the movement of fog.

The importance of fog drip in the hydrology of a site can also be demonstrated by the effects of harvesting cloud forests on the water budget. Streamflow increases that are often anticipated after tree removals on a watershed can be limited and flows might even decrease following timber harvesting on sites in fog belts (Azevedo and Morgan 1974, Lovett et al. 1982). On the Bull Run municipal watershed in western Oregon, for example, where timber harvesting significantly reduced fog drip inputs to the watershed, low flows observed in the dry season were reduced to levels lower than expected (Harr 1982). In these instances, cloud forest should be managed to sustain vigorous growth of mature stands for purposes of sustaining water yield.

Estimating Changes in Water Yield

Changes in water yield that accompany changes in vegetative cover can be estimated by several methods. A discussion of three general approaches follows.

Regional Relationships

Mathematical relationships have been developed to predict water yield response to vegetative changes. Such expressions are often the result of localized experiments or regression relationships and, therefore, should be used only in the region or area from which they were developed. Relationships can be used as approximations if climate, soils, topography, and vegetative types are similar in the area of application to the area for which they were developed (Box. 6.2). Previously discussed methods by Douglas (1983) and Hibbert (1983) are examples of regional relationships.

Water Budget Approach

A water budget approach, such as that of the USDA Natural Resources Conservation Service, can be used to estimate water yield for watersheds with deep soils and high infiltration capacities. Water yield from such watersheds is governed primarily by soil moisture storage characteristics. Water yield changes associated with changes in the types of vegetative cover can be estimated by water budget analyses using different effective rooting depths (Box. 6.3).

The basic method can be modified at the discretion of the user and as more detailed information becomes available. For example, the resolution can be reduced from monthly to daily accounting. If seasonal *ET* relationships are known, the *ET-PET* relationship can be modified accordingly. Edwards and Blackie (1981) reported several *ET*:*PET* ratios for different plant-soil systems in East Africa; these ratios could be used in that region to improve estimates of *ET* changes. Likewise, functions like Equation 3.23 [$ET = (PET)f(AW/AWC)$] can be used. The *AW* and *AWC* terms are related to the effective rooting depth of the respective vegetation types. Knowledge of transpiration response to soil moisture conditions is assumed.

Computer Simulation Models

Computer models for hydrologic simulation include simplified empirical relationships at one extreme and detailed process-oriented models at the other extreme. Computer simulation models can contain derivations or elements of the previously discussed methods. More intricate and complex relationships can be considered and sensitivity analyses performed where data or assumptions are weak. Nevertheless, we are often constrained by a lack of basic knowledge relating vegetation and other land use changes to hydrologic response. A further discussion of hydrologic and watershed management computer simulation models is found in Chapter 18.

Upstream-Downstream Considerations

Increasing water yield from upland watersheds does not necessarily result in a significant increase in water yield at downstream reservoir sites. If a small portion of the watershed is clearcut, there will generally be little effect on streamflow. As the distance

Box 6.2

Estimating Water Yield Changes Associated with the Conversion from Aspen to Red Pine in the Lake States

The paper industry in the Lake States required softwood species for making high quality paper. Mixed stands of conifers (softwood) and hardwoods and extensive aspen (*Populus* spp.) forests occur in the region. Projections of deficiencies in softwood supplies in the 1970s led to many stands of aspen being converted to conifers in the region. The implications of widespread conversion on water yield can be approximated with the method of Verry (1976). Differences in interception between aspen forests and red pine forests, including overstory and understory species, were estimated and transformed into a gross precipitation–net precipitation relationship (see figure). With this relationship, changes in net precipitation caused by changes in forest type can be approximated. For example, if annual gross precipitation is 756 mm, the conversion of an aspen stand (23.0 m^2/ha) to a red pine (*Pinus resinosa*) stand of the same basal area would decrease net precipitation by 66 mm/yr (633 mm – 567 mm). Because transpiration differences are not included in the method, the changes in net precipitation would provide conservative estimates of water yield changes. Annual streamflow for north central Minnesota averages about 190 mm; for the area in which a conversion took place, annual streamflow would be reduced 35%. If only a portion of a watershed undergoes conversion, the changes in annual streamflow would be proportional to the percentage of the watershed affected. If 20% of a watershed was converted as explained above, annual water yield would be reduced by about 175 mm or 7%. The utility of this method is its simplicity and its use of data that are normally available to resource managers.

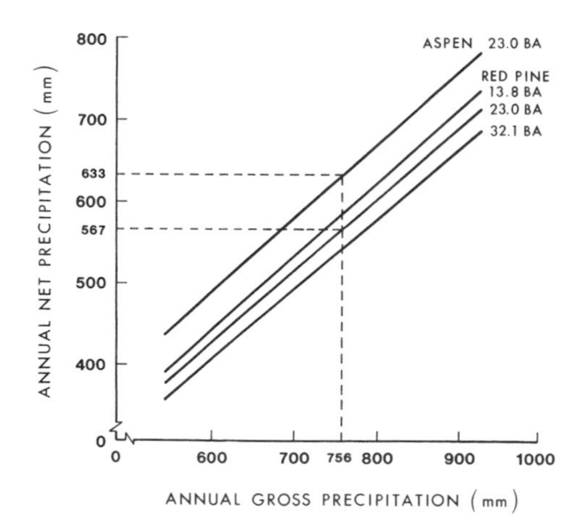

Relationship between net and gross annual precipitation for aspen and red pine (from Verry 1976).

increases between treated watersheds and the storage reservoir, opportunities for water losses increase. Riparian-phreatophyte vegetation along stream courses can transpire large amounts of water. Likewise, channel infiltration losses, called *transmission losses*, can exceed any water yield increases from upstream areas, particularly in the case of ephemeral streams in dryland regions.

Transmission losses can be estimated by (1) estimating the hydraulic conductivity of stream bottom material, (2) applying the hydraulic conductivity to the total area wetted by flow, and (3) applying the above for the duration of flow. Clean gravel and coarse sand bed materials can have hydraulic conductivities in excess of 127 mm/hr. At the other extreme, consolidated bed material with a high silt-clay content can have hydraulic conductivities of 0.03 mm/hr. Transmission losses as high as 62,060 m^3/km have been reported for channels in the southwestern United States (Lane 1983).

Water yield improvement schemes should also take into account the evaporative losses from the reservoir pool. In arid regions, reservoir evaporation can represent a large percentage of annual streamflow at the site. Todd (1970) reports annual lake evaporation to vary from 405 mm/yr in Maine to 2500 mm/yr in Arizona. The relationship between incremental increases in storage and corresponding increases in surface area of the reservoir pool largely determines whether any water yield increase at the reservoir site will be available for later use.

Several methods are available to estimate lake evaporation. One of the simplest and most widely used is the pan evaporation method. Lake evaporation can be estimated by multiplying pan evaporation (E_p) by a pan coefficient (C_e) (see Chapter 3).

Another consideration is that of the timing of the increased water yield. For example, if water yield is increased during the season when the reservoir is normally full, any additional water supply will be of little value.

Although many studies have quantified the effects of vegetation changes on streamflow at upland watersheds, few have determined the net effects of the changes in downstream user areas. An Arizona study estimated that less than one-half of the streamflow increase attributed to vegetation management in the Verde River basin would reach water users in Phoenix, approximately 150 km downstream (Brown and Fogel 1987).

Vegetation Management and Streamflow Pattern

Many water resource problems are related to the timing of water yield. Droughts and floods are the two extremes of streamflow that result from meteorological events. Solving problems of such streamflow extremes involves a variety of nonstructural and structural engineering approaches, as well as people management. For example, reservoirs can be used to augment streamflow during droughts, and vegetation manipulation can either increase or decrease flows into the reservoir, thereby affecting reservoir management (Box 6.4). But, water conservation measures by consumers and agricultural enterprises require sociopolitical approaches or economic incentives.

In the case of flooding, several approaches can be taken either to minimize the effects of floods or to prevent them. Structural solutions include reservoirs and levees. However, as discussed later in Chapter 10, changes in velocities, patterns, and quantities of streamflow caused by such projects can influence stream channel morphology and

Box 6.3

Application of a Water Budget Method to Estimate Changes in Water Yield due to Clearcutting a Mature Hardwood Deciduous Forest

A clearcut of 190 ha of mixed hardwoods is to be considered on a watershed that drains into a water supply reservoir. The city that receives water from this reservoir wants to determine how much water yield increase can be expected from such a cut. If sufficient water yield increases can be expected, the city may implement a sustainable forest management operation in which portions of their municipal watershed are maintained in clearcut or young-growth conditions.

To provide a conservative estimate of water yield expectations, precipitation and temperature records corresponding to a relatively dry 14 mo period were used to perform a water budget analysis for existing conditions—a mature, mixed hardwood forest (Table A). Soils were clay-loam textured with a plant available soil moisture content of 164 mm/m. Plot studies indicated that the mature forest had an effective rooting depth of 1.7 m, which means that 279 mm of soil moisture could be used to satisfy evapotranspiration demands (1.7 m × 164 mm/m). Based on this water budget analysis, the 14 mo water yield was 242 mm.

A. Water budget for a hardwood-covered watershed before clearcutting

	Year 1									Year 2				
	Apr	May	Jun	Jul	Aug	Sep	Oct	Nov	Dec	Jan	Feb	Mar	Apr	May
	(mm)													
Average precipitation[a]	27	4	31	42	36	12	50	120	140	105	90	95	65	20
Initial soil moisture[b]	279	248	163	67	0	0	0	0	100	240	279	279	279	279
Total available moisture	306	252	194	109	36	12	50	120	240	345	369	374	344	299
Potential *ET*[c]	58	89	127	173	157	107	57	20	0	0	3	13	58	89
Actual *ET*[d]	58	89	127	109	36	12	50	20	0	0	3	13	58	89
Remaining available moisture	248	163	67	0	0	0	0	100	240	345	366	361	286	210

Final soil moisture[e]	248	163	67	0	0	0	0	100	240	279	279	279	279	210
Water yield[f]	0	0	0	0	0	0	0	0	0	66	87	82	7	0

[a]Average over the watershed for each month of record.
[b]At start of each month. Same as "final soil moisture" of previous month.
[c]Average annual values for the month, as estimated by Thornthwaite's method.
[d]Total available moisture, or potential *ET*, whichever is smaller.
[e]At end of month. Same as "initial soil moisture" for next month. This value cannot be larger than the available soil water-holding capacity determined for the watershed, for this watershed 279 mm.
[f]Water is yielded when the remaining available moisture exceeds the water holding capacity for the soil in the watershed (279 mm).

To estimate the effects of clearcutting, the same initial conditions were used, but the effective rooting depth of the remaining plants was assumed to be 0.8 m, which corresponds to a herbaceous-shrub plant cover, a condition similar to that of a clearcut. The resulting available soil moisture capacity for the clearcut condition was 131 mm (0.8 m × 164 mm/m). Water yield for the clearcut condition (Table B) was 390 mm for the same 14 mo period. Water yield, therefore, was increased by 148 mm or 273,600 m^3. Of course, one must recognize that the 148 mm increase would be expected at the clearcut site, and water can be lost before reaching the reservoir site.

B. Water budget for a clearcut hardwood forest

	Year 1									Year 2				
	Apr	May	Jun	Jul	Aug	Sep	Oct	Nov	Dec	Jan	Feb	Mar	Apr	May
	(mm)													
Average precipitation	27	4	31	42	36	12	50	120	140	105	90	95	65	20
Initial soil moisture	279	248	163	148	148	148	148	148	248	279	279	279	279	279
Total available moisture	306	252	194	190	184	160	198	268	388	384	369	374	344	299
Potential *ET*	58	89	127	173	157	107	57	20	0	0	3	13	58	89
Actual *ET*	58	89	46[a]	42	36	12	50	20	0	0	3	13	58	89
Remaining available moisture	248	163	148[a]	148	148	148	148	248	388	384	366	361	286	210
Final soil moisture	248	163	148	148	148	148	148	248	279	279	279	279	279	210
Water yield	0	0	0	0	0	0	0	0	109	105	87	82	7	0

[a]Actual *ET* is restricted by the available soil water capacity of the reduced rooting zone (279 mm – 131 mm = 148 mm); the final soil water content must still exceed 279 mm before any water is yielded.

Box 6.4

Water Resource Problems Resulting from Vegetative Changes

To develop a wood-based industry on the Fiji Islands, 60,000 ha of *Pinus caribaea* were planted on the country's two largest islands, Viti Levu and Vanua Levu. Plantations were established on the dry, leeward zones of both islands. On Viti Levu, a water supply dam with hydroelectric power stations was developed coincident with afforestation. The project was intended to supply water to the two largest "dry zone" towns on the island for 30 yr. Although annual rainfall on the windward sides of the mountains can exceed 4800 mm/yr, rainfall during the dry season (May through October) on the leeward slopes varies from 300 to 500 mm with prolonged dry periods. As forest cover replaced mission grass cover, dry-season streamflow diminished, a cause for concern to the water supply project. Streamflow reductions of 50–60% were observed from watersheds that had 6 yr old pine stands. Greater reductions were expected once pine forests became mature. In this instance, the afforestation project was at cross-purposes with the water resource project (Drysdale 1981).

dynamics, sometimes resulting in unwanted effects. Floodplain management and zoning represent a nonstructural alternative. Vegetation management of upland watersheds and along stream channels should be included as part of either approach.

Forest vegetation has long been thought to influence the timing of streamflow by storing water during wet periods and releasing water during dry periods. Such relationships are based largely on myth but have provided the impetus for forest conservation movements in Europe and the United States. By not properly accounting for the effects of forest vegetation on the amount and timing of water yield, water resource management objectives can become compromised (Box 6.4).

The following sections examine the extent to which watersheds can be managed to help solve problems of flooding and droughts.

Stormflow-Flooding Relationships

Questions dealing with the influence of vegetation and land use activities on flooding must be addressed with precise and consistent terminology. As pointed out by Hewlett (1982), some confusion and misconceptions have arisen because of terminology problems. In popular usage, *flooding* usually means a high flow of water that causes economic loss or loss of life. A technical definition of flooding, and the one used here, is streamflow that rises above the stream banks and exceeds the capacity of the channel. A second point of confusion arises from streamflow-frequency analyses in which the annual maximum discharge is commonly called the *annual flood.* In most years the bankfull stage of a stream may be exceeded one or more times, as discussed in Chapter 10. Not all of these floods cause serious damage and may not be considered to be a flood in the eyes of the general public. During dry years, the annual maximum discharge may not rise above the stream banks and, therefore, technically is not a flood. Floods of greater magnitude occur less frequently and become the focus of flood prevention and/or flood avoidance programs. Nevertheless, one will often encounter discussions of "annual floods," "5 yr recurrence interval floods," "100 yr recurrence interval floods," and so forth, when only some of these floods have significant impacts on humans.

To be more precise, streamflow events should be defined in terms of probability and in terms of hydrographic characteristics. For example, events should be described in terms such as *the maximum annual peak discharge*; *the 0.02 probability, or 50 yr recurrence interval, peak discharge; the 0.05 probability, or 20 yr recurrence interval, stormflow volume*; and so on. By using this admittedly cumbersome terminology, we are accurately defining the event. The occurrence of flooding and consequences to humans must then be determined by subsequent study.

Determining the effects of upstream watershed disturbances on flooding requires that the on-site effects on hydrographic characteristics be identified and that these effects be viewed within the context of a watershed. Determining the impacts on flooding is difficult because of the routing and combining of all flows to downstream points of interest. Most experimental watershed studies consider the impacts of land use on stormflow events only at the outlet of the watershed in which the land use change was imposed. However, the effects of land use on the magnitude and timing of peak discharges can be diminished and lengthened, respectively, as the stormflow moves downstream. To understand the flooding implications of watershed changes better, consider first the stormflow response of first-order, or headwater, watersheds.

Stormflow-flooding analyses are facilitated by studying the streamflow hydrograph and relating the factors that govern its makeup. As discussed in Chapter 4, the streamflow hydrograph represents the integrated hydrologic response of a watershed to a given sequence of precipitation. Hydrographic characteristics of interest include stormflow volume and the magnitude and timing of peak discharge. Watershed size, shape, land slope, soils, and geologic strata are characteristics that remain relatively constant through time; each influences the streamflow response for a given precipitation event. The soil-plant system on the watershed and the stream channel system are the dynamic components of a watershed that, when altered, can affect the streamflow response to precipitation.

Changes in watershed and stream channel conditions affect stormflow volume, peak magnitude, and timing of the peak from rainfall and snowmelt runoff events. These effects are more pronounced with events that are not of an extreme magnitude. As the amount and duration of precipitation increase, the influence of human-induced changes on the watershed and channel system on the magnitude of peak flow and volume diminishes. Therefore, the influence of vegetative cover and soil compaction, for example, is minimal for extremely large precipitation events that usually are associated with major floods.

For events other than the extreme, stormflow characteristics change in relation to the severity of disturbance of the soil-plant system and the stream channel and to the percentage of watershed area affected. The percentage of area of large watersheds or river basins disturbed severely by fires, timber harvesting, road construction, and urbanization is normally small. Changes in streamflow that are detected in first-order watersheds become less evident downstream and can become confounded and difficult to predict because of the combined changes in peak flow discharge and timing at different

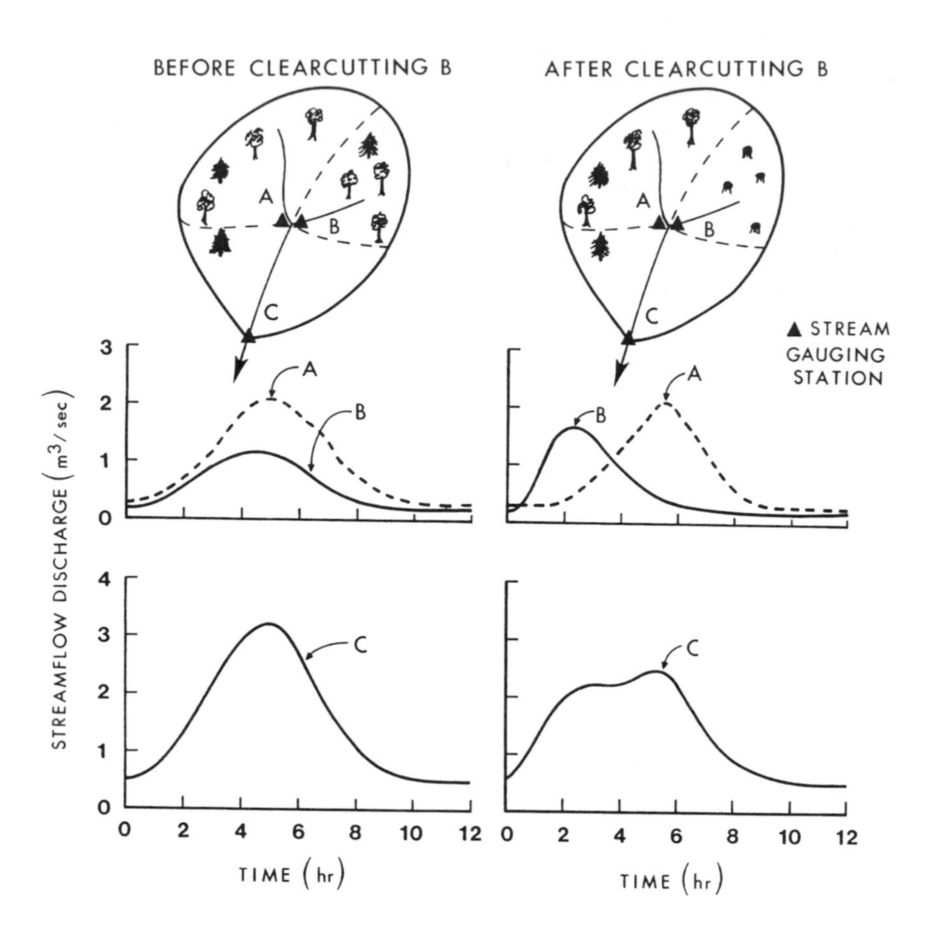

FIGURE 6.1. Effects of forest removal on upstream and downstream stormflow hydrographs where desynchronization of stormflow hydrographs occur.

locations in the watershed. For example, increased peak flows from one tributary watershed can increase as a result of watershed disturbance, but the magnitude of the peak discharge downstream can be reduced if the timing of upstream peak flows are desynchronized (Fig. 6.1). On the other hand, when tributaries of a watershed experience increased stormflow volumes, there is a cumulative and *additive effect on downstream peaks*. That is, when a significant area of a watershed has increased flow volumes, there is little desynchronization effect. It is not surprising, therefore, that reports of stormflow response to land treatments vary considerably. A few stormflow–land treatment studies of rainfall and stormflow events are examined in the following paragraphs to illustrate some of the points made above.

Rainfall Events

Several watershed experiments have shown increases in rainfall-caused stormflow volume and peak discharge following forest cutting. Some controlled experiments, however, have indicated little effect or even reductions in stormflow characteristics, which limits generalizations (Table 6.4). Although many studies indicate that the greatest increases in peaks and stormflow volumes occur under wet antecedent conditions, a tropical rain forest example in Australia indicated little effect of forest clearing on stormflow parameters but an annual water yield increase of 10% following forest clearing (Gilmour et al. 1982). The mechanism of stormflow production helps explain this discrepancy. Soils in the watershed have a hardpan with a marked reduction in hydraulic conductivity at depths from 0.2 to 0.5 m, restricting rapid percolation. The zone of low permeability in the soil, not vegetative cover, dominates the stormflow process during wet seasons. During the monsoon and postmonsoon seasons, widespread surface runoff occurs from undisturbed forests as well as from cleared areas.

A study indicating a reduction in peak discharge following forest clearcutting in British Columbia, Canada (Table 6.4), also showed that peak discharges were delayed by several hours (Cheng et al. 1975). The delayed peaks were attributed to soil disturbance that resulted in a rough surface with greater retention storage. Velocities of flow were also reduced by debris in the channel following logging. The previous experiments clearly show the need to understand all factors governing the rainfall-stormflow process on the watershed and in the stream channel.

Snowmelt Events

Forest removal affects streamflow from snowmelt by changing the spatial deposition of snow and changing the energy budget at the snowpack surface. Changes in snowmelt runoff are often attributed to changes in the timing of snowmelt; this led to the idea that the timing of snowmelt runoff can be managed by manipulating forest cover to desynchronize snowmelt from different parts of a watershed (see Chapter 15). This effect was observed in Minnesota, where a watershed was partially cleared in two successive years.

Clearing patches of forest cover in mountainous areas would not be as effective in reducing peak discharges as in flat terrain. Snowmelt runoff from mountainous watersheds is normally desynchronized as a result of differences in elevation, slope, and aspect. However, for relatively flat topography, there is some ability to change snowmelt rates and, therefore, influence snowmelt peak discharges by manipulating forest cover. As with rainfall runoff, the influence of vegetative changes diminishes as the magnitude of precipitation (snowmelt) increases.

In cold continental climates, such as the northern United States, Canada, and northern Europe, changes in forest cover can also affect snowmelt runoff by altering soil frost. Soils are usually wetter in the fall after clearcutting, and they tend to freeze deeper, with a greater occurrence of concrete frost. Snowmelt on frozen soil runs off rapidly. If large areas within a watershed are cleared at any point in time or are converted to agricultural croplands, higher peak discharges and volumes of runoff over frozen soil can lead to more frequent flooding. Rain or snow events, particularly when the soil is frozen, can cause severe flooding.

Concluding Thoughts on Flooding

Land use activities on the watershed affect stormflow response and flood peaks by changing *storage*, *conveyance*, or both. These changes are the result of:

- The extent of the change in vegetative cover as it relates to changes in interception and antecedent soil moisture condition. Removal of vegetation or conversion from plants with high annual transpiration and interception losses to those with low

TABLE 6.4. Changes in rainfall-produced stormflow following forest cover removal

Location, Climate, Zone	Vegetation and Soils	Treatment and Percent of Area Affected	Changes in Stormflow: Peak (%)	Changes in Stormflow: Volume (%)	Changes in Stormflow: Timing (%)	References
North Carolina, USA, humid, temperate	Mixed hardwood, moderately deep soils, gravelly loam	Clearcut, 100% (no log removal)	+6	+11	0	Hewlett and Helvey 1970
New Hampshire, USA, humid, temperate	Mixed hardwood, sandy loam soils, average depth o.s.m.	Clearcut, 100% Regrowth prevented by herbicides for 3 yr (no log removal)	+100 to +200	+30	0	Hornbeck 1973
Oregon, USA, humid, coastal	Conifers, shallow to deep sandy soils	Commercial clearcut, 100% (logging, road construction)	+100 or less	+10 or less	0	Harr et al. 1975
Minnesota, USA, cold, continental	Mixed hardwood, glacial till soils, medium depth	Commercial clearcut, 70%	+170	+200	0	Verry et al. 1983
British Columbia, Canada, humid, temperate	Conifers, gravelly sandy loam soils from glacial till	Commercial clearcut, 100% (logging)	–22	NR[a]	Delayed	Cheng et al. 1975
North Queensland, Australia, humid, tropics	Rainforest, deep clay soils, shallow hardpan	Clearcut, 100% (pasture development)	0	0	0	Gilmour et al. 1982

[a]NR means not reported.

losses can increase stormflow volumes and the magnitude of peak flows. Such practices can also expand the source areas of flow. After a given precipitation event, antecedent soil moisture and water tables will tend to be higher; consequently, less storage is available to hold precipitation from the next event, and source areas are expanded.

- Changes in the mechanisms of stormflow production through reductions in infiltration capacities and alterations of source areas of the variable source area. Changes in the hydraulic properties of soils and in the presence of water-impeding layers in the soil can affect the pathways and volumes of runoff. Intensive grazing, road construction, logging, and urbanization can increase surface runoff. As the proportion of precipitation that occurs as surface runoff increases, the volume of runoff and the rate of flow to the watershed outlet (conveyance) increase, resulting in higher peak discharges. Activities that increase infiltration capacities would be expected to have the opposite effect.
- The development of roads, drainage ditches, and skid trails and their orientation with respect to land slope and proximity to stream channels and alterations of the stream channel can change the conveyance system in a watershed. Changes that reduce the time of concentration shorten the travel time of flow to the watershed outlet, which increases peak discharge.
- Changes in detention and retention storage associated with channels, ponds, wetlands, and reservoirs. Drainage of wetlands or ponds will reduce storage on the watershed with the potential of increasing stormflow or runoff from snowmelt. In contrast the construction of a reservoir or raising the height of the spillway of a reservoir will increase storage and diminish rainfall or snowmelt volumes from a watershed.
- The extent of surface erosion, gully erosion, and mass movements (mudslides, landslides, etc.) in relation to detention storage on the watershed and in the conveyance system. Increased sediment deposition can reduce the capacity of stream channels at both upstream and downstream locations. Channel storage is diminished, and flows that would have remained within the stream banks previously can now flood.
- Stream channel modifications that are intended to improve conveyance and restrict localized flooding, result in less storage and increase the velocity of flow downstream.

Flooding concerns are often in the more developed downstream areas, sometimes far removed from areas where land use and vegetative cover is undergoing change. Peak discharges and associated flood stages along major streams and rivers represent the accumulated flows from many watersheds of diverse topography, vegetation, soils, and land use. Increases in peak discharges from any headwater watershed can have little effect on downstream peaks because of the routing and desynchronization that normally occur. However, when stormflow volumes are increased from upland watersheds, they are not damped to the extent that peaks are, resulting in a cumulative effect on downstream volumes and peak discharges. The combination of increased stormflow volumes and increased amounts of sediment deposited in channels can increase the frequency with which streamflow exceeds channel capacity.

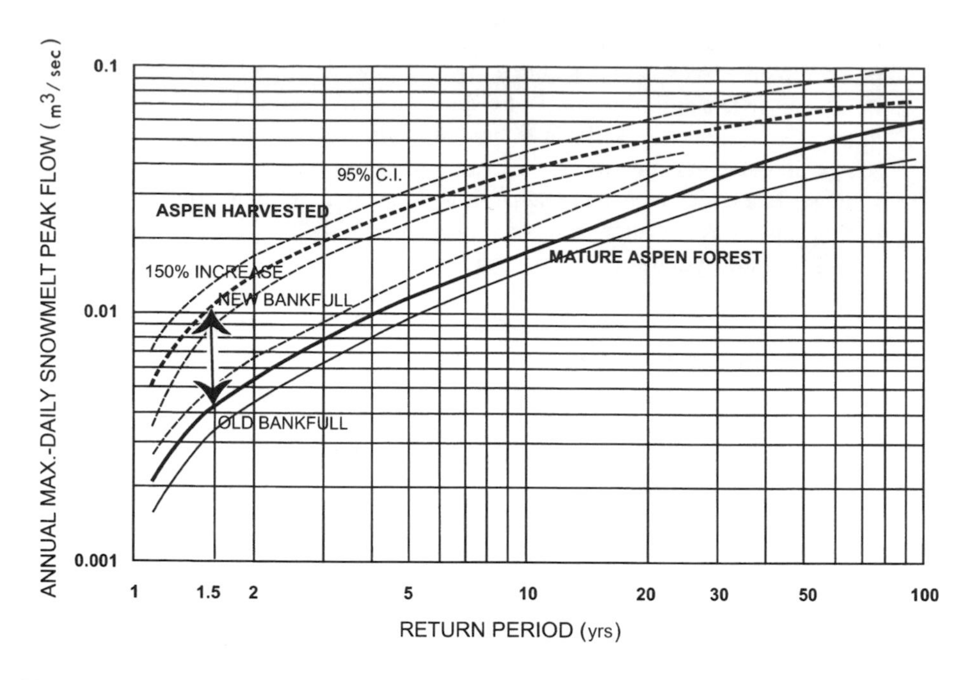

FIGURE 6.2. Changes in the magnitude and recurrence interval of annual peak flows following aspen clearcutting in northern Minnesota (from Lu 1994 as presented by Verry 2001).

Changes in land use, particularly changes in forest cover, will more likely affect smaller floods with return periods of from 5 to 20 yr, for example, than major floods with return periods of 50 yr or greater. For example, clearcutting mature aspen on a small watershed in northern Minnesota was shown to increase the magnitude of annual snowmelt peaks associated with return periods up to about 25 yr (Fig. 6.2). Note in this figure that peak flows associated with the 1.5 yr return period were increased 150%. The 1.5 yr return period peak flow for many stream channels is associated with bankfull conditions; increasing the magnitude of this flow has important implications for stream channel morphology changes (discussed in Chapter 10).

Roads and culverts in rural areas, campgrounds, and small upland communities will generally be affected by watershed changes more than large urban centers and agricultural areas along major rivers (Box 6.5). Floods of major rivers are affected more by meteorological factors than by land use activities in upland watersheds. Therefore, watershed management should be viewed in terms of complementing other means of achieving flood control or protection and not as a means by itself.

Low Streamflow

Dry-season streamflow is a concern to water resource managers because it often coincides with periods of greatest need. When precipitation is lacking, groundwater and reservoir storage are needed to provide water for irrigation and municipal requirements. Low streamflows can concentrate pollutants in the water, and streams become more sensitive to perturbations and temperature fluctuations (Hornbeck et al. 1997).

Box 6.5

Determining Effects of Watershed Changes on Stormflow Peak and Culvert Design for Small Catchments

Several upland forested watersheds were cleared and converted to pastures. The original forest roads in the area were designed on the basis of a 10% risk of failure over a 5 yr design life (thus, culverts were sized to accommodate the 50 yr return period peak flow). After the vegetation conversion, are the culverts now underdesigned?

For a 60 ha watershed that has been cleared, estimate the change, if any, that might be expected in the magnitude of the design stormflow peak, that is, the adequacy of the existing culvert system. Soils are medium heavy clays with good structure, and the 50 yr rainfall for a 12 min storm over 60 ha is 74 mm/hr.

Using the Rational Method (Chapter 17),

$$Q_p = (CP_gA)/360$$

where, Q_p = peak discharge in m³/sec; C = runoff constant (0.3); P_g = rainfall intensity (74 mm/hr); and A = area of watershed (60 ha).

$$Q_p = \frac{(0.3)(74)(60)}{360} = 3.7 \text{ m}^3/\text{sec}$$

Under a pasture condition, the C value is estimated to be 0.4; thus, the peak associated with the above design criteria is now equal to:

$$Q_p = \frac{(0.4)(74)(60)}{360} = 4.9 \text{ m}^3/\text{sec}$$

In this example, the existing culverts are now underdesigned, and the risk of having roads wash out is greater than 10% over the 5 yr period. An economic analysis would be needed to evaluate the benefits and costs of putting a new culvert system in place.

Therefore, low flows can place aquatic ecosystems under considerable stress. Because of the consequences of low flows, water resource management can be aimed at increasing streamflows during dry periods, or at least not diminishing low flows further.

Early conservationists pointed to the apparent relationships between forest cover and high streamflow, and arid nonforested areas and low streamflow (Box 6.6). Forests were thought of as reservoirs that could store water during wet periods and release water during dry periods. Our present knowledge about vegetation–*ET*–water yield relationships conflicts with this notion.

Most experimental evidence in rainfall-dominated regimes suggests that forest removal, or conversion from plants that are high users of water to plants that are low

Box 6.6

Vegetation, Land Use, and Hydrologic Response

A multiple regression analysis was performed to evaluate the influence of forest cover, soils, climatic variables, and physical watershed features on flood peaks in West Virginia (Frye and Runner 1970). The following relationship was developed for flood peaks:

$$Q_p = aK_i(F)^b$$

where Q_p = peak flow (m^3/sec), a,b = constants for a given flood frequency, F = percent forest cover, and K_i = influence of all other factors. This relationship shows that the magnitude of flood peaks increases with forest cover. How can you explain this finding?

Lee (1980) uses this example to show the types of problems one can encounter when using regression analyses to explain cause and effect relationships. The regression was developed for a large enough area to include forested, mountainous watersheds with shallow soils, high precipitation, and steep topography in contrast to the nonforested watersheds with lower precipitation and streamflow. A valid comparison should have included watersheds that only differed in forest cover and that were under similar precipitation regimes.

users of water, increases recession flow and sometimes dry-season flow. Wetter soil conditions resulting from reduced *ET* make the watershed system more responsive to small rainfall amounts and apparently lead to a longer period of soil water drainage and subsurface flow; continued precipitation inputs are needed, however, to sustain flow. Factors such as topography and geologic strata can also influence the time response of subsurface flow to rainfall events. Streamflow during droughts is sustained primarily by groundwater flow; therefore, any effects that changes in vegetative cover and soils have on streamflow during droughts would be attributed largely to changes in groundwater. In any case, changes in vegetative cover will not substantially alter low flows resulting from extended dry periods or droughts.

Streamflow influenced by snowmelt is usually characterized by a long period of recession flows. Since forest cover can be manipulated to affect snow accumulation and melt, it has been suggested that snowpacks can be managed to enhance dry-season flows. Kattelmann et al. (1983) stated that the greatest contribution watershed management can make in meeting future water demands in California is by manipulating forest cover to delay snowmelt runoff. Any significant delay in streamflow response would be beneficial for the operation of reservoirs. As in rain-dominated regions, delayed and increased flows caused by vegetative manipulations would normally extend over short periods and would not affect flows during lengthy droughts.

Although early conservationists might not have recognized the correct cause-and-effect relationships of rainfall-runoff processes, they did observe that forest cover generally promotes high-quality water that flows less erratically than that from most other watershed cover conditions.

Fire and Streamflow Pattern

The frequent occurrence of wildfire and prescribed burning throughout history has led to an extensive body of knowledge concerning the effects of fire on hydrologic processes and watershed lands. Depending on its severity, fire can change vegetative structures, modify soil properties, and impact components of the hydrologic cycle on watersheds. Lightning-caused wildfire requiring a suppression response by people can result in modification or replacement of the prefire vegetation on a watershed and, in doing so, alterations of hydrologic processes. A controlled application of fire as a watershed management tool, that is, a prescribed fire, can also change the behavior of hydrologic processes.

When vegetation and litter accumulations and other decomposed organic matter on the soil surface of a watershed are destroyed by fire, reductions in interception and *ET* can result, along with decreases in infiltration and increases in overland flow, subsurface flow, and streamflow pattern (DeBano et al. 1998). Responses of a streamflow regime to fire are reflected by changes in streamflow discharge, peak flow, timing of flow, and baseflow.

Streamflow Discharge

To determine the responses of streamflow discharge and, more generally, the responses of a streamflow regime to fire requires that on-site fire effects be determined initially and that these effects then be evaluated within the context of the entire watershed. However, determining effects of fire on a watershed scale is difficult, because of the complexity of routing water within and throughout a watershed and combining of all of the pathways of flow to the outlet of the watershed and other downstream points of interest. It is known that the effects of fire on the magnitude and timing of streamflow discharge can be diminished and lengthened, respectively, as a postfire streamflow event moves downstream.

There is little doubt that high severity wildfire often has an influence on streamflow discharge. The combined effects of a loss of vegetative cover, a decrease in litter accumulations and other decomposed organic matter on the soil surface, and the possible formation of water-repellent soils are some of the causative mechanisms for the increase in streamflow discharge. Increases in streamflow discharge following wildfire are highly variable (Box 6.7), although they are generally greater in regions with high *ET* losses.

Streamflow responses to prescribed fire are usually smaller in magnitude, or almost nonexistent in instances of low fire severity, in comparison to streamflow responses to wildfire. It is generally not the purpose of prescribed burning to completely consume extensive areas of litter accumulations and other decomposed organic matter on the soil surface, with resultant large changes in streamflow regimes.

A burn that was prescribed to reduce fuel loads on a 180 ha watershed in the Cape Region of South Africa caused a 15% increase (80 mm) in average annual streamflow discharge (Scott 1993). The watershed was vegetated with fynbos shrubs, most of which were not damaged by the fire. Effectiveness of the prescribed fire was less than anticipated, however, because of the unseasonably high rainfall at the time of burning.

Box 6.7

Increases in Streamflow Discharge Following Wildfire

Washington

Annual streamflow discharge from a 564 ha watershed in the Cascade Range of eastern Washington, on which a wildfire killed nearly 100% of the mixed conifer forest vegetation, increased dramatically relative to a prefire streamflow relationship between the burned watershed and an unburned control. Differences between measured and predicted streamflow discharge varied from 107 mm in a dry year (1977) to 477 mm in a wet year (1972). Soil water storage remained high for the period of streamflow record largely because of abnormally high precipitation (rain and snow) inputs (Helvey 1980). As a consequence, the burned and control watersheds became more sensitive to precipitation.

Cape Region of South Africa

Increases of 12% (70 mm) in average annual streamflow discharge and more than 60% (almost 6.5 mm) in stormflow discharge were observed after a wildfire occurred on a 201 ha watershed in the Cape Region that had been converted from indigenous fynbos shrubs and grasslike plants to a *Pinus radiata* plantation (Scott 1993). While hydrologic effects of the fire were the result of changes on the watershed itself, reductions in interception and *ET* were the dominant factors (Scott and Van Wyk 1990).

There were no changes in annual streamflow or stormflow discharges after a wildfire swept across a second watershed 132 ha in size. This fire killed most of the planted *Eucalyptus festigala* trees, consumed all of the surface fuels, and completely exposed the soil beneath dying trees (Scott 1993). However, the unusually low rainfall that followed the burn could have limited postfire interpretations of the statistically weak calibration relationship with an unburned control watershed.

Southern France

There was a 30% (nearly 60 mm) increase in streamflow discharge the first year after a wildfire on a 146 ha watershed in southern France, near the Mediterranean Sea (Lavabre et al. 1993). Prefire vegetation on the watershed was a mixture of maquis, cork oak, and chestnut trees. The increase in streamflow discharge was largely related to the reduction in *ET* due to destruction of this vegetative cover by the fire.

Southwestern United States

Annual streamflow discharge from watersheds in the fire-prone chaparral shrublands of the southwestern United States have increased by varying magnitudes, at least temporarily, as a result of wildfire (DeBano et al. 1998). The combined effects of loss of vegetative cover, decreased litter accumulations, and formation of water-repellent soils following burning are the presumed reasons for these streamflow increases.

TABLE 6.5. Effect of wildfire on annual peak flows

Location	Vegetation Type	Increase (%)	References
Eastern Oregon	Ponderosa pine	45	Anderson et al. 1976
Central Arizona	Ponderosa pine	500–1500	Rich 1962
Cape Region of South Africa	Monterey pine	290	Scott 1993
Southwestern United States	Chaparral	200–45,000	Sinclair and Hamilton 1955, Glendening et al. 1961

Annual streamflow discharge was not significantly changed relative to prefire levels in 6 yr following a prescribed fire on 43% of a 471 ha watershed in eastern Arizona supporting a ponderosa pine forest (Gottfried and DeBano 1990). The burning plan specifying a 70% reduction in fine fuels and a 40% reduction in heavy fuels was satisfactorily met. Damage to the residual stand of trees was minimal.

Burning of logging residues following timber-harvesting operations, burning of competing vegetation to prepare a site for tree planting, and burning of forests in the process of clearing land for agricultural production are common practices in many parts of the world. Depending on their intensity and extent, these prescribed burns cause changes in streamflow discharge from watersheds on which these burns are conducted. However, it is difficult to isolate effects of these burning treatments from the accompanying hydrologic impacts of timber-harvesting operations, site preparation, and clearing of forest vegetation in analyzing the responses of streamflow discharge to fire.

Peak Flow

Destruction of vegetation and reductions in litter accumulations and other decomposed organic matter by fire can also increase peak flow. However, the magnitude of the increase is variable (Table 6.5). Increases in peak flow following fire are generally related to the occurrence of intense and short-duration rainfall inputs, slope steepness of the watershed, and a formation of soil water repellency after burning. Postfire streamflow events with excessively high peak flows are often characteristic of flooding regimes.

Timing of Flow

Another concern of hydrologists and watershed managers regarding the effects of fire on a streamflow regime is the likelihood of changes in the timing of flow. While information is limited, it has been observed that streamflow regimes from burned nearby watersheds often respond to rainfall inputs faster than watersheds supporting protective vegetative cover, producing streamflow events where time-to-peak is earlier (DeBano et al. 1998). Earlier time-to-peak, coupled with higher peak flow, can increase the frequency of flooding.

Timing of snowmelt in the spring can also be advanced by fire. Earlier snowmelt is initiated by lower snow albedo caused by blackened trees and increased surface exposure with elimination of vegetative cover (Helvey 1973).

Baseflow

Fire in combination with other management activities can increase baseflow. For example, while the hydrologic mechanisms involved are largely unclear, Berndt (1971) observed immediate increases in baseflow following a wildfire on a 564 ha watershed in eastern Washington. In addition, removal of riparian vegetation by the fire eliminated the diurnal fluctuations of baseflow. Increased baseflow persisted above prefire levels 3 yr after the fire.

Following a wildfire in central Arizona, streamflow from watersheds dominated by chaparral shrub species became perennial at the watershed outlets after the residual shrubs were chemically treated to suppress resprouting, even though upstream channels were frequently dry (Hibbert et al. 1982). Streamflow was largely absent in the immediate prefire years, but it became intermittent shortly after the wildfire and remained so until the chemical treatment was imposed, at which time the streamflow regime became perennial.

Crouse (1961) reported increased baseflows from burned watersheds on the San Dimas Experimental Forest in southern California. While these watersheds had been cleared of chaparral vegetation by burning, seeded to grass, and maintained in a grass cover with herbicides to induce higher streamflow discharges, the author felt that wildfire had made a significant contribution to the increased dry-season flow.

Cumulative Watershed Effects

The cumulative watershed effects (CWEs) of changing watershed conditions and stream channel characteristics present a challenge to the managers and users of the land and stream system. It is important that watersheds be managed in such a way that the combination of management activities within a watershed does not significantly impact other beneficial uses (Reid 1993). Managers must be able to predict the environmental effects of these activities if they hope to prevent unwanted downstream effects. Before this prediction is possible, however, it is necessary to know how CWEs come about and how watershed systems respond to them.

Most land management activities change the character of vegetation and soils; import and remove water, sediment, and chemicals; and introduce pathogens and heat. As these environmental parameters change, the processes associated with the transport of water through a watershed change in response and alter the production of water, sediment, chemicals, and other watershed products. A comprehensive review of how changes in environmental parameters affect the generation and transport of these watershed products can be found in Reid (1993), Rees (1995), Sidle (2000), and other references. We describe these processes below and in subsequent chapters.

Runoff volume, its mode and timing, and its transport through a watershed system into a stream channel are all affected, to some extent, by environmental changes brought about by vegetation management practices. The changes in vegetative composition, relative density, and age structure that occur through management affect the rates of ET and, as a consequence, runoff volume and timing as discussed in this chapter. Depending on the extent of the watershed affected and antecedent soil moisture conditions, these changes can increase storm peaks.

As discussed earlier in this chapter, in some cases vegetation can be intentionally modified to increase water yield. However, altered water yields can also be unintended by-products of vegetative management practices implemented primarily for other purposes, such as harvesting timber for processing into wood products or converting a forest overstory to herbaceous plants to increase the yields of livestock forage. Conversion from perennial vegetative cover to annual crops can increase water yield. Converting from annual crops to tree plantations will have the opposite effect. The processes of *ET* and runoff generation in relation to vegetation management practices have been described to the point where computer simulation models of these processes have been constructed (see Chapter 19). These simulators are useful in estimating the effects of altered vegetative covers, whether planned or unplanned, on the hydrology of a watershed and larger river basin and, therefore, in relation to CWEs.

It is also important to understand the relation between other aspects of land management activities and CWEs. Unsurfaced roads and construction sites, which are associated with many vegetative management programs, are often highly compacted and, as a result, generate runoff and ultimately streamflow (Dickerson 1976; Johnson and Beschta 1980). Roads have also been shown to be problematic for erosion because of exposure to erodible soil and subsoil by construction, reduced infiltration on the road surface, increased gradients on cut and fill slopes, and the concentration of overland flow from precipitation excess and interception of subsurface flow (Burroughs and King 1989, Megahan and Hornbeck 2000).

Trampling by livestock and people also compacts the soil and alters its hydrologic properties. Physical disturbance by scarification can increase soil permeability and its infiltration capacity. The accompanying increases in soil surface roughness can slow surface runoff and allow it more time to infiltrate. Increased infiltration, in turn, decreases runoff peaks, increases low-flow discharges, and decreases runoff volume by providing more water for *ET*.

As discussed in Chapter 4, burning of some plant species releases volatile oils that can coat soil particles to form hydrophobic (water-repellent) layers in soil (DeBano 1981, DeBano et al. 1998). Runoff rates generally increase if the hydrophobic effects are widespread on a watershed. In addition, flood peaks can be increased, low-flow discharges lowered, and available soil moisture decreased. Water catchments such as stock tanks and surface cavities created by uprooting trees and shrubs alter runoff volumes leaving watersheds by increasing the area of water surface susceptible to evaporation.

By considering CWEs, we can better understand the spatial and temporal relationships of various types of land use and their effects on the production of water, sediment, and chemicals in a watershed. To understand CWEs, however, requires knowledge of how land management activities influence the environmental parameters that in turn alter the controlling watershed processes and all of the interactions therefrom (Reid 1993, MacDonald 2000, Sidle 2000). Many of these relationships have been studied and, as a consequence, some of the interactions are qualitatively and, in some cases, quantitatively predictable. An interdisciplinary approach incorporating hydrology, geomorphology, and ecology is needed to better understand and appreciate cumulative watershed effects.

Summary

The relationship of forests and other vegetative covers to water supply has been the subject of considerable controversy, partly due to early misconceptions about the role of forests and vegetation in the hydrologic cycle. Verry (1986) responded as follows to earlier assumptions that were taken as certain by Zon (1927) in light of the information existing in the 1920s. These responses remain valid at the beginning of the 21st century. Zon's original statements and Verry's responses are:

- The forest lowers air temperature inside and above it. [TRUE]
- Forests increase the abundance and frequency of precipitation. [usually FALSE]
- The destruction of forests affects the climate. [FALSE, but TRUE for microclimates]
- On level terrain, forest transpiration drains marshy land. [mostly FALSE]
- In hill and mountain country, forests conserve water for streamflow. [usually FALSE, sometimes TRUE]
- Forests retard snowmelt. [usually TRUE, sometimes FALSE]
- Forests prevent erosion. [TRUE, to a point]
- Forests regulate the flow of springs. [usually FALSE, sometimes TRUE]
- The total discharge of large rivers depends on climate. [TRUE]
- Forests tend to equalize streamflow throughout the year by making the low stages higher and the high stages lower. [sometimes FALSE, sometimes TRUE]
- Forests cannot prevent floods produced by exceptional precipitation, but they can mitigate their destructiveness. [TRUE]

Notice that most of the above answers are qualified; there are exceptions to most of the "rules." An understanding of the cause-and-effect relationships and basic hydrologic principles helps to explain such exceptions.

Key points that you should be able to explain after completing this chapter are phrased in the following questions:

- What changes in vegetative cover usually result in an increase in the quantity of water yield?
- What are the exceptions to the above?
- What methods are available to estimate changes in water yield caused by changes in vegetative cover?
- What factors are important in determining how much of a given change in water yield from an upstream watershed becomes realized downstream?
- How can land use practices affect streamflow during periods of high flows and of low flows? Give examples of changes in a watershed that can result in:
  - higher flows during the dry season,
  - higher flows during the wet season, and
  - lower flows during both the dry and the wet season
- Does clearcutting of forests cause flooding to increase? If so, explain.
- How do wildfire and prescribed fire differ in their effects on streamflow pattern?
- In what way do land use activities and environmental change affect hydrologic processes on watersheds and the ultimate streamflow response? What are the relations between hydrologic processes on a watershed and cumulative watershed effects?

REFERENCES

Anderson, H.W., M.D. Hoover, and K.G. Reinhart. 1976. *Forests and water: Effects of forest management on floods, sedimentation, and water supply*. USDA For. Serv. Gen. Tech. Rep. PSW–18.

Azevedo, J., and D.L. Morgan. 1974. Fog precipitation in coastal California forests. *Ecology* 55:1135–1141.

Baker, M.B., Jr. 1986. *Effects of ponderosa pine treatments on water yield in Arizona. Water Resour. Res.* 22:67–73.

Baker, M.B., Jr., compiler. 1999. *History of watershed research in the Central Arizona Highlands.* USDA For. Serv. Gen. Tech. Rep. RMRS–GTR–29.

Berndt, H.W. 1971. *Early effects of forest fire on streamflow characteristics.* USDA For. Serv. Res. Note PNW–148.

Bosch, J.M., and J.D. Hewlett. 1982. A review of catchment experiments to determine the effect of vegetation changes on water yield and evapotranspiration. *J. Hydrol.* 55:3–23.

Bowie, J.E., and W. Kam. 1968. *Use of water by riparian vegetation, Cottonwood Wash, Arizona.* U.S. Geol. Surv. Water Supply Pap. 1858.

Brown, T.C., and M.M. Fogel. 1987. Use of streamflow increases from vegetation management in the Verde River basin. *Water Resour. Bull.* 23:1149–1160.

Burroughs, E.R., Jr., and J.G. King. 1989. *Reduction of soil erosion on forest roads.* USDA For. Serv. Gen. Tech. Rep. INT–264.

Cheng, J.D., T.A. Black, J. DeVries, R.P. Willington, and B.C. Goodell. 1975. The evaluation of initial changes in peak streamflow following logging of a watershed on the west coast of Canada. *Int. Assoc. Hydrol. Sci. Publ.* 117:475–486.

Crouse, R.P. 1961. *First-year effects of land treatment on dryseason streamflow after a fire in southern California.* USDA For. Serv., Pacific Southwest Forest and Range Experiment Station, Res. Note 191.

Dawson, T.E. 1998. Fog in the California redwood forest: Ecosystem inputs and use by plants. *Oecologia* 117:476–485.

DeBano, L.F. 1981. *Water repellent soils: A state-of-the-art.* USDA For. Serv. Gen. Tech. Rep. PSW–46.

DeBano, L.F., D.G. Neary, and P.F. Ffolliott. 1998. *Fire's effects on ecosystems.* New York: John Wiley & Sons, Inc.

Dickerson, B.P. 1976. Soil compaction after tree-length skidding in northern Mississippi. *Soil Sci. Soc. Amer. J.* 40:965–966.

Douglas, J.E. 1983. The potential for water yield augmentation from forest management in the eastern United States. *Water Resour. Bull.* 19:351–358.

Drysdale, P.J. 1981. Status of general and forest hydrology research in Fiji. In *Country papers on status of watershed forest influence research in Southeast Asia and the Pacific.* Working Pap. Honolulu, HI: East-West Center.

Dunin, F.X., and S.M. Mackay. 1982. Evaporation of eucalypt and coniferous forest communities. In *First national symposium on forest hydrology*, ed. E.M. O'Loughlin and L.J. Bren, 18–25. Melbourne, Australia: National Committee on Hydrology and Water Resources of the Institution of Engineers.

Edwards, K.A., and J.R. Blackie. 1981. Results of the East African catchment experiments, 1958–1974. In *Tropical agricultural hydrology*, ed. R. Lal and E.W. Russell, 163–188. New York: John Wiley & Sons.

Frye, P.M., and G.S. Runner. 1970. *A proposed streamflow data program for West Virginia.* Charleston, WV: U.S. Dept. of Interior, Geological Survey.

Gilmour, D.A., D.S. Cassells, and M. Bonell. 1982. Hydrological research in the tropical rainforests of North Queensland: Some implications for land use management. In *First national symposium on forest hydrology*, ed. E.M. O'Loughlin and L.J. Bren, 145–152. Melbourne, Australia: National Committee on Hydrology and Water Resources of the Institution of Engineers.

Glendening, G.E., C.P. Pase, and P. Ingebo. 1961. Preliminary hydrologic effects of wildfire in chaparral. In *Proceedings of the 5th annual arizona watershed symposium*, 12–15. Phoenix, Arizona.

Gottfried, G.J., and L.F. DeBano. 1990. Streamflow and water quality responses to preharvest prescribed burning in an undisturbed ponderosa pine watershed. In *Effects of fire management of southwestern natural resources*, tech. coord. J.S. Krammes, 222–228. USDA For. Serv. Gen. Tech. Rep. RM–191.

Gregersen, H.M., K.N. Brooks, J.A. Dixon, and L.S. Hamilton. 1987. *Guidelines for economic appraisal of watershed management projects.* FAO Conserv. Guide 16. Rome.

Harr, R.D. 1982. Fog drip in the Bull Run municipal watershed, Oregon. *Water Resour. Bull.* 18:785–789.

Harr, R.D. 1983. Potential for augmenting water yield through forest practices in western Washington and western Oregon. *Water Resour. Bull.* 19:383–393.

Harr, R.D., W.C. Harper, and J.T. Krygier. 1975. Changes in storm hydrographs after road building and clear cutting in the Oregon coast range. *Water Resour. Res.* 11:436–444.

Helvey, J.D. 1973. Watershed behavior after forest fire in Washington. In *Agriculture and urban considerations in irrigation and drainage: Irrigation and drainage specialty conference*, 403–433. New York: American Society of Civil Engineers.

Helvey, J.D. 1980. Effects of a north central Washington wildfire on runoff and sedimentation. *Water Resour. Bull.* 16:627–634.

Hewlett, J.D. 1982. Forests and floods in the light of recent investigations. In *Proceedings of the Canadian hydrology symposium '82 on hydrological processes of forested areas*, 543–559. Ottawa: National Research Council, Canada.

Hewlett, J.D., and J.D. Helvey. 1970. Effects of forest clear-felling on the storm hydrograph. *Water Resour. Res.* 6:768–782.

Hibbert, A.R. 1979. *Managing vegetation to increase flow in the Colorado River*. USDA For. Serv. Gen. Tech. Rep. RM–66.

Hibbert, A.R. 1983. Water yield improvement potential by vegetation management on western rangelands. *Water Resour. Bull.* 19:375–381.

Hibbert, A.R., E.A. Davis, and O.D. Knipe. 1982. Water yield changes resulting from treatment of Arizona chaparral. In *Dynamics and management of Mediterranean-type ecosystems*, tech. coords. C.E. Conrad, and W.C. Oechel, 382–389. USDA For. Serv. Gen. Tech. Rep. PSW–58.

Hornbeck, J.W. 1973. Storm flow from hardwood-forested and cleared watersheds in New Hampshire. *Water Resour. Res.* 9:346–354.

Hornbeck, J.W., M.B. Adams, E.S. Corbett, E.S. Verry, and J.A. Lynch. 1993. Long-term impacts of forest treatments on water yield: A summary for northeastern USA. *J. Hydrol.* 150:323–344.

Hornbeck, J.W., S.W. Bailey, D.W. Buso, and J.B. Shanley. 1997. Streamwater chemistry and nutrient budgets for forested watersheds in New England: Variability and management implications. *For. Ecol. and Manage.* 93:73–89.

Horton, J.S., and C.J. Campbell. 1974. *Management of phreatophyte and riparian vegetation for maximum multiple use values*. USDA For. Serv. Res. Pap. RM–117.

Huff, D.D., B. Hargrove, M.L. Tharp, and R. Graham. 2000. Managing forests for water yield: The importance of scale. *J. For.* 98(12):15–19.

Ingraham, N.L., and R.A. Matthews. 1988. Fog drip as a source of groundwater recharge in northern Kenya. *Water Resour. Res.* 24:1406–1410.

Ingraham, N.L., and R.A. Matthews. 1995. The importance of fog-drip water to vegetation: Point Reyes Peninsula, California. *J. Hydrol.* 164:269–285.

Johnson, M.G., and R.L. Beschta. 1980. Logging, infiltration capacity, and surface erodibility in western Oregon. *J. For.* 78:334–337.

Jones, J.A. 2000. Hydrologic processes and peak discharge response to forest removal, regrowth, and roads in 10 small experimental basins, western Oregon. *Water Resour. Res.* 39:2621–2642.

Kattelmann, R.C., N.H. Berg, and J. Rector. 1983. The potential for increasing streamflow from Sierra Nevada watersheds. *Water Resour. Bull.* 19:395–402.

Lane, L.J. 1983. Transmission losses. In *National engineering handbook*, Sect. 4, Hydrology, Chap. 19. Washington, DC: USDA Soil Conservation Service.

Lavabre, J., D.S. Torres, and F. Cernesson. 1993. Changes in the hydrological response of a small Mediterranean basin a year after a wildfire. *J. Hydrol.* 142:273–299.

Lee, R. 1964. Potential isolation as a topoclimatic characteristic of drainage basins. *Bull. Int. Assoc. Hydrol. Sci.* 9:27–41.

Lee, R. 1980. *Forest hydrology*. New York: Columbia Univ. Press.

Lovett, G.M., W.A. Reiners, and R.K. Olson. 1982. Cloud droplets deposition in sub-alpine forests: Hydrological and chemical inputs. *Science* 218:1303–1304.

Lu, S.-Y. 1994. Forest harvesting effects on streamflow and flood frequency in the Lake States. Ph.D. Thesis. St. Paul, MN: University of Minnesota.

Lull, H.W., and K.G. Reinhart. 1972. *Forests and floods in the eastern United States*. USDA For. Serv. Res. Pap. NE–226.

Marques, J., J.M. Santos, N.A. Villa Nova, and E. Salati. 1977. Precipitable water and water flux between Belem and Manaus. *Acta Amazonica* 7:355–362.

MacDonald, L.H. 2000. Predicting and managing cumulative watershed effects. In *Watershed 2000: Science and engineering for the new millennium*, eds. M. Flug and D. Frevert. Reston, VA: American Society of Civil Engineers. [CD-ROM] Windows

Megahan, W.F., and J. Hornbeck. 2000. Lessons learned in watershed management: A retrospective view. In *Land stewardship in the 21st century: The contributions of watershed management*, tech. coords. P.F. Ffolliott, M.B. Baker, Jr., C.B. Edminster, M.C. Dillon, and K.L. Mora, 177–188. USDA For. Serv. Proc. RMRS–P–13.

Pilgrim, D.H., D.G. Boran, I.A. Rowbottom, S.M. Mackay, and J. Tjendana. 1982. Water balance and runoff characteristics of mature and cleared pine and eucalypt catchments at Lidsdale, New South Wales. In *First national symposium on forest hydrology*, ed. E.M. O'Loughlin and L.J. Bren, 103–110. Melbourne, Australia: National Committee on Hydrology and Water Resources of the Institution of Engineers.

Rees, W.E. 1995. Cumulative environmental assessment and global change. *Environ. Assessment Rev.* 15:295–309.

Reid, L.M. 1993. *Research and cumulative watershed effects*. USDA For. Serv. Gen. Tech. Rep. PSW–GTR–141.

Rich, L.R. 1962. *Erosion and sediment movement following a wildfire in a ponderosa pine forest in central Arizona*. USDA For. Serv., Rocky Mountain Forest and Range Experiment Station, Res. Note 76.

Rowe, L.K., and A.J. Pearce. 1994. Hydrology and related changes after harvesting native forest catchments and establishing *Pinus radiata* plantations. Pt. 2, The native forest water balance and changes in streamflow after harvesting. *Hydrological Processes* 8:281–297.

Salati, E., and P.B. Vose. 1984. Amazon basin: A system in equilibrium. *Science* 225:129–138.

Satterlund, D.R., and P.W. Adams. 1992. *Wildland watershed management*. New York: John Wiley & Sons, Inc.

Scott, D.F. 1993. The hydrological effect of fire in South African mountain catchments. *J. Hydrol.* 150:409–432.

Scott, D.F., and D.B. Van Wyk. 1990. The effects of wildfire on soil wettability and hydrologic behaviour of an afforested catchment. *J. Hydrol.* 121:239–256.

Sidle, R.C. 2000. Watershed challenges for the 21st century: A global perspective for mountainous terrain. In *Land stewardship in the 21st century: The contributions of watershed management*, tech. coords. P.F. Ffolliott, M.B. Baker, Jr., C.B. Edminster, M.C. Dillon, and K.L. Mora, 45–56. USDA For. Serv. Proc. RMRS–P–13.

Sinclair, J.D., and E.L. Hamilton. 1955. Streamflow reactions to a fire-damaged watershed. In *Proceedings of the Hydraulic Division, American Society of Civil Engineers*, np. New York.

Swank, W.T., and D.A. Crossley, Jr., eds. 1988. *Forest hydrology and ecology at Coweeta*. New York: Springer-Verlag.

Todd, D.K. 1970. *The water encyclopedia*. Port Washington, NY: Water Information Center, Inc.

Van Hylckama, T.E.A. 1970. Water use by salt cedar. *Water Resour.* Res. 6:728–735.

Verry, E.S. 1976. *Estimating water yield differences between hardwood and pine forests: An application of net precipitation data*. USDA For. Serv. Res. Pap. NC–128.

Verry, E.S. 1986. Forest harvesting and water: The Lake States experience. *Water Res. Bull.* 22:1039–1047.

Verry, E.S. 2001. Land fragmentation and impacts to streams and fish in the central and upper Midwest. Presentation in: *Hydrologic Science: Challenges for the 21st Century.* Annual Meeting of the American Institute of Hydrology. October 14–17, 2001. Bloomington, MN.

Verry, E.S., J.R. Lewis, and K.N. Brooks. 1983. Aspen clearcutting increases snowmelt and storm flow peaks in north central Minnesota. *Water Resour. Bull.* 19:59B67.

Whitehead, P.G., and M. Robinson. 1993. Experimental basin studies—An international and historical perspective of forest impacts. *J. Hydrol.* 145:217–230.

Zadroga, F. 1981. The hydrological importance of a montane cloud forest area of Costa Rica. In *Tropical agricultural hydrology,* ed. R. Lal and F.W. Russell, 59–73. New York: John Wiley & Sons.

Zon, R. 1927. *Forests and water in the light of scientific investigation.* USDA For. Serv. Rep. unclassified.

Stream channels adjust to excessive sediment loads; a tributary of the Betsiboka River, Madagascar.

PART 2

Erosion, Sediment Yield, and Channel Processes

The hydrologic processes discussed in Part 1 directly and indirectly affect soil erosion, the transport of eroded sediments, the deposition of sediments downstream, and the fluvial processes that define the stream channel system. Land use and watershed management practices also directly affect erosion, sedimentation, and channel processes via changes in the hydrologic processes and physical changes in the channel system itself. These topics are the subject of Part 2.

Soil erosion affects the productivity of upland areas and can adversely impact downstream areas. In an overview of global erosion and sedimentation, Pimentel et al. (1995) stated that > 50% of the world's pasture lands and about 80% of agricultural lands suffer from significant erosion. Furthermore, they indicated that in the United States alone, erosion results in losses of land productivity amounting to $27 billion/yr (1992 dollars) and $17 million/yr in downstream or offsite impacts (damages to streams, dredging costs, etc.). Some would argue that these cost estimates are too high; nevertheless, the fact remains that excessive soil erosion and sedimentation represent significant costs to people throughout the world.

There is recent evidence that erosion rates are being reduced through improved land use practices in some parts of the world. Trimble (1999) reported that soil conservation practices in a 360 km^2 watershed in Wisconsin have substantially reduced erosion and sedimentation rates that were at a peak in the 1920s and 1930s. Rates of sediment storage in this watershed that were 405,000 Mg/yr from 1853 to 1938 have been reduced to 80,000 Mg/yr from 1975 to 1993. In contrast, sedimentation rates in the Minnesota River Basin have increased in spite of improved soil conservation on farmlands. Here, part of the problem may be attributed to accelerated levels of streamflow that have resulted from extensive tile drainage of farm fields and loss of wetlands.

A basic understanding of erosion processes and the factors affecting erosion and streamflow-sediment relationships is needed by managers of watersheds. A thorough knowledge of erosion processes and their control probably is not reasonable for all resource managers and engineers, but they should have a working knowledge of the principal causes of erosion.

Soil erosion is the process of dislodgement and transport of soil particles by wind and water. Climate, topography, soil characteristics, vegetative cover, and land use all affect soil erosion. Chapter 7 is concerned with surface erosion from water and wind and methods of soil control. Gully erosion and control and soil mass movement, including sand dune formation, are discussed in Chapter 8.

The downstream impacts of erosion depend on the factors that govern sediment transport from watershed surfaces and through the stream channel. Channel systems adjust to alterations in water flow and sediment delivery. Sediment transport and deposition and channel processes are the focus of Chapter 9. The channels of rivers and tributary streams are in a state of dynamic equilibrium and are formed by the flows of water and sediment over time. These flows and hence, the morphology of stream channels can be altered by human interventions on the watershed, in the riparian zone and in the channels themselves. The dynamics of stream channels and their response to alterations, and the role of stream classification are presented in Chapter 10.

References

Pimentel, D., C. Harvey, P. Resosudarmo, K. Sinclair, D. Kurz, M. McNair, S. Crist, L. Shpritz, L. Fitton, R. Saffouri, and R. Blair. 1995. Environmental and economic costs of soil erosion and conservation benefits. *Science* 267:1117–1123.

Trimble, S.W. 1999. Decreased rates of alluvial sediment storage in the Coon Creek Basin, Wisconsin, 1975–93. *Science* 285:1244–1246.

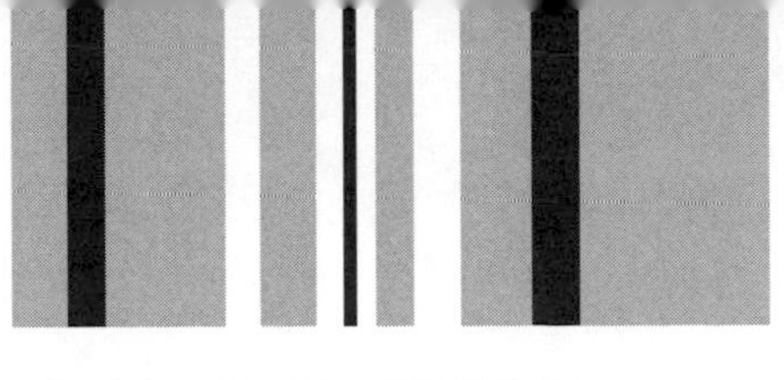

CHAPTER 7

Surface Erosion and Control of Erosion on Upland Watersheds

INTRODUCTION

In general, there are three erosion processes on upland watersheds: surface erosion, gully erosion, and soil mass movement. *Surface erosion* involves the detachment and subsequent removal of soil particles and small aggregates from land surfaces by water or wind. This type of erosion is caused by the action of raindrops, thin film flows, concentrated overland flows, or wind. While less serious in forested environments, surface erosion can be an important source of sediment from rangelands and cultivated agricultural lands. *Gully erosion* is the detachment and movement of material, either individual soil particles or large aggregates, in a well-defined channel. This kind of erosion is a major form of geologic erosion that can be accelerated greatly under poor land management. *Soil mass movement* includes erosion in which cohesive masses of soil are displaced. Movement can be rapid, as with landslides, or it can be quite slow, as with soil creep and certain soil slumps. A detailed discussion of the processes involved in gully erosion and soil mass movement is presented in Chapter 8.

All of the above erosion processes can occur singly or in combination. Human activities, such as construction, road building, forest removal, intensive livestock grazing, and agriculture, can accelerate these processes. At times, it is difficult to distinguish the basic types of erosion and to determine whether they are natural geologic processes or have been accelerated by poor land use practices. Factors that affect soil erosion and sediment movement from a watershed are summarized in Figure 7.1.

THE EROSION PROCESS

Soil erosion is the process of dislodgement and transport of soil particles from the surface by water and wind. The soil particles can be dislodged by the energies expended at

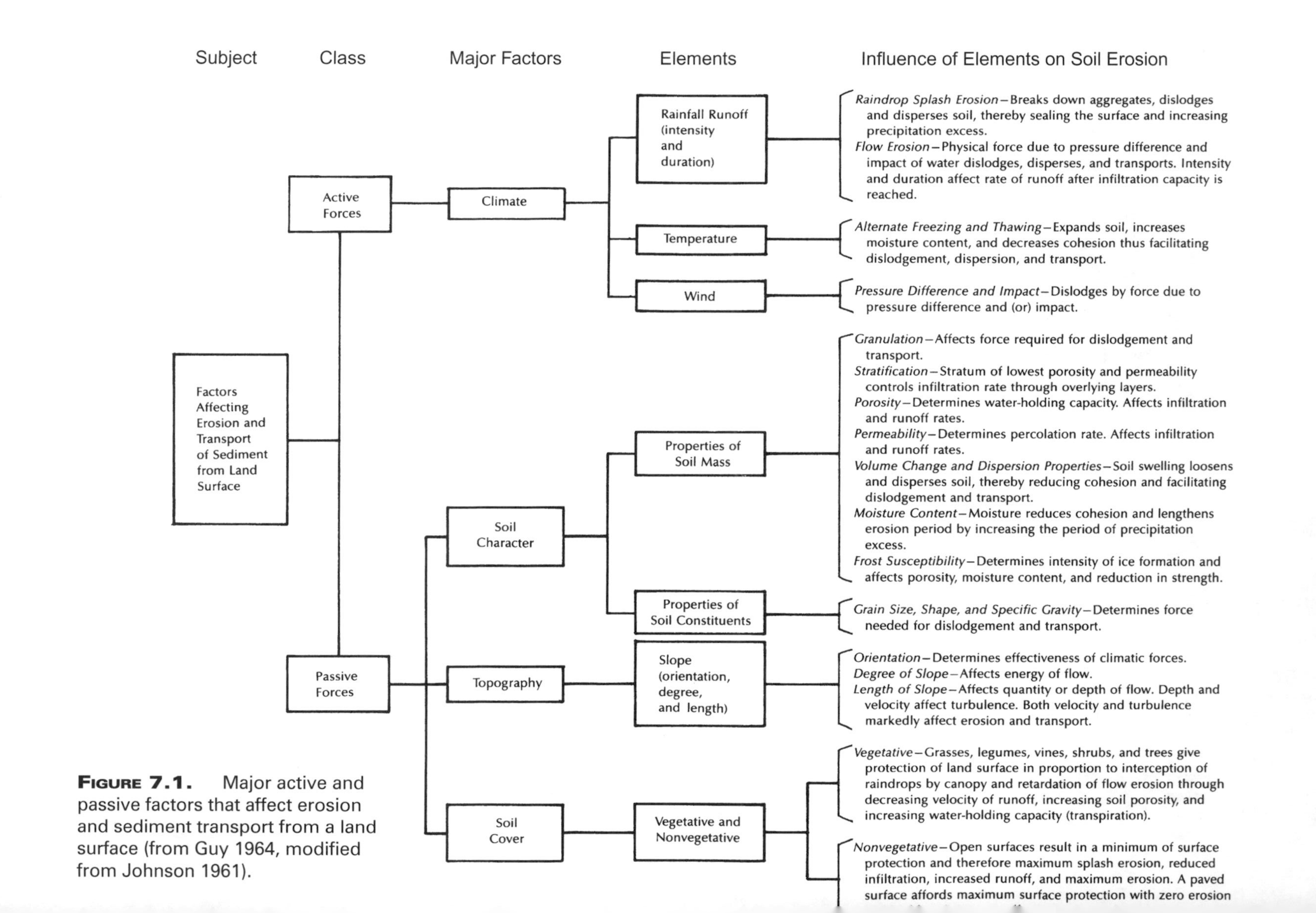

FIGURE 7.1. Major active and passive factors that affect erosion and sediment transport from a land surface (from Guy 1964, modified from Johnson 1961).

the soil surface by raindrops or by the eddies in surface runoff and wind, and then transported by water or wind or the force of gravity. Therefore, erosion is a process in the physical sense that work requires the expenditure of energy. The energy is imparted to the soil surface by forces resulting from impulses produced by the momentum (mass × velocity) of falling raindrops or by the momentum of eddies in the turbulent flows of runoff or wind. It is these forces that cause work to be done in both the dislodgement and transport phases of the process.

Water Erosion

The dislodgement of soil particles at the soil surface by energy imparted to the surface by falling raindrops is a primary agent of erosion, particularly on soils with sparse vegetative cover (Table 7.1). The energy released at the surface during a large storm is sufficient to splash > 200 t of soil into the air on a single hectare of bare and loose soil. Individual soil particles can be splashed > 0.5 m in height and 1.5 m sideways.

It has often been reported that the energy released at the soil surface by a rainstorm is greater than that released by the runoff produced. However, these calculations did not take into account the energy of turbulent eddies in runoff (Fig. 7.2). Furthermore, rainfall is

Table 7.1. Kinetic energy (K_e) associated with different intensities of rainfall

	Rainfall Intensity (mm/hr)	Kinetic Energy[a] (MJ/ha·mm)[b]
Drizzle	1	0.12
Rain	15	0.22
Cloudburst	75	0.28

Source: Calculated from Diesmeyer and Foster 1980.
[a] K_e = 1/2 (mass)(velocity)^2.
[b] Units are megajoules per hectare millimeter.

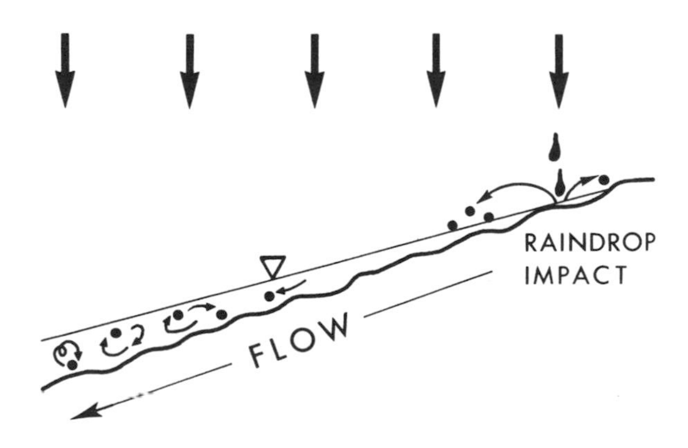

Figure 7.2. Surface soil erosion as a result of raindrop impact and turbulent surface runoff.

often distributed more or less uniformly over an area, whereas runoff quickly becomes concentrated in rills and channels, where its erosive power becomes magnified.

The major result of the impulses imparted to the soil surface by raindrops is the deterioration of soil structure by the breakdown of soil aggregates. The subsequent splashing of finer soil particles tends to puddle and close the soil surface and thereby increase surface runoff.

Surface runoff takes place when the rate of rainfall exceeds the infiltration rate on slopes. Just as with rainfall, the kinetic energy of surface runoff depends upon the mass (depth) of water and its velocity. In addition, surface runoff is turbulent; eddies in the flow make up the turbulence. These eddies are random in size, orientation, and velocity and provide the impulses to dislodge and entrain soil particles. The intensity of the turbulence in surface runoff depends upon the velocity and depth of runoff and the roughness of the surface over which the water flows.

Surface runoff combined with the beating action of raindrops causes rills to be formed in the soil surface. *Rill erosion* is the form of erosion that produces the greatest amount of soil loss worldwide. *Sheet erosion* takes place between rills and thus is also called *inter-rill erosion*. Sheet erosion is the movement of a semisuspended layer of soil particles over the land surface. However, minute rills are formed almost simultaneously with the first detachment and movement of particles. The constant meanders and changes in position of these small rills obscure their presence from normal observation, hence the concept of "sheet erosion."

As runoff becomes concentrated in rills and small channels and moves downslope, the velocity and mass of the suspension, as well as the intensity of the turbulence in the flow, increase. When the depth of runoff is shallow, raindrops striking the water surface can add to the turbulence. This increase of kinetic energy results in an even greater increase in the ability of the flow to dislodge and transport larger soil particles. If the flow carries a large load of sediment, the abrasive action of the load adds to the erosive power of the runoff. On steep, unobstructed slopes and with heavy rains, soil loss in this manner can be dramatic; such losses are also common on drylands where normally sparse vegetative cover has been disturbed by poor land practices.

The momentum gained by surface runoff on a sloping area and, consequently, the amount of soil that can be lost from the area depend upon both the inclination and the length of unobstructed slope. With increasing slope length, soil loss per unit length initially is accelerated but then approaches a constant rate. As the inclination of the slope increases, soil loss increases. Slope angle and slope length allow the buildup in momentum in flowing water and are major factors in accelerating rill erosion; the steeper and longer the slope, the greater the problems of control. Once it becomes channelized, uncontrolled surface runoff is capable of creating the more spectacular gully erosion (see Chapter 8). Gullies are common features of sparsely vegetated lands.

Wind Erosion

In dry regions, erosion by both water and wind is a natural feature. Such erosion is an inevitable consequence of the environment, largely because rainfall is inadequate to support a protective cover of vegetation. Any use of drylands that further reduces vegetation cover tends to accelerate erosion beyond that which is a natural consequence of the environment. As a rule, watersheds that have natural vegetative cover and that receive precipitation of > 400 mm/yr experience little wind erosion. However, when soils are exposed, excessive wind erosion can occur even in regions with > 800 mm/yr annual precipitation. In either case, wind erosion diminishes with increasing annual precipitation.

The actions of water or wind are often complementary in their roles of removing soil in dry regions. For example, a soil stripped of vegetation by the abrasive action of wind-blown sand is rendered vulnerable to erosion by water; likewise, a barren outwash sediment deposit is subject to erosion by wind. Sometimes it is difficult to determine which is the dominant agent on a particular site; generally, wind erosion is a long-term process of gradual removal, and water erosion is shorter term and can be very rapid. The conservation of soil and water in arid zones often must address both processes simultaneously. Fortunately, many principles of the erosion process and most of the methods of controlling erosion apply to either process.

The action of wind on the soil surface is analogous to the action of flowing water. Wind also exhibits turbulent flow, having a net velocity in the horizontal direction but with strong eddying in both upward and downward directions. The air is compressed randomly by this turbulence, which produces gusts. Hot desert soils create thermal updrafts, which increase the turbulence. By their great velocity and upward eddies, gusty winds are able to dislodge small soil particles, lift them, and carry them away, much like suspended sediment in flowing water.

Wind can move larger soil particles by making them "jump" along the ground. The jumping particles also apply energy to the soil surface each time they hit the ground and, in doing so, dislodge other particles so that they too can be moved by the wind. This process is called *saltation* and is also a major process in the movement of bed load, the larger particles that move along the bottom of a stream channel.

The largest soil particles that wind can move to any extent are about 1 mm. Thus, where the size of the soil particles extends over a wide range, wind has a sorting effect on the soil. Very fine clay and silt particles (< 0.02 mm) are lifted into the air and carried away as wind-blown dust. Sand-size particles are carried along in the air layer near the ground by saltation until they reach an obstruction, where they can pile up into drifts and (under extreme conditions) into dunes. Just as gullies are advanced stages of water erosion, sand dunes are severe stages of wind erosion.

The erosive power of wind, like that of water, increases exponentially with velocity; but, unlike water, it is not affected by the force of gravity. Therefore, slope inclination is not a factor in wind erosion, except where sloping or hilly terrain forms barriers or influences wind direction. However, similar to the effect on the erosive power of water, the length of unobstructed terrain (*fetch*) over which the wind flows is important in allowing the wind to gain momentum and to increase its erosive power. Winds with velocities less than about 12–19 km/hr at 1 m above the ground seldom impart sufficient energy at the soil surface to dislodge and put into motion sand-size particles.

Measurement of Surface Erosion

Surface erosion from a relatively small area can be measured or approximated by several field methods. The most common methods include the use of plots or stakes and measuring natural landscape features such as pedestals. Surface erosion over larger areas such as a drainage basin can be measured by repeated reservoir surveys of designated transects or through the use of tracers (Morgan 1995). The emphasis in this discussion is placed on the former.

Erosion Plots

The most widely used method of quantifying surface erosion rates is to measure the amount of soil that washes from plots. Collecting troughs are sunk along the width of the bottom of the plots. Walls made of plastic, sheet metal, plywood, or concrete (called fabric dams) are inserted at least 10 cm into the soil surface and form the boundaries of the plot. The collecting trough empties into a tank, in which sediment and runoff are measured. Sometimes, these tanks have recording instruments that can measure rates of flow. In other cases, the total volume of sediment and water is measured after a rainfall event.

Plots can vary in size from microplots of 1–2 m^2 to the standard plot of approximately 2 m $\times$ 22 m used for the Universal Soil Loss Equation, discussed later. Techniques used and objectives of the experiment dictate plot size. For multiple comparisons of vegetation, soils, and land use practices, microplots are less expensive and more practical for experiments that use rainfall simulators. Larger plots provide more realistic estimates of erosion because they better represent the effects of increasing runoff and velocity downslope. Plots larger than the standard can produce large volumes of runoff and sediment that are difficult to store. In such cases, devices that split or sample a portion of total water and sediment flow are preferred. Such fractioning devices allow larger plots to be used to quantify the effects of larger-scale land use practices.

Erosion Stakes and Other Methods

The insertion of stakes or pins into the soil can be used to estimate soil losses and sediment deposition that occur along hillslopes. Commonly, a long metal nail with a washer is inserted into the soil and the distance between the head of the nail and the washer is measured. This distance increases as erosion occurs because the soil that supports the washer is washed away. If the washer causes a pedestal to form immediately beneath the washer (because of protection from rainfall impact), measurements should be made from the nail head to the bottom of the pedestal. A benchmark should be established in close proximity to the stakes as a point of reference, and stakes should be clearly marked so that original stakes can be accurately relocated on subsequent surveys.

Normally, erosion stakes are arranged in a grid pattern along hillslopes. Repeated measurements of stakes over time provide information on the changes in soil surface that result from soil loss and deposition. This method is inexpensive compared with the plot method but presents more difficulty in converting observations into actual soil losses in tons per hectare.

Using the same principles as the stake method, erosion estimates can sometimes be made from natural landscape features. Pedestals often form beneath clumps of bunchgrass, dense shrubs, stones, or other areas that are protected from rainfall. As erosion removes soil from around them, the distance between the pedestal top and bottom increases. Repeated measurements of the height of residual soil pedestals provide estimates as described above. The key is to relate measurements to a common point of reference or benchmark. Sometimes, soil that has eroded away from the base of trees can be estimated with repeated measurements of soil surface and a point on exposed tree roots or from a nail driven into the tree trunk.

Prediction of Soil Loss

Because land management practices create a variety of conditions that influence the magnitude of surface erosion, land managers frequently want to predict the amount of soil loss by surface erosion. Several models are available for predicting erosion caused by the action of water, including the universal soil loss equation (USLE), modified soil loss equation (MSLE), revised universal soil loss equation (RUSLE), and the water erosion prediction project (WEPP) model. The most familiar method available for predicting soil loss is the USLE in both its original and its modified and revised forms.

Universal Soil Loss Equation

Prior to the development of the USLE, estimates of erosion rates were made from site-specific data on soil losses. As a result, these estimates were limited to particular regions and soils. However, the need for a more widely applicable erosion prediction technique led to the development of the USLE by the USDA Agricultural Research Service. The original USLE of 1965 was based on the analysis of 10,000 plot-yr of source data, mostly collected from agricultural plots under natural rainfall. Subsequently, because of the high costs of collecting data from plots under natural rainfall, erosion research has been conducted on plots with simulated rainfall. Rainfall-simulator data are employed in the revised USLE of 1978 (not the same as the RUSLE of the 1980s, which is discussed later) to describe soil erodibility and to provide values for the effectiveness of conservation tillage and construction practices for controlling soil erosion.

In the following description of the USLE, the English units employed in its original development are used in the presentation rather than the corresponding metric units. The basic USLE (Wischmeier and Smith 1965, 1978) is:

$$A = RK\,(LS)\,CP \tag{7.1}$$

where A = computed soil loss in tons per unit area (acre); R = a rainfall erosivity factor for a specific area, usually expressed in terms of average erosion index (EI) units; K = a soil erodibility factor for a specific soil horizon; LS = topographic factor, a combined dimensionless factor for slope length and slope gradient, where L is expressed as the ratio of soil loss from a given slope length to soil loss from a 72.6 ft length under the same conditions (it is not the actual slope length), and S is expressed as the ratio of soil loss from a 9% slope under the same conditions (it is not the actual slope steepness); C = a dimensionless cropping management factor, expressed as a ratio of soil loss from the condition of interest to soil loss from tilled continuous fallow (condition under which K is determined); and P = an erosion control practice factor, expressed as a ratio of the soil loss with the practices (e.g., contouring, strip-cropping, or terracing) to soil loss with farming up and down the slope.

Equation 7.1 provides an estimate of sheet and rill erosion from rainfall events on upland areas. It does not include erosion from stream banks, snowmelt, or wind, and it does not include eroded sediment deposited at the base of slopes and at other reduced-flow locations before runoff reaches the streams or reservoirs.

The term *universal* was given to the USLE to indicate that, in contrast to earlier erosion prediction equations that applied only to specific regions, the USLE applied initially in 1965 to the area of the United States east of the Rocky Mountains, and with the 1978 revision, to all of the United States.

By 1970, the USLE also was being applied to nonagricultural situations, such as construction sites and undisturbed lands, including forests and rangelands. However, as extensive baseline data were not available for these applications, a subfactor method was developed to estimate values for the C factor. In essence, the subfactor method employs a set of relationships for canopy cover, ground cover, and internal soil effects to estimate a composite value for C. This development allows the use of source data collected from more basic studies to be utilized in applications of the USLE.

Rainfall Erosivity Factor

The rainfall erosivity factor (R) is an index that characterizes the effect of raindrop impact and rate of runoff associated with the rainstorm. It is determined by calculating the EI for a specified period, usually 1 yr or one season within the year. The EI averaged over a number of these periods (n) equals R:

$$R = \frac{\sum_{i}^{n} EI_i}{n} \tag{7.2}$$

The energy of a rainstorm depends on the amount of rain and all the component rainfall intensities of the storm. For any given mass in motion, the energy is proportional to velocity squared; therefore, rainfall energy is related directly to rain intensity by the relationship:

$$E = 916 + 331\,(\log I_i) \tag{7.3}$$

where E = kinetic energy per inch of rainfall in ft-tons/acre and I = rainfall intensity in each rainfall intensity period of the storm (in./hr).

The total kinetic energy of a storm (k_e) is obtained by multiplying E by the depth in inches of rainfall in each intensity period (n), and summing:

$$k_e = \sum_{i}^{n} [916 + 331\,(\log I_i)] \tag{7.4}$$

The EI for an individual storm is calculated by multiplying the total kinetic energy (k_e) of the storm by the maximum amount of rain falling within 30 consecutive minutes (I_{30}), multiplying by 2 to obtain in./hr, and dividing the result by 100 (to convert from hundreds of ft-tons/acre to ft-tons/acre):

$$EI\,(\text{storm}) = \frac{2k_e I_{30}}{100} \tag{7.5}$$

The EI for a specific period (year or season) is the sum of the individual storms' EI values computed for all significant storms during that period. Usually, only storms > 0.5 in. are selected. The R is then determined as the sum of the EI values for all such storms that occurred during a 20–25 yr period, divided by the number of years (Eq. 7.2).

Studies conducted on rangelands in the southwestern United States have shown that storm runoff is correlated highly with the R value for the storm. Therefore, although runoff could have been a parameter for inclusion in the USLE, the use of R is considered a better index for runoff and precipitation-induced erosion.

Runoff events associated with snowmelt and thawing soils are common on many watersheds. Erosion during these events can be appreciable. Therefore, to apply the USLE to these situations, a use for which it was not originally intended, the R factor must be adjusted. This adjustment, which is based on only limited data and is general in nature, is 1.5 times the winter precipitation (measured in inches of water), a value that is added to the R erosivity values for winter storms. Additional research is necessary to further verify and, if required, modify this adjustment procedure for wider use.

Soil Erodibility Factor

The soil erodibility factor (K) indicates the susceptibility of soil to erosion and is expressed as soil loss per unit of area per unit of R for a unit plot. By definition, a unit plot is 72.6 ft long, on a uniform 9% slope, maintained in continuous fallow, with tillage when necessary to break surface crusts and to control weeds. These dimensions are selected because they coincide with the erosion research plots used in early work in the United States. Continuous fallow is selected as a base because no single cropping system is common to all agricultural areas, and soil loss from any other plot conditions would be influenced, to a large extent, by residual and current crop and management effects, both of which vary from one location to another. The K value can be determined as the slope of a regression line through the origin for source data on soil loss (A) and erosivity (R), once the ratios for L, S, C, and P have been adjusted to those of unit conditions. When the K value was originally determined with natural rainfall data, it covered a range of storm sizes and antecedent soil moisture conditions. Results of later studies conducted with rainfall simulators were used to produce a soil erodibility nomograph based on soil texture and structure (Wischmeier et al. 1971). This nomograph, illustrated in Figure 7.3, is now used to obtain the K factor.

Slope Length Factor and Slope Gradient Factor

The topographic factors L and S indicate the effects of slope length and steepness, respectively, on erosion. Slope length refers to overland flow, from where it originates to where runoff reaches a defined channel or to where deposition begins. In general, slopes are treated as uniform profiles. Maximum slope lengths are seldom longer than 600 ft or shorter than 15–20 ft. Selection of a slope length requires on-site inspection and judgment.

The slope length factor (L) is defined as:

$$L = \left(\frac{\lambda}{72.6}\right)^m \tag{7.6}$$

where λ = field slope length (ft) and m = exponent, affected by the interaction of slope length with gradient, soil properties, type of vegetation, etc. The exponent value ranges from 0.3 for long slopes with gradients < 5% to 0.6 for slopes > 10%. The average value of 0.5 is applicable to most cases.

The maximum steepness of agricultural cropland plots used to derive the S factor was 25%, which is less than many forested and rangeland watershed slopes. Recent investigations on rangelands suggest that the USLE can overestimate the effect of slope on noncrop situations. Consequently, the S factor likely will be adjusted downward in future revisions of the USLE.

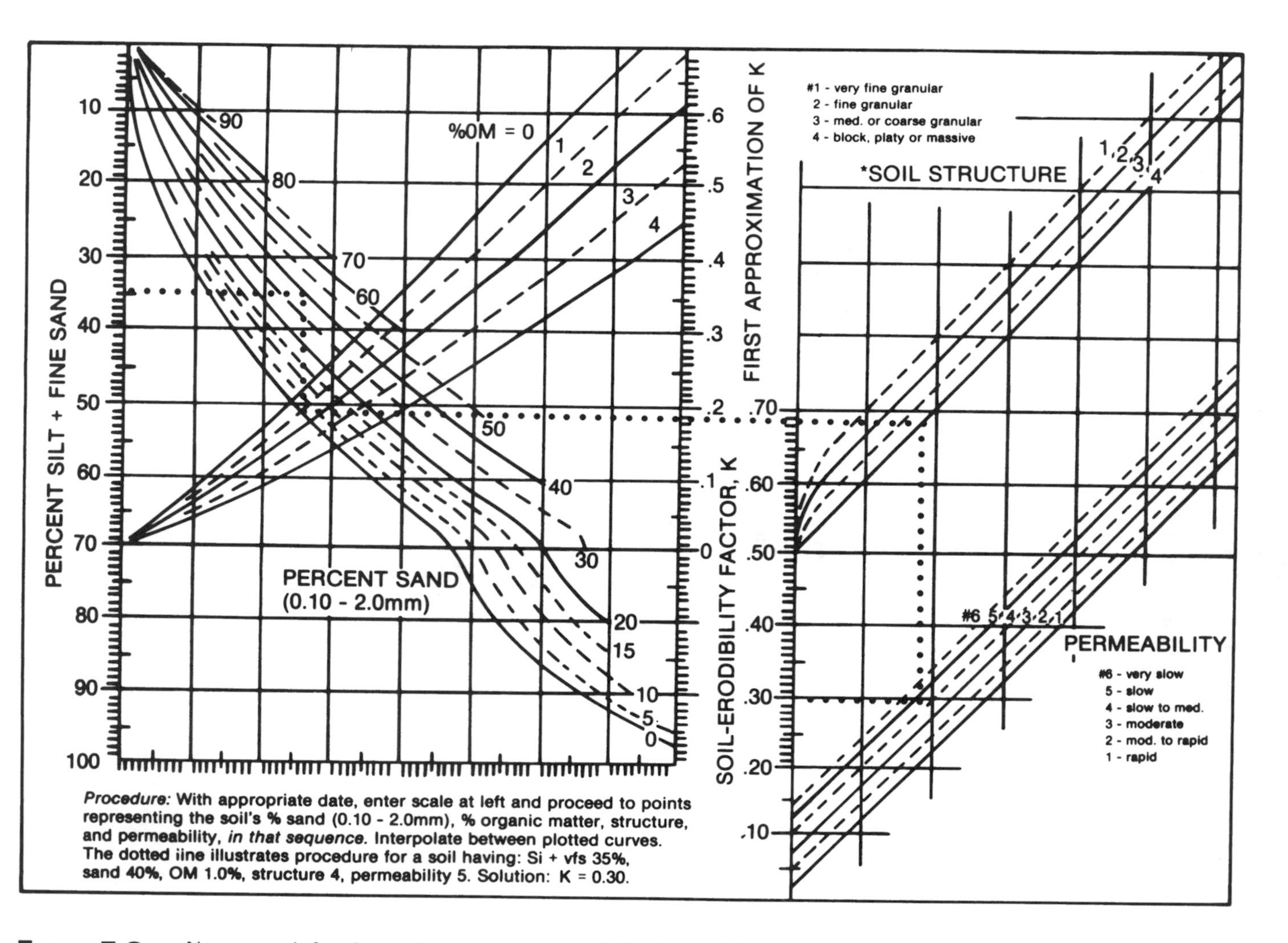

FIGURE 7.3. Nomograph for determining the soil erodibility factor (K) in English units (from USDA Forest Service 1980, adapted from Wischmeier et al. 1971).

The slope gradient factor (S) is defined as:

$$S = \frac{0.43 + 0.30s + 0.043s^2}{6.613} \tag{7.7}$$

where s = slope gradient (in percent).

Foster and Wischmeier (1973) adapted the LS factors for use on irregular slopes; this is especially useful on wildland sites, which rarely have uniform slopes. They describe the combined factor as:

$$LS = \frac{1}{\lambda_e} \sum_{j=1}^{n} \left[\left(\frac{S_j \lambda_j^{m+1}}{(72.6)^m} - \frac{S_j \lambda_{j-1}^{m+1}}{(72.6)^m} \right) \left(\frac{10{,}000}{10{,}000 + s_j^2} \right) \right] \tag{7.8}$$

where λ_e = overall slope length; j = sequence number of segment from top to bottom; n = number of segments; λ_j = length (ft) from the top to the lower end of the jth slope segment; λ_{j-1} = the slope length above segment j; S_j = S factor for segment j (Eq. 7.7); and s_j = slope (%) for segment j.

For uniform slopes, LS is determined as:

$$LS = \left(\frac{\lambda_e}{72.6} \right)^m S \left(\frac{10{,}000}{10{,}000 + s^2} \right) \tag{7.9}$$

Cropping Management Factor

The cropping management factor (C) of the USLE represents an integration of several factors that affect erosion, including vegetative cover, plant litter, soil surface, and land management. Embedded in the term is a reflection of how intercepted raindrops that are reformed on a plant canopy affect splash erosion. Also, the binding effect of plant roots on erosion and how the properties of soil change as it lies idle are considered. Unfortunately, the manner in which grazing animals and other plant cover manipulations change the magnitude of C is not well defined. Studies to define these cause-and-effect relationships are needed to better understand the role of the C factor in calculating annual erosion rates.

In most cases, the value of C is not constant over the year. Although treated as an independent variable in the equation, the true value of this factor probably is dependent upon all other factors. Therefore, the value of C should be established experimentally. Runoff plots and fabric dams (filter fences) are useful for this purpose. One simple procedure is:

1. Install runoff plots on the cover complexes of interest, measure soil loss with fabric dams for each storm event, and record rainfall intensity and amount for at least a 2 yr calibration period.
2. Identify rainfall events with a threshold storm size great enough to produce soil loss.
3. Calculate the R value for each storm greater than the threshold storm (usually $>$ 0.5 in.) for each year of record.
4. Determine the K and LS factors from the nomographs and equations given in the text.
5. Solve the USLE for C for each year of record; that is, $C = A/RK(LS)$, and calculate an average value for C.

Because of the variability in storms from year to year, an average value or an expected range in R values is useful. Such estimates can be made by developing a regression relationship between the calculated R values and storm amounts measured during the calibration period. By using a historical record of daily rainfall from the nearest weather station, the R value can be determined for each storm on record, and the average R for the watershed can be calculated.

In areas of the world for which there are no guidelines for the establishment of C values for field crops, it is easiest to correlate soil loss ratio with the amount of dry organic matter per unit area or with percent ground cover. For permanent pasture, rangelands, idle lands, and woodlands, C values can be estimated from published tables.

Certain subfactors should be considered when developing this factor for forest conditions: soil consolidation, surface residue, canopy, fine roots, residual effect of fine roots after tillage, contour effect, roughness, weeds, and grasses. Surface residue (litter, slash, and live vegetation) is the predominant element. Tables and nomographs for estimating the C factor for forestry practices are given in Dissmeyer and Foster (1985).

Erosion Control Practice Factor

The effect of erosion control practice (P) measures is considered an independent variable; therefore, it has not been included in the cropping management factor. The soil loss ratios for erosion control practices vary with slope gradient. Practices characterized by P, including strip-cropping and terraces, are not applicable to most forested and rangeland watersheds. Experimental data to quantify the P factor for noncrop management practices on forested and rangeland watersheds are not available.

Although the USLE was designed and has been field-tested for specific agricultural applications, it is frequently used in other ways because it seems to meet the need better than other tools available. Examples of applying the USLE to a variety of conditions and geographic regions are summarized in Table 7.2. In addition, the USLE or parts thereof, in original or modified forms, have been incorporated into other erosion prediction technologies (Toy and Osterkamp 1995).

Modifications of the Universal Soil Loss Equation

The USLE has been modified for use in rangeland and forest environments. The cropping management (C) factor and the erosion control practice (P) factor used in the USLE have been replaced by a vegetation management (VM) factor to form the modified soil loss equation (MSLE):

$$A = RK\,(LS)\,(VM) \tag{7.10}$$

where VM = the vegetation management factor, the ratio of soil loss from land managed under specified conditions of vegetative cover to that from the fallow condition on which the K factor is evaluated.

Vegetative cover and soil surface conditions of natural ecosystems, whether undisturbed or disturbed, are accounted for with the VM factor. Three different kinds of effects are considered as subfactors: (1) canopy cover effects; (2) effects of low-growing vegetative cover, mulch, and litter; and (3) bare ground with fine roots. Relationships have been

TABLE 7.2. Application of USLE under a variety of conditions

Application	Location	Reference	Comment
Rangelands	Arizona	Osborn et al. 1977	Evaluation of USLE for use on rangelands
	California	Singer et al. 1977	Evaluation of USLE for use on rangelands
	Various, in western United States	Renard and Foster 1985	Evaluation of USLE for use on rangelands (includes literature summary)
Forest lands	Manual for use of USLE on forest lands	Dissmeyer and Foster 1980	Guidance for use of USLE under forest vegetation with adjustments to factors and subfactors
Surface mining (coal)	Alabama	Shown et al. 1982	Equation used to evaluate erosion for assumed various phases of mining and reclamation
	Wyoming	Frickel et al. 1981	Equation used to evaluate erosion for assumed various phases of mining and reclamation
	Pennsylvania	Khanbilvardi et al. 1983	Equation used to estimate soil loss from inter-rill areas on mined and reclaimed lands
	Manual for use of USLE on surface-mined lands of western United States	USDA Soil Conservation Service 1977	Guidance for application of USLE on western mined lands
Watershed	Arizona	Fogel et al. 1977	MUSLE as a basis for stochastic sediment yield measurements
	Mississippi	Murphree et al. 1976	Comparison of equation estimates with sediment yield measurements
	Idaho, Colorado, Arizona	Jackson et al. 1986,	MUSLE used to estimate sediment yield from rangeland drainage basins
	Maryland	Stephens et al. 1977	Equation used for erosion and sediment inventory of basin
	Tennessee	Dyer 1977	Equation used in comprehensive resource planning of river basin

Source: Modified from Toy and Osterkamp 1995.

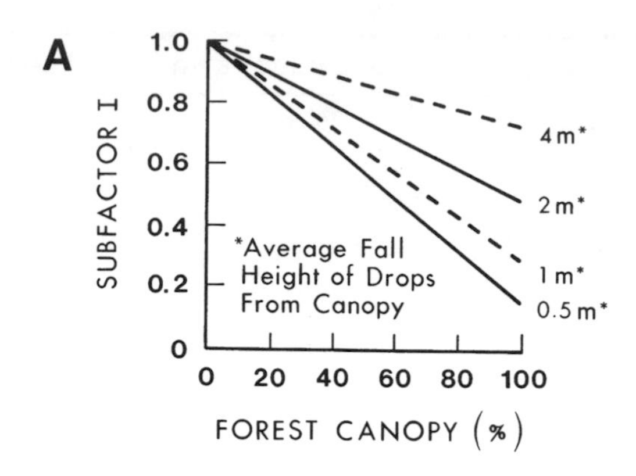

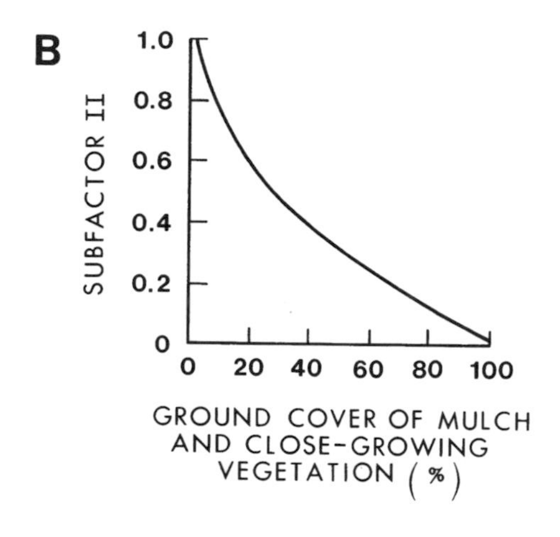

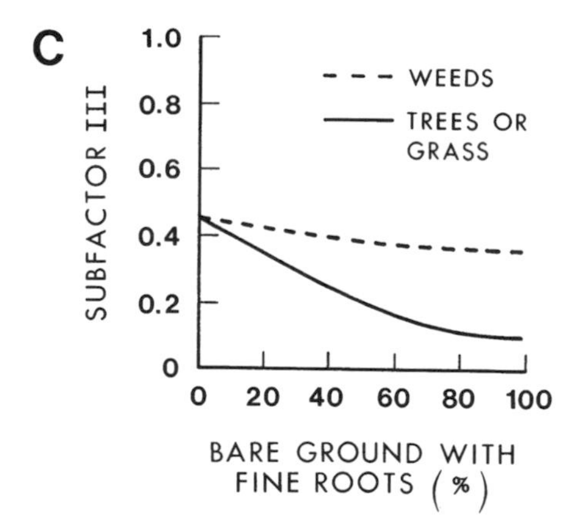

SUBFACTOR I x SUBFACTOR II
x SUBFACTOR III = VM FACTOR

Figure 7.4. Relationships of forest canopy cover (A), ground cover (B), and fine roots in the topsoil (C) used to determine subfactors I, II, and III, respectively, for the *VM* factor (from Wischmeier 1975).

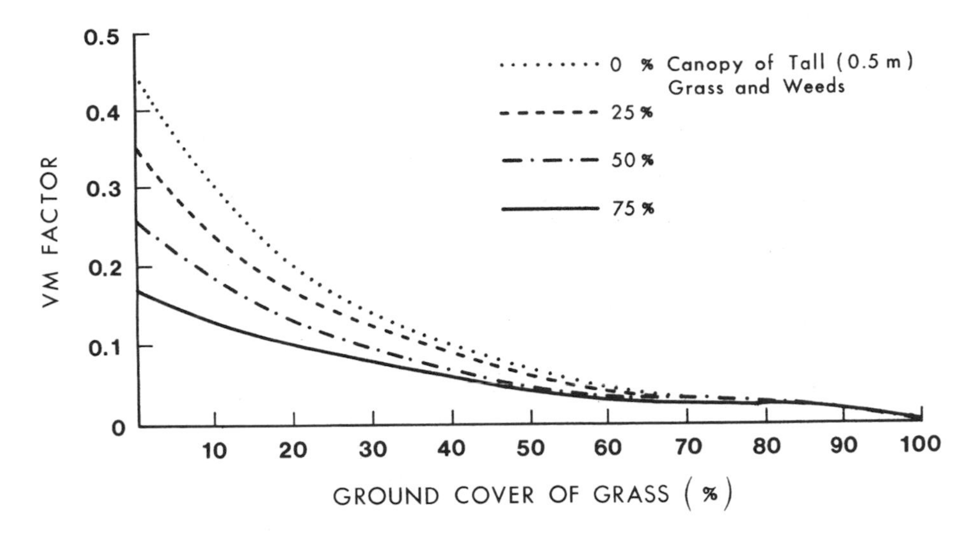

Figure 7.5. Relationship between ground cover conditions and the *VM* factor for the modified soil loss equation (adapted from Clyde et al. 1976).

developed for each of the three subfactors (Fig. 7.4). The three subfactors are multiplied together to obtain the *VM* value. When a forest canopy is not present (as, for example, in many rangeland conditions), relationships such as the one illustrated in Figure 7.5 can be used directly. The MSLE procedure is applied in Box 7.1.

More work has been carried out determining *C* values for the USLE than for *VM* factors; therefore, there are more numerous tables of relationships for *C* than for *VM* factors. Published values of *C* can be used as a substitute for *VM* if they account for the three effects described in Tables 7.3 and 7.4.

The MSLE procedure can be used as a guide for quantifying the potential erosion of different land use and land management strategies *only* if the principal interactions on which the equation is based are thoroughly understood. Failure to understand the equation can lead to invalid interpretations. If the underlying assumptions do not represent the actual processes in the forest environment, then predicted erosion values can be far different from actual erosion.

Both the USLE and the MSLE require an estimated value for the *R* factor. General estimates can be obtained for the United States from publications by the USDA Soil Conservation Service (now the U.S. Natural Resources Conservation Service). For other parts of the world and specific conditions in the United States, long-term rainfall-intensity records must be analyzed. Often, these records are not available.

Williams (1975) modified the USLE by replacing the *R* factor with a runoff factor. The modification is based on the assumption that the total discharge and peak discharge rate resulting from a storm on the watershed depend upon the duration, amount, and intensity of the storm. The modified equation is:

$$Y_s = 95 q_p Q K\,(LS)\,CP \tag{7.11}$$

where Y_s = sediment yield (tons); Q = volume of storm runoff (acre-ft); q_p = peak discharge (ft^3/sec); and $K(LS)CP$ = as defined in the USLE.

Box 7.1

Existing Surface Erosion Rates (Sheet and Rill Erosion) Estimated for a 12,000 ha Watershed in the Loukos Basin, Northern Morocco, by Applying the Modified Soil Loss Equation (MSLE)

Based on previous studies in the area, the values determined were $R = 400$, $K = 0.15$, $LS = 10$, and $VM = 0.13$ (based on a 30% ground cover of grass with a 25% canopy of tall weeds).

$A = RK\,(LS)\,(VM)$
$A = (400)\,(0.15)\,(10)\,(0.13) = 78$ t/ha/yr (estimated annual soil loss)

The total erosion rate from the watershed was 936,000 t/yr. Revegetation measures, as part of a watershed rehabilitation project, are anticipated to result in the following:

1/2 watershed area = 25% forest canopy with a 60% ground cover of grass and
1/2 watershed area = 80% grass cover with a 25% short shrub cover.

These changes in vegetative cover would be expected to reduce the surface erosion (estimated by the MSLE with the modified *VM* factors determined from Table 7.4) as follows:

A (1/2 area) = (400) (0.15) (10) (0.041) = 24.6 t/ha
A (1/2 area) = (400) (0.15) (10) (0.012) = 7.2 t/ha

The erosion rate for the rehabilitated watershed would be 190,800 t/yr.

TABLE 7.3. ***C* factors for undisturbed woodlands**

Effective Canopy[a] (% of area)	Forest Litter[b] (% of area)	C factor[c]
100–75	100–90	0.0001–0.001
70–40	85–75	0.002–0.004
35–20	70–40	0.003–0.009

Source: From USDA Soil Conservation Service 1977.
[a]When effective canopy is less than 20%, the area will be considered as grassland or idle land for estimating soil loss. Where woodlands are being harvested or grazed, use Table 7.4.
[b]Forest litter is assumed to be at least 5 cm deep over the percent ground surface area covered.
[c]The range in *C* values is due in part to the range in the percent area covered. In addition, the percent of effective canopy and its height has an effect. Low canopy is effective in reducing raindrop impact and in lowering the *C* factor. High canopy, over 13 m, is not effective in reducing raindrop impact and will have no effect on the *C* value.

The equation was developed to directly estimate sediment yield at the outlet of a watershed, rather than soil loss, on a storm-by-storm basis. Peak and total discharge can be estimated by the methods described in Chapter 17 if runoff data are not available. Satisfactory results have been obtained with Equation 7.11 when tested for a wide range

TABLE 7.4. *C* or *VM* factors for permanent pasture, rangeland, idle land, and grazed woodland

Type and Height of Raised Canopy[a]	Canopy Cover[b] (%)	Type[c]	Cover That Contacts the Surface (% ground cover) 0	20	40	60	80	95–100
No appreciable canopy		G	.45	.20	.10	.042	.013	.003
		W	.45	.24	.15	.090	.043	.011
Canopy of tall weeds or short brush (0.5 m fall ht)	25	G	.36	.17	.09	.038	.012	.003
		W	.36	.20	.13	.082	.041	.011
	50	G	.26	.13	.07	.035	.012	.003
		W	.26	.16	.11	.075	.039	.011
	75	G	.17	.10	.06	.031	.011	.003
		W	.17	.12	.09	.067	.038	.011
Appreciable brush or bushes (2 m fall ht)	25	G	.40	.18	.09	.040	.013	.003
		W	.40	.22	.14	.085	.042	.011
	50	G	.34	.16	.085	.038	.012	.003
		W	.34	.19	.13	.081	.041	.011
	75	G	.28	.14	.08	.036	.012	.003
		W	.28	.17	.12	.077	.040	.011
Tress but no appreciable low brush (4 m fall ht)	25	G	.42	.19	.10	.041	.013	.003
		W	.42	.23	.14	.087	.042	.011
	50	G	.39	.18	.09	.040	.013	.003
		W	.39	.21	.14	.085	.042	.011
	75	G	.36	.17	.09	.039	.012	.003
		W	.36	.20	.13	.083	.041	.011

Source: From USDA Soil Conservation Service 1977.
Note: All values assume (1) random distribution of mulch or vegetation, and (2) mulch of appreciable depth where it exists. Idle land refers to land with undisturbed profiles for at least a period of three consecutive years. Also to be used for burned forest land and forest land that has been harvested less than 3 yr.
[a]Average fall height of water drops from canopy to soil surface.
[b]Portion of total area surface that would be hidden from view by canopy in a vertical projection (a bird's-eye view).
[c]G = cover at surface is grass, grasslike plants, decaying compacted duff, or litter at least 2 in. deep; W = cover at surface is mostly broadleaf herbaceous plants (as weeds with little lateral-root network near the surface), and/or undecayed residue.

of watershed sizes and slopes. However, it tends to overestimate sediment yield from small storms and to underestimate sediment yield from large storms.

Soil Loss and Conservation Planning

For conservation planning, the *soil loss tolerance* needs to be established. Soil loss tolerance, sometimes called *permissible soil loss*, is the maximum rate of soil erosion that will still permit a high level of crop productivity to be sustained economically and ecologically. Soil loss tolerance (T_e) values of 2.5–12.5 t/ha/yr are often used. The numbers represent the permissible soil loss where food, forage, and fiber plants are to be grown. Values are not applicable to construction sites but can be used for forest and other wildland sites.

A single T_e value is normally assigned to each soil series. A second T_e value can be assigned to certain kinds of soil where erosion has reduced the thickness of the effective root zone significantly, diminishing the potential of the soil to produce biomass over an extended period of time. The following criteria are used to assign T_e values to a soil series:

- An adequate rooting depth must be maintained in the soil for plant growth. For shallow soils overlying rock or other restrictive layers, it is important to retain the remaining soil; little soil loss is tolerated. The T_e should be less on shallow soils or those with impervious layers than for soils with good soil depth or for soils with underlying soil materials that can be improved by management practices.
- Soils that have significant yield reductions when the surface layer is removed by erosion are given lower T_e values than those where erosion has little impact on yield.

A T_e of 11.2 t/ha/yr has been used for agricultural soils in much of the United States. This maximum value has been selected for the following reasons:

- Soil losses in excess of 11.2 t/ha/yr affect the maintenance, cost, and effectiveness of water control structures, such as open ditches, ponds, and other structures affected by sediment.
- Excessive surface erosion is accompanied by gully formation in many places, causing added problems for tillage operations and increasing sedimentation of ditches, streams, and waterways.
- Plant nutrients are lost; Pimentel et al. (1995) estimated that erosion rates on U.S. agricultural croplands average about 17 t/ha/yr and that the cost of replacing the associated loss of nutrients is $30/ha/yr.
- Numerous practices are known that can be used successfully to keep soil losses below this maximum tolerable level.

After having established the soil loss tolerance, the USLE can be written as:

$$CP = \frac{T_e}{RK\,(LS)} \tag{7.12}$$

By choosing the right cropping management system and appropriate conservation practices, a value for the combined effect of C and P (or VM) that fits the equation can be established. To do so, it is helpful to consult the erosion index distribution curve of the area to select the most critical stages as far as rainfall erosivity is concerned.

Revised Universal Soil Loss Equation

The USLE was a state-of-the-art tool for predicting soil erosion when it was released for application. Subsequently, its use revealed significant weaknesses in terms of the conditions for which it was applicable and the nature of the results obtained. The USDA Agricultural Research Service and their cooperators initiated the development of the revised universal soil loss equation (RUSLE) in the 1980s to address these weaknesses by accounting for temporal changes in soil erodibility and plant factors that were not originally considered (Weltz et al. 1998). RUSLE has a snowmelt-erosion component and can be applied to single events. Improvements were also made to the rainfall, length, slope, and management practice factors of the original USLE model.

The RUSLE technology is computer based, replacing the tables, figures, and often tedious USLE calculations with keyboard entry. The improvements of the RUSLE over the USLE are detailed by Renard et al. (1997). In the most general terms, the RUSLE is a software version of a greatly improved USLE and draws heavily on the USLE data and documentation. Specific improvements in the RUSLE relative to the USLE include

(1) more data from different locations, for different crops and cropping systems, and for forest and rangeland erosion have been incorporated into the RUSLE; (2) corrections of errors in the USLE analysis of soil erosion have been made and gaps in the original data filled; and (3) increased flexibility of the RUSLE, which allows for predicting soil erosion for a greater variety of ecosystems and management alternatives.

The RUSLE has been replacing the USLE as a soil conservation planning tool in many areas, largely because of its improved accuracy and flexibility over the USLE. Until more powerful process-based tools and the data they require for application become readily available, Yoder and Lown (1995) suggest that the RUSLE model is one of the better soil erosion prediction technologies currently available.

Water Erosion Prediction Project Model

One effort to formulate a new generation of technologies for predicting soil erosion by water has been undertaken by the USDA Agricultural Research Service and their cooperators to develop the process-based water erosion prediction project (WEPP) model. The WEPP model has been developed to estimate soil erosion from single events, long-term soil loss from hillslopes, and soil detachment and deposition in small channels and impoundments within a small watershed (Weltz et al. 1998). The goal of the WEPP effort is a process-oriented model, or family of models, that are conceptually superior to the lumped RUSLE model and more versatile as to the conditions that can be evaluated. The WEPP technology is expected to replace RUSLE sometime in the future (Renard et al. 1997).

The WEPP model operates on a daily time-step, allowing the incorporation of temporal changes in soil erodibility, management practices, above- and below-ground biomass, litter biomass, plant height, canopy cover, and ground cover into the prediction of soil erosion on agricultural and rangeland watersheds. Linear and nonlinear slope segments and multiple soil series and plant communities on a hillslope are represented. The WEPP technology is intended to apply to all situations where erosion occurs, including those resulting from rainfall, snowmelt, irrigation, and ephemeral gullies. Once it has been thoroughly tested and becomes operational (Box 7.2), the model is intended to evaluate the impact of management practices (livestock grazing, cropping practices, soil conservation methods, etc.) on surface runoff, sedimentation, and crop or forage production.

Wind Erosion Equation

Prediction of soil erosion by wind is limited by the complexity of the processes and factors involved. However, researchers and managers have used a Wind Erosion Equation (WEE) originally proposed by Woodruff and Siddoway (1965) with its various modifications (Skidmore 1994) for the past 35 yr to obtain this prediction. The form of the WEE is

$$E = f(IKCLV) \tag{7.13}$$

where E = annual soil loss in tons per unit area; I = the soil erodibility index; K = the soil-ridge roughness factor; C = the climatic factor; L = the unsheltered median travel distance of wind across a field; and V = the equivalent vegetative cover.

The WEE is somewhat analogous to the more widely used USLE. Unlike the USLE, however, the wind erosion factors in Equation 7.13 are not multiplied together but require graphical and tabular solutions that consider the interactions of the process of wind erosion. Details on obtaining these solutions are presented by Hagen (1991), Skidmore (1994), and others.

Box 7.2

A Test of the Accuracy of the WEPP Technology

A Log-Pearson Type III (LP III) distribution was fitted to measured and WEPP-predicted soil loss values from runoff plots on 15 agricultural watersheds at six locations in the United States for periods ranging from 6 to 10 yr to test the accuracy of the WEPP technology in reproducing the statistical distributions of daily soil loss (Baffaut et al. 1998). Cumulative soil loss expressed as a function of storm recurrence interval was also used to show the relative contributions of large and small storms to the total soil loss at each site. The comparisons showed that both the measured and WEPP-predicted frequency curves fell within the 95% confidence range of the LP III distributions. This finding was true using either weather data from the site for the period of monitoring or synthetically generated weather series. Large storms contributed a major portion of the soil erosion for conditions where vegetative cover was high, but not necessarily under conditions of low cover. The results of this test were encouraging in terms of using WEPP to generate long-term daily soil loss for measured soil erosion.

Preventing Soil Erosion

Avoiding erosion-susceptible situations and inappropriate land uses in the first place is the most economical and effective means to combat soil erosion and to maintain the productivity of watersheds. The guiding rule is that the land user should consider carefully the principles of water and wind action in relation to each management decision, whether this concerns soil conservation techniques, water resource development, range management, forest management, or agriculture.

Situations that are particularly susceptible to soil loss include (1) sloping ground, particularly hills with shallow soils; (2) soils with inherently low permeabilities; and (3) sites where denudation of vegetation is likely.

Maintaining a vegetative cover is the best means of reducing erosion hazard, but the occurrence of an adequate cover cannot be relied upon in drylands (Box 7.3). Instead, the importance of reducing the energy of flowing water or wind suggests preventive measures that can be applied using simple land use practices. In preventing water erosion, the key is to maintain the surface soil in a condition that readily accepts water; the more water that infiltrates the soil, the better the chance of sustaining plant growth and reducing the erosive effects of surface runoff. Guidelines for preventing water and wind erosion are outlined in Table 7.5.

In areas where accelerated erosion is occurring, remedial soil conservation techniques will be necessary. These techniques should be designed to reduce the impact of the erosion processes and be applied first to those areas that have the highest production potential.

Controlling Soil Erosion

An understanding of the erosion processes suggests several broad actions that can be undertaken to control accelerated surface soil erosion. Actions that protect the soil surface against the energy of rainfall impact and increase the roughness of the surface, and

Box 7.3

Surface Erosion in Dryland Africa Increases Drastically after Removal of Vegetative Cover (Harrison 1987)

What protects dryland Africa's vulnerable soils from the erosive actions of water and wind is vegetation. Trees, shrubs, and herbaceous plants break the force of raindrops and hold the soil in place in the face of blowing wind. With respect to water, the roots of these plants plus the activities of the earthworms and termites they foster create thousands of pores and channels through which water can infiltrate into and percolate through the soil. But when the vegetative cover is removed, soil becomes exposed to the erosive power of water and wind. The increase in surface erosion after the removal of vegetation can be spectacular. In one series of studies, the annual rate of soil loss from forests was nil—a mere 30-200 kg/ha. However, annual losses from agricultural croplands was nearly 90 t/ha. From bare soil, a common situation on agricultural lands at the beginning of the rainy season, the rates of annual soil loss ranged from 10 to a massive 170 t/ha. Soil that formed over hundreds of years would wash away in only 1 yr at this latter rate of surface erosion.

Table 7.5. Guidelines for preventing water and wind erosion

Water Erosion

Avoid land use practices that reduce infiltration capacity and soil permeability,

Encourage grass and herbaceous cover of the soil for as long as possible each year,

Locate livestock watering facilities to minimize runoff production to water bodies,

Avoid logging and heavy grazing on steep slopes,

Conduct any skidding of logs on steep slopes in upward directions to counteract drainage concentration patterns,

Lay out roads and trails so that runoff is not channelized on steep, susceptible areas,

Apply erosion control techniques on agricultural fields and promote infiltration, and

Remember that the more water that goes into the soil, the better is the chance of sustaining plant growth and reducing the erosive effects of surface runoff.

Wind Erosion

Avoid uses that will lead to the elimination of shrubs and trees over large areas,

Avoid locating livestock watering facilities on erodible soils,

Protect agricultural fields and heavy use areas with shelterbelts,

Manage animals and plants in your area to maintain a good balance between range plants, woody trees, and shrubs, and

When planting shrubs and trees on grazing lands, locate and space them to reduce wind velocity.

thus the tortuosity of the flow path, reduce the energy of rainfall and surface runoff. Mechanical treatments that shorten the slope length and reduce the slope inclination lessen the energy of overland flow and can reduce the quantity and velocity of surface runoff. Any actions that prevent the channelization of surface runoff will reduce the opportunity for gully formation. Often, strips of vegetation perpendicular to the slope can slow and reduce surface runoff.

When controlling wind erosion, actions should be taken that reduce the length of fetch to reduce the momentum of wind and that increase soil cohesiveness or armor the soil surface to prevent the lifting of soil particles by wind. The key is to reduce wind velocity near the ground and deflect the wind direction.

The most effective techniques are those that combine several of these actions. For example, nearly all the actions can be accomplished with an effective vegetation cover. In addition to reducing rainfall impact, increasing surface roughness, and thus reducing the velocity of surface runoff, maintenance of a vegetation cover can improve infiltration rates. Soil erodibility is decreased by the activity of roots and the improvement of soil structure by the addition of organic matter. Also, evapotranspiration by vegetation reduces soil water content between rainfall events and, consequently, provides more storage for rainfall and lessens runoff.

Surface Erosion Control on Forestlands

A minimal amount of surface erosion is expected in natural forests (Fig. 7.6). Undisturbed forests rarely experience erosion rates in excess of 0.04 t/ha/yr. Activities that remove vegetative cover and, most important, expose mineral soil lead to high rates of surface erosion. Silvicultural treatments in which the destruction of lesser vegetation and litter accumulations is minimized will help to control erosion by reducing the raindrop impact on a soil surface and by maintaining high infiltration rates. Residual strips of vegetation alternated with clearcuts and aligned perpendicularly to the slope can function as barriers to flowing water and the downslope movement of soil particles. Retaining strips of vegetation can also be employed to protect channel banks and streambeds during timber-harvesting operations. However, leaving residual strips of vegetation can have minimal effect in controlling surface erosion in mountainous watersheds because of rapid channeling of surface runoff.

The nature of the logging operation used in timber harvesting can affect the magnitude of surface erosion on a watershed. Watersheds undergoing logging operations and associated road development often have erosion rates in excess of 15 t/ha/yr. Logging operations that have minimal effect on the compaction and disturbance of surface soils should be favored. Small cable systems can be used to remove felled timber on sites where tractors would cause excessive soil disturbance, for example, on slopes of > 30–35%. Other methods, such as cable systems with intermediate supports to attain the necessary lift and extend the yarding distance, can be utilized on flatter terrain. Also, double-tired low-ground-pressure vehicles with torsion suspension might replace crawler tractors. Limiting timber-harvesting operations to the dry season, adjusting them to the soil type, and minimizing the disturbance to the litter layer will lessen soil compaction and consequent surface erosion. A variety of Best Management Practices (BMPs) have been devised to cope with surface erosion from timber-harvesting activities (Martin and Hornbeck 1994, NCASI 1994).

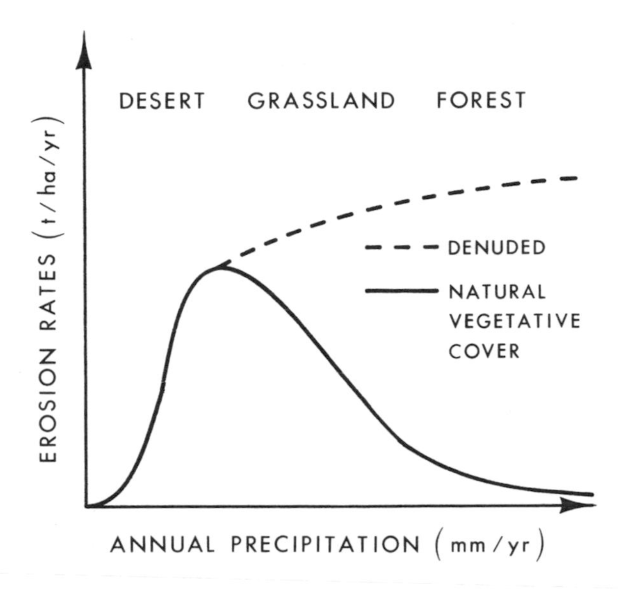

FIGURE 7.6. Relationship between erosion rates and annual precipitation for vegetation types and watershed cover conditions (modified from Hudson 1981 and others).

Establishing and maintaining a vegetative cover to protect cutbanks and fill slopes along roads, on landings in timber-harvesting operations, and in other critical areas of exposed mineral soil will, in many instances, help to control surface erosion. Individual plant species have different values for erosion control, different site requirements, and different cultural requirements. Therefore, knowledge of available plant species for erosion control is important. Guides that specify plant species, site preparation, and seeding or planting techniques for local areas in the United States are available from federal and state land management agencies.

Most erosion problems on forest lands are attributed to improper road and skid trail design, location, and layout. Roads and trails are necessary for many activities on watersheds, but their potential impact upon erosion usually exceeds that of all other management activities considered. Erosion rates from road construction sites commonly exceed 95 t/ha/yr. No other activity results in such intensive and concentrated soil disturbance.

Many potential erosion problems can be eliminated in the planning stage, before road or trail construction. Emphasis should be placed on the development of a system, not a randomly located network. A thoughtfully devised system of all feeder and hauling roads will minimize the amount of mineral soil exposed. Good planning will also minimize the investment in, and maintenance of, the system. One of the more important decisions to be made in planning is the allowable width and grade of the roads, as these standards will affect the area of disturbance within a watershed.

Some erosion will result from roads and trails no matter how careful the layout; therefore, the objective is to minimize the magnitude of this erosion. To do so, a road or trail system should be located to minimize the extent of exposed soil and disturbed, unstable areas on a watershed; also, the system should be kept away from stream channels to the extent possible. A satisfactory location is the product of both considerations. Steep gradients should be avoided because roads or trails on these sites often are less stable than on lesser slopes, and they expose more soil due to excessive cut-and-fill requirements. Also, steeper slopes concentrate water more quickly and impart a higher velocity to the flow than less steep slopes. However, sufficient slope is needed at each location to facilitate drainage.

Other factors to consider in road construction are culvert size, spacing, and maintenance; adequate compaction of fill materials; and minimization of the amount of sidecast materials. Roadside ditch turnouts interrupt flow and divert runoff onto adjacent land. Temporary roads and trails should be stabilized immediately after use by seeding grasses or herbaceous cover mixtures and appropriate follow-up management. The USDA Forest Service and other agencies have established construction guidelines and BMPs that, when followed, should reduce the chances of serious erosion problems.

Megahan (1977), Burroughs and King (1989), and others have identified basic principles to follow in reducing impacts of road construction on erosion and sedimentation. These principles include: (1) keep the area of roads on a watershed to a minimum by reducing mileage and disturbance, (2) do not locate roads in high-erosion-hazard areas, (3) apply erosion control measures in areas that are disturbed by road construction, and (4) reduce sediment delivery from roads to stream channels.

Surface Erosion Control on Rangelands

The greatest amount of erosion worldwide occurs in dry regions (Fig. 7.6). These regions are generally too dry for productive rain-fed agriculture; commonly, grazing is the only economical use of the land. In many developing countries, grazing is uncontrolled and excessive. The establishment of appropriate range management practices must be a priority under these conditions; otherwise, any other soil conservation practice will fail. Often, simply controlling livestock density and the grazing practice is sufficient to restore depleted and eroding rangelands. The key is to maintain a healthy and extensive vegetative cover and not to reduce infiltration capacities of rangelands, which if overgrazed, are characterized by low plant density, compacted soils, surface runoff, and excessive erosion.

Fire is commonly used as a management tool to increase forage production of rangelands by removing woody vegetation. Controlled burns, if properly managed, should not adversely impact the hydraulic properties of soils (DeBano et al. 1998). Uncontrolled fires can reach temperatures that are high enough to reduce infiltration capacities. In any event, fires leave large areas of exposed mineral soil, which are vulnerable to rainfall impact, surface runoff, and erosion.

If rangelands are in poor condition, reseeding and other measures can be necessary. Reseeding may require temporary mechanical treatments to conserve water and, thereby, aid vegetation establishment. Control of weed species can also be necessary.

Vegetative Measures

Reseeding is expensive and can be justified economically only if it provides returns in forage and erosion protection over many years. Successful reseeding efforts depend upon rainfall patterns and site conditions and require the following conditions:

- Soils and precipitation must be adequate to provide a reasonable probability for success. Higher precipitation is usually required to establish plants on clay soils than on more sandy soils.
- Indigenous or site-adapted species must be used. Selecting adapted ecotypes within a species also can provide an extra measure of success on some sites.
- Seedbeds should be prepared to control unwanted vegetation and provide rapid infiltration. A loose, irregular-surface seedbed is often desirable.
- An adequate quantity of seed is needed to ensure a successful stand of vegetation, but not to the point of being wasteful. Seeding approximately 2–3 million viable seeds per hectare is a rough estimate of proper seeding intensity.
- Seeds must be planted at the proper depth. The smaller the seed, the shallower it should be seeded. Broadcasting small seeds on loose seedbeds frequently is adequate, but drilling is recommended for large seeds and on crusted seedbeds.
- Seeding should occur when favorable moisture and temperature conditions combine to give the longest possible period for germination and early growth.
- Seedlings should be protected from grazing until they have become well established; this can require three or more growing seasons on dry sites.
- Site preparation by mechanical techniques can be needed to prepare seedbeds and conserve water.

These techniques are costly and in many cases cannot be justified economically.

Mechanical Techniques

Because they are expensive, mechanical treatments to improve rangelands can be justified only if runoff and sediment from the watershed threaten important downstream developments, if reclamation is essential to the survival of people in the area who have no alternative means of livelihood, or if the value of the increased production equals or exceeds the cost of treatment.

The purpose of mechanical treatments is to reduce surface runoff and soil loss by retaining water on the site until a vegetative cover can become established. Some of the more common treatments include contour furrows, contour trenches, fallow strips, pits, and basins. The type of treatment chosen depends upon the runoff potential of the site. Often, a combination of treatments may be desirable, but planting should follow treatment as soon as possible.

Mechanical treatments have a life expectancy limited by the amount of runoff and sediment produced on the site. Therefore, it is important to evaluate the site for its potential for maintaining a vegetative cover after the treatment has lost its effectiveness. A livestock management strategy must usually be formulated and enforced if the treatment is to have a lasting effect.

Contour Furrows. Contour furrows are constructed to break slope length and provide depression storage for surface runoff. They are small ditches 20–30 cm deep that follow the contour; they usually are constructed with a single-blade, furrowed plow that forms a berm on the downslope side. Furrows that form miniature terraces that hold the water in place until it infiltrates the soil are suitable for plant establishment and usually are seeded after construction. Studies indicate that furrows are effective if the spacing between them is < 2 m. At greater spacings, there is little effect except along the furrows.

Many early trials with furrows failed because of the difficulty of following the contour and the lack of seeding and follow-up maintenance. If not placed along contours, furrows become drainage ditches that concentrate runoff and can accelerate erosion rather than prevent it. Following the contour is a difficult process and one of the disadvantages of the method. The effectiveness of furrows can be increased and the misalignment of furrows with the contours can be corrected somewhat by constructing crossbars across the furrows at intervals of 1.5–10 m. The furrows then become small basins, and water from adjacent sections will be held in place if one section should break.

Contour Trenches. Trenches are nothing more than large furrows and usually are required on slopes too steep for contour furrows. There are two types: a shallow outside type (for slopes up to 70%) where the excavated material forms the barrier to overland flow; and the deeper inside type where the excavation retains most of the overland flow. Both types are expensive, usually require machinery, and must be designed to handle large stormflows. The failure of an upper trench could result in a domino effect on trenches downslope. If such a situation occurs, the resulting erosion can be much greater than that which would have occurred in the first place.

Fallow Strips. Revegetated strips have proven successful on level to gently rolling land to break the slope length until desirable vegetation can become established. The strips usually are about 1 m wide and parallel to the contour. Initially, the strips are cultivated to destroy unwanted vegetation, loosen the soil surface, and prepare a seedbed. The original vegetation is left between the strips until the newly planted vegetation has become established. New strips then are tilled and planted and the process continued until the range has been rehabilitated. Conventional tillage equipment can be used for this technique.

Pitting. Pitting is a technique of digging or gouging shallow depressions into the soil surface to create depression storage for surface runoff. The treatment of extensive areas requires a tractor and pitter. A conventional disk plow can be modified into a pitter, although a special disk plow is made for range use. Every alternate disk can be removed and the axle holes of the remaining disks offset from the center. Alternatively, rather than offsetting, a half-moon section can be cut from each of the remaining disks. The disks are arranged to strike the surface at different times, creating an alternate pattern of pits as the plow is pulled across the surface. The furrow is broken by the missing disk or by the missing section of the cut disk, which then forms a discontinuous furrow. A standard heavy 51 cm disk plow with holes 8 cm off center will produce pits about 20–30 cm wide by 45–60 cm long, 15 cm deep, and 40 cm apart. Under favorable conditions of soil type, soil moisture, and machine weight, the pits will have a storage capacity of about 0.013 m^3. The number of pits required per hectare can be estimated by dividing the estimated runoff produced from a design storm by the storage capacity per pit. Design storms are established from criteria set by respective agencies with management responsibilities. Pitting is effective on slopes of up to 30%, and it has been estimated that the technique can produce as much as four times the amount of forage produced on untreated areas.

Basins. Basins are larger pits, usually about 2 m long, 1.8 m wide, and 15–20 cm deep. They store a greater amount of water and can help create pockets of lush vegetation. Basins generally are more costly to construct and are not as widely used as pitting methods.

SUMMARY

Erosion of the soil by water can occur as surface erosion, gully erosion, and soil mass movement from upland watersheds. After completing this chapter, you should have a basic understanding of soil surface erosion and should be able to:

- Describe the process of detachment of soil particles by raindrops and transport by surface runoff. How does this differ for wind erosion?
- Explain how land use practices and changes in vegetative cover influence the processes of soil detachment described above.
- Compare the advantages and limitations of the USLE, RUSLE, and WEPP models as predictors of soil loss from the actions of water.
- Explain and apply the modified soil loss equation (MSLE) in estimating soil erosion under different land use and vegetative cover conditions.
- Explain how surface erosion can be controlled, or at least maintained at acceptable levels. What types of watershed management practices and guidelines are appropriate on forest lands? On rangelands?

REFERENCES

Agassi, M., ed. 1995. *Soil erosion, conservation, and rehabilitation*. New York: Marcel Decker.

Baffaut, C., M.A. Nearing, and G. Govers. 1998. Statistical distributions of soil loss from runoff plots and WEPP model simulations. *Soil Sci. Soc. Amer*. 62:756–763.

Barrow, C.J. 1994. *Land degradation*. Cambridge, UK: Cambridge University Press.

Burroughs, E.R., Jr., and J. King. 1989. *Reduction of soil erosion on forest roads*. USDA For. Serv. Gen. Tech. Rep. INT-264.

Clyde, C.F., C.E. Israelsen, and P.E. Packer. 1976. Erosion during highway construction. In *Manual of erosion control principles and practices*. Logan: Utah Water Research Laboratory, Utah State Univ.

DeBano, L.F., D.G. Neary, and P.F. Ffolliott. 1998. Fire's effects on ecosystems. New York: John Wiley & Sons, Inc.

Dissmeyer, G.E., and G.R. Foster. 1980. *A guide for predicting sheet and rill erosion on forest lands*. USDA Tech. Pub. SA-TP 11. Atlanta, GA.

———. 1985. Modifying the universal soil loss equation for forest lands. In *Soil erosion and conservation*, ed. S.A. El Swaify, W.L. Moldenhauer, and A. Lo, 480–495. Ankeny, IA: Soil Conservation Society of America.

Dyer, E.B. 1977. Use of the universal soil loss equation: A river basin experience. In *Soil erosion: Prediction and control*, ed. G.R. Foster, 292–297. Ankeny, IA: Soil Conservation Society of America.

Fogel, M.M., L.H. Hekman, and L. Duckstein. 1977. A stochastic sediment yield model using the modified universal soil loss equation. In *Soil erosion: Prediction and control,* ed. G.R. Foster, 226–233. Ankenny, IA: Soil Conservation Society of America.

Foster, G.R., and W.H. Wischmeier. 1973. *Evaluating irregular slopes for soil loss prediction.* Am. Soc. Agri. Eng. Pap. 73227.

Frickel, D.G., L.M. Shown, R.F. Hadley, and R.F. Miller. 1981. *Methodology for hydrologic evaluation of a potential surface mine: The Red Rim site, Carbon and Sweetwater Counties, Wyoming*. USGS Survey—Water-Resources Investigation, Open File Rep. 81B75.

Guy, H.P. 1964. *An analysis of some storm-related variables affecting stream sediment transport*. U.S. Geol. Surv. Prof. Pap. 462-E.

Hagen, L.J. 1991. A wind erosion prediction system to meet user needs. *J. Soil Water Conserv*. 46:106–111.

Harrison, P. 1987. *The greening of Africa: Breaking through in the battle for land and food*. New York: Penguin Books.

Hudson, N.W. 1981. *Soil conservation*. 2nd ed. New York: Cornell Univ. Press.

Jackson, W.L., K. Gebbardt, and B.P. Van Haveren. 1986. Use of the modified universal soil loss equation for average annual sediment yield estimates on small rangeland drainage basins. In *Drainage basin sediment delivery*, ed. R.F. Hadley, 413–422. Int. Assoc. Hydrol. Sci. Publ. 159. Washington, DC: IAHS.

Johnson, A.W. 1961. Highway erosion control. *Am. Soc. Agric. Eng. Trans*. 4:144–152.

Khanbilvardi, R.M., A.S. Rogowski, and A.C. Miller. 1983. Predicting erosion and deposition on a stripmined and reclaimed area. *Water Resour. Bull*. 19:585–593.

Martin, C.W., and J.W. Hornbeck. 1994. Soil disturbance by logging in New England. *Northern J. Applied For*. 11:17–23.

Megahan, W.F. 1977. Reducing erosional impacts of roads. In *Guidelines for watershed management*, ed. S.H. Kunkle and J.L. Thames, 237–261. FAO Conserv. Guide 1. Rome.

Morgan, R.P.C. 1995. *Soil erosion and conservation*. Essex, England: Longman.

Murphree, C.E., C.K. Mutchler, and L.L. McDowell. 1976. Sediment yields from a Mississippi delta watershed. In *Proceedings of the Third Interagency Sedimentation Conference,* 1–99 to 1–109. PB-245-100. Washington, DC: Water Resource Council.

NCASI. 1994. *Forests as nonpoint sources of pollution and effectiveness of best management practices*. Natl. Coun. Air and Stream Improve. Tech. Bull. 672.

Osborn, H.B., J.R. Simanton, and K.G. Renard. 1977. Use of the universal soil loss equation in the semiarid Southwest. In *Soil erosion: Prediction and control*, ed. G.R. Foster, 19–41. Ankeny, IA: Soil Conservation Society of America.

Pimentel, D., C. Harvey, P. Resosudarmo, K. Sinclair, D. Kurz, M. McNair, S. Crist, L. Shpritz, L. Fitton, R. Saffouri, and R. Blair. 1995. Environmental and economic costs of soil erosion and conservation benefits. *Science* 267:1117–1123.

Renard, K.G., and G.R. Foster. 1985. Managing rangeland soil resources: The Universal Soil Loss Equation. *Rangelands* 7:118–122.

Renard, K.G., G.R. Foster, D.C. Yoder, and D.K. McCool. 1994. RUSLE revisited: Status, questions, answers, and the future. *J. Soil Water Conserv*. 49:213–220.

Renard, K.G., G.R. Foster, G.A. Weeies, D.K. McCool, and D.C. Yoder. 1997. *Predicting soil erosion by water: A guide to conservation planning with the revised universal soil loss equation (RUSLE)*. USDA Washington, DC: Agric. Handb. 703.

Savabi, M.R., D.C. Flanagan, B. Hebel, and B.A. Engel. 1995. Application of WEPP and GIS-GRASS to a small watershed in Indiana. *J. Soil Water Conserv*. 50:477–483.

Shown, L.M., D.G. Frickel, R.F. Miller, and F.A. Branson. 1982. *Methodology for hydrologic evaluation of a potential surface mine: Loblolly Branch Basin, Tuscaloosa County, Alabama*. USGS Survey, Water-Resources Investigation, Open File Rep. 82–50.

Singer, M.J., G.L. Huntington, and H.R. Sketchley. 1977. Erosion prediction on California rangelands. In *Soil erosion: Prediction and control*, ed. G.R. Foster, 143–151. Ankeny, IA: Soil Conservation Society of America.

Skidmore, E.L. 1994. Wind erosion. In *Soil erosion research methods*, ed. R. Lal, 265–293. Delray, FL: St. Lucie Press.

Stephens, H.V., H.E. Scholl, and J.W. Gaffney. 1977. Use of the universal soil loss equation in wide-area soil loss surveys in Maryland. In *Soil erosion: Prediction and control,* ed. G.R. Foster, 277–282. Ankeny, IA: Soil Conservation Society of America.

Toy, T.J., and W.R. Osterkamp. 1995. The applicability of RUSLE to geomorphic studies. *J. Soil Water Conserv*. 50:498–503.

Troch, F.R., J.A. Hobbs, and R.L. Donahue. 1999. *Soil and water conservation*. 3rd ed. Englewood, NJ: Prentice Hall.

USDA Forest Service. 1980. *An approach to water resources evaluation, nonpoint silvicultural sources*. Environ. Res. Lab., EPA-600/880012. Athens, GA: Environmental Protection Agency.

USDA Soil Conservation Service. 1977. *Preliminary guidance for estimating erosion on area disturbed by surface mining activities in the interior western United States.* Interim Final Rep., EPA-908/4-77-005.

Weltz, M.A., M.R. Kidwell, and H.D. Fox. 1998. Influence of abiotic and biotic factors in measuring and modeling soil erosion on rangelands: State of knowledge. *J. Range Manage.* 51:482–495.

Williams, J.R. 1975. Sediment yield prediction with universal equation using runoff energy factor. In *Present and prospective technology for predicting sediment yields and sources,* 244–252. USDA-ARS-S-40.

Wischmeier, W.H. 1975. Estimating the soil loss equation's cover and management factor for undisturbed areas. In *Present and prospective technology for predicting sediment yields and sources,* 118–124. USDA-ARS-S-40.

Wischmeier, W.H., and D.D. Smith. 1965. *Predicting rainfall-erosion losses from cropland east of the Rocky Mountains.* USDA Agric Handb. 282.

———. 1978. *Predicting rainfall-erosion losses—A guide to conservation planning.* USDA Agric Handb. 537.

Wischmeier, W.H., C.B. Johnson, and B.V. Cross. 1971. A soil erodibility nomograph for farmland and construction sites. *J. Soil Water Conserv.* 26:189–193.

Woodruff, N.P., and F.H. Siddoway. 1965. A wind erosion equation. *Soil Sci. Society of America Proc.* 29:602–608.

Yoder, D., and J. Lown. 1995. The future of RUSLE: Inside the new Revised Universal Soil Loss Equation. *J. Soil Water Conserv.* 50:484–489.

CHAPTER 8 *Gully Erosion and Soil Mass Movement*

INTRODUCTION

This chapter discusses the processes and control of gully erosion, the recognition of hazardous situations that can create massive earth movements, and the control of wind-transported sands. A fluvial system, of which a gully is a special case, is the product of interaction among many variables: relief above a base level, climate, lithology, the area and shape of the drainage basin, hillslope morphology, soils, vegetation, human and animal activity, the slope of a channel, channel pattern and roughness, the bed load it moves, and its discharge of water and sediment. Whenever any of these variables are changed, the others (except such independent variables as climate and lithology) also shift in response to the altered system. Basically, gully erosion occurs when the force of concentrated flowing water exceeds the resistance of the soil over which it is flowing. Gully stabilization occurs in a natural system moving toward a state of dynamic equilibrium.

Gullies are severe stages of water erosion. Steep mountain watersheds are particularly prone to *mass wasting* and gully cutting. Mass wasting is the movement of massive amounts of soil by gravity and is one of several natural processes that have occurred to a greater or lesser extent long before the advent of humans. However, poor land use practices and poorly planned development activities can accelerate erosion processes. Gullies, for example, can be created within a small, badly managed farm unit or by excessive livestock grazing. Development activities, such as road building, often trigger massive earth movements. The greatest economic loss from these processes frequently does not occur in the area of origin but downstream, downwind, or downslope from the problem area. Sediment from severely gullied uplands can bury fertile bottomland soils, sand dune migration can cover more productive lands or human developments, and massive earth movements can cause destruction, sometimes of human life.

Gully Erosion

A gully is a relatively deep, recently formed channel on valley sides and floors where no well-defined channel previously existed. The flow in these channels is almost always ephemeral. Gully development can be triggered by mass wasting or tectonic movements causing a change in base level. Base levels are the longer-term topographic elevations on a landscape that are in dynamic equilibrium with their environment. More often, gullies occur on areas that have a low-density vegetative cover and highly erodible soils. Development of new gullies or the rapid expansion and deepening of older gullies can often be traced to removal of vegetative cover through some human activity.

A gully develops when surface runoff is concentrated at a *nickpoint*, which is an area where an abrupt change of elevation and slope gradient and a lack of protective vegetation occur (Fig. 8.1). The fall of water over this nickpoint causes it to be undermined and to migrate uphill (headcuts). Simultaneously, the force of the falling water dislodges sediment below the fall and transports it downhill, lengthening and deepening the gully in the downhill direction (downcuts). In a *discontinuous gully*, where both processes are equally important, the gully bed has a stairstep configuration. A *continuous gully* generally gains depth rapidly from the headcut and then maintains a relatively constant gradient to the mouth, where the most active changes take place. Frequently, a series of discontinuous gullies will coalesce into a continuous gully (Box 8.1).

If conditions on a watershed are not improved, the gully will continue to deepen, widen, and lengthen until a new equilibrium is reached. The process of deposition can then begin at the gully mouth and proceed upstream until the slope of the gully sides and bottom is shallow enough to permit vegetation to become established. Plants growing in a gully signal the end of its active phase, barring disequilibration of its downcutting and depositional forces.

Gully erosion is most prevalent in drylands and in disturbed humid areas where soil compaction and vegetation removal have resulted in surface runoff. However, subsurface flow can also dissolve, dislodge, and transport soil particles. When large subterranean voids are present in the soil, matric flow can be minimal. When subsurface flow becomes turbulent (as opposed to laminar, matric flow), it is called *pipeflow*. Although common in humid, deep-soil areas (particularly in old tree root cavities, animal burrows, etc.), pipeflow is also common in dryland regions. In some cases, pipes can reach diameters of > 1 m. Soil pipes can form and grow in diameter until the soil above them collapses. This process can lead to the formation of gullies, which most likely results in greater soil erosion than the actual pipeflow itself.

Discontinuous and Continuous Gullies

Gullies can be classified as *discontinuous* or *continuous*. As described by Heede (1976, p. 4), "Discontinuous gullies can be found at any location on a hillslope. Their start is signified by an abrupt headcut. Normally, gully depth decreases rapidly downstream. A fan forms where the gully intersects the valley." Discontinuous gullies can occur singly or in a system of downslope steps in which one gully follows the next. Fusion with a tributary can form a continuous system. Gullies can also become captured by a continuous stream when shifts of streamflow on an alluvial fan divert flow from a discontinuous gully into a parallel secondary gully. At the point of overflow, a headcut that advances upstream into the discontinuous channel will develop. Here, it will form a

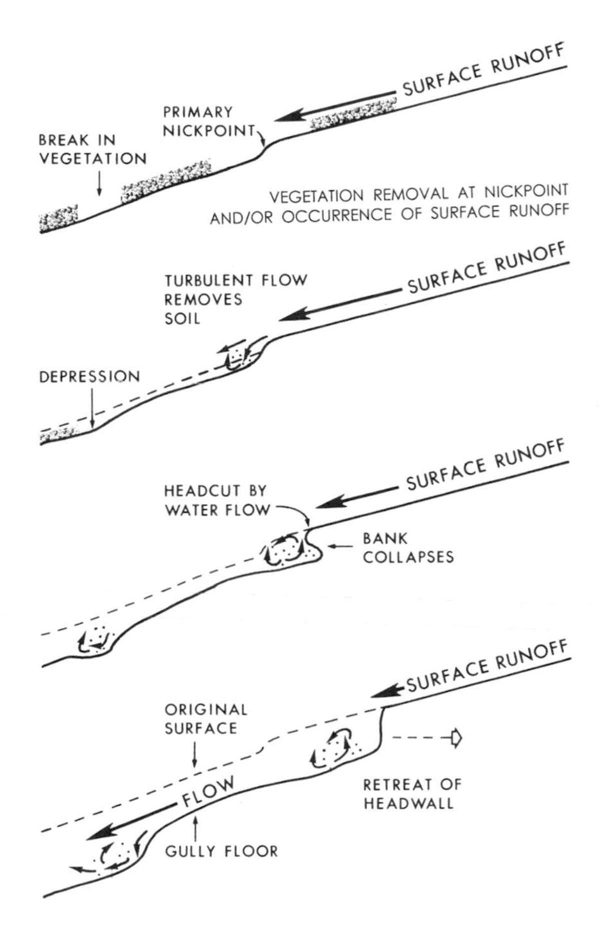

FIGURE 8.1. Illustration of gully formation and headwall retreat over time (adapted from Heede 1967 and Harvey et al. 1985).

nickpoint and intercept all flow; the gully deepens with the upstream advance of the nickpoint.

Heede (1976, p. 8) describes the formation of continuous gullies as beginning "with many fingerlike extensions into the headwater area. It gains depth rapidly in the downstream direction and maintains approximately this depth to the gully mouth. Continuous gullies nearly always form systems (stream nets)." Continuous gullies are found in different vegetation types but are prominent in dryland regions. Removal of any vegetative cover can lead to gully and gully stream net formation if topography and soils are conducive to gully initiation.

At an early stage, gully processes proceed toward the attainment of dynamic equilibrium. As gullies become older, they become more like a river or normal stream. Active gully development can be reinitiated by changes in runoff or vegetative cover and by physical changes in the landscape such as land uplift. Gully development is not necessarily an orderly process that proceeds from one condition to the next-advanced one.

Box 8.1

Use of Field Surveys and Aerial Photo Interpretation to Estimate Gully Erosion and Sources of Sediment (from Stromquist et al. 1985)

Interpretation of aerial photos at a 1:30,000 scale and field surveys were used to identify sediment sources, transport mechanisms, and storage elements to help estimate erosion rates from a 6.15 km^2 watershed in southwestern Lesotho (see the figure below). Gully-eroded areas increased by 200,000 m^2 and produced about 300,000 t of sediment from 1951 to 1961. From 1961 to 1980, the gully-eroded areas increased by 60,000 m^2 and contributed about 90,000 t of sediment.

Sediment production at the reservoir site was estimated to be 300,000 t from gully erosion and about 80,000 t from surface erosion. The volume of sediment at the reservoir at full supply level was 267,000 m^3, which corresponds to 360,000 t of trapped sediment.

Many of the discontinuous gullies became continuous gullies from 1951 to 1961. This rapid gully expansion was explained partly by unusually high rainfall rates from 1953 to 1961, which averaged 100 mm/yr more than normal. This period was preceded by an extreme drought beginning in 1944, throughout which the loss of vegetative cover made the area susceptible to erosion during the high-rainfall period of 1953 to 1961.

The use of aerial photography helped quantify the loss of productive area and, as used in this study, helped identify sources of sediment. In this case, only a fraction of the watershed contributed the bulk of sediments to the reservoir site.

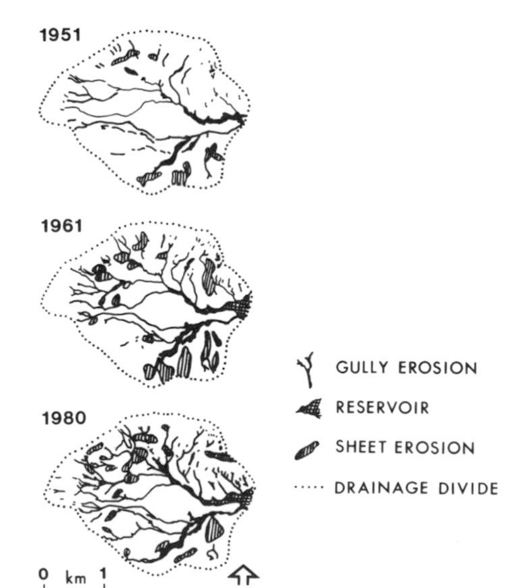

Gully and sheet over time on a 6.15 km^2 watershed in southwestern Lesothos (Stromquist et al. 1985, © *S. African Geol. J.,* by permission).

Gully Control

Gully erosion is the result of two main processes: *downcutting* and *headcutting.* Downcutting is the vertical lowering of the gully bottom and leads to gully deepening and widening. Headcutting is the upslope movement that extends the gully into headwater areas and increases the number of tributaries. Gully control must stabilize both the channel gradient and channel nickpoints to be effective.

Once it has been allowed to develop, gully erosion is difficult and expensive to control. However, severely gullied lands can threaten valuable on-site agricultural fields or cultural improvements, such as buildings or roads. Gullied watersheds can also be a source of sediment and floodwater that threaten the production of valley farmland or the effective lifetime of irrigation works and reservoirs. In cases such as these, extensive gully control projects should be undertaken. Even then, the costs of control should be weighed carefully against probable benefits.

Permanent gully control can be obtained only by returning the site to a good hydrologic condition. Obtaining this condition usually requires the establishment and maintenance of an adequate cover of vegetation and plant litter, not only on the eroding site but also on the area where runoff originates. Expensive mechanical structures in the form of check dams and lined waterways can be necessary to stabilize a gully channel temporarily to allow vegetation to become established. These structures, however, require continual monitoring and maintenance over time in order to assure their initial effectiveness. The time required for revegetation varies from 1 to 2 yr in humid areas and up to 20–25 yr in dryland environments; therefore, the structures have to be built accordingly. In no case should mechanical structures be considered an end in themselves or permanent solutions, no matter how well they are constructed.

Prioritizing Gully Control Treatments

Generally, the funds for controlling gullies are limited, and therefore, a procedure is needed to give the highest return for the least investment in gully control activities. A gully treatment strategy developed by Heede (1982) uses physical factors and parameters to establish priorities for treatment (Box 8.2). Effective treatment is based largely upon whether the gullies are continuous or discontinuous, and where the most critical points are located. The critical locations in continuous gullies are at the gully mouth, where (unless an unusual flow deepens the channel) deposition continues and the gully widens. This channel widening creates an inherently unstable situation. Discontinuous gullies have two critical locations: the headcut at the upstream end and the alluvial fan formed below the gully mouth. Classifying gullies into continuous or discontinuous types not only allows determination of network types but also identifies critical locations that must be considered when designing treatments.

Objective of Controls: Vegetation Establishment and Mechanical Structures

The long- and short-term objectives of gully control must be recognized because sometimes reaching the long-term goal of revegetation directly is difficult, particularly in areas of low rainfall. Stabilization of the gully channel is the first objective.

Where a vegetative cover can be established, channel gradients can sometimes be stabilized without resorting to mechanical or engineering measures. However, vegetation

Box 8.2

Procedure for Analyzing Continuous and Discontinuous Gully Networks to Devise a Prioritized Treatment Plan (from Heede 1982)

The five steps in the procedure are:

1. Determine the type of network based on gully types.
2. Determine the stream ordering of the network gullies.
3. Tally the tributaries of each gully.
4. Analyze the stage of development of each gully.
5. Rank treatment priorities.

Aerial photos are useful for determining the gully networks and the relationship of individual gullies to each other in step 1. Ordering of gullies in step 2 can be done by commonly used stream-ordering techniques. Ordering of gullies, like streams, is the arrangement of channel segments in which first-order channels are the smallest unbranched tributaries; second-order channels are initiated by the confluence of two first-order channels; and so forth. The individual gullies can be classified into youthful, mature, and old in step 4. Ranking the streams in step 5 utilizes the information gained in the first four steps and prioritizes gullies into a network hierarchy based on the overall situation. The number of tributaries, stage of development, and expected treatment returns for the total network must be used in the final ranking. The potential for vegetative rehabilitation of the watershed must also be considered in the final ranking.

alone can rarely stabilize headcuts because of concentrated flow at these locations. Vegetation types that grow rapidly and establish a high plant density and deep, dense root systems are most effective. Tall grasses that lie down on the gully bottom under flow conditions provide a smooth interface between flow and original bed and can increase flow velocities; these plants are not suitable for gully stabilization. The higher flow velocities with such plant cover can widen the gully even though the gully bottom is protected. Trees and shrubs can restrict high flow volumes and velocities and cause diversion against the bank. Where diverted flows are concentrated, new gullies can develop, and new headcuts can form where the flow reenters the original channel. However, on low gradients and in wide gullies, especially at the mouth of such gullies, trees and shrubs can be planted to form live dams to build up sediment deposits by reducing flow velocities.

If climate or site conditions do not permit the establishment of vegetation, mechanical measures or control structures will be required. Structures are usually required at critical locations along a gully channel such as nickpoints on the gully bed, headcuts, and gully reaches close to the gully mouth. At the gully mouth, changes in flow cause frequent changes in deepening, widening, and deposition. Normally, critical locations are identified in the field.

An effective control structure design must help vegetation to become established and survive. Once the gully gradient is stabilized, vegetation can become established on the gully bottom; stabilized gully bottoms will then lead to the stabilization of banks because the toe of the gully side slopes is at rest (Heede 1976). Gully banks that are too steep for vegetation establishment can be stabilized more quickly by physical sloughing. The gully bottom should be stable before banks are sloughed.

Vegetation can be established rapidly if substantial deposits of sediment accumulate in the gully above control structures. Such deposits can store soil moisture and decrease channel gradients. The net effect of vegetation establishment in the channel and the reduced channel gradient is a decrease in peak discharge.

Gullies that are undergoing active headcutting and downcutting are difficult to stabilize and revegetate. Most often, mechanical treatments are needed to provide the short-term stability necessary for vegetation establishment. Other considerations such as soil type, rate of sedimentation, hydraulics, and the logistics needed to manage a watershed also enter into the development of a solution. For example, if gullies are wide and deep, the construction of large dams across the gullies to accumulate sufficient sediment is necessary to allow equipment to move across the gully. Large check dams may be undesirable or uneconomical. Other mechanical or structural controls must then be considered.

Check Dams. A check dam is a barrier placed in an actively eroding gully to trap sediment carried down the gully during periodic flow events. Sediment backed up behind the check dam (1) develops a new channel bottom with a gentler gradient than the original gully bottom and hence reduces the velocity and the erosive force of gully flow; (2) stabilizes the side slopes of the gully and encourages their adjustment to their natural angle of repose, reducing further erosion of the channel banks; (3) promotes the establishment of vegetation on the gully slopes and bottom; and (4) stores soil water so that the water table can be raised, enhancing vegetative growth outside the gully.

There are two general types of check dams: nonporous dams (without weep holes) and porous dams, which release part of the flow through the structure. A *weep hole* is a porous material located in an otherwise impermeable structure; it allows water to seep through and drain the structure.

Nonporous dams, such as those built from concrete, sheet metal, wet masonry, or earth, receive heavy impact from the hydrostatic forces of gully flow. These forces require strong anchoring of the dam into the gully banks, where most of the pressure is transmitted. *Earthen dams* should be used for gully control only in exceptional cases, for it generally was the failure of the earth material in the first place that caused the gully. However, earthen dams constructed at the mouth of gullies (*gully plugs*) can be effective in regions where the watershed can be revegetated quickly and where the storage upstream from the plug is adequate to contain the larger stormflows. The plug may have no central reinforced spillway but must have an emergency spillway to discharge water from extreme flow events. The flow released by the emergency spillway should not be concentrated but should spread out over an area stabilized by an effective vegetation cover or by some other type of protection such as a gravel field.

Porous dams transmit less pressure to the banks of gullies than do nonporous dams. Because gullies generally form in erodible, soft soils, constructing porous dams is easier, cheaper, and often more effective. Loose rocks, rough stone masonry, gabions, old car tires, logs, and brush have all been used successfully to construct porous dams. Rocks, which are abundant in many areas, are superior to most other materials for constructing inexpensive but durable porous check dams. It is important not to have large

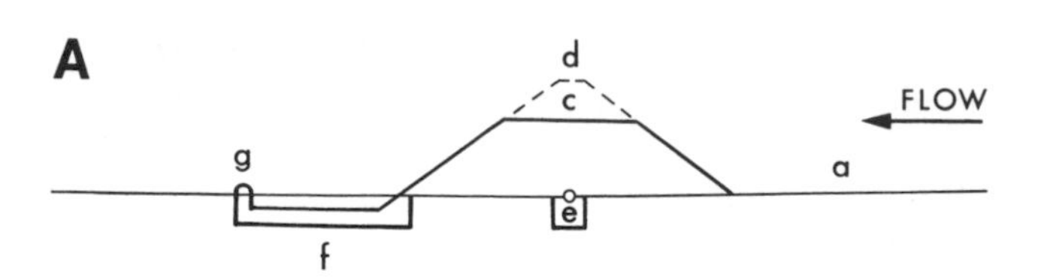

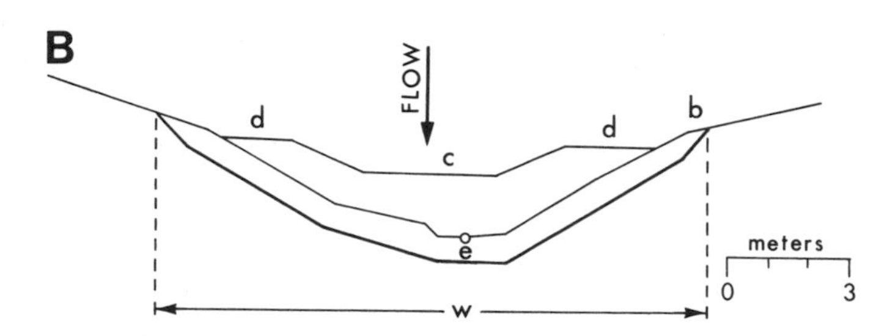

FIGURE 8.2. Cross section of a check dam. A: section of the dam parallel to the centerline of the gully; B: section of the dam at the cross section of the gully. a = original gully bottom, b = original gully cross section, c = spillway, d = crest of freeboard, e = excavation for anchoring key, f = apron, g = end sill, w = width of bank (Heede 1976, p. 14).

voids that allow jetting through the structure because this decreases the dam's effectiveness in trapping sediment. For loose rock dams, a range of rock sizes can be used to avoid this.

With the exception of gully plugs, an effective dam has three essential elements, the omission of any one of which will cause the structure to fail:

1. A spillway adequate to carry a selected design flow.
2. A key that anchors the structure into the bottom and sides of the gully.
3. An apron that absorbs the impact of water from the spillway and prevents undercutting of the structure.

Two additional components help to ensure the life of the structure:

1. A sill at the lower end of the apron to provide a hydraulic jump that reduces the impact of falling water on the unprotected gully bottom (Fig. 8.2).
2. Protective armoring on the gully banks on the downstream side of the structure to help prevent undercutting on the sides of the dam.

The purpose of the spillway is to direct the flow of water to the center of the channel, preventing cutting around the ends of the dam. Spillways of check dams can be considered broad-crested weirs, and the discharge relationship (Heede 1976) is:

$$Q_p = C_f L H^{3/2} \tag{8.1}$$

where Q_p = the peak discharge of the design flow (m³/sec); C_f = the coefficient of the spillway; L = the effective length of the spillway (m); and H = the total depth of flow above the spillway crest (m).

The value of C_f varies with the roughness, breadth, and shape of the spillway. Because check dams are not constructed to precise engineering standards, a mean value of 1.65 for C_f is acceptable. Generally, trapezoidal spillways are preferred, since the effective length of the spillway becomes larger with increasing depth of flow.

Where vegetation is slow to establish, spillways should be designed to carry the peak discharge of the 20 or 25 yr recurrence interval peak discharge. The length of the spillway relative to the width of the gully bottom is important for the protection of the channel and the structure. Normally, spillways should be designed with a length not greater than the gully bottom to reduce splashing of water against the sides of the gully.

Most gullies have either trapezoidal, rectangular, or V-shaped channels. For broad rectangular and trapezoidal channels, Equation 8.1 can be used to determine spillway depth (H):

$$H = \left(\frac{Q_p}{C_f L}\right)^{2/3} \quad \textbf{(8.2)}$$

where L can be any length that does not exceed the width of the gully bottom.

In narrow rectangular and V-shaped gullies, the length of the spillway must be adjusted to prevent the water overfall from striking the gully sides. Heede and Mufich (1973) developed the following equations for calculating spillway dimensions for these gully shapes:

$$H_v = \left(\frac{Q_p}{C_f Las}\right)^{2/3} \quad \textbf{(8.3)}$$

where:

$$Las = \frac{W}{D}(H_e - f_b) \quad \textbf{(8.4)}$$

where H_e = effective dam height (m); H_v = spillway depth (m); Las = spillway length (m); W = bank width of the gully measured from brink to brink (m); D = depth of the gully (m); and f_b = a constant referring to the length of the freeboard. In gullies with a depth of 1.5 m or less, the f_b value should not be < 0.15; in gullies deeper than 1.5 m, the maximum value should be at least 0.30.

The stability of a check dam is increased by keying the dam into the bottom and sides of the gully. The purpose of extending the key into the gully sides is to prevent water from flowing around the dam, making the structure ineffective. Keying the dam into the gully bottom also prevents undercutting at the downstream side.

Dam construction begins with filling the key trench with loose rock, the size distribution of which should prevent large voids that allow flow to reach velocities that can lead to washouts. It is generally recommended that smaller materials be used, with 80% of material smaller than 14 cm.

Aprons are installed on the gully bottom to prevent flows from undercutting the structures at the downstream side. A general rule of thumb is that the length of the apron should be about 1.5 times the height of the structure in channels with gradients < 15% and 1.75 times for those with gradients steeper than 15% (Heede 1976). These lengths prevent the waterfall from the spillway from hitting the unprotected gully bottom.

The apron should be embedded in the gully bottom so that its surface is roughly level at about 0.3 m below the original gully bottom. Where flows are high, aprons are endangered by the so-called *ground roller* that develops where the hydraulic jump hits the gully bottom. These rollers rotate upstream, and if the hydraulic jump is too near the apron, they can undermine the apron. Therefore, at the downstream end of the apron, a loose rock sill should be built about 0.15 m in height above the gully bottom. The sill creates a pool, which cushions the impact of the waterfall.

Check dams will fail if flows in the channel scour the gully side slopes below the structures, creating a gap between the dam and the bank. Turbulent flow below a check dam creates eddies that move upstream along the gully sides and erode the bank. Loose rock is effective for bank protection but should be reinforced with wire mesh secured to posts on all slopes steeper than 80–100%.

Channel banks should be protected along the entire length of the apron, and banks should be protected to the height of the dam if the channel bottom is sufficiently wide; waterfall from the spillway should strike only on the channel floor. This height can be reduced away from the structure. If the waterfall strikes against the sides of a narrow gully, the height of bank protection should be maintained for the entire length of the apron.

Headcut Control. Different types of structures can be used to stabilize headcuts. All of the types should be designed with sufficient porosity to prevent excessive pressures and thus eliminate the need for large structural foundations. Also, some type of reverse filter is needed to promote gradual seepage from smaller to larger openings in the structure. Reverse filters can be constructed if the slope of the headcut wall is sufficient to layer material, beginning with fine to coarse sand and on to fine and coarse gravel. Erosion cloth can also be effective.

Loose rock can provide effective headcut control, but flow through the structure must be controlled. The shape (preferably angular) and the size distribution of the rock must again be selected to avoid large openings that allow flow velocity to become too great. Care must be taken to stabilize the toe of the rock fill to prevent the fill from being eroded. Loose rock dams can dissipate energy from chuting flows and can trap sediment, which can facilitate vegetation establishment, helping to stabilize the toe of the rock fill.

Spacing between Dams. The purpose of check dams is to stabilize the gully bottom to prevent further downcutting and subsequent headcutting and extension of the gully. Therefore, each dam should be spaced upstream at the toe of the expected sediment wedge formed by the dam below. The first dam should be constructed in the gully where downcutting does not occur (i.e., where sediment has been deposited at the mouth of the gully, where there may be a rock outcrop or a maintained road crossing, or where the gully enters a stream system). The spacing of subsequent dams constructed upstream from the base dam depends upon the gradient of the gully floor, the gradient of the sediment wedges deposited upstream of the dams, and the effective height of the dams as measured from the gully floor to the bottom of the spillway.

Sediment will be deposited behind a dam on a gradient less than the gradient of the gully bottom. The gradient of the deposit depends upon the velocity of the flows and the size of the sediment particles (Fig. 8.3). The ratio of the gradient of sediment deposits to the gradient of the original gully bottom has been estimated at between 0.3 and 0.6 for sandy soils and 0.6–0.7 for fine-textured soils; the steeper the original gully

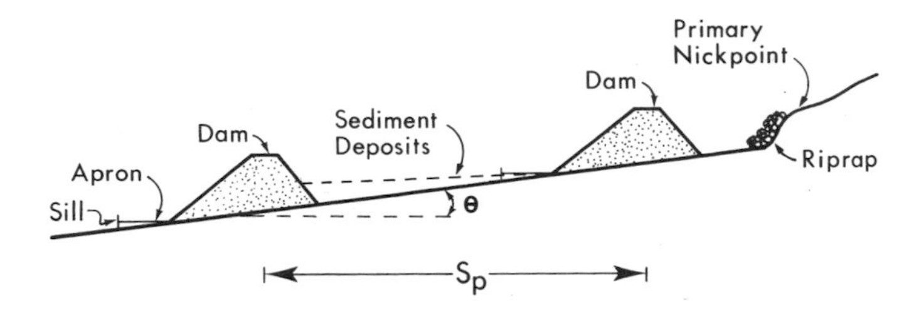

FIGURE 8.3. Diagram of placement of check dams: S_p = spacing, θ = angle of gully gradient.

gradient, the smaller the ratio of aggraded slope to original slope. Headcut areas above the uppermost dam should be stabilized with loose rock or riprap material (Fig. 8.3).

Heede and Mufich (1973) developed a calculation for the spacing of check dams:

$$S_p = \frac{H_e}{K_c G \cos\theta} \quad \textbf{(8.5)}$$

where S_p = spacing; H_e = effective height of the dam, from gully bottom to spillway crest; θ = angle corresponding to gully gradient; G = gully gradient as a ratio ($G = \tan\theta$); and K_c = a constant, related to the gradient of the sediment deposits (S_s), which is assumed to be $(1 - K_c)\,G$.

Sample values of K_c for clay-rich soils in Colorado are $K_c = 0.3$ for $G \leqq 0.2$ and $K_c = 0.5$ for $G > 0.2$. A K_c value can be determined for a particular area by measuring sediment deposits backed up behind 10 yr old structures and solving for K_c:

$$K_c = 1 - \frac{S_s}{G} \quad \textbf{(8.6)}$$

Spacing of dams calculated by the above formula is only a guide. The choice of actual sites should be made in the field and should take into consideration local topography and other conditions such as (1) placing the dam at a constriction in the channel rather than at a widened point if there is a choice of one or the other within a short distance of the calculated position; (2) placing the dam so that it does not receive the impact of flow of the tributary, where a tributary gully enters the main gully; and (3) placing the dam below the meander, where the flow in the gully has meandered within the channel.

The spacing and effective height chosen for the dams depend not only on the gradient and local conditions in the gully but also on the principal objective of the gully control. When the intention is to achieve the greatest possible deposition, the dams should have a relatively greater effective height and be spaced farther apart. If the main concern is to stabilize the gully gradient and sediment deposits are not of interest, the dams could be lower and closer together.

Vegetation-Lined Waterways. The gully control measures described above are designed to reduce flow velocity within the channel and aid in the establishment of vegetation. Waterways are designed to reduce the flow in the gully by modifying the topography; to lengthen the watercourse, resulting in a gentler bed gradient; and to increase the cross section of flow, resulting in gentle channel side slopes. Shallow flows over a rough surface with a large wetted perimeter reduce the erosive power of flowing water.

The rapid establishment of vegetation lining the waterway is essential for successful erosion control. Adequate precipitation, favorable temperature, and soil fertility are all necessary for quick plant growth. Other requisites include the following (Heede 1976, p. 34):

- The gully should not be larger than the available fill volumes.
- The valley bottom must be wide enough to accommodate a waterway that is longer than the gully.
- The soil mantle must be deep enough to permit shaping of the topography.
- The topsoil must be deep enough to permit later spreading on all disturbed areas.

Waterways are more susceptible to erosion immediately following construction than are check dams, and vegetation-lined waterways require careful attention and maintenance during the first years after construction.

Cumulative Effects on Gully Erosion

Gully formation generally results in a transport of water, soil, and chemicals from a watershed. It is important, therefore, that the manager appreciate the relationships between gully formation and land use practices on the watershed. For example, Melton (1965) hypothesized that arroyo formation in the southwestern United States might have been promoted by decreased swale vegetation. Topographic modifications from road construction or skid-trail use are a common cause for gullying in logged areas (Reid 1993, Prosser and Soufi 1998). Gullies often form where drainage is diverted onto unprotected slopes by roadside ditches and culverts, where ditches and ruts concentrate the flow of water or where culverts block and divert the flow over roadbeds.

Most of the changes affecting gully erosion involve altered hydrology (Reid 1993). Either surface runoff is increased and increases erosion power, or channel networks are modified and expose susceptible sites to erosion.

In assessing gully erosion and developing management solutions to control gully erosion, one must consider the cumulative effects across the watershed.

Soil Mass Movement

Soil mass movement refers to the instantaneous downslope movement of finite masses of soil, rock, and debris that gravity drives. Examples of these movements include landslides, debris avalanches, slumps and earthflows, creep, and debris torrents (Fig. 8.4). Such movement occurs at specific sites where hillslopes (and alterations to hillslopes) experience conditions in which *shear-stress* factors become large compared with *shear-strength* factors. Because of the strong influence of gravity, these conditions are pronounced in steep, mountainous areas, particularly in humid zones that experience high-intensity rainfall events (Box 8.3) or rapid snowmelt. A general classification of hillslope failures is presented in Table 8.1.

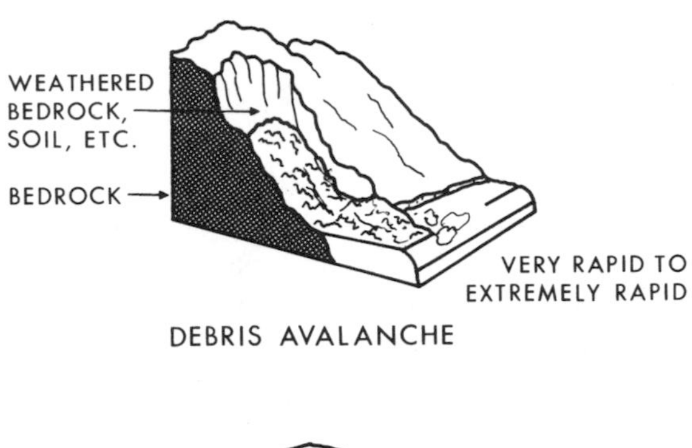

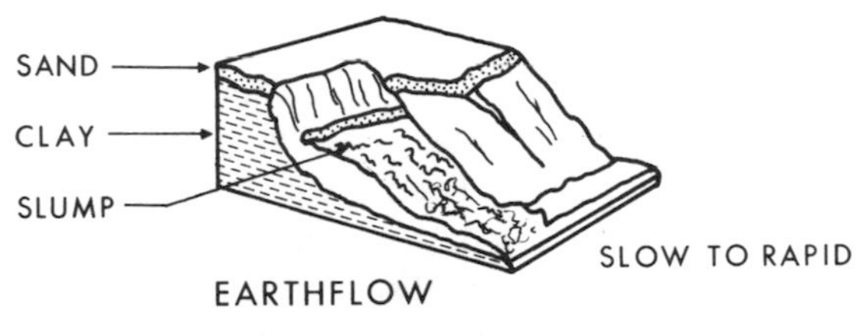

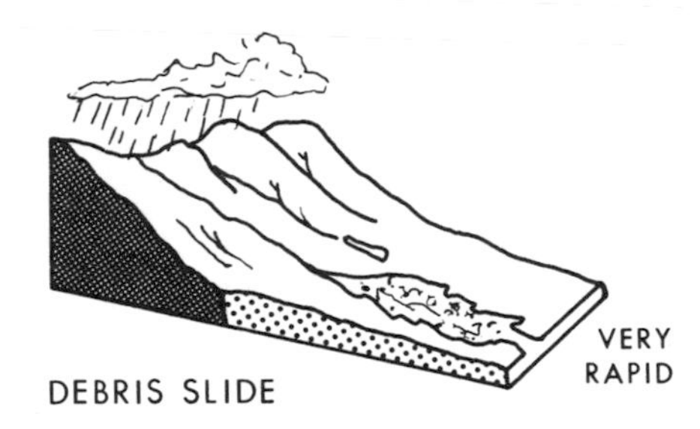

FIGURE 8.4. Illustration of soil mass movements (adapted from Varnes 1958 and Swanston and Swanson 1980).

Processes of Soil Movement

The stability of soils on hillslopes is often expressed in terms of a safety factor (F):

$$F = \frac{\text{resistance of the soil to failure (shear strength)}}{\text{forces promoting failure (shear stress)}} \quad \textbf{(8.7)}$$

A value of $F = 1$ indicates imminent failure; large values indicate little risk of failure. The factors affecting shear strength and shear stress are illustrated in Figure 8.5.

Shear stress increases as the inclination (slope) increases or as the weight of the soil mass increases. The presence of bedding planes and fractures in underlying bedrock can result in zones of weakness. Earthquakes or blasting for construction can augment stress. The addition of large amounts of water to the soil mantle and the removal of downslope material by undercutting (for road construction, for example) are common causes of movement due to increased stress.

Box 8.3

Soil Mass Movement and Debris Flows in Taiwan

Few places in the world experience the severity of landslides and debris flows that frequently occurs in Taiwan. This country is about 75% mountainous areas, much of which have shallow soils overlying weak, fractured, and intensively weathered geologic formations. More than 50% of Taiwan's land area of 35,900 km^2 has slopes steeper than 20 degrees, and the island has over 100 peaks that exceed 3000 m in elevation. Population density is greater than 600 people/km^2. Every typhoon season of May to October brings torrential rainfall resulting in frequent landslides, debris torrents, and flooding of severe magnitudes. The consequent losses of life, injury, and damage to property and infrastructure are acute. People are particularly vulnerable to landslides and debris flows in the less habitable upland areas of the country, where houses are built along hillslopes and small rural communities are often clustered at the mouth of small drainages or within the floodplain itself.

Lee et al. (1990) summarized the surveys of 9900 landslides with a cumulative slide area of 16,171 ha that had occurred on 40 selected watersheds of a total area of 20,428 km^2 between 1963 and 1977. Many of these landslides occurred during the typhoon season and triggered damaging debris flows in upland drainages of these watersheds. Impacts on life and property of debris flows caused by selected typhoons in central Taiwan are summarized below to illustrate their destructive effects (Cheng et al. 1997).

Location	Dates	Rainfall Event	Impacts on Life and Property
Tung-Men, Hualin	June 23, 1990	475 mm/3 hr	29 deaths, 7 injured, 6 missing, 24 houses destroyed, severe road damage
Er-Bu-Keng, Nantou	July 31–August 1, 1996	> 700 mm in less than 2 days	5 deaths, 10 houses and 3.8 ha of fruit orchards destroyed
Tung-Fu, Nantou	July 31–August 1, 1996	> 1300 mm in less than 2 days	2 deaths, 18 houses damaged or destroyed
Shen-Mu Village, Nantou	July 31–August 1, 1996	> 1600 mm in less than 2 days	5 deaths, 6 injured, 8 houses destroyed, 3 ha of fruit orchards destroyed

High intensity rainfall was the cause of all of these debris flows. The July 31 to August 1 1996 event was the result of Typhoon Herb, one of the most destructive typhoons to strike Taiwan in recent decades (see Box 1.2 for a discussion of Typhoon Herb). One of the highest-elevation rain gauges near the resulting debris flow site measured 1987 mm of rainfall in 42 hr during this typhoon event.

TABLE 8.1. Classification of hillslope failures

Kind	Description	Favored by	Cause
Falls	Movement through air; bouncing, rolling, falling; very rapid	Scarps or steep slopes, badly fractured rock, lack of retaining vegetation	Removal of support, wedging and prying, quakes, overloading
Slides (avalanches)	Material in motion not greatly deformed, movement along a plane; slow to rapid	Massive overweak zone, presence of permeable or incompetent beds, poorly cemented or unconsolidated sediments	Oversteepening, reduction of internal friction
Flows	Moves as viscous fluid (continuous internal deformation); slow to rapid	Unconsolidated material, alternate permeable, impermeable fine sediment on bedrock	Reduction of internal friction due to water content
Creep	Slow downhill movement, up to several cm per year	High daily temperature ranges, alternate rain and dry periods, frequent freeze and thaw cycles	Swaying of trees, wedging and prying, undercutting or gullying
Debris torrents	Rapid movement of water-charged soil, rock, and organic material in stream channels	Steep channels, thin layer of unconsolidated material over bedrock within channel; layered clay particles (lacustrine clays), which form slippage plane when wet	High streamflow discharge, saturated soils, often triggered by debris avalanches; deforestation accelerates occurrence

Source: From Swanston and Swanson 1980.

Shear strength is determined by complex relationships between the soil and slope and the strength and structure of the underlying rock. Cohesion of soil particles and frictional resistance between the soil mass and the underlying sliding surface are major factors affecting shear strength. Frictional resistance is a function of the angle of internal friction of the soil and the effective weight of the soil mass. Pore water pressure in saturated soil tends to reduce the frictional resistance of the soil. Rock strength is affected by structural characteristics such as cleavage planes, fractures, jointing, bedding planes, and strata of weaker rocks.

Plants exert a pronounced influence on many types of soil mass movement. The removal of soil water by transpiration results in lower pore water pressures, reduced chemical weathering, and reduced weight of the soil mass. Tree roots, which add to the

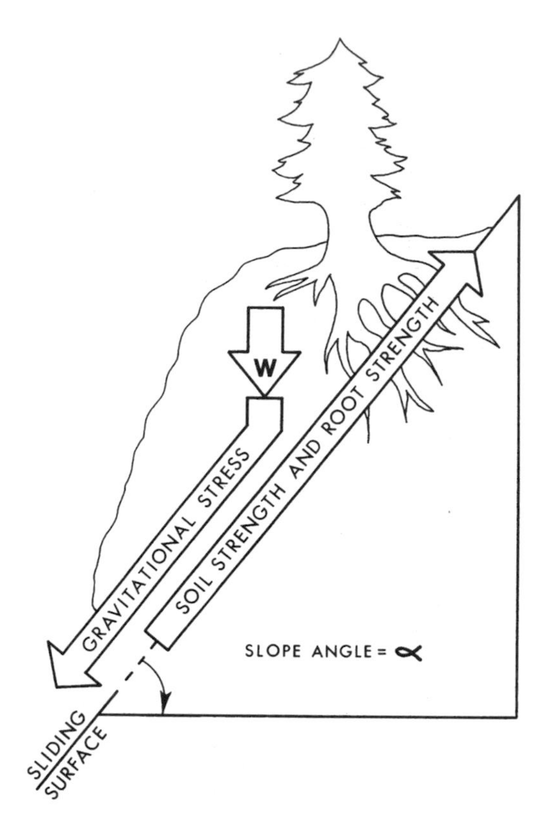

Figure 8.5. Simplified diagram of forces acting on a soil mass on a slope (adapted from Swanston 1974).

frictional resistance of a sloping soil mass, can effectively stabilize thin soils, generally up to 1 m in depth, by vertically anchoring into a stable substrate. Medium to fine root systems can provide lateral strength and also improve slope stability.

Factors Affecting Slope Stability

Several important physical and biological factors influence slope stability and erosion (O'Loughlin 1985, Sidle et al. 1985). The more important factors that act singly or in combination include:

- Climate—These factors include rainfall intensity and duration and temperature changes.
- Soil—Soil factors affecting slope stability include soil strength, particle size distribution, clay content, clay type, infiltration capacity, soil drainage condition, porosity, organic content and depth, stratification, and lithic contacts.
- Physiography—These features include factors such as slope steepness, slope length, and slope roughness.
- Vegetation—Key vegetation factors include cover density and type, forest litter thickness, tree root distributions, and strength of tree roots.

- Water erosive forces—These factors include surface runoff and the flow of water from snowmelt.
- Human factors—Human activities include timber harvesting (and especially tree felling), clearing of forests for other land uses, road construction, and loading slopes with fill material.
- Animal factors—Most important are overgrazing or overbrowsing by domestic livestock or wild animal populations.

Evaluating the Stability of Hillslopes

Procedures have been developed to assess soil mass movement hazards and the potential for sediment delivery to channels (Hicks and Smith 1981, Fannin et al. 1997). At the landscape level, terrain evaluation procedures that utilize topographic and geologic information have been developed to provide broad categories of landslide hazard related to potential timber-harvesting, road construction, and other management activities (Sidle 2000). Many of these procedures are based on the factors responsible for slope stability and erosion presented above. A discussion of the methods for evaluating the stability of hillslopes is beyond the scope of this book; however, key factors that must be considered for hazard assessment are identified in Box 8.4.

Reducing Impacts of Soil Mass Movement

Natural events such as large and high intensity rainfall events, earthquakes, and wildfire have a profound effect on soil mass movement. By increasing soil saturation on steep slopes that are normally unsaturated, these events can trigger landslides and debris flows on undisturbed forestlands. Wildfire can also increase the risk of landslides and debris flow. Benda and Dunne (1997) found that severe wildfire was a major cause of landslides on prehistoric landscapes in the coastal mountain ranges of Oregon. Relatively little can be done to prevent soil mass movement when natural events occur on terrain that is susceptible to slopes failures.

Occurrences of landslides and debris flows can increase when poorly conceived timber-harvesting activities, road construction, and vegetative conversions are carried out on these sensitive sites; these actions need not take place, however. Careful planning and implementation of these management activities helps to mitigate the human-induced activities compounding the occurrence of soil mass movement (see Chapter 19). Guidelines to achieve this goal are considered below in the context of cumulative effects on soil mass movement.

Cumulative Effects on Soil Mass Movement

Processes of soil mass movement attack the entire soil profile. Plant roots inhibit these processes by increasing soil cohesion. The modification of the vegetative cover, the soil system, or the inclination of a hillslope can affect soil mass movement. The impacts of land use can be estimated by relating them to factors affecting shear strength and resistance to shear. Most commonly, road construction and forest removal activities have the greatest effect on soil mass movement. Undercutting a slope and improper drainage are major factors that accelerate mass movement (Hagans and Weaver 1987, Fannin and Rollerson 1993). Proper road layout, design and control of drainage, and minimizing cut-and-fill (earthwork) can help prevent problems. Areas that are naturally susceptible to soil mass movement should simply be avoided (Burroughs and King 1989). In terms of logging practices on steep

Box 8.4

Factors to Consider When Making Hazard Assessments of Hillslope Failure (from Swanston and Swanson 1980)

The stability of hillslopes can be judged by evaluating the following:

Land Features

- Landforms—qualitative indicator of potentially unstable landforms, for example, fracturing and bedding planes parallel to slopes, steep U-shaped valleys.
- Slope configuration—convex or concave.
- Slope gradient.

Soil Characteristics

- Present soil mass movement and rate.
- Parent material—cohesive characteristics; for example, colluvium, tills, and pumice soils possess little cohesion.
- Occurrence of cemented, compacted, or impermeable subsoil layer—identify principal planes of failure.
- Evidence of concentrated subsurface drainage—indications of local zones of high soil moisture, springs, seeps, and so forth.
- Soil characteristics—depth, texture, clay mineralogy, angle of internal friction, cohesion.

Bedrock Lithology and Structure

- Rock type—volcanic ash, breccias, and silty sandstone are susceptible to earthflows, and so forth.
- Degree of weathering.
- Bedding planes or dips parallel to slope.
- Jointing and fracturing—locations, directions, and relationship to slope.

Vegetative Characteristics

- Root distribution and degree of root penetration in the subsoil.
- Vegetative type and distribution—cover density, age, and so forth.

Hydrologic characteristics

- Saturated hydraulic conductivity.
- Pore water pressure.

Climate

- Precipitation occurrence and distribution.
- Temperature fluctuations—frost heaving, and so forth.

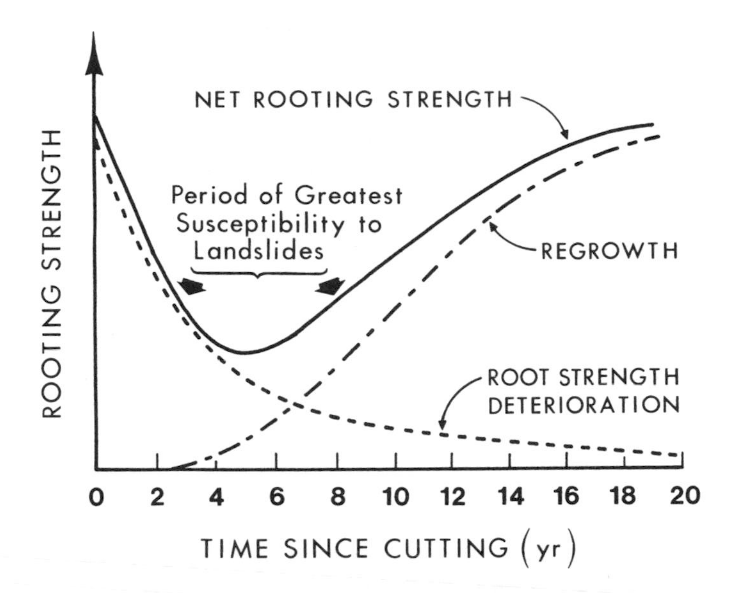

FIGURE 8.6. Hypothetical relationship of root-strength deterioration after timber harvesting and root-strength improvement with regenerating forest (from Sidle 1985).

slopes, full-suspension yarding, cable yarding, balloon logging, and other alternatives to skid roads should be used.

The removal of trees from steep slopes, and particularly the permanent conversion from forest to pasture or agricultural crops, can also result in accelerated mass movement. The reduced evapotranspiration with such activities can lead to wetter soils. Shear resistance is reduced by the loss and deterioration of tree roots, particularly in areas where roots penetrate and are anchored into the subsoil. Accelerated soil mass movement following conversion from forest to pasture has been pronounced in steep, mountainous areas of New Zealand (Trustrum et al. 1984). In many instances, therefore, maintaining tree cover on steep slopes to reduce the hazard of soil mass movement is desirable.

Routinely implemented forest-harvesting activities and regeneration practices periodically can leave hillslopes susceptible to mass movement. Root strength deteriorates rapidly as roots decay following timber harvesting (Fig. 8.6). In the Pacific Northwest, several years are required before the regrowing forest exhibits root strength that is equivalent to that of mature forests (Sidle 1985, 1991). As a consequence, there is generally a 3–8 yr period when net root strength is at a minimum. However, the first several years after timber harvesting also coincide with the period of maximum water yield increases caused by reduced evapotranspiration (see Chapter 6). The result is more frequent occurrences of shallow landslides on steep slopes for several years following logging. The period of susceptibility is somewhat species dependent; that is, it is affected by the rate of root decay and the rate of regrowth of the tree species being managed.

SUMMARY

Gully erosion and soil mass movement can reduce the productive area and productive capacity of a watershed and can cause large quantities of sediment to be moved from the upland areas to downstream channels or downwind areas. A basic understanding of the processes involved and the factors that affect gully erosion and soil mass movement should be gained after reading this chapter. Specifically, you should be able to:

- Understand how gullies are formed.
- Explain the role of structural and vegetative measures in controlling gully and wind erosion.
- Explain how you would prioritize the treatment of gullies.
- Describe the different types of soil mass movement and explain the causes of each.
- Explain shear stress and shear strength as they pertain to soil mass movement.
- Explain how different land use impacts, including road construction, timber harvesting, and conversion from deep-rooted to shallow-rooted plants, affect both gully erosion and soil mass movement.

References

Benda, L.E., and T. Dunne. 1997. Stochastic forcing of sediment supply to channel networks from landsliding and debris flow. *Water Resour. Res.* 33:2849–2863.

Burroughs, E.R., Jr., and J.G. King. 1989. *Reduction of soil erosion on forest roads.* USDA For. Serv. Gen. Tech. Rep. INT–264.

Cheng, J.D., H.L. Wu, and L.J. Chen. 1997. A comprehensive debris flow hazard mitigation program in Taiwan. In *Proceedings of the First International Conference on Debris Flow Hazards Mitigation: Mechanics, Predictions, and Assessment*, ed. C. Chen, 93–102. New York: American Society of Civil Engineers.

Fannin, R.J., and T.P. Rollerson. 1993. Debris flows: Some physical characteristics and behaviour. *Canadian Geotech. J.* 30:71–81.

Fannin, R.J., M.P. Wise, J.M.T. Wilkinson, B. Thompson, and E.D. Hetherington. 1997. Debris flow hazard assessment in British Columbia. In *Proceedings of the First International Conference on Debris—Flow Hazards Mitigation*, 197–206. San Francisco, CA.

Hagans, D.K., and W.E. Weaver. 1987. Magnitude, cause, and basin response to fluvial erosion, Redwood Creek basin, northern California. In *Erosion and sedimentation in the Pacific Rim*, 419–428. Int. Assoc. Hydrol. Sci. Publ. 165. Washington, DC: IAHS.

Harvey, M.D., C.C. Watson, and S.A. Schumm. 1985. *Gully erosion.* USDI Bur. Land Manage. Tech. Note 366.

Heede, B.H. 1967. The fusion of discontinuous gullies: A case study. *Bull. Int. Assoc. Hydrol. Sci.* 12(4): 4250.

_________. 1971. *Characteristics and processes of soil piping in gullies.* USDA For. Serv. Res. Pap. RM–68.

_________. 1976. *Gully development and control: The status of our knowledge.* USDA For. Serv. Res. Pap. RM–169.

_________. 1977. *Gully control structures and systems.* FAO Conserv. Guide 1, 181–222. Rome.

_________. 1982. Gully control: Determining treatment priorities for gullies in a network. *Environ. Manage.* 6:441–451.

Heede, B.H., and J.G. Mufich. 1973. Functional relationships and a computer program for structural gully control. *Environ. Manage.* 1:321–344.

Hicks, B.G., and R.D. Smith. 1981. Management of steeplands impacts by landslide hazard zonation and risk evaluation. *J. Hydrol. N.Z.* 20:63–70.

Lee, S.W., J.D. Cheng, C.W. Ho, Y.C. Chiang, H.H. Chen, and K.J. Tsai. 1990. An assessment of landslide problems and watershed management in Taiwan. *IAHS–AISH Publ.* 192:238–246.

Melton, M.A. 1965. The geomorphic and paleoclimate significance of alluvial deposits in southern Arizona. *J. Geol.* 73:1–38.

O'Loughlin, C.I. 1985. *The effects of forest land use on erosion and slope stability*. Rep. Seminar. Honolulu, HI: East-West Center.

Prosser, I.P., and M. Soufi. 1998. Controls on gully formation following forest clearing in a humid temperate environment. *Water Resour. Res.* 34:3661–3671.

Reid, L.M. 1993. *Research and cumulative watershed effects*. USDA For. Serv. Gen. Tech. Rep. PSW–GTR–141.

Sidle, R.C. 1985. Factors influencing the stability of slopes. In *Proceedings of the workshop on slope stability: Problems and solutions in forest management*, ed. D. Swanston, 17–25. USDA For. Serv. Gen. Tech. Rep. PNW–180.

________. 1991. A conceptual model of changes in root cohesion in response to vegetation management. *J. Environ. Qual.* 20:43–52.

________. 2000. Watershed management challenges for the 21st century: A global perspective for mountainous terrain. In *Land stewardship in the 21st century: The contributions of watershed management*, tech. coords. P.F. Ffolliott, M.B. Baker, Jr., C.B. Edminster, M.C. Dillon, and K.L. Mora, 45–56. USDA For. Serv. Proc. RMRS–P–13.

Sidle, R.C., A.J. Pierce, and C.L. O'Loughlin. 1985. *Hillslope stability and land use*. Washington, DC: Water Resour. Monogr. 11.

Stromquist, L., B. Lunden, and Q. Chakela. 1985. Sediment sources, sediment transfer in a small Lesotho catchment: A pilot study of the spatial distribution of erosion features and their variation with time and climate. *S. African Geol. J.* 67:3–13.

Swanston, D.N. 1974. *The forest ecosystem of southeast Alaska*. USDA For. Serv. Gen. Tech. Rep. PNW–7.

Swanston, D.N., and F.J. Swanson. 1980. Soil mass movement. In *An approach to water resources evaluation, nonpoint silvicultural sources*, V.1–V.49. Environ. Res. Lab., EPA–600/880012. Athens, GA: Environmental Protection Agency.

Toy, T.J., and R.F. Hadley. 1987. *Geomorphology and reclamation of disturbed lands*. Orlando, FL: Academic Press.

Trustrum, N.A., V.J. Thomas, and M.G. Lambert. 1984. Soil slip erosion as a constraint to hill country pasture production. *Proc. N.Z. Grassl. Assoc.* 45:66–76.

Varnes, D.J. 1958. Landslide types and processes. In *Landslides and engineering practice*, ed. E.B. Eckel, 20–47. Highway Res. Board Spec. Rep. 29. Washington, DC: National Academy of Sciences.

CHAPTER 9 *Sediment Yield and Channel Processes*

INTRODUCTION

The amount of sediment in a stream, river, lake, or reservoir is of keen interest in watershed management. Excessive sediment levels in streamflow can be indicative of high levels of soil erosion from a watershed, an unstable stream channel, or both. Excessive sediment can adversely affect water quality and aquatic habitat and is one of the primary targets of the total maximum daily load (TMDL) provisions of the Clean Water Act in the United States that are aimed at reducing nonpoint pollution from agricultural and silvicultural practices (see Chapter 12). In addition, excessive sediment deposition can adversely impact stream channel morphology, reservoirs, lakes, and associated infrastructures. Higher than anticipated levels of sediment deposition in a reservoir can reduce its economic life. Similarly, high levels of suspended sediment can damage turbines that provide hydroelectric power. Consequently, there are both environmental and financial costs of excessive sediment in water bodies.

Determining what constitutes excess sediment in a stream is not straightforward. It is important to recognize that erosion and sedimentation processes occur naturally and that levels of sediment in streamflow can vary considerably from one location to another and from one time period to another. The relationships between upland erosion and downstream sedimentation involve complex channel processes and their dynamics, many of which are poorly understood. A full theoretical treatment of this subject is beyond the scope of this book. Instead, this chapter examines basic processes affecting sediment transport and deposition as related to channel dynamics. Methods of measurement and analysis of sediment data are also described. Land use and resource management actions that directly and indirectly affect sedimentation and stream channels are discussed.

Processes of Sediment Transport

Sources of Sediment

Sediment is the product of erosion whether it occurred as surface erosion, gully erosion, soil mass movement, or stream channel erosion. The amount of sediment contributed to a stream channel from the surrounding watershed is dependent on many factors, including proximity of the erosional feature to the channel, characteristics of sediment particles, and the efficiency by which sediment particles are transferred from one part of the landscape to another. Only a portion of eroded soil is passed through and out of a watershed during most storm events. Most sediment is deposited at the base of hillslopes, in riparian buffer strips (see Chapter 13), in floodplains following high flows or flood events, and within river channels.

Although the types of erosion occurring on a watershed can be determined, it is difficult to determine how much of the eroded material will be transported and how much will be deposited into a channel over time. The following section focuses attention on the processes of sediment transport in the channel—that is, after eroded material reaches the channel.

Fundamental Energy Relationships of Streams

Unsaturated flow of water in response to energy gradients in the soil-plant-atmosphere system was the focus of Chapter 3. Here, attention is placed on the flow of unbound water and sediment in a channel. Water that is flowing from a watershed is moving from a state of higher energy status to one of lower energy status, the base level being sea level. Evaporation, wind, and precipitation redistribute water and can return it back to the watershed and to an elevated energy status by virtue of its elevation above sea level. In responses to gravity, much of the water in the form of excess precipitation is returned to the oceans, and the energy status is reduced through the process of channel flow.

Once water reaches a stream channel, the rate and type of flow are determined by gravity and resistance forces of friction (recall Manning's n in Chapter 4). Gravity forces are expressed as a continuous energy gradient, called the *hydraulic gradient*. The hydraulic gradient is a potential energy gradient that exists by virtue of the elevation of water above sea level. Although the overall hydraulic gradient is determined by the change in elevation from the highest to the lowest elevations in the watershed, the gradient is not uniform throughout the stream channel. Hydraulic gradients are generally steepest at the uppermost part of the watershed and diminish as the channel nears sea level.

Water that is ponded at upper portions of a watershed has a high potential energy but cannot perform work until it is released. Water drains from the soil and otherwise reaches a channel under the force of gravity and flows from its higher energy state to the downstream, lower energy state. In this process, the stream's potential energy is converted to kinetic energy by virtue of the velocity of flowing water:

$$\text{Energy} = 0.5 \times \text{mass} \times \text{velocity}^2 \quad \textbf{(9.1)}$$

Water flow in a channel is governed by energy relationships that are based upon the *Bernoulli equation*, which states:

$$\frac{P}{\rho g} + \frac{V^2}{2g} + z = \text{constant} \quad \textbf{(9.2)}$$

where P = pressure (bars or newtons/m^2), ρ = density of fluid (kg/m^3), g = acceleration due to gravity (m/s^2), V = velocity (m/s), and z = elevation above some datum (m).

The three components in Equation 9.2 have units of length (m) and can be considered as pressure head, velocity head, and elevational head.

For a given discharge value (Q), we know from the conservation of mass principle that even though channel dimensions can change from one section to another, the products A_1V_1 at section 1 = A_2V_2 at section 2. Therefore, if we consider the first term in Equation 9.2 to be equivalent to water depth in the channel, the velocity and depth of flow will change for a given discharge in response to changing channel dimensions of width and bottom configurations. The overall change in the energy status of a stream, therefore, can be accounted for by the change in the water surface elevation of the stream, which is the slope.

The *Bernoulli equation* is a one-dimensional energy equation that illustrates the above relationships for a section of stream channel from its upstream location (subscript 1) to its downstream location (subscript 2):

$$z_1 + D_1 + \frac{V_1^2}{2g} = z_2 + D_2 + \frac{V_2^2}{2g} + h_L \quad \textbf{(9.3)}$$

where D = mean water depth (m), and h_L = the head loss due to energy losses associated largely with friction.

The specific energy for a given channel section with a small slope and a given discharge is a function of water depth:

$$E_s = D + \frac{V^2}{2g} \quad \textbf{(9.4)}$$

The above energy relationships in a channel have important effects on fluvial processes and are the basis for several relationships and terms that are defined in Table 9.1.

Subcritical and Supercritical Flow

At and below *critical flow*, a more or less stable relationship exists between a given depth of flow and the ensuing rate of streamflow. It is at or near critical flow, therefore, that the best measurements of streamflow are generally obtained (e.g., such relationships are used in the design of flumes). Subcritical flow is tranquil and exerts relatively low energies on the channel banks and beds. In contrast, supercritical flow has high energy and, as a consequence, exerts considerable stress on the bed and sides of channels. The *Froude number* (Fr) is a dimensionless parameter that is used as a quantitative measure of whether subcritical or supercritical flow will occur:

$$Fr = \frac{V}{(gD)^{1/2}} \quad \textbf{(9.5)}$$

where V = the average velocity in the cross section of measurement, g = the acceleration due to gravity, and D = the average water depth.

If $Fr < 1$, subcritical flow occurs. Supercritical flow occurs when $Fr > 1$.

Laminar and Turbulent Flow

Streamflow can also be characterized by the movement of individual fluid elements with respect to each other. This movement results in either laminar or turbulent flow. In

Table 9.1. Terminology used in describing channel dynamics and processes

Term	Definition
Uniform flow	Flow conditions in which the water surface is parallel to the streambed; depth and velocity of flow are constant over the channel reach.
Varied flow	In contrast to uniform flow, flow velocity and depth change over the channel reach.
Steady flow	Depth and velocity of flow do not change over a given time interval at a point within the stream channel.
Unsteady flow	Conditions of flow in which depth and velocity change with time at a point within the stream channel (e.g., when a wave passes a point in the channel).
Laminar flow	Flow of fluid elements in a stream in parallel layers past each other in the same direction but at different velocities.
Turbulent flow	Flow of fluid elements in a stream past each other in all directions with random velocities (e.g., eddies).
Critical flow	Occurs when the Froude number = 1 (see Eq. 9.5), which is a condition of minimum specific energy (E_s) for a given discharge and channel condition.
Critical depth (D_c)	For any given channel section and discharge, this is a depth of water that corresponds to the minimum specific energy (E_s); if water depth is $> D_c$, the flow is *subcritical*, if water depth is $< D_c$, the flow is *supercritical*.
Aggradation	A process by which sediments collect in streambeds, floodplains, and other water bodies and become deposited, raising their elevations.
Degradation	A process by which streambeds and floodplains are lowered by erosion and removal of material from the site.
Dynamic equilibrium	A channel system that is sufficiently stable so that compensating changes (either aggradation or degradation) will not alter the equilibrium.

laminar flow, each fluid element moves in a straight line with uniform velocity. There is little mixing between the layers, or elements, of flow and, therefore, no turbulence. In contrast, *turbulent flow* has a complicated pattern of eddies, producing random velocity fluctuations in all directions. The constant changes of flow lines during turbulent flow lead to surges of flow against stream banks and structures, increasing shear stress. Turbulent flow is the normal condition in streams. The *Reynolds number* is a dimensionless measure used to distinguish quantitatively between laminar and turbulent flow. The Reynolds number is:

$$R_e = \frac{(VD)}{\nu} = \frac{(\text{inertial force})}{(\text{viscous force})} \quad \textbf{(9.6)}$$

where V = the average velocity in the cross section of measurement (m or ft/sec), D = the average water depth (m or ft), and ν = the kinematic viscosity (m^2 or ft^2/sec).

In natural stream channels, normally a Reynolds number less than 2000 indicates laminar flow; one over 2000 indicates turbulent flow.

Sediment Movement

Both water and sediment make up the flow in a stream. As water and sediment move from a higher to a lower energy state, energy must be released. The stream dissipates energy (1) as heat from friction and (2) by performing work on the channel and sediments. It is this process of work by the flow of water and sediment that forms the stream channel and changes the slope of the stream channel.

Channel erosion is not the only way that a stream can dissipate excess energy. Energy can be internally dissipated through turbulent flow—the eddies and boils in a stream. When a stream loses energy to its channel, momentum is transferred from the flowing water to the channel; this change in momentum is a *force*.

When sufficient force is applied to the channel, sediment will be eroded and carried away. The energy that is lost in a section of channel (h_L) is a function of channel roughness (n), the hydraulic radius (R_h), and velocity (V) of the stream as shown below:

$$h_L = n\left(\frac{1}{4R_h}\right)\left(\frac{V^2}{2g}\right) \quad \textbf{(9.7)}$$

As discussed in Chapter 4, channel roughness includes channel vegetation, woody debris, bottom forms, and the morphological features of the channel.

The resisting force exerted by the channel is a *shear stress* (τ), referred to here as *shear resistance*. The shear force is that shear stress generated on the bed and banks (wetted perimeter) of a channel in the direction of flow. Expressed as a per unit of wetted area, shear stress (τ_o) can be defined as:

$$\tau_o = \gamma\,(R_h)\,s_b \quad \textbf{(9.8)}$$

where γ is the specific weight of the fluid and s_b is the channel bed slope.

Types of Sediment Transported in a Stream

The *sediment discharge* of a stream is defined as the mass rate of transport through a given cross section of the stream and is usually measured in milligrams per liter (mg/l) or parts per million (ppm). Sediment discharge contains fine particles that are transported in suspension. Part of the *suspended load* is called the wash load. The *wash load* is made up only of silt and clay, whereas suspended sediment also includes sand-sized particles. The bed load consists of sand, gravel, or larger cobbles and is transported along the stream bottom by traction, rolling, sliding, or saltation (Fig. 9.1). Particles are moved when eddies formed by turbulent flow dissipate part of their kinetic energy into mechanical work.

Suspended Load. Particles can be transported as suspended load if their *settling velocity* is less than the buoyant velocity of the turbulent eddies and vortices of the water. Settling velocity primarily depends upon the size and density of the particle. In general, the settling velocity of particles < 0.1 mm in diameter is proportional to the square of the particle diameter, while the settling velocity of particles larger than 0.1 mm is proportional to the square root of the particle diameter. Once particles are in suspension, little energy is needed for transport. A heavy suspended load

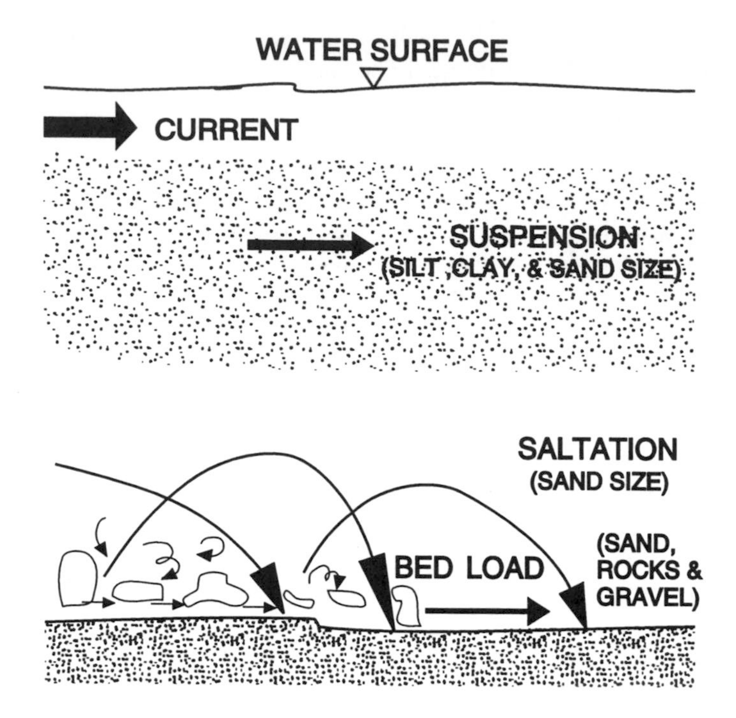

FIGURE 9.1. Transportation of particles of sediment in running water.

decreases turbulence and makes the stream more efficient. Concentrations are highest in shallow streams where velocities are high.

As one would suspect, the concentration of sediment in a stream is lowest near the water surface and increases with depth. Silt and clay particles < 0.005 mm in diameter generally are dispersed uniformly throughout the depth, but large grains are more concentrated near the bottom.

For most streams there is a correlation between suspended load and stream discharge. During stormflow events, the rising limb of the hydrograph is associated with higher rates of sediment transport and degradation (Fig. 9.2). As the flood peak passes and the rate of discharge drops, the amount of sediment in suspension also diminishes rapidly and aggradation occurs. If sufficient measurements of discharge and sedimentation are available, a relationship can be developed for use as a *sediment rating curve*. The relationship will most often take a power function form, such as:

$$SS = kq^m \tag{9.9}$$

where SS = suspended-sediment load (mg/l), q = rate of stream discharge (m^3/sec), and k and m = constants for a particular stream.

The variability in the suspended-sediment discharge relationship for a given stream relates to channel stability and sources of sediment. Streams draining undisturbed forested watersheds are characteristically stable, with low levels of suspended sediment. However, both suspended-sediment and bed load are discharged from streams

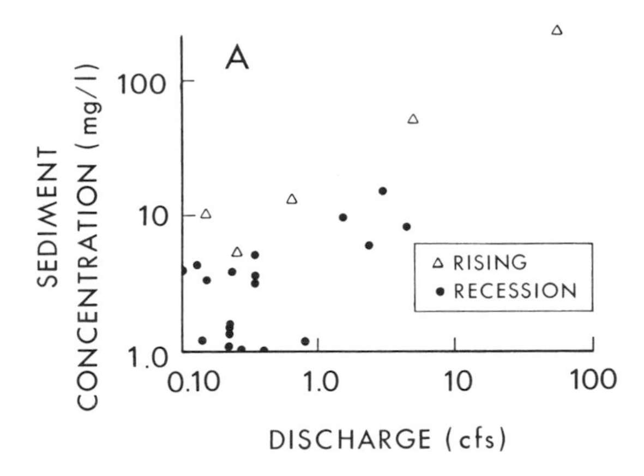

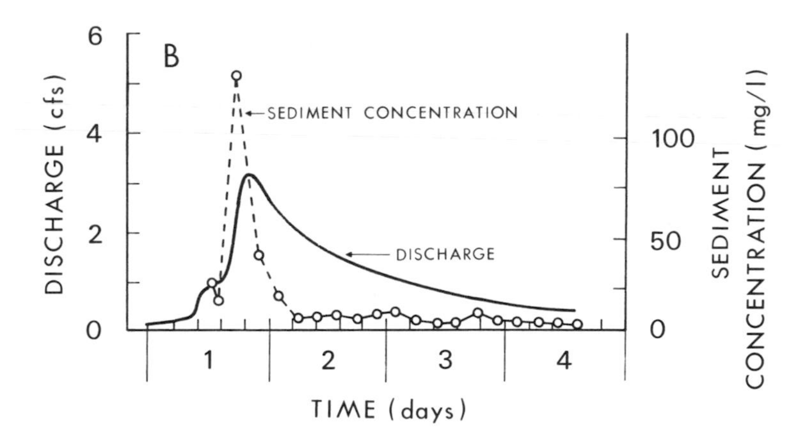

Figure 9.2. Example of the relationship between streamflow discharge and suspended sediment for a small stream in northeastern Minnesota. A: the relationship for several storm events; B: a single storm event (included in A).

draining forested watersheds, depending on stream type. As changes occur in the magnitude of sediment contributed from the watershed, the discharge regime, or the stream channel morphology, the suspended-sediment discharge relationship for a channel reach changes. For example, forest fires can temporarily cause increases in suspended sediment downstream, and the increase in the amount of sediment and any increase in streamflow associated with the fire can shift the relationship (Fig. 9.2). Likewise, floods can change the relationship by bank overflow and the cutting of new channel segments that provide new sources of sediment. Newly cut stream channels have suspended-sediment relationships different from those of a well-armored, stable stream. Studies have shown that in most instances, the suspended sediment relationships following such disturbances will eventually (sometimes it takes several years) adjust back to the original sediment rating curve.

A rating-curve relationship such as Equation 9.9 has been suggested as a method for estimating the effects of land use and management activities on suspended sediment. To use such an equation, stable relationships must be developed from field data, and any changes in the relationship caused by natural phenomena must be taken into account. Any significant shift in the relationship following some action such as logging or conversion of vegetative cover on a watershed could then be quantified. The major difficulty, however, is separating changes in sediment rating curves caused by natural phenomena from those caused by human activities. Obtaining a representative sample of suspended sediment for measurement is also difficult, for concentrations can vary considerably with time and within a cross section of a stream. The usefulness of sediment rating curves can often be improved by separating the data into streamflow generation mechanisms (rainfall events, snowmelt-runoff events, etc.), rising and falling stages of the hydrograph, or combinations thereof (Box 9.1).

Bed Load. Bed load particles can be transported in groups or singly and can be entrained if the vertical velocity of eddies creates sufficient suction to lift the particle from the bottom. These particles can also be started in motion if the force exerted by the water is greater on the top of the grain than on the lower part. Particles can move by saltation if the hydrodynamic lift exceeds the weight of the particle (Fig. 9.1). They will be redeposited downstream if not reentrained. Large as well as small particles can roll or slide along the stream bottom; rounded particles are more easily moved. Generally the largest particles are moved in the steeply sloping channels of headwater streams.

The largest grain size that a stream can move as bed load determines *stream competence*. The competency of a stream varies greatly throughout its length and with time at any given point along its length. Stream competence is increased during high peak discharges and flood events.

The force required to entrain a given grain size is called the *critical tractive force*. The velocity at which entrainment takes place is called the *erosion velocity*. DuBoy's equation generally is used to calculate the tractive force for low velocities and small grains as a function of stream depth and gradient:

$$T_f = W_w DS \tag{9.10}$$

where T_f = tractive force; W_w = specific weight of water; D = depth of water; and S = stream gradient.

For high velocities and large particles, stream velocity is more important than depth and slope; this has given rise to the sixth-power law:

$$\text{competence} = CV^6 \tag{9.11}$$

where C is a constant.

Doubling the stream velocity means that particles 64 times larger can be moved. However, the exponent is only approximate and varies with other conditions of flow.

Stream power, the rate of doing work, is used to express the ability of a stream to transport bed load particles. It is the product of streamflow discharge, water surface slope, and the specific weight of water. Relationships similar to those shown in the sediment rating curve discussed previously can be developed between unit stream power and unit bed load transport rate for a given stream, similar to the sediment rating curve discussed previously.

Box 9.1

Sediment Rating Curves for Watersheds of Arizona

Sediment rating curves were developed to analyze the effects of vegetative management practices on suspended-sediment discharge from watersheds in ponderosa pine forest and pinion-juniper woodlands of north central Arizona. The sediment rating curves were derived from measurements of suspended-sediment concentration and streamflow discharge on the Beaver Creek watersheds (Lopes et al. 2001) and are expressed in the form of power equations (Eq. 9.9). One example of these curves is presented below. Disturbances from vegetation management practices generally increased suspended-sediment transport above those of control watersheds. Completely cleared and strip-cut ponderosa pine watersheds produced higher sediment concentrations than did a control watershed. Likewise, cabled and herbicide-treated pinion-juniper watersheds yielded higher sediment-laden streamflows than did a control.

Sediment transport regimes were also related to streamflow generation mechanisms and hydrograph stages. Although about 85% of the data analyzed represented snowmelt-runoff events in both vegetative types, derivation of sediment rating curves based on streamflow generation mechanisms (snowmelt-runoff events, high-intensity and short-duration rainfall events, and low-intensity and relatively long rainfall events) improved the sensitivity of the analysis. Sediment data collected during rising and falling stages of a hydrograph varied between the two vegetative types. Sediment concentrations were generally higher in the rising stage than in the falling stage for ponderosa pine watersheds. However, there was no clear evidence of higher sediment concentrations in the rising stage of the hydrograph as compared to the falling stage in the pinion-juniper watersheds.

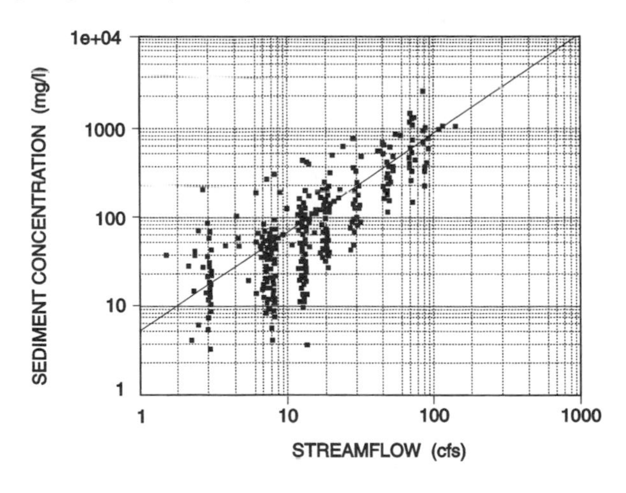

Relationship between streamflow discharge and suspended sediment from a clearcut ponderosa pine watershed in north central Arizona (from Lopes and Ffolliott 1993, © *Water Resources Bulletin*, by permission).

Another important concept in sediment transport is *stream capacity*, which is the maximum amount of sediment of a given size and smaller that a stream can carry as bed load. Increased channel gradient and discharge rate result in increased stream capacity. It has been found that if small particles are added to predominantly coarse streambed material, the stream capacity for both large and small particles is increased, but if large particles are added to small-size grain material, the stream capacity is reduced. Small particles increase the density of the suspension and, therefore, the carrying capacity. Capacity also decreases with increasing grain size.

All of the variables affecting stream capacity are interrelated and vary with channel geometry. Streams that carry large bed loads (such as those found in dryland regions where sediment sources are great) have shallow, rectangular, or trapezoidal cross sections, because there is a steep velocity gradient near the streambed in such cross sections (Morisawa 1968). The typically parabolic cross sections of channels in humid regions, where sediment loads are relatively small, usually do not have steep velocity gradients near the streambed.

Aggradation and Degradation

The amount of sediment carried by a stream depends largely upon the interrelationships between the supply of material to the channel, characteristics of the channel, the physical characteristics of the sediment, and the rate and amount of streamflow discharge. The supply of material and streamflow depend upon the climate, topography, geology, soils, vegetation, and land use practices on the watershed. Channel characteristics of importance are the morphological stage of the channel, roughness of the channel bed, bed material, and steepness of the channel slope. Soils and geological materials of the watershed and stream channels and the state of their weathering largely determine the physical characteristics of the sediment particles.

The interrelationships of these factors determine the type and amount of sediment and the amount of energy available for the stream to entrain and transport the particles. When stream energy exceeds the sediment supply, channel *degradation* occurs (Fig. 9.3). Localized removal of channel bed material by flowing water is called channel scour. On the other hand, when sediment supply exceeds stream energy, *aggradation* occurs within the channel. For a particular stream and flow condition, a relationship between transport capability or capacity and supply can be developed (Fig. 9.4). The wash load, illustrated in Figure 9.4, consists of silts and clays, generally 0.0625 mm or smaller. Sediment supply generally limits total sediment transport for smaller particles. As material gets larger, total sediment transport is more likely to be limited by transport capability.

The processes of aggradation and degradation are important when considering stream dynamics because these are the main mechanisms for sediment storage and release, respectively, in a stream channel (Fig. 9.3). When aggradation occurs, excess material is deposited; eventually a new slope is established that equals the upstream slope. The new equilibrium slope established by aggradation can carry the incoming sediment, but the downstream slope has not adjusted and deposition occurs. This obstructs the flow, and deposition also occurs above the reach. As a result, deposition occurs above and below the reach, raising the bed parallel to the new equilibrium slope. The rate of this movement decreases with time or with downstream advance, and so does the rate of aggradation. The channels upstream and downstream from the

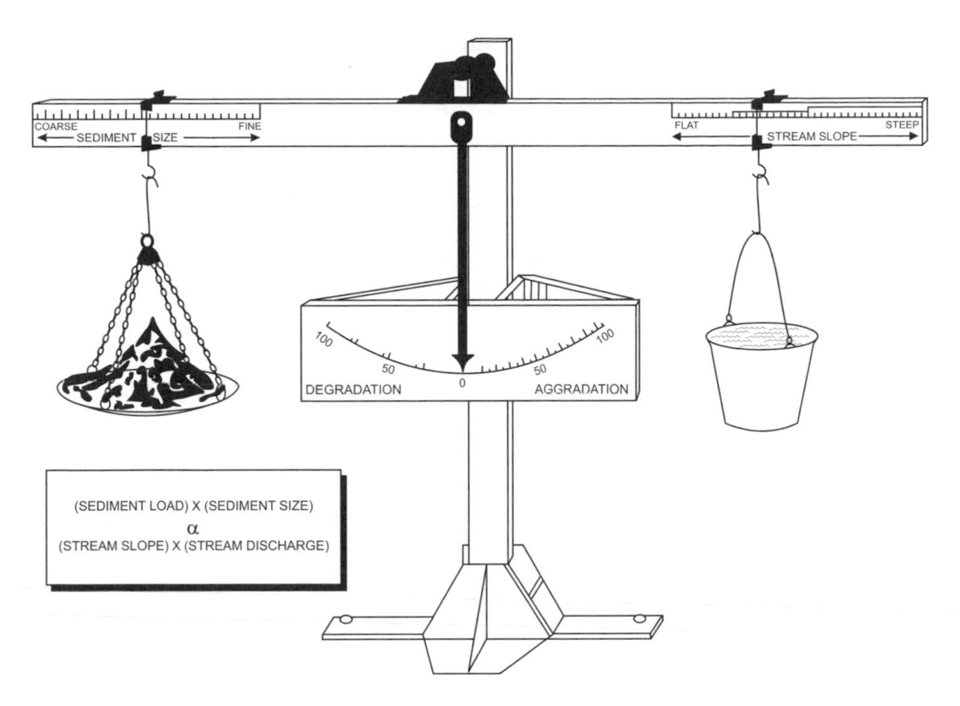

FIGURE 9.3. Relationships involved in maintaining a stable channel balance (from Lane 1955, based on Rosgen 1980).

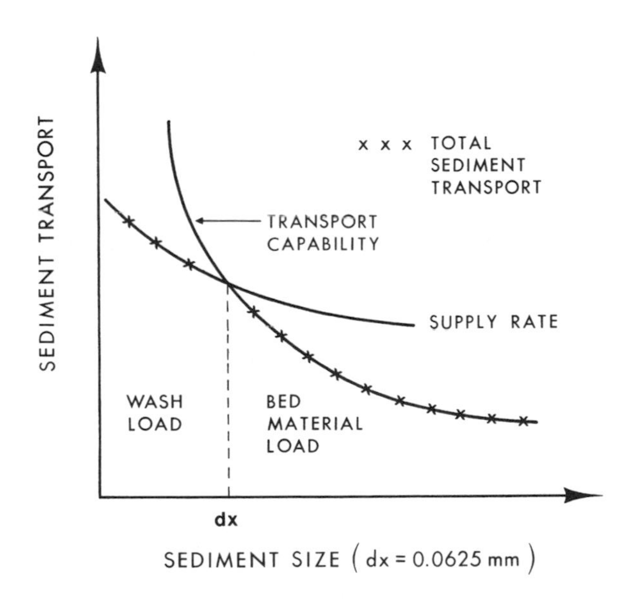

FIGURE 9.4. Rate of sedimentation as affected by transport capability and supply rate for different-sized particles for a particular stream and flow condition (from Shen and Li 1976, as presented in Rosgen 1980).

aggrading front behave like two different reaches with different flows until aggradation ceases and dynamic equilibrium (see below) is restored.

Aggradation generally creates a somewhat convex longitudinal profile in a stream system, and coarse material is deposited first while the finer material moves farther downstream. As a consequence, particle size decreases downstream. When aggradation occurs, the streambed rises slowly and there is a tendency for the water to flow over the banks. This process can lead to natural levee formation. The continual deposition and aggradation ultimately lead to a braided river.

In contrast to aggradation, *degradation* involves sediment removal. Relatively speaking, natural degradation is usually a slow process. When the dynamic equilibrium of a stream system is disturbed, rapid degradation can occur. This initially rapid degradation following disturbance diminishes slowly over time. The profile of a degrading stream system is concave, and the channel cross sections tend to be V-shaped. During degradation, soil particles are picked up from the bed until the load limits are reached. The formation of V-shaped cross sections occurs because of the variations in flow resistance across the channel. Normally, the carrying capacity for sediment is lower near banks than in the center of the stream channel because of bank roughness. Therefore, more material is picked up in the center of the stream, causing the V-shaped profile.

Aggradation and degradation processes also lead to shifts in channel morphology, as discussed later.

Dynamic Equilibrium

Stream channels are in a constant state of change because of the processes of aggradation and degradation, as illustrated in Figure 9.3. The concept of *dynamic equilibrium* is useful in describing stream systems and their stages of development (Heede 1980). When a channel system is in dynamic equilibrium, it is sufficiently stable so that compensating changes can occur without significantly altering this equilibrium. This resilience, or resistance to rapid change, results from internal adjustments to change in flow or sediment movement that are made by several factors (vegetation, channel depth, stream morphology, etc.) operating simultaneously in the system.

Several visual features indicate when streams are not in equilibrium. These features include channel headcuts, underdeveloped drainage nets (such as those having channelized watercourses on only one-half or less of the watershed area), frequent bed scarps, and the absence of a concave longitudinal profile where watershed conditions are relatively constant. Channel headcuts are sources of local erosion and indicate that the stream length and gradients have not allowed an equilibrium condition to develop. Bed scarps develop at nickpoints and indicate marked changes in longitudinal gradients. These scarps proceed upstream until a smooth transition between upstream and downstream gradient is attained.

Streams in dynamic equilibrium do not have headcuts. Watercourses that begin high on the watershed form a smooth transition between the unchannelized area and the defined channels. Bed scarps are not developed and the longitudinal profile is concave. In general, flow discharge, depth, and width increase downstream, and gradient and sediment particle size decrease, if watershed conditions are relatively constant over long stream reaches. As a consequence, sediment production is negligible.

MEASUREMENT OF SEDIMENT

Suspended Sediment

Various techniques are available to estimate suspended sediment concentrations. The collection of grab samples is a common procedure, especially in small streams. However, this method might not be reliable because of the variability in sediment concentrations. Single-stage samplers consisting of a container with an inflow and outflow tube at the top are used on small, fast-rising streams. A single-stage sampler begins its intake when the water level exceeds the height of the lower inflow tube and continues until the container is full. Because only the rising stage of the hydrograph is sampled, the use of such data is limited.

Depth-integrating samplers minimize the sampling bias involved with single-stage samplers. A depth-integrating sampler (such as the DH-48 or DH-49) has a container that allows water to enter as the sampler is lowered and raised at a constant rate. Consequently, a relatively uniform sample for a given vertical section of a stream is obtained. Depending upon the size of the stream, a number of these samples can be taken at selected intervals across the channel. Each suspended sediment measurement should be accompanied by a measurement of streamflow discharge through the channel cross section.

After a suspended-sediment sample is obtained, the liquid portion is removed by evaporating, filtering, or centrifuging, and the amount of sediment is weighed. The dry weight of suspended sediment is commonly expressed as a concentration in milligrams per liter or parts per million (ppm) of water. Usually, measurements of suspended sediment and bed load are made separately because of differences in the sizes of the particles and in the distribution of particles in a stream.

Bed Load Measurement

Bed load is more difficult to measure than suspended load. No single device for measuring bed load is reliable, economical, and easy to use. While many bed load samplers exist, the Helley-Smith bed load sampler is the most widely used in the western United States (Leopold and Emmett 1976). Estimates can also be obtained by measuring the amount of material deposited in sediment traps, reservoirs, settling basins, or upstream of porous sediment-collecting dams. These volumetric measurements can be partitioned into sands, gravels, and cobbles to determine the contributions by particle size.

SEDIMENT YIELD

Sediment yield is the total sediment outflow from a watershed or drainage basin measured for a specific period and at a defined point in a stream channel. Sediment yield is normally determined by sampling sediment and relating the results to streamflow discharge or by performing sediment deposit surveys in reservoirs. Estimated sediment yields from major river basins of the world are presented in Table 9.2.

Streams discharging large quantities of sediment annually are those that drain areas undergoing active geologic erosion or improper land use. The degree of aridity also affects sediment yields. Because of lower vegetation densities on arid watersheds, sediment yields

TABLE 9.2. Average annual sediment yield from selected river basins that are among the 21 largest sediment-yielding rivers in the world

River, Country	Drainage Area ($\times 10^6$ km^2)	Average Annual Streamflow Discharge ($\times 10^9$ m^3/yr)	Average Annual Sediment Yield (10^6 t/yr)[a]
Ganges/Brahmaputra, India	1.48	971	1670
Yellow (Huangho), China	0.77	49	1080
Amazon, Brazil	6.15	6300	900
Mississippi, USA	3.27	580	210
Mekong, Vietnam	0.79	470	160
Nile, Egypt	2.96	30	110
La Plata, Argentina	2.83	470	92
Danube, Romania	0.81	206	67
Yukon, USA	0.84	195	60

Source: From Milliman and Meade 1983, by permission.
[a]Metric tons per year.

in relation to streamflow discharge generally are higher. For example, the Mississippi River watershed is 4 times larger than that of the Yellow River and the annual discharge is 12 times greater. The ratio of sediment load to discharge for the Yellow River is 22.04; for the Mississippi, the ratio is 0.36 (Table 9.2). The Mississippi traverses a humid zone, much of the watershed is vegetated, and soil conservation is practiced over extensive farm areas. In contrast, the Yellow River traverses a semiarid region of north central China; it drains an area of deep loess, highly susceptible to geologic erosion, that also has been denuded by centuries of primitive agriculture.

Undisturbed forested watersheds generally yield the lowest amount of sediment of any vegetative cover or land use condition. Examples of annual sediment yields from small forested watersheds compared to larger watersheds with mixed land uses in the United States are presented in Table 9.3.

Sediment Budgets

As suggested earlier, the transport, or routing, of sediment from source areas where active erosion takes place to downstream channels involves many complex processes. A *sediment budget* is an accounting of sediment input, output, and change in storage for a particular stream system or channel reach. It is a simplification of the processes that affect sediment transport and includes consideration of (1) sediment sources (e.g., surface erosion, gully erosion, soil mass movement, bank erosion), (2) the rate of movement from one temporary storage area to another, (3) the amount of sediment and time in residence in each storage site, (4) linkages among the processes of transfer and storage sites, and (5) any changes in material as it moves through the system. In essence, the sediment budget is a quantitative statement of rates of production, transport, and discharge of sediment. To account properly for the spatial and temporal variations of

TABLE 9.3. Sediment yield from small forested watersheds and larger watersheds of mixed land use in the United States

Region	Number of Watersheds	Sediment Yield (t/ha/yr)[a]	
		Mean	Range
East			
Forested	65	0.17	0.02–2.44
Mixed use	226	0.35	0.02–4.42
West			
Forested	80	0.16	0.02–1.17
Mixed use	312	0.42	0.02–13.38
Pacific Coast			
Forested	26	3.93	0.04–43.56
Mixed use	103	10.37	0.13–111.86

Source: From Patric et al. 1984.
[a]Metric tons per year.

transport and storage requires rather sophisticated models. A discussion of such models is beyond the scope of this chapter.

Sediment Delivery Ratio

A commonly used method for relating erosion rates to sediment transport is the *sediment delivery ratio* (D_r), defined as:

$$D_r = \frac{Y_s}{T_e} \tag{9.12}$$

where Y_s = sediment yield at a point (weight/area/yr) and T_e = total erosion from the watershed above the point at which sediment yield is measured (weight/area/yr).

The sediment delivery ratio is affected by the texture of eroded material, land use conditions, climate, local stream environment, and general physiographic position. Generally, as the size of the drainage area increases, the sediment delivery ratio decreases (Fig. 9.5). Such relationships should only be used to provide rough approximations. As discussed previously, erosion and sediment concentrations can vary greatly for any given watershed.

Before sediment delivery ratios or more detailed sediment routing models can be developed, erosion and sediment data must be collected. Often, these data are not available for upland watersheds.

CUMULATIVE WATERSHED EFFECTS ON SEDIMENT YIELD

Sediment yield can be altered by any changes on a watershed that influence sediment deposition to a channel, streamflow rates, or both. Likewise, any changes to the channel itself that affect stream slope, channel roughness, and channel morphology can alter aggradation-degradation processes and sediment delivery. Among the array of human

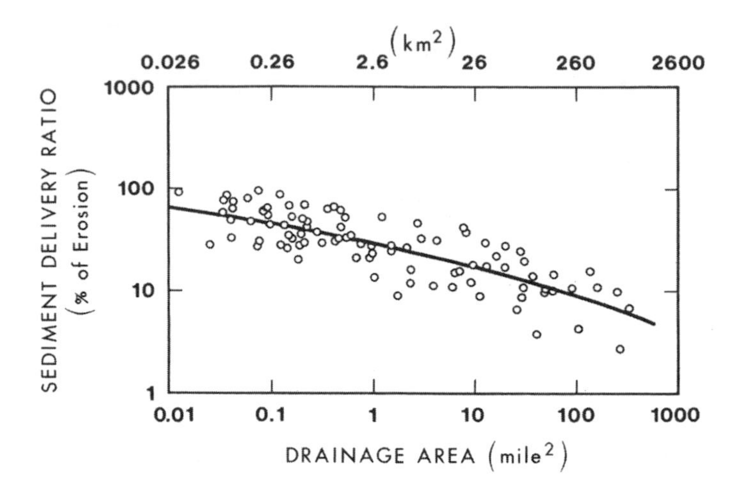

FIGURE 9.5. Sediment delivery ratio determined from watershed size (from Roehl 1962).

activities on watersheds that can affect these relationships are altering vegetative cover, agricultural crop cultivation, urbanization, road construction and maintenance, wetland drainage, and ditching and stream channelization (Dissmeyer 2000). Often, such changes occur piecemeal on a watershed and are implemented incrementally over time. However, the cumulative effect of such changes can impact the dynamic equilibrium of streams (Box 9.2).

Many studies describe the effects of various land use activities on sediment yields without identifying the sediment source. Sediment yields from logging activities and roads are widely documented (Reid 1993, Megahan and Hornbeck 2000). These studies generally show a 2- to 50-fold increase in sediment yields, with most of the increase associated with improperly aligned roads. Increases in sediment input can be much larger on sites where landslides are common. Sediment yields decrease relatively rapidly once road use is discontinued and logged areas regenerate.

Studies that considered only the effects of roads on sediment delivery in the Oregon Coast Range indicated that road density alone could not explain the rates of sediment yield from surface erosion (Luce et al. 2001). These studies suggested several direct and indirect effects that roads can have on sediment delivery including:

- Road slope, length, and condition of the surface as affected by traffic and maintenance.
- Distance from the stream channel and the road drainage system (culverts, ditches, etc.) that affects the volume and rate of sediment and water entering stream channels.
- Gullies and mass wasting initiated by road construction and drainage on hillslopes.
- Failures of culverts due to blockage or poor design.
- Effects of roads on peak flows, for example, as caused by intercepting subsurface flow and/or concentrating surface runoff from the road surface.
- Roads located in the riparian zone or floodplain; cut and fill activities; use of riprap that changes the channel form, and so forth.

Box 9.2

Cumulative Watershed Effects on Stream Sediment: Two Examples

Minnesota River Basin

The 43,264 km^2 Minnesota River Basin is the largest tributary to the Mississippi River in Minnesota. The 558 km river flows through some of the state's richest agricultural land. Like many basins in the midwestern United States, the basin has undergone major changes since the arrival of European settlers. Conversion from prairie and savannah vegetation to agricultural cropping, drainage of wetlands, urbanization, construction of roads, construction of levees, river channelization, removal of native riparian forests, and development on floodplains have changed the landscape and hydrologic characteristics of the basin. Over time, the rates of erosion and the magnitude of streamflow volumes and peaks have increased, resulting in greater sediment levels in the river. Thirteen years of data at Mankato indicated a median suspended sediment concentration of 92 mg/l (Magner et al. 1993) and an annual sediment load of 26 metric tons/km^2 (MPCA 1994). Improved agricultural cultivation practices and conservation practices within the Conservation Reserve Program (see Chapter 12) that promoted perennial vegetation in erosive areas were implemented in the 1980s to reduce erosion and sediment loads in the river. Consequently, surface erosion has been reduced in much of the basin. However, higher sediment loads have been observed in the 1980s when compared with the 1970s; these can be partly accounted for by higher river flows (Klang et al. 1996). It is suspected that the extensive loss of wetlands (see Chapter 14) and the associated ditching and drain tiling in the basin have cumulatively increased average annual peak flows (Miller 1999). As a result, even though surface erosion rates have been reduced, with the river degrading, much of the sediment is now derived from channel sources.

Nemadji River

The Nemadji River, which drains an area of about 1120 km^2 that straddles the Minnesota-Wisconsin border south of Duluth, Minnesota, has the highest suspended sediment load per square kilometer of all monitored rivers in either Minnesota or Wisconsin (NRCS 1998). Over 118,000 t of suspended sediment and 4000 t of bed load are discharged into Lake Superior annually. Hillslopes are unstable and soil mass movement is common in this area that is dominated by clay deposits and clayey glacial till. Since the middle 1880s, sediment deposition from the Nemadji River into Lake Superior has increased. Changes in watershed conditions that possibly explain the increased sedimentation include (1) intensive timber harvesting of mostly white (*Pinus strobus*) and red pine (*P. resinosa*) forests in the middle to late 1800s and the associated logging activities; (2) channel cleaning, straightening, and dam building that occurred

(continues)

(continued)

in conjunction with logging; (3) replacement of the pine forests with pioneer aspen (*Populus* spp.) forests and agricultural cropping; (4) road construction and maintenance; and (5) degradation of riparian areas.

The cumulative effects of these activities include increased streamflow volumes and peaks, wetter soils and accelerated rates of soil slumping along hillsides adjacent to channels, unstable stream channels, and consequently, increased channel erosion (Riedel 2000) and increased sediment transport into Lake Superior. Reversing these cumulative effects will require a comprehensive restoration program that includes efforts to (1) restore pine forest cover, (2) improve road maintenance and drainage, (3) manage riparian areas, and (4) restore stream channels (as discussed in Chapter 10).

Urbanization also increases sediment yield in many instances. Walling and Gregory (1970) reported 2- to 100-fold increases in suspended sediment below building construction sites. Wolman and Schick (1967) estimated increases in sediment yields of 700–1800 t/1000 population due to construction activities in the metropolitan area of Washington, DC.

While sediment yields are often an indicator of land use impact, they are also difficult to interpret in terms of causal effects without information on how the sediment was produced. Van Sickle (1981) summarized annual sediment yields from many Oregon watersheds and showed that year-to-year variation was large. To arrive at an accurate long-term average value, therefore, it is necessary to measure sediment yields over a long time period. Roels (1985) and others have stated that many results obtained from plot-sized experiments cannot be used generally to estimate erosion rates and sediment yields on a watershed basis because of the inherent limitations in the study designs. All sediment measurement and monitoring programs should be carried out within the framework of statistically valid sampling schemes.

The diversity of land use changes that occurs spatially and temporally on a watershed can add up to significant changes in sediment yield from the watershed. Reductions in riparian vegetation and streambank alterations that increase streambank erosion can increase sediment yield. Increases in streamflow resulting from reduced time of concentration of runoff (see Chapter 17) can lead to changes in stream channel dimensions to accommodate the modified flow. Channel erosion, primarily lateral extension, which increases channel width, adds to sediment supply. Reductions in flow can also lead to channel adjustments and alterations of sediment yield.

Summary

Surface erosion, gully erosion, soil mass movement, streambank erosion, and channel scour combine to produce sediment in stream channels. Streams transport sediment as suspended sediment and bed load. The channel processes involved in sediment transport and deposition in a stream system are dynamic and complex. Several important fluvial processes and hydraulic variables are involved in the aggradation and degradation of stream channels during the storage and release of sediment in channels. To obtain a detailed and thorough understanding of these processes would require the studying of separate texts. The purpose of this chapter was to introduce the subject and to provide sufficient information so that you should be able to:

- Explain the fundamental energy relationships of streamflow in a channel that affect streamflow velocity, stream channel erosion, and sediment transport.
- Describe the different types of sediment transport.
- Explain the relationships between stream capacity and sedimentation.
- Explain the conditions under which aggradation and degradation occur in a stream channel.
- Discuss the relationships that are normally found when the hydrograph characteristics of stormflow events are compared to the corresponding suspended-sediment loads.
- Explain the relationships between upland erosion and downstream sediment delivery, and identify the factors affecting a sediment delivery ratio.
- Explain the differences between laminar and turbulent flow, and between subcritical and supercritical flow.
- Discuss why dynamic equilibrium is a useful concept when describing stream systems and their stages of development.
- Provide examples of how cumulative watershed effects can alter sediment transport and delivery.

References

Dissmeyer, G.E., ed. 2000. *Drinking water from forests and grasslands: A synthesis of the scientific literature.* USDA For. Serv. Gen. Tech. Rep. SRS-39.

Dunne, T., and L.B. Leopold. 1978. *Water in environmental planning.* San Francisco: W.H. Freeman and Co.

Hansen, D., and D.I. Bray. 1993. Single-station estimates of suspended sediment loads using sediment rating curves. *Can. J. Civ. Eng.* 20:133–142.

Heede, B.H. 1980. *Stream dynamics: An overview for land managers.* USDA For. Serv. Gen. Tech. Rep. RM-72.

Klang, J., S. McCollor, B. Davis, and J. Porter. 1996. *Sediment concentration reductions in the Minnesota River.* St. Paul: Minnesota Pollution Control Agency.

Lane, E.W. 1955. The importance of fluvial morphology in hydraulic engineering. *Proc. Am. Soc. Civ. Eng.* 81(745): 117.

Leopold, L.B., and W.W. Emmett. 1976. Bedload measurements, East Fork River, Wyoming. *Proc. Nat. Acad. Sci.* 73:1000–1004.

Lopes, V.L, and P.F. Ffolliott. 1993. Sediment rating curves for a clearcut ponderosa pine watershed in northern Arizona. *Water Res. Bull.* 29:369–382.

Lopes, V.L., P.F. Ffolliott, and M.B. Baker, Jr. 2001. Impacts of vegetative practices on suspended sediment from watersheds in Arizona. *J. Water Resour. Plan. Manage.* 127:41–47.

Luce, C.H., B.E. Reiman, J.B. Dunham, J.L. Clayton, J.G. King, and T.A. Black. 2001. Incorporating aquatic ecology into decisions on prioritization of road decommissioning. *Water Resour. Impact* 3(3): 8–14.

Magner, J.A., G.D. Johnson, and T.J. Larson. 1993. The Minnesota River Basin: Environmental impacts of basin-wide drainage. In *Industrial and Agricultural Impacts of the Hydrologic Environment*, eds. Y. Eckstein and A. Zaporozek. Vol. 5:147–162, Alexandria, VA: Water Environment Federation.

Megahan, W.F., and J. Hornbeck. 2000. Lessons learned in watershed management: A retrospective view. In *Land stewardship in the 21st century: The contributions of watershed management*, tech. coords. P.F. Ffolliott, M.B. Baker, Jr., C.B. Edminster, M.C. Dillon, and K.L. Mora, 177–188. USDA For. Serv. Proceed. RMRS-13.

Miller, R.C. 1999. *Hydrologic effects of wetland drainage and land use change in a tributary watershed of the Minnesota River Basin: A modeling approach.* Masters Thesis. University of Minnesota, St. Paul.

Milliman, J.D., and R.H. Meade. 1983. World-wide delivery of river sediments to the oceans. *J. Geol.* 91:1–21.

Minnesota Pollution Control Agency (MPCA). 1994. *Minnesota River Assessment Project Report.* Report to the Legislative Commission of Minnesota Resources. St. Paul: MPCA.

Morisawa, M. 1968. *Streams: Their dynamics and morphology.* Earth and Planet. Sci. Ser. New York: McGraw-Hill.

Natural Resource Conservation Service (NRCS). 1998. *Erosion and sedimentation in the Nemadji River Basin.* Nemadji River Basin Project Final Report. Duluth, MN: NRCS.

Patric, J.A., J.O. Evans, and J.D. Helvey. 1984. Summary of sediment yield data from forested land in the United States. *J. For.* 82:101–104.

Reid, L.M. 1993. *Research and cumulative watershed effects.* USDA For. Serv. Gen. Tech. Rep. PSW-GTR-141.

Riedel, M.S. 2000. Geomorphic impacts of land use on clay channel streams. Ph.D. Thesis. University of Minnesota, St. Paul.

Roehl, J.W. 1962. Sediment source area delivery ratios and influencing morphological factors. *Int. Assoc. Hydrol. Sci. Publ.* 59:202–213.

Roels, J.M. 1985. Estimation of soil loss at a regional scale based on plot measurements—Some critical considerations. *Earth Surface Process. and Landform* 10:587–598.

Rosgen, D.L. 1980. Total potential sediment. In *An approach to water resources evaluation, nonpoint silvicultural sources.* VI.1–VI.43, Environ. Res. Lab., EPA-600/880012. Athens, GA: Environmental Protection Agency.

Shen, H.W., and P.Y. Julien. 1993. Erosion and sediment transport. In *Handbook of Hydrology*, Chapter 12, ed. D.M. Maidment, 12.1–12.61. New York: McGraw-Hill.

Shen, H.W., and R.M. Li. 1976. Water sediment yield. In *Stochastic approaches to water response*, ed. H.W. Shen, 21–68. Fort Collins: Colorado State Univ.

Simons, D.B., and F. Senturk. 1977. *Sediment transport technology.* Fort Collins, CO: Water Resources Publications.

Swanson, F.J., R.J. Janda, T. Dunne, and D.N. Swanston. 1982. *Sediment budgets and routing in forested drainage basins.* USDA For. Serv. Gen. Tech. Rep. PNW-141.

Toy, T.J., ed. 1977. *Erosion: Research techniques, erodibility, and sediment delivery.* Norwich, CT: Geo. Abstracts Ltd.

Toy, T.J., and R.F. Hadley. 1987. *Geomorphology and reclamation of disturbed lands.* Orlando: Academic Press.

Van Sickle, J. 1981. Long-term distribution of annual sediment yields from small watersheds. *Water Resour. Res.* 17:659–663.

Walling, D.E., and K.J. Gregory. 1970. The measurement of the effects of building construction on drainage basin dynamics. *J. Hydrol.* 11:129–144.

Wolman, M.G., and A.P. Schick. 1967. Effects of construction on fluvial sediment, urban and suburban areas of Maryland. *Water Resour. Res.* 3:451–464.

CHAPTER 10 *Stream Channel Morphology and Stream Classification*

INTRODUCTION

Over geologic time there is an ongoing struggle on the landscape between the uplifting of land masses and the weathering processes of erosion and sediment transport that work to lower land masses to sea level. These processes define the form of the landscape and delineate its watersheds and subsequent stream channels. Flowing water alters the topography and, in the process, forms stream channels that are the conduits for moving water and sediment. The study of these processes and relationships has been the purview of fluvial geomorphology. With few exceptions, it has only been in recent decades that hydrologists and watershed managers have recognized the importance of fluvial geomorphology in hydrology and watershed management. This perspective helps us to better understand how land use activities affect the flow of water and sediment. Through field observations of land and channel forms, it is possible to link geomorphic features with hydrologic processes and, further, to link human-induced changes on the watershed with changes in the stream channel.

The previous chapter discussed basic energy relationships and hydraulic principles that help explain geomorphic processes. The intent of this chapter is to link these hydraulic principles and fundamental energy relationships with stream channel form and function. Furthermore, the application of geomorphic stream classification approaches is discussed in terms of better understanding the hydrologic and cumulative land use effects on stream channels. At the conclusion of this chapter, the implications for watershed management should be apparent.

BASIC CONCEPTS OF FLUVIAL GEOMORPHOLOGY

A stream channel is formed by the energy of flowing water and factors that affect that flow. Topographic relief, geologic strata, parent material and soils, vegetation, and characteristics of the riparian zone all play a role. Water initially flows over the land surface and/or through the soil systems's micro- and macropore system. Such flow, for the most part, is episodic. At the uppermost parts of the watershed, where flow becomes concentrated and has sufficient kinetic energy to pick up and transport soil particles, it begins to cut its channel. At first this channel may be a small rivulet that soon combines with other rivulets, gaining greater velocity. The channel can be fed by subsurface flow from the hillslope and channel bank, overland flow, and groundwater flow as it continues downstream.

A river network can be viewed as having an upper erosion zone, a transition zone, and a deposition zone (Schumm 1977). The erosion zone is the upper reach of a river basin and has the steepest hydraulic gradient. With the coincident increase in mass of streamflow, the channel deepens, widens, and begins its course to the ocean, in accordance with the energy relationships of water and sediment flow discussed in Chapter 9. The main stem of the river is formed and is considered a transition zone between the upper erosion zone and the lower deposition zone at the river estuary. In the transition zone, there is little gain or loss of sediment over the long term. The deposition zone has the lowest hydraulic gradient and is evident by the vast deltas formed by major rivers (e.g., the Nile River and the Mississippi River deltas).

During storm events or snowmelt episodes, the magnitude of flow increases, resulting in increased energy that can pick up, move, and transport sediment. The larger the stormflow event, the greater the energy and thus the work on the channel and its sediments. Over many years of episodic stormflow events followed by longer periods of baseflow, channels in the various reaches of the river basin take on their form. Depending on their position in the basin and the corresponding hydraulic gradient, the type of material present, and other factors discussed below, the channels will take on various forms.

Rivers, Floods, and Floodplains

Streams and rivers form *channel banks* as they cut into the earth in their journey to the ocean. These banks form over time and become established at a height that confines the stream for all but the larger streamflow events in a year. When flow increases to the point where the elevation of the stream exceeds the elevation of the streambanks, part of the stream leaves its channel and a flood occurs. Except in steep mountainous terrain, or severely entrenched streams, when flooding occurs, the overbank flow spreads out onto the *floodplain*. Here, the velocity is reduced because of the increased roughness that is encountered (riparian vegetation, debris deposits from previous floods, etc.). As a result, sediment and debris are deposited, forming a relatively flat feature that is adjacent to the channel called the *floodplain*. The elevation at the top of the streambank where water begins to overflow onto the floodplain is called the *bank-full stage*. For most rivers, water flows onto the floodplain about once every year. Over time the river lowers its elevation, and over time new floodplains develop. Terraces are old floodplains that have been abandoned as the stream channel lowers its elevation. These features are illustrated in Figure 10.1.

Over time, a river will experience floods of various magnitudes. Major flood events with deeper flow on the floodplain and higher velocities can cause considerable erosion and sediment transport. A new channel may be cut into the floodplain at certain locations, and as the flood recedes, more sediment and debris are deposited. As the

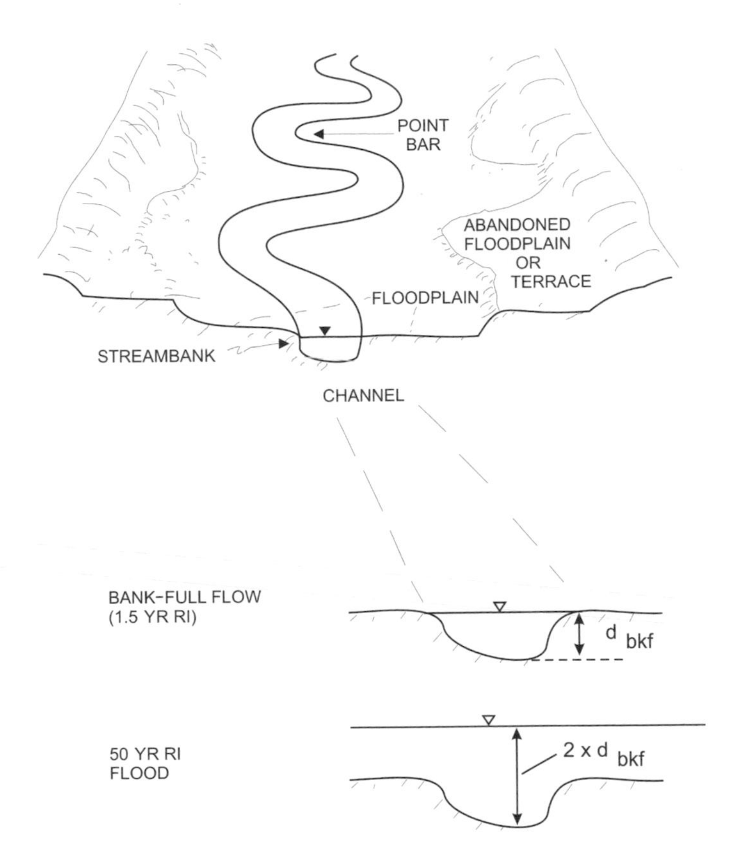

FIGURE 10.1. Landforms in a river valley and flow conditions in a channel for bank-full and the 50 yr recurrence Interval (RI) flood (modified from Fitzpatric et al. 1999 and Verry 2000).

hydraulic gradient is reduced (in contrast to the steepest gradients in mountain streams), streams meander—that is, they move laterally throughout the valley floor by eroding one bank and depositing sediment (point bar) on the opposite side of the channel. As a result of both flood events and natural channel processes, streams migrate laterally within the floodplain. As pointed out by Dunne and Leopold (1978), in a flat valley floor over a long period of time, eventually the stream channel will have occupied all positions in the valley. As the channel meanders, it maintains its depth and width dimensions, unless there is a substantial shift in either flow regimes and/or sediment deposition in the channel. Over time, new sections of channel become formed and other sections of the previous channel become abandoned—this is how most valleys are formed.

Bank-Full Stage

The *bank-full stage* is an important feature of stream channels. As stated by Rosgen (1996), "the bank-full stage and its attendant discharge serve as consistent morphological indices [that] can be related to the formation, maintenance, and dimensions of the channel as it exists under the modern climatic regime." For most streams, the discharge associated with the bank-full stage has a recurrence interval that averages between 1.4 and 1.6 yr. This discharge is the most effective at doing the work that maintains the channel's morphological characteristics. The reason for this is best illustrated in Figure 10.2. Although the largest streamflow discharges that have occurred historically in a channel have the greatest power to form a channel and transport the most sediment, they occur infrequently. The product of the transport rate and the frequency of occurrence is at a maximum at some point between the most frequent discharge level of the stream and the highest levels of flow; this maximum point is the effective discharge (Wolman and Miller 1960). This effective discharge is generally associated with that at the bank-full stage. This tells us that the flows at or near the annual maximum peak flow discharge are the most important in working the channel and its sediments. As a result, the bank-full stage has become the key, or standard, for describing stream channel morphological features (e.g., width, depth, entrenchment, etc.), and the basis for morphological classification (described later). Does the bank-full discharge for all streams coincide with a 1.5 recurrence interval? Surprisingly, this seems to be consistent. However, as Rosgen (1996) points out, stream channels with bed material that consists of large cobbles and boulders require discharges of a larger magnitude (and with a longer return period) than flows near the average annual event. Nevertheless, most stream channels seem to be formed and maintained by the more frequently occurring flood events rather than the extreme, large floods.

The Floodplain

The relationship between a river or stream channel and its floodplain is an important one. During flood events, the floodplain becomes an integral part of the river. As indicated earlier, this is not a rare occurrence, but this fact is often ignored or not well understood by those who insist upon cultivating, urbanizing, and/or otherwise developing the floodplain. It is important to note here the distinction between the floodplain and terraces, which represent the old, abandoned floodplain (Fig. 10.1). While the active floodplain experiences frequent inundation, the higher elevation terrace will be flooded less frequently, if at all. As pointed out by Dunne and Leopold (1978), it is therefore important for developers to recognize the difference between the two.

When *overbank flow* occurs and moves onto natural (unaltered) floodplains, the flow encounters a higher level of resistance or roughness that is due to natural riparian vegetation and debris from previous floods. The reduced velocity has important implications on the floodplain and on downstream areas. In the immediate vicinity of the floodplain, a given discharge will have a higher stage as a result of the reduced velocity. This means that the area inundated by the flood is larger than it would otherwise be with a higher velocity. The accumulation of sediment and debris can further add to the height of the stage and increases roughness. For floods with a more frequent occurrence, such as the average annual flood (2 yr recurrence interval), the 5 yr recurrence interval flood, etc., considerable deposition of sediment can occur. More extreme floods can scour a new channel and accelerate channel migration in the floodplain, as mentioned previously.

By reducing the velocity of flow over the floodplain and by providing storage, floodplains have a cumulative effect of reducing flood peaks downstream. Furthermore, there is evidence the saturated soils and residual pools of water retained in the floodplain can reduce nitrogen levels that would otherwise be transported downstream (Cirmo and

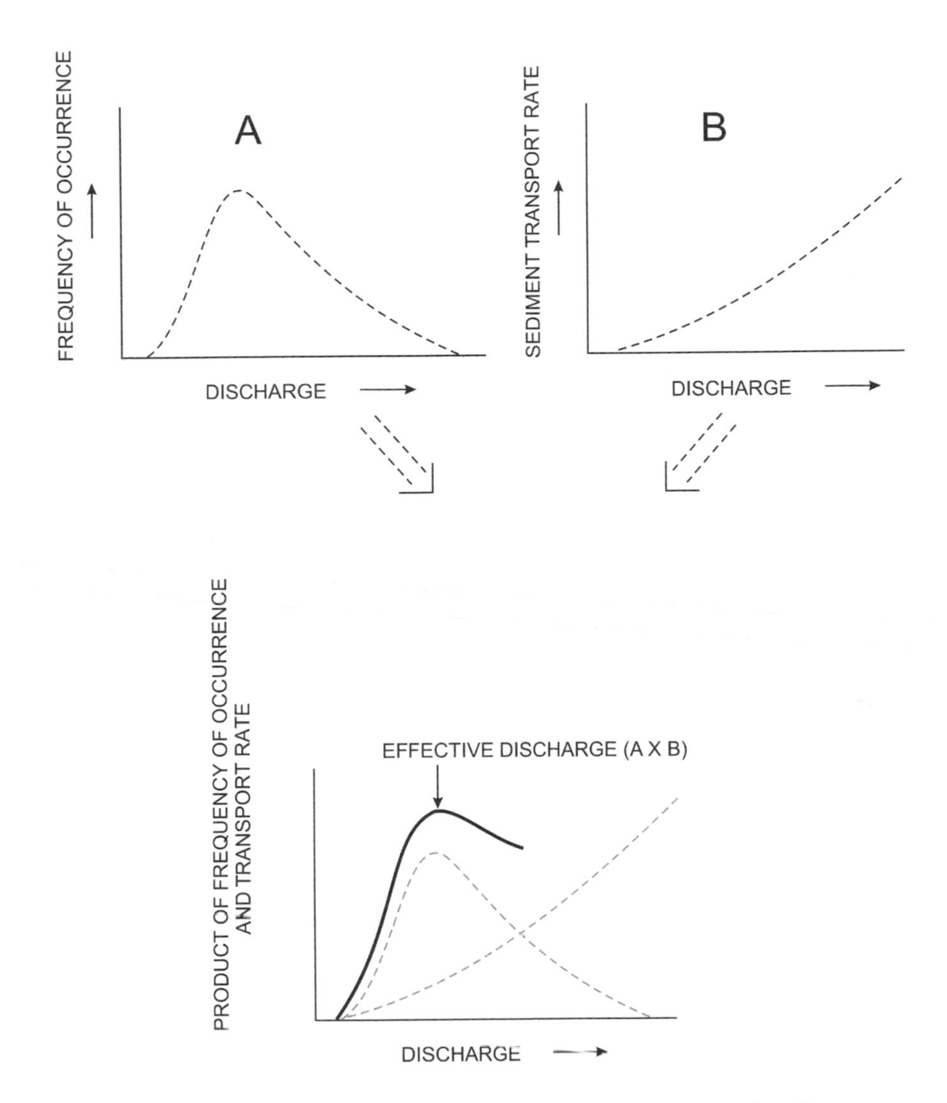

FIGURE 10.2. Effective discharge as determined by the product of sediment transport rate and frequency of occurrence of streamflow discharge (from Wolman and Miller 1960 as presented by Rosgen 1996).

McDonnell 1997). Uptake by riparian vegetation and denitrification (Chapter 11) can reduce nitrogen levels in the floodplain. Finally, fish yield and production are strongly related to the amount of floodplain accessible by a water body (Junk et al. 1989).

Floods, Flood Damage, and Human Interventions

There appears to be a widespread misconception held by the public about floods, their occurrence, and their causes. The above discussion emphasizes the fact that floods are naturally occurring phenomena that are integral to the formation and maintenance of river channels and floodplains. Yet when floodplains are flooded, lives lost, and/or property damaged, there is a tendency by the public to view such events as unexpected, extraordinary, and extreme. Following large floods, there is often an outcry for an explanation, which translates into someone to blame.

Land use changes can affect the magnitude of flood discharges up to a point. Any changes on the watershed and/or channel that either (1) reduce storage or (2) facilitate the conveyance of flow have the potential to increase flood peaks. As discussed in Chapter 6, changes in land use that reduce storage or improve conveyance are generally thought to have a greater influence on peak flow events of a smaller magnitude, such as the average annual flood peak (2 yr recurrence interval), than on the extremely large events with recurrence intervals greater than 100 yr. Many studies have suggested that effects of land use can influence events associated with 25 to 50 yr recurrence intervals. As the amount of rainfall and/or snowmelt becomes greater, the storage capacity of any watershed system will be exceeded, and discharge will become excessive. Eventually the effects of human activities on peak flow discharge will diminish dramatically. It is not clear at what magnitude or flood frequency land use no longer influences peak flow discharge. Intuitively, one would think that the most extreme events, such as the 500 yr recurrence interval flood, would be minimally affected by land use conditions. Some have suggested that for substantial land use change on a watershed, such as clearcutting a forest (Lu 1994) or widespread drainage of wetlands with channel modifications (Miller 1999), frequency curves for conditions before and after the change converge near the 100 yr recurrence interval.

Quite separate from anthropogenic influences on the magnitude of flood flows are the effects of human occupancy in flood prone areas that exacerbate flood damages and loss of life. Encroachment onto the floodplain and flood-prone areas is the single most important cause of flood damages. The causes of such encroachment can be many, including land scarcity, the desire to live near streams and rivers, and so forth. In addition, the construction of levees, flood control dams and diversions, and channel modifications aimed at reducing flood frequency encourage floodplain development. As described in Box 10.1, flood control measures themselves can have unwanted effects. A false sense of security can accompany various flood control measures; people believe that dams and levees will protect them from all flooding, even if they occupy the floodplain. Commonly, there is another important factor that influences redevelopment on floodplains: short memories. As a result, efforts to reduce flood damages today emphasize floodplain zoning and management that restrict human activities in hazardous areas rather than trying to control the occurrence and magnitude of floods.

Stream Channel Changes in Response to Human Activities

Changes in streamflow regime, channel modifications, and floodplain alterations not only affect flooding, as discussed above, but can collectively cause channel morphology to change. Streams are in a state of dynamic equilibrium as discussed in Chapter 9. The form of the stream channel is determined by, and characterized by, eight interrelated variables: total discharge, stream velocity, hydraulic gradient, roughness of the bed and bank, concentration of sediment, size of sediment, width, and depth (Leopold et al. 1964). Changes in any of these variables will affect degradation or aggradation, and thereby the channel form.

Land use on the watershed affects the quantity of streamflow and streamflow regime, which in turn can alter the degradation-aggradation process. For example, urbanization (Box 10.2), conversions from natural forests to agricultural crop lands,

Box 10.1

Assessment of the 1993 Flood on the Upper Mississippi River, USA

The Mississippi River flood of 1993 caused more than $10 billion in damages (Tobin and Montz 1994). In the aftermath of this flood, many questioned the extent of flood damages, given the extensive flood control management and engineering works in the basin. Abnormally heavy rainfall in the fall of 1992 persisted in the spring and early summer of 1993 over large areas of the Upper Mississippi River Basin. Record flood stages were recorded in many areas. Some suggested that the drainage of wetlands and extensive use of drain tiles in farmlands added to the flooding. A forum of hydrologists determined that agricultural drainage was not an important contributor to this flood, given the extreme nature of the event (University of Minnesota 1997). Others pointed out that the extensive use of levees and river channelization measures that were aimed at reducing flood damages, may actually be offset by the loss of flood reduction benefits of natural meandering rivers, riparian systems, and floodplains.

Leopold (1994) compared flood stages that occurred during the 1993 flood with those that would have occurred for the same discharges, but under the conditions prior to 1927, the year that intensive construction of levees began on the river (see below). River stages at several locations in 1993 were substantially higher as a result of levees that confine the river and prevent it from overflowing onto the floodplain.

Location	1993 Stage (m)	Pre-1927 Stage (m)	Stage Difference (m)
St. Louis, MO	14.8	11.8	3
Chester, IL	14.8	10.0	4.8
Keokuk, IA	8.2	7.0	1.2

Two points need to be made here: (1) as long as levees are not washed out or breached, the increased height of flood stages in the channel does not cause damage. However, when levees fail or are breached, as occurred in 1993, the increased stages result in greater devastation than would have occurred without levees. There is the potential for a cumulative effect of increasing flood stages downstream as the intensity of levee construction and channelization increase.

removal of perennial vegetative cover, and drainage of wetlands can increase streamflow discharge and, importantly, peak flow discharges associated with the 1.4–1.6 yr recurrence interval. A higher discharge for the bank-full recurrence interval, or viewed another way, a more frequent occurrence of bank-full flow, can tip the balance toward

Box 10.2

Urbanization Effects on Stream Channel Stability

The process of urbanization, which increases the impervious area of a watershed, brings about some of the most dramatic changes in streamflow of any land use conversion. The extent to which urbanization subsequently affects downstream channels depends on several factors, including: (1) the original watershed condition that has undergone urbanization; (2) the percentage of the watershed that is urbanized; (3) the extent of soil erosion and sediment delivery during construction phases; and (4) the extent to which urbanized areas on the watershed are directly connected to the stream channel.

Bledsoe and Watson (2001) summarized the hydrologic effects of urbanization that impact stream channel stability. They concentrated on changes in the magnitude of the median annual peak flows, stating that they are closely associated with effective channel forming flows normally with a 1.5–2 year recurrence interval. The median annual peak flow increases with increasing imperviousness for watersheds in six different states. For most areas, there was a doubling of median annual peaks with a 10–20% increase in impervious area. The magnitude of response of any given watershed varied according to the pre-urbanization watershed storage conditions, and the connectivity and conveyance systems of the urban areas. The combination of the increased power of streamflow and increased sediment loading translates into greater risk of channel instability. The extent to which stream channel morphology changes to these new flow regimes depends on the resilience of the respective channels, riparian conditions, erodibility of stream beds and banks, mode of sediment transport, and status of stream conditions to geomorphic thresholds.

degradation. Higher discharges mean higher velocities and accelerated scouring of the channel bed and banks. In contrast, land use that adds excessive sediment to the channel from surface, gully, or mass soil erosion can result in channel aggradation.

Floodplain occupancy and the accompanying stream channel modifications (sometimes called "channel improvement projects") and construction of levees or flood walls directly impact channel form. Deepening, dredging (cleaning), and straightening of channels have been carried out with the intent of increasing the velocity of flow for any given discharge and lowering the elevation of streamflow levels that were historically related to flooding. Channel width and depth ratios are significantly changed. Levees confine the flow and reduce upstream storage and consequently increase discharge rates and velocities downstream. The cumulative effects of these activities often result in increased stream gradient, loss of channel roughness and floodplain storage, and the resulting increase in velocity for any given discharge (compared to conditions prior to channel changes).

Stream Evaluation and Classification

The form and functions of a stream channel, the pattern of streamflow, the types of habitats present, and the quality of water are all indicators of watershed and stream channel conditions. It should also be clear that human activities on the watershed and in stream channels can affect the stability, form, and function of streams; the quality of water (see Chapter 11); and the aquatic ecosystem. Over time, there can be a stream reach or even an entire river system that becomes degraded or has undergone changes that indicate some type of management action is needed. Such changes can include excessive streambank erosion, sediment deposition in pools, river widening, increased suspended sediment concentrations, loss of fish habitat, and so forth. In response to such situations, managers need to be able to evaluate, classify, and set priorities for streams.

As stream channels adjust to natural or human-induced perturbations, there is often interest in evaluating the status of streams and ultimately developing some type of remedy to restore or improve the form, habitat, and function of the stream. This interest has lead to numerous approaches for evaluating and classifying stream channels in terms of their morphology, habitat, and related biological conditions. Because of the diverse nature of physical and biological relationships in a stream and the diverse interests of managers (water quality constituents, fish habitat, riparian management, channel restoration, etc.), there is no single classification system that applies for all interests. Some approaches focus on the biological aspects of stream habitats and aquatic organisms (Hilsenhoff 1987, Hawkins et al. 1993). There are certain fundamentals that should be observed in adopting a particular classification system including (Mosley 1994):

- The purpose and application of the classification must be specific and clear.
- Objects (stream channels) that differ in kind should not fit into the same classification.
- Classifications are not absolute and can change with new information.
- Classifications should be exhaustive and exclusive (e.g., any stream channel should be assigned to only one class).
- Differentiating characteristics must be relevant to the purpose of the classification.

A detailed discussion of classification and evaluation methods is not within the scope of this book. Therefore, we present and discuss selected classification and evaluation methods that have been widely applied.

Stream Channel Stability

Because streams and rivers are dynamic systems, the concept of *stream channel stability* may seem hypothetical. Streams are constantly adjusting, as indicated earlier, yet they do assume some features and forms that reflect their natural setting and habitat. In this regard, stream classification provides a systematic approach that yields reproducible and consistent results with respect to some established reference so that streams can be evaluated over time and with one another. Departures from expected conditions due to human interventions can yield obvious indicators that the stream channel is degrading. The challenge, therefore, is to develop some type of framework in which stream channel stability can be evaluated.

A channel evolution model was developed for west Tennessee by Schumm et al. (1984) that provides a basis for studying the stability of alluvial channels (Fig. 10.3). It is suggested that alluvial channels that become destabilized, whether due to human actions or natural disturbance, "pass through a sequence of channel changes with time" (Simon and Rinaldi 2000). The effect of riparian vegetation removal on this process and, conversely, the effect of channel degradation on riparian vegetation is apparent. Beginning with a stable channel (Stage I), removal of riparian vegetation and/or increased stream discharges begins channel entrenchment. As the channel deepens, banks become steepened, slope failures begin to occur, and riparian trees begin to tilt toward the channel (Stages II and III). Channel banks are now destabilized, and mass wasting occurs. As soil material fills into the channel, the stream channel widens (Stage IV) and begins aggradation (Stage V). Degradation proceeds upstream as the reduced gradient in this section can no longer transport sediment from the upstream degraded channel. Eventually, as riparian vegetation becomes reestablished on the channel banks, the banks flatten, and the hydraulic gradient is reduced through meander extension and elongation, and the channel achieves a new dynamic equilibrium (Stage VI). A new, more narrow, floodplain is then formed and the old floodplain becomes a terrace. Under this new situation, flood flows are constrained by the boundaries of the new terraces, which can result in higher velocities and greater erosive power. These unstable channels can cause structural failures of culverts, bridges, and so on, at road crossings (Box 10.3).

Pfankuch (1975) presented a field procedure for inventorying mountainous streams and evaluating their stability. With this procedure, several indicators are used in differ-

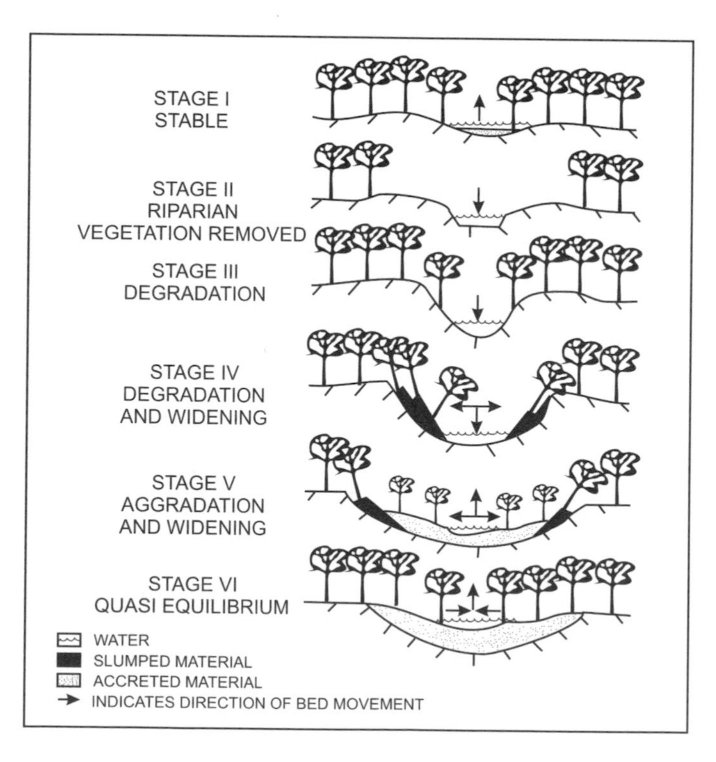

FIGURE 10.3. Channel evolution model developed in west Tennessee (modified from Simon and Hupp 1986 as presented by Simon and Rinaldi 2000).

Box 10.3

Channel Instability: Loess Area of the Midwestern United States (Simon and Rinaldi 2000)

Unstable stream channels are common in the loess area of the Midwest as a result of human activities on watersheds and in stream channels. This area extends south to north from Mississippi to Minnesota and Wisconsin and from east to west from Ohio to Kansas. Since 1910 channels have been enlarged and straightened to reduce flooding along their floodplains. These activities, accompanied by intensive agricultural development, urbanization, road construction, and other land use changes on watersheds have resulted in system wide channel instability. Channel degradation has caused over $1.1 billion damages to bridges, pipelines, and to adjacent lands since the early 1900s. Channels have become "canyon-like" in some areas, while in others they have over widened as a result of accelerated soil mass movement. In western Iowa, nearly 95% of the stream channels were considered unstable (Stages II-V as illustrated in Fig. 10.3).

Land use activities that have contributed to channel instability include:

- Clearing of the land for cultivation, which was accelerated following the Civil War, and which increased runoff, peakflow discharges, erosion of uplands, and gullying of the landscape.
- The removal of woody vegetation along streambanks and riparian areas, including floodplains, which reduced the hydraulic roughness, increased flow velocities and stream power to erode stream beds and banks and transport sediment.
- Channelization projects that dredged and straightened channels, which caused streams to become entrenched.

Channel degradation is widespread in the loess area, but differences in channel instability were pronounced between streams in sand beds and those in silt beds. Silt-bed streams downcut for long periods of time, as much as 70 years in western Iowa. Without coarse-grained sediment for downstream aggradation, energy is dissipated through channel widening. In the silt-bed streams, channel widening is sustained by the high streambanks and low cohesive strength of soils. Where streams cut into sands, aggradation occurs more quickly and channel widening occurs at more moderate rates. Mass wasting into stream channels is a dominant process throughout the region and extends meanders. Some riverside residents with homes and property on bluffs overlooking rivers have become victims of this process.

ent positions in a channel cross section to determine the "resistive capacity of streams to the detachment of bed and bank materials and to provide information about the capacity of streams to adjust and recover from potential changes in flow and/or increase in sediment production." There is a rating score for different indicators in the channel bottom, lower bank (the top of which is bank-full), and upper bank (Fig. 10.4). Each indicator is given a numerical score; lower scores indicate greater stability. The rating form and numerical scores for each indicator are presented in Table 10.1.

Based on the total numerical score, each channel segment is rated excellent, good, fair, or poor. Although somewhat subjective, these categories provide managers with a means to set priorities. A stream reach with a poor rating should get more immediate attention for upstream watershed management and/or channel modifications than should one rated as good or excellent. To be more specific, Pfankuch (1975) points out that stability ratings can be useful in evaluating needs for improvements at culvert and

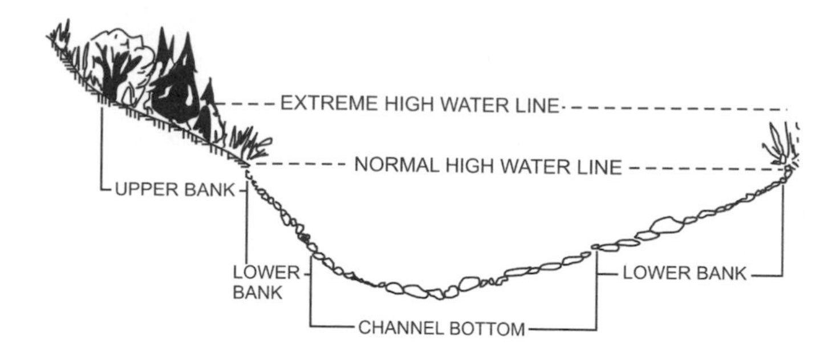

FIGURE 10.4. Stream channel and bank locations used in the Pfankuch (1975) method of streambank stability rating.

bridge sites, campgrounds, or where livestock are concentrated near stream channels. This method has been incorporated into the Rosgen stream classification and assessment approach described in the following section.

Stream Classification—A Morphological Approach

Stream systems are complex. To understand these systems and place them into a management framework requires knowledge of the processes that influence the pattern and character of the systems (Rosgen 1994). However, there is often more information available on stream systems than can be efficiently applied in a management context. Part of the problem confronting watershed managers in this regard is the large number of "pieces" making up the data sets describing the stream form and fluvial processes involved. Definitions of terminology used in stream classification are presented in Table 10.2.

Stream classification is one way in which these "pieces" of data can be brought together, along with the disciplines represented, into a common and usable format. Several methods of classifying streams and rivers have been developed based on hydrologic and geomorphic characteristics (Gordon et al. 1992). Any stream classification scheme risks oversimplification, but it is helpful in organizing the relevant data in a logical and reproducible way. Such a morphological approach has been developed for mountain streams by Montgomery and Buffington (1997). A widely applied method, and one that has been adopted by the USDA Forest Service (Stream Systems Technology Center 2001) is that of Rosgen (1994, 1996). Streams and rivers are classified in this system to:

- Predict a stream's behavior from its appearance.
- Develop specific hydraulic and sediment relations for a given morphological channel type and state.
- Provide a mechanism to extrapolate site-specific data collected on a particular stream reach to those of similar character.
- Provide a consistent and reproducible frame of reference for communication among watershed managers working with stream systems in a variety of disciplines.

The channel width, depth, and slope, the roughness of the channel materials, the stream discharge and velocity, and the sediment load and sediment size all define and determine channel morphology (Leopold et al. 1964). A change in one of these variables sets up a series of channel adjustments that, inevitably, lead to changes in the others, resulting in channel pattern alteration. Because stream morphology is the product of this integrative process, the variables that are measurable should be used as stream classification criteria. This is the basis of the *Rosgen Stream Classification System (RSCS).*

The Rosgen Stream Classification System

The level of classification presented by Rosgen (1994) is commensurate with many planning-level objectives in watershed management. Because these objectives vary, a hierarchy of stream classification and inventories is desirable. Such a hierarchy allows an organization of stream inventory data into levels of resolution ranging from very broad morphological characteristics (Level I) to reach-specific descriptions (Levels II, III, and IV), as described in Table 10.3. Level I classifies the stream from type Aa to G. The pattern of streams is classified as relatively straight and steep (Aa and A stream types), low sinuosity and moderate gradient (B stream types), or low gradient and meandering (C and E stream types). Complex multiple-channel stream patterns are the braided (D) and anastomotic (DA) stream types. Types F and G represent entrenched, unstable streams that characteristically exhibit high bank erosion rates. The Aa and A streams are steep, entrenched, straight reaches with rapidly flowing water typical of high elevation, mountainous streams. As the hydraulic gradient diminishes, streams meander and their sinuosity increases.

Level II further subdivides streams according to slope ranges and dominant channel-material particle sizes. The stream types are assigned numbers related to the median diameter of particles: 1 = bedrock, 2 = boulder, 3 = cobble, 4 = gravel, 5 = sand, and 6 = silt/clay. These separations initially produce 41 major stream types, as shown in Figure 10.5. A range of values for each criterion is presented in the key to classification for the 41 major stream types (Fig. 10.6). The values selected for each criterion were obtained from data for a large assortment of streams throughout the United States, Canada, and New Zealand.

The RSCS is widely applicable under a variety of conditions. It can be applied to either ephemeral or perennial channels with little modification. An important feature in classification, the bank-full stage, where water completely fills the stream channel, can be identified in most perennial channels through field indicators. These bank-full indicators are often more elusive in ephemeral channels. Although originally developed for application in streams of the Rocky Mountains, the RSCS has been successfully applied in low relief terrain and low gradient stream channels in Wisconsin (Savery et al. 2001) and moderate gradient streams in clay-bed channels in Minnesota (Riedel 2000).

The morphological variables considered in the classification often change in short distances along a stream channel, due largely to changes in geology and the tributaries. As a consequence, the morphological descriptions are based on observations and measurements made on selected, representative reaches of the stream (that is, reaches of only a few meters to several kilometers) so that the stream types classified by the system apply only to these selected reaches of the channel. The observations and measurements from selected reaches are not averaged over entire watersheds to classify stream systems.

Continuum Concept

The ranges in slope, width/depth ratio, entrenchment ratio, and sinuosity shown in Figure 10.5 span the most commonly observed values. The entrenchment ratio is

TABLE 10.1 Stream reach stability evaluation

		Stability Indicators by Classes			
	Key#	Excellent		Good	
Upper Banks	1	Bank slope gradient < 30%.	(2)	Bank slope gradient 30–40%	(4)
	2	No evidence of past or any potential for future mass wasting into channel.	(3)	Infrequent and/or very small. Mostly healed over. Low future potential.	(6)
	3	Essentially absent from immediate channel area.	(2)	Present but mostly small twigs and limbs.	(4)
	4	90%+ plant density. Vigor and variety suggests a deep, dense, soil-binding root mass.	(3)	70–90% density. Fewer plant species or lower vigor suggests a less dense or deep root mass.	(6)
Lower Banks	5	Ample for present plus some increases. W/D ratio < 7.	(1)	Adequate. Overbank flows rare. W/D ratio 8–15.	(2)
	6	65%+ with large, angular boulders 12+ in. numerous.	(2)	40–65%, mostly small boulders to cobbles 6–12 in.	(4)
	7	Rocks and old logs firmly embedded. Flow pattern without cutting or deposition. Pools and riffles stable.	(2)	Some present, causing erosive cross currents and minor pool filling. Obstructions and deflectors newer and less firm.	(4)
	8	Little or none evident. Infrequent raw banks less than 6 in. high generally.	(4)	Some, intermittently at outcurves and constrictions. Raw banks may be up to 12 in.	(6)
	9	Little or no enlargement of channel or point bars.	(4)	Some new increase in bar formation, mostly from coarse gravels.	(8)
Bottom	10	Sharp edges and corners, plane surfaces roughened.	(1)	Rounded corners and edges, surfaces smooth and flat.	(2)
	11	Surfaces dulll, darkened, or stained. Generally not "bright."	(1)	Mostly dull, but may have up to 35% bright surfaces.	(2)
	12	Assorted sizes tightly packed and/or overlapping.	(2)	Moderately packed with some overlapping.	(4)
	13	No change in sizes evident. Stable materials 80–100%.	(4)	Distribution shift slight. Stable materials 50–80%.	(8)
	14	Less than 5% of the bottom affected by scouring and deposition.	(6)	5–30% affected. Scour at constrictions and where grades steepen. Some deposition in pools.	(12)
	15	Abundant. Growth largely mosslike, dark green, perennial. In swift water, too.	(1)	Common. Algal forms in low velocity and pool areas. Moss here, too, and swifter water.	(2)
		Total	____	Total	____

Source: From Pfankuch 1975.

Note: Stability ratings: Add column scores. Total score < 38 = excellent; 39–76 = good; 77–114 = fair; 115+ = poor. Key number on field cards: 1 = landform slope; 2 = mass wasting (existing or potential); 3 = debris jam potential; 4 = vegetative bank protection; 5 = channel capacity; 6 = bank rock content; 7 = obstructions, flow deflectors, sediment traps; 8 = cutting; 9 = deposition; 10 = rock angularity; 11 = brightness; 12 = consolidation or particle packing; 13 = bottom size distribution and percent stable materials; 14 = scouring and deposition; 15 = clinging aquatic vegetation (moss and algae).

Stability Indicators by Classes			
Fair		**Poor**	
Bank slope gradient 40–60%	(6)	Bank slope gradient 60%+	(8)
Moderate frequency and size, with some raw spots eroded by water during high flows.	(9)	Frequent or large, sediment nearly yearlong or imminent danger of same.	(12)
Present, volume and size are both increasing.	(6)	Moderate to heavy amounts, predominantly larger sizes.	(8)
50–70% density. Lower vigor and still fewer species form a somewhat shallow and discontinuous root mass.	(9)	< 50% density plus fewer species and less vigor indicate poor, discontinuous, and shallow root mass.	(12)
Barely contains present peaks, occasional overbank floods. W/D ratio 15–25.	(3)	Inadequate. Overbank flows common. W/D ratio > 25.	(4)
20–40%, with most in the 3–6 in. diameter class.	(6)	< 20% rock fragments of gravel size, 1-3" or less.	(8)
Moderately frequent, moderately unstable obstructions and deflectors move with high water causing bank cutting and filling of pools.	(6)	Frequent obstructions and deflectors cause bank erosion yearlong. Sediment traps full, channel migration occurring.	(8)
Significant. Cuts 12–24 in. high. Root mat overhangs and sloughing evident.	(12)	Almost continuous cuts, some over 24 in. high. Failure of overhangs frequent.	(16)
Moderate deposition of new gravel and coarse sand on old and some new bars.	(12)	Extensive deposits of predominantly fine particles. Accelerated bar development.	(16)
Corners and edges well rounded in two dimensions.	(3)	Well rounded in all dimensions, surfaces smooth.	(4)
Mixture, 50–50% dull and bright, ± 1%, i.e., 35–65%.	(3)	Predominantly bright, 65%+, exposed or scoured surfaces.	(4)
Mostly a loose assortment with no apparent overlap.	(6)	No packing evident. Loose assortment, easily moved.	(8)
Moderate change in sizes. Stable materials 20–50%.	(12)	Marked distribution change. Stable materials 0–20%.	(16)
30–50% affected. Deposits and scour at obstructions, constrictions, and bends. Some filling of pools.	(18)	More than 50% of the bottom in a state of flux or change nearly yearlong.	(24)
Present but spotty, mostly in backwater areas. Seasonal blooms make rocks slick.	(3)	Perennial types scarce or absent. Yellow-green, short-term bloom may be present.	(4)
Total	_____	Total	_____

Table 10.2. Terminology used in describing stream morphology and classification

Term	Definition
Anastomosis	A stream channel with connections between the parts of its branching system.
Bank-full stage	A stage where water completely fills the channel system of a stream without spreading onto the floodplain; flow at this stage is often assumed to control the form of alluvial stream channels and is associated with a recurrence interval averaging 1.5 yr.
Braided	Multiple stream channels woven together.
Meander	A stream channel with a winding or indirect course.
Pool	A section of a stream channel with deeper, slow-moving water with fine bed material.
Riffle	A section of a stream channel with shallow depth and fast-moving water; bed materials are coarse.
Sinuosity	A measure of the number of bends, curves, and meanders in a stream channel compared to a straight channel (channel length/straight-line distance from upstream to downstream point).
Sinuous	A stream channel of many curves, bends, or turns; a stream channel which is intermediate between straight and meandering.
Thalweg	The deepest part of the channel

defined by Rosgen as "the ratio of the width of the flood-prone area to the bank-full surface width of the channel" (1994, p. 181). Exceptions in which the value of one of these variables is outside the range for a given stream type occur infrequently.

The *RSCS* recognizes a continuum of stream morphology within and between stream types. This continuum is applied where values outside the "normal range" are encountered but do not warrant classification of the stream as a unique type. The general appearance of a stream and the associated dimensions and patterns of the stream do not generally change with a minor change in one of the delineative criteria.

Streams do not usually change instantaneously. Rather, streams more frequently undergo a series of channel adjustments through time to accommodate any change in the "driving" variables. Their dimensions, profiles, and patterns reflect these adjustment processes, which are largely responsible for the observed stream form. The rate and direction of channel adjustment are functions of the nature and magnitude of the change and the stream type involved. Some streams change rapidly, while others are slow in their response to change.

Processes of aggradation and degradation, discussed earlier, can cause a change in stream type. For example, aggradation can trigger changes in width/depth ratio, slope, and sinuosity; type E streams can change to type C or type G (Fig. 10.7). Conversely, degradation would involve moving from C to E or F to G type streams.

TABLE 10.3. Levels of the Rosgen Stream Classification system (from Rosgen 1994, 1996)

Level	Description of Inventory	Objectives	Information Required
I	Broad morphological classification	For generalized description of major types of streams	Landform, lithology, soils, climate, basin relief, deposition history, valley morphology, river profile morphology and pattern
II	Morphological description of stream types	Provide delineation and detailed interpretation of homogeneous stream reaches based on reference reach measurements	Channel patterns, entrenchment ratio, width/depth ratio, sinuosity, channel material, slope
III	Stream state or condition	Determine existing condition or stability of streams; estimate departure from their potential	Channel stability index, riparian vegetation, deposition patterns, debris occurrence, bank erosion, confinement features, fish habitat indices, river size category
IV	Verification	Verify the stream condition, stability and potential as predicted from Levels I, II and III	Streamflow, suspended sediment and bedload measurements, bank erosion rates, aggradation-degradation of channel beds, biological data, riparian vegetation measurements, hydraulic geometry

Applications of Stream Classification—Management Interpretations

The ability to predict a stream's behavior from its appearance and to extrapolate information from similar stream types is helpful to watershed managers in applying interpretive information (Rosgen 1994). These interpretations can evaluate stream types in relation to their sensitivity to disturbance, recovery potential, sediment supply, the controlling influence of vegetation, and streambank erosion potential (Box 10.4). Applications of these interpretations can be useful in impact assessment, risk analysis, and management direction by stream type.

Levels III and IV provide more detailed descriptions and characteristics of stream channels that provide indications of stream potential, stability, and the status of stream channels in comparison with their potential. Such detail is required to understand and link past land use with channel changes and to help guide stream restoration efforts.

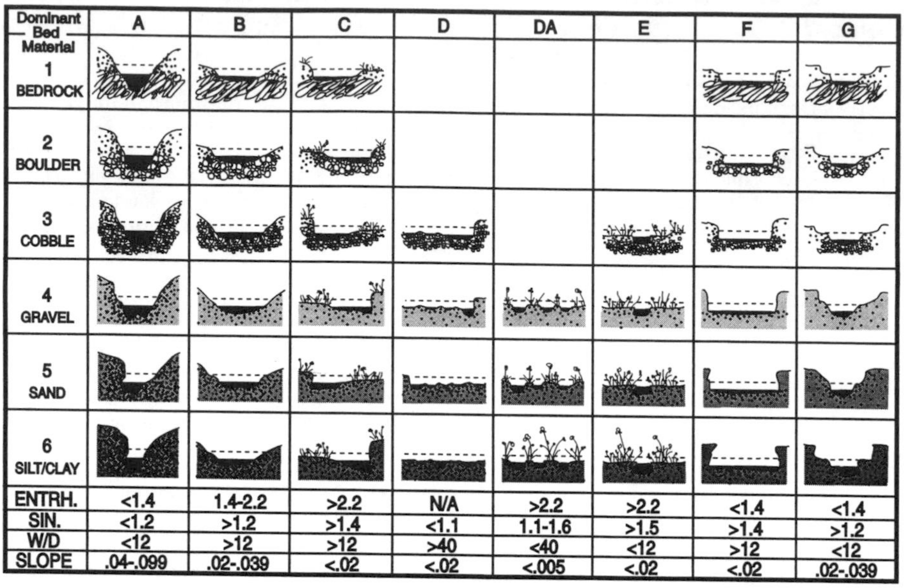

Dominant Bed Material	A	B	C	D	DA	E	F	G
1 BEDROCK								
2 BOULDER								
3 COBBLE								
4 GRAVEL								
5 SAND								
6 SILT/CLAY								
ENTRH.	<1.4	1.4-2.2	>2.2	N/A	>2.2	>2.2	<1.4	<1.4
SIN.	<1.2	>1.2	>1.4	<1.1	1.1-1.6	>1.5	>1.4	>1.2
W/D	<12	>12	>12	>40	<40	<12	>12	<12
SLOPE	.04-.099	.02-.039	<.02	<.02	<.005	<.02	<.02	.02-.039

FIGURE 10.5. Major stream types with their cross-sectional configurations and physical characteristics (from Rosgen 1994, © *Catena,* by permission).

Interpretive information by stream type can also be used in establishing guidelines for watershed management practices, silvicultural standards, and riparian and floodplain management and in analyzing possible cumulative effects.

SUMMARY

An appreciation of how energy relationships, hydraulic principles, and geomorphic processes are linked with stream channel form and function is crucial to effective watershed management. Therefore, this chapter considers these linkages and the application of geomorphic stream classification systems in the context of the hydrologic and cumulative land use effects on stream channels. After reading this chapter, you should be able to:

- Define a flood, a floodplain, and the bank-full flow condition.
- Discuss the relevance of bank-full flow in terms of stream channel morphology.
- Describe factors that determine the form of a stream channel and indicate how human activities can affect stream channel form, function, and stability.
- Discuss the importance of stream classification in helping to predict a stream's behavior from its appearance.
- Explain the role of stream classification in establishing general guidelines for watershed management, silvicultural standards, riparian management, and floodplain management and in analyzing cumulative effects.

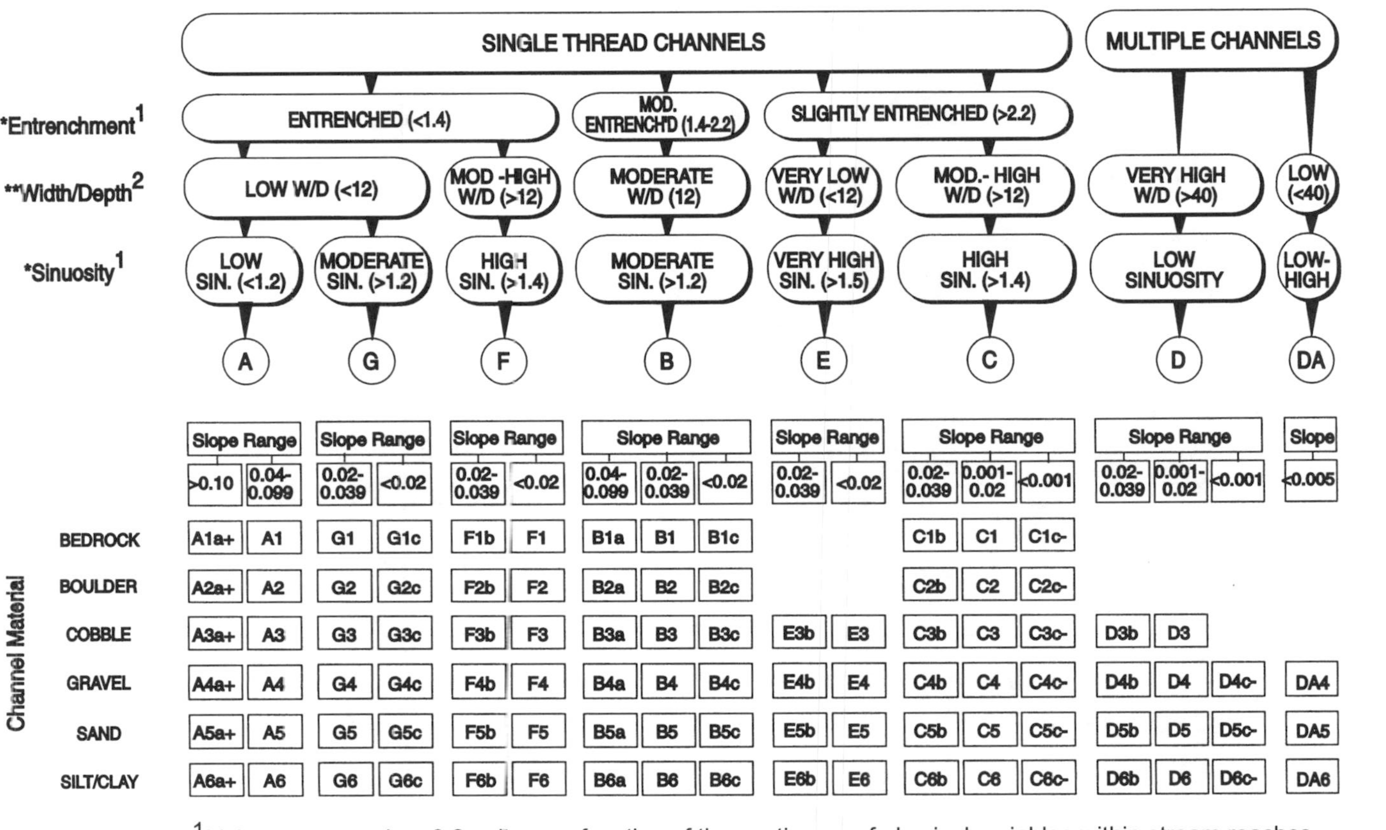

[1] Values can vary by ±0.2 units as a function of the continuum of physical variables within stream reaches.
[2] Values can vary by ±2.0 units as a function of the continuum of physical variables within stream reaches.

FIGURE 10.6. Classification system for natural rivers (from Rosgen 1994, © *Catena,* by permission).

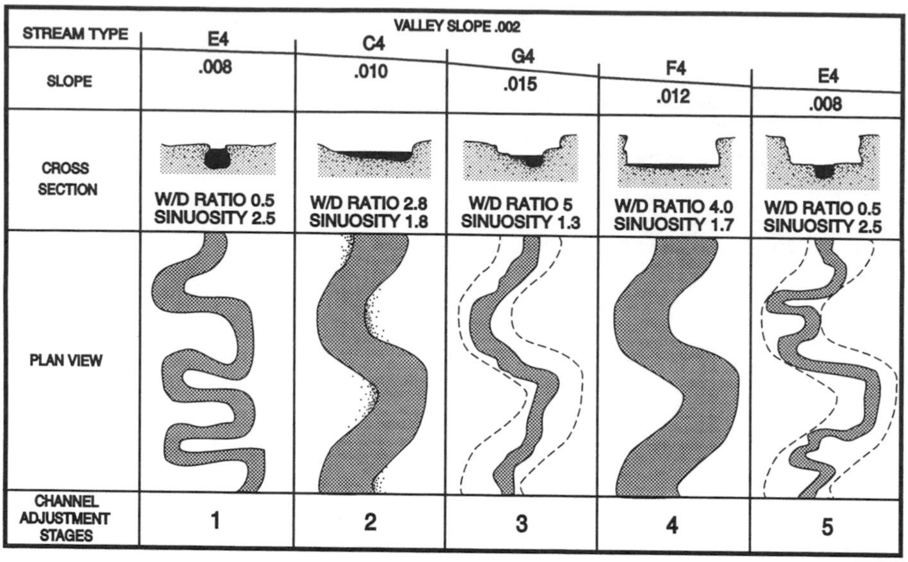

Figure 10.7. Example of evolutionary adjustments in stream channels (from Rosgen 1994, © *Catena,* by permission).

Box 10.4

Management Interpretations of Stream Types Classified by Rosgen (1994)

Stream Type	Sensitivity to Disturbance[a]	Recovery Potential[b]	Sediment Supply[c]	Streambank Erosion Potential	Vegetation Controlling Influence[d]
A1	Very low	Excellent	Very low	Very low	Negligible
A3	Very high	Very poor	Very high	High	Negligible
A4	Extreme	Very poor	Very high	Very high	Negligible
A6	High	Poor	High	High	Negligible
B1	Very low	Excellent	Very low	Very low	Negligible
B3	Low	Excellent	Low	Low	Moderate
B4	Moderate	Excellent	Moderate	Low	Moderate
B6	Moderate	Excellent	Moderate	Low	Moderate
C1	Low	Very good	Very low	Low	Moderate
C3	Moderate	Good	Moderate	Moderate	Very high
C4	Very high	Good	High	Very high	Very high
C5	Very high	Fair	Very high	Very high	Very high
C6	Very high	Good	High	High	Very high
D3	Very high	Poor	Very high	Very high	Moderate
D5	Very high	Poor	Very high	Very high	Moderate

Stream Type	Sensitivity to Disturbance[a]	Recovery Potential[b]	Sediment Supply[c]	Streambank Erosion Potential	Vegetation Controlling Influence[d]
D6	High	Poor	High	High	Moderate
DA4	Moderate	Good	Very low	Low	Very high
DA6	Moderate	Good	Very low	Very low	Very high
E3	High	Good	Low	Moderate	Very high
E4	Very high	Good	Moderate	High	Very high
E6	Very high	Good	Low	Moderate	Very high
F1	Low	Fair	Low	Moderate	Low
F3	Moderate	Poor	Very high	Very high	Moderate
F4	Extreme	Poor	Very high	Very high	Moderate
F6	Very high	Fair	High	Very high	Moderate
G1	Low	Good	Low	Low	Low
G2	Moderate	Fair	Moderate	Moderate	Low
G3	Very high	Poor	Very high	Very high	High
G4	Extreme	Very poor	Very high	Very high	High
G6	Very high	Poor	High	High	High

[a]Includes increases in streamflow magnitude and timing and/or sediment increases.
[b]Assumes natural recovery once cause of instability is corrected.
[c]Includes suspended and bed load from channel-derived sources and/or from stream-adjacent slopes.
[d]Vegetation that influences width/depth ratio stability.

References

Bevenger, G.S., and R.M. King. 1995. *A pebble count procedure for assessing watershed cumu lative effects*. USDA For. Serv. Res. Pap. RM-RP-319.

Bledsoe, B.P., and C.C. Watson. 2001. Effects of urbanization on channel instability. *J. American Water Resources Association* 37(2): 255–270.

Cirmo, C.P., and J.J. McDonnell. 1997. Linking the hydrologic and biochemical controls of nitrogen transport in near-stream zones of temperate-forested catchments: A review. *J. Hydrol.* 199:88–120.

Dunne, T., and L.B. Leopold. 1978. *Water in environmental planning*. San Francisco: W.H. Freeman and Company.

Fitzpatric, F.A., J.C. Knox, and H.E. Whitman. 1999. *Effects of historical land-cover changes on flooding and sedimentation, North Fish Creek, Wisconsin*. U.S. Geological Survey Water Resources Investigations Rep. 99–4083.

Gordon, N.D., T.A. McMahon, and B.L. Finlayson. 1992. *Stream hydrology—An introduction for ecologists*. New York: John Wiley & Sons.

Harrelson, C.C., C.L. Rawlins, and J.P. Potyondy. 1994. *Stream channel reference sites: An illustrated guide to field technique*. USDA For. Serv. Gen. Tech. Rep. RM-245.

Hawkins, C.P., J.L. Kershner, P.A. Bisson, M.D. Bryant, L.M. Decker, S.V. Gregory, D.A. McCullough, C.K. Overton, G.H. Reeves, R.J. Steedman, and M.K. Young. 1993. A hierarchal approach to classifying stream habitat features. *Fisheries* 18(6): 3–11.

Heede, B.H., and J.N. Rinne. 1990. Hydrodynamic and fluvial morphologic processes: Implications for fisheries management and research. *North Amer. J. Fish. Manage.* 10:249–268.

Hilsenhoff, W.L. 1987. An improved biotic index of organic stream pollution. *The Great Lakes Entomologist* 20(1): 31–39.

Junk, W.J., P.B. Bayley, and R.E. Sparks. 1989. The flood pulse concept in river-floodplain systems. *Can. Spec. Publ. Fish. Aquat. Sci.* 106:110–127.

Leopold, L.B. 1994. Flood hydrology and the floodplain. *Water Resources Update Issue* 94:11–14. The Universities Council on Water Resources.

Leopold, L.B., M.G. Wolman, and J.P. Miller. 1964. *Fluvial processes in geomorphology*. San Francisco: W.H. Freeman and Co.

Lu, S.Y. 1994. Forest harvesting effects on streamflow and flood frequency in the northern Lake States. Ph.D. Dissertation. St. Paul: University of Minnesota.

Meiman, J., and L.J. Schmidt, comp. 1994. *A research strategy for studying stream processes and the effects of altered streamflow regimes*. Fort Collins, CO: USDA For. Serv.

Miller, R.C. 1999. Hydrologic effects of wetland drainage and land use change in a tributary watershed of the Minnesota River Basin: A modeling approach. M.S. Thesis. St. Paul: University of Minnesota.

Montgomery, D.R., and J.M. Buffington. 1997. Channel-reach morphology in mountain drainage basins. *GSA Bulletin* 109:596–611.

Mosley, M.P. 1994. The classification and characterization of rivers. Chapter 12 in *River channels: Environment and process*. Oxford, UK: B. Blackwell.

Pfankuch, D.J. 1975. *Stream reach inventory and channel stability classification—A watershed management procedure*. USDA For. Serv. R1-75-002. Government Printing Office #696-260/200, Washington, DC.

Riedel, M.S. 2000. Geomorphic impacts of land use on clay channel streams. Ph.D. Thesis. St. Paul: University of Minnesota.

Rosgen, D. 1994. A classification of natural rivers. *Catena* 22:169–199.

Rosgen, D. 1996. *Applied river morphology*. Pagosa Springs, CO: Wildland Hydrology.

Savery, T.S., G.H. Belt, and D.A. Higgins. 2001. Evaluation of the Rosgen stream classification system in Chequamegon-Nicolet National Forest, Wisconsin. *J. American Water Resources Association* 37(3): 641–654.

Schumm, S.A. 1977. *The fluvial system*. New York: Wiley and Sons.

Schumm, S.A., M.D. Harvey, and C.C. Watson. 1984. *Incised channels, morphology, dynamics and control*. Littleton, CO: Water Resources Publications.

Simon, A., and C.R. Hupp. 1986. Channel evolution in modified Tennessee channels. In Proc. Fourth Federal Interagency Sedimentation Conference, Las Vegas, Nevada, March 24–27, 1986. Vol. 2, 5.71–5.82.

Simon, A., and M. Rinaldi. 2000. Channel instability in the loess area of the midwestern United States. *J. American Water Resour. Association* 36:133–150.

Stream Systems Technology Center. 2001. Forest Service stream classification: Adoption of a first approximation. *Stream Notes* (April):1–4. Fort Collins, CO: Rocky Mountain Research Station.

Tobin, G.A., and B.E. Montz. 1994. *The great Midwestern floods of 1993*. New York: Saunders College Publ., Harcourt Brace College Publ.

University of Minnesota. 1997. *Hydrologic impacts of drainage*. Report of a Technical Forum. June 4, St. Paul, MN.

Verry, E.S. 2000. Water flow in soils and streams: Sustaining hydrologic function. In *Riparian management of forests*, ed. E.S. Verry, J.W. Hornbeck, and C.A. Dolloff, 99–124. Boca Raton: Lewis Pub.

Wolman, M.G., and L.B. Leopold. 1957. *River flood plains—Some observations on their formations*. U.S. Geological Survey Professional Paper 282-C, 87–109.

Wolman, M.G., and J.P. Miller. 1960. Magnitude and frequency of forces in geomorphic processes. *J. Geol.* 68:54–74.

The use and management of surrounding watersheds affect the quality of water collected in reservoirs. The Roosevelt Reservoir in central Arizona, one of the oldest reclamation projects in the United States, provides water to the rapidly growing Phoenix metropolitan area.

PART 3

Water Quality

Concerns of water quality and nonpoint pollution associated with land use are of major concern to resource managers. Alterations in the chemical, physical, and biological characteristics of streams affect aquatic ecosystems and impact the quality of water for downstream users. Water quality characteristics and the influence of land use on water quality are the subjects of Chapter 11.

Actions can be taken to prevent or minimize adverse effects of land use on water quality. *Best management practices (BMPs)* are examples of such preventative measures. *Total maximum daily loads (TMDLs)* represent another approach in which different water bodies are examined, problems identified, and standards established that are intended to reduce loading of pollutants to acceptable limits. The management of water quality in the context of current water quality issues and regulatory considerations is the focus of Chapter 12.

CHAPTER 11 *Water Quality Characteristics*

INTRODUCTION

Before we can intelligently discuss water quality, we must identify how the water is to be used. Water quality involves a long list of individual components and chemical constituents. A *water quality standard* refers to the physical, chemical, or biological characteristics of water in relation to a specified use. For example, water quality standards for irrigation are not necessarily acceptable for drinking water. Changes in water quality due to watershed use can make water unusable for drinking, but it can be acceptable for fisheries, irrigation, or other uses. In some instances, we may be required by law to prevent water quality characteristics from being degraded from natural, or background, conditions. The objective of such laws or regulations is to maintain the quality of water for some possible unforeseen future use. The term *pollution* means that water has been degraded or defiled in some way by human actions. Pollution that is the result of poorly planned or implemented land use is the focus of this discussion. It should be recognized, however, that water quality degradation can also result from natural events, such as large rainstorms, fires, or volcanic eruptions.

This chapter deals primarily with water quality characteristics of naturally occurring *surface water* in "wildland settings" and land use impacts on these characteristics. The discussion focuses on nonpoint rather than point pollution. *Point pollution* is associated with industries and/or municipalities and the discharge of pollutants through a pipe or ditch, while *nonpoint pollution* refers to pollution that is associated with land use activities such as agricultural cultivation, grazing of livestock, and forest management practices and occurs over wide areas. Management responses to nonpoint pollution problems are presented in the following chapter. Groundwater quality in terms of its usefulness for drinking and irrigation is also considered in this chapter. We begin this discussion of water quality characteristics with a brief look at the chemistry of the precipitation that falls onto watershed lands and, therefore, influences the quality of water flowing from these lands.

Chemistry of Precipitation

Air pollution and its influence on precipitation chemistry and dry deposition additions to terrestrial and aquatic systems are of considerable interest to farmers, natural resource managers, and society as a whole. The potential additions of higher concentrations of chemicals to land and water and other changes in water chemistry can affect land productivity and water quality, with significant environmental implications globally.

Acid Precipitation and Chemical Constituents

Acid precipitation became a major environmental issue in the late 1970s and early 1980s in the industrial countries of the Northern Hemisphere. This issue prompted a global program of monitoring the chemistry of precipitation. In the United States, the National Atmospheric Deposition Program (NADP) was created in 1977 to determine atmospheric chemical deposition and its effects. The NADP sites all measured concentrations of five ions: sulfate, nitrate, ammonium, calcium and hydrogen (Kapp et al. 1988). Such programs have provided considerable insight into the chemical characteristics of both rainfall and snowfall.

Naturally occurring precipitation is slightly *acidic* (pH of 5.6–5.7) because of the reaction of water with normal levels of atmospheric carbon dioxide (Environmental Protection Agency 1980). The industrial northeastern United States has experienced increasing acidity, with large areas experiencing rainfall pH values of 4.5 and below. In northern Europe, pH values as low as 2.4 have been reported—this is close to the acidity of lemon juice. The increasing acidity of precipitation is caused by the atmospheric inputs of sulfur oxides and nitrogen oxides from the burning of fossil fuels, such as coal, gas, and oil. Of major concern are the impacts of acid deposition—both liquid and solid atmospheric particulates—on aquatic and terrestrial ecosystems.

Examples of precipitation chemistry in the United States are presented in Box 11.1.

The potential effects of acid deposition are widespread because of the long-range transport of acid by the atmosphere. The actual impacts of acid deposition on streams and lakes are largely a function of the buffering capacity of soil surrounding them and the size of the watershed. If soils are alkaline or contain sufficient calcium, acids become neutralized and waters become acidified slowly, if at all. However, lakes and streams that occur on infertile, shallow soils over dense bedrock and that have a low watershed area to water surface area ratio are susceptible to acidification. Once the pH

Definitions of Acidity and pH

Acidity of water—the capacity of water to neutralize a strong base to a designated pH.

pH of water—the negative log, base 10, of the hydrogen ion (H^+) activity in moles per liter; a value of 7 is neutral, < 7 is acidic water and > 7 is basic or alkaline water.

Box 11.1

Precipitation Chemistry in the United States: Some Patterns (Kapp et al. 1988)

Between 1979 and 1984, the temporal and spatial patterns of precipitation chemistry in the United States indicated that many watersheds in the United States are exposed to large numbers of low concentration events and a few events with extremely high concentrations. The highest concentration events occurred most frequently during the growing seasons in most sites; this has implications for sensitive agricultural crops and native plants.

The highest concentrations of hydrogen ions (highest acidity) occurred in the states of Pennsylvania, New York, Ohio, and West Virginia, while the lowest concentrations of hydrogen ions occurred in central Minnesota. The spatial patterns of sulfate and nitrate concentrations were similar to those of hydrogen ion concentrations. There was a striking east-west differentiation in pH values, with all sites east of a line from northwestern Wisconsin to the Gulf Coast of Texas having pH < 5.0; west of this line all but one site had pH > 5.0. In contrast, calcium concentrations in precipitation decreased from west to east across the Great Lakes region.

Patterns of ammonium were different from the above and appeared to be influenced by extensive cattle feedlot operations (e.g., eastern Nebraska) and areas with high emissions (e.g., New York).

Average annual precipitation is generally higher in the majority of eastern states than in most western states. With higher concentrations of hydrogen, sulfate, and nitrate in precipitation also occurring in the east, the result is a higher total amount of these ions being deposited on terrestrial and aquatic systems in the east.

of such streams and lakes begins to drop much below 6.0, fish food organisms and fish fauna are affected. In Sweden, 50% of the lakes have a pH of 6.0 or less; half of those are below pH 5.0. When lakes reach a pH of < 4.5, they are considered to be critically acid and cannot support fish life. Also of importance to watershed management is the impact of acid deposition on forest and other vegetation. Atmospheric pollution can inhibit nitrogen fixation in the soil, cause calcium, magnesium, and potassium to be leached from the soil, and inhibit bacterial decomposition. Air pollutants, including acid deposition, caused a $1.2 billion loss of trees in (former) West Germany during the summer of 1983 alone (Postel 1984). In severe cases around smelters, forests can become nonproductive, and a loss of plant cover can lead to the same serious hydrologic problems encountered with other barren landscapes.

Snowpacks in the northern hemisphere have also been observed to be acidic. Snowflakes can collect a variety of impurities by scavenging as they descend through the atmosphere. Dry atmospheric fallout between snowstorms and local contamination can also add to the impurities found in snowpacks. There is no general agreement on how low

Box 11.2

Sources of Contaminants in Snowpacks: An Example in Montana, USA (Pagenkopf 1983)

The low pH values of snowpacks observed in the southwest corner of Montana were initially puzzling because there is little industry or population in the vicinity. Research indicated that the source of the acid oxides originated from five distinct areas outside Montana: the Seattle-Tacoma area of Washington, the Portland area of Oregon, the San Francisco Bay area, the greater Los Angeles basin, and the Wasatch front in Utah. Apparently, storms that generally track from west or southwest to east cross the mountains of western Montana and deposit the accumulated sulfuric and nitric acids with the falling snow.

the pH of a snowpack must be before it should be considered *acid deposition*, although studies in the western United States and Canada suggest that snowpacks with pH values of 5.0 and above represent "normal" snow and those with pH values below 5.0 represent acidic snow. Snowpack acidity is generally not a problem in the western United States and Canada, with the possible exceptions of "hot spots" that are found downwind of major industrial developments (Box 11.2). However, snowpack acidity, the frozen version of the acid rain problem, often occurs in highly industrialized regions in northeastern North America, Scandinavia and northern Europe, and northern Japan and China.

Generalizations relative to concentrations of chemical pollutants in snowpacks are difficult to make. There is always a degree of variability in the chemical makeup of snow, although increased contamination is likely to occur as a result of human activities, such as the development of industrialized urban areas. Calcium is often a dominant cation in many areas unaffected by industrial emissions. Although found in areas unaffected by industrial pollution, SO_4^{-2} is a dominant anion in snowpacks that are influenced by industrial air pollution. Other chemical constituents found in varying concentrations include sodium, nitrate, fluorite, and chloride.

Regional differences in pH and the concentrations of chemical constituents, when they occur, are attributed to some combination of the "age" of the snowpack when it was sampled; the snowpack layer sampled; pH of the snowpack, which can correlate with sulfate, nitrate, and other constituents in some cases; the portion of the snowmelt period sampled; and the sampling techniques and laboratory procedures used in their detection and analysis.

Snowpack solute contents and melting patterns can affect subsequent snowmelt runoff chemistry and ionic concentrations in streamwater (Stottlemyer 1987). The pathway of flow of snowmelt water to the stream channel also influences the ultimate chemistry of snowmelt water that reaches the stream. It has been suggested that sudden increases of acidity and elevated solute concentrates from outflow at the base of the snowpacks have the potential to adversely affect fish populations and inhibit amphibian reproduction (Fahey 1979, Hutchinson and Haras 1980, Johannessen and Henriksen 1978). However, in most cases, the concentrations of chemical pollutants in snowmelt runoff do not exceed the levels of acceptability for aquatic life, irrigation, or public water supplies. Long-term monitoring is needed to link the relative contributions of the chemical constituents in snowpacks to the chemistry of subsequent snowmelt runoff and streamwater.

Atmospheric Deposition of Mercury

One of the main sources of mercury (Hg) to watersheds, streams, and lakes is atmospheric deposition (Kolka 1996). As a trace element in geologic formations, Hg is naturally released by the slow process of weathering. This process does not yield high concentrations of Hg to terrestrial and aquatic systems. However, there are a variety of sources of Hg resulting from natural events and human activities in which Hg becomes volatile. Volcanic eruptions and forest fires expel Hg to the atmosphere. Human activities, including metal production, waste handling and treatment, and the burning of fossil fuels, peat, and wood contribute to atmospheric Hg.

Once in the atmosphere, Hg becomes widespread, and it has been found in the most remote wilderness areas. Mercury has a strong affinity for organic matter in soils and surface water (Kolka 1996). As a result, Hg transport and cycling in a watershed is closely tied to the processes and movement of organic carbon. Boreal forests and peatlands in the northern hemisphere have been shown to be a source for organic carbon that leaves watersheds in this region. Atmospheric Hg apparently accumulates on the forest canopy and reaches the soil, which acts as a sink. Watershed studies in northern Minnesota indicate that from 50 to 80% of the Hg export was associated with particulate organic carbon in runoff from uplands and peatlands (Kolka 1996). In this region, this represents the main source of Hg that is reaching streams and rivers and affecting the aquatic food chain. The risks of accelerated Hg levels to wildlife, fish, and humans are reasonably well known.

Cumulative Effects

The cumulative effects of accelerated levels of atmospheric deposition on both terrestrial and aquatic components of watersheds are difficult to quantify or even anticipate. In nutrient poor systems, additional nutrients (within reasonable ranges of pH) can increase biomass productivity of land and water systems for a period. For some lakes and streams, even small increases in nutrient loading can be detrimental—leading to eutrophication. The impacts of such deposition over time depend on soils, vegetation types, and other factors affecting the buffering capacity of a given system.

Atmospheric deposition of pollutants, such as mercury (Hg), has serious implications for watersheds, streams, and lakes. Mercury accumulates in the food chain and is toxic to wildlife, fish, and humans. The effects of atmospheric contamination can be widespread and can accrue over a long period of time. Contamination of fish from atmospheric Hg has led to fish consumption advisories in 26 states in the United States, as well as in Canada and Sweden (Glass et al. 1992).

Even though atmospheric deposition of pollutants and acid is not affected by watershed boundaries, the direct and indirect effects on aquatic and terrestrial ecosystems can be more clearly identified by using the watershed as a unit for study. The solution to such problems, however, is not within the realm of watershed management.

Physical Characteristics of Surface Water

Among the more important physical characteristics of surface water are suspended sediment concentrations, thermal pollution, and the level of dissolved oxygen. Suspended sediments, which consist largely of silts and colloids of various materials, affect water quality in terms of domestic and industrial uses and can adversely affect aquatic organisms and their environments. Thermal pollution also has many direct and indirect impacts on aquatic organisms and water quality. Although dissolved oxygen is one physical measure of water quality, it is also an important determinant of chemical and biological processes in water.

Suspended Sediment

The physical quality of naturally occurring surface water is determined largely by the amount of sediment that it carries. As discussed in Chapter 9, the *total sediment load* in streamflow comprises *suspended sediment* and *bed load.* Of the two, suspended sediment has a greater effect on water quality because it restricts sunlight from reaching photosynthetic plants (measured as turbidity), and it can affect aquatic ecosystems adversely by smothering benthic communities and by covering gravels that are often important spawning habitat for fish. Sediment also carries many nutrients and heavy metals that affect water quality.

Sources

Throughout the world, an increase in the sediment load of streams is one of the most widespread causes of degradation in the quality of water from agricultural, grazed, and forested watersheds. Surface water from undisturbed forested upland watersheds has relatively low suspended-sediment concentrations (usually < 20 ppm). Much of this sediment comes from the stream channel itself, although limited amounts can be contributed from surface runoff during large storm events. Where organic soils are prevalent, much of the suspended material in streams is organic particles rather than mineral particles. Higher concentrations of suspended sediment are often the result of accelerated erosion caused by disturbances in drainage areas, such as urban expansion, road construction, agricultural cultivation, overgrazing by livestock, logging operations, or natural catastrophes (including large floods, landslides, or fires).

Some of the most important ecological impacts of land use practices involve physical changes in stream structure, such as increased content of fine particles in gravel beds, erosion of streambanks, increases in stream width, decreases in stream depth, and fewer deep pools (MacDonald et al. 1991). Roads are the major contributors of sediment in streams; therefore, proper road design and maintenance are critical to minimizing sediment problems (Binkley and Brown 1993). These problems are compounded when disturbances take place on steep terrain and near stream channels.

Nutrient and Heavy-Metal Transport Capacities of Sediment

Nutrient and heavy-metal losses from upland watersheds are usually measured by dissolved ion concentrations. However, a potentially important source of nutrient and heavy-metal loss, and one that is often ignored, is transportation by sediment. The losses occur as a result of the weathering forces of the physical and biological environment, the latter represented by the vegetation type on the watershed acting upon the parent bedrock. Pesticides (such as atrazine) are also known to adsorb to soil particles and can be transported by the water system in this manner.

Transported sediment from watersheds composed of different bedrock and vegetation combinations can carry high levels of nutrients and heavy metals. Sediment from upland watersheds with limestone, granite, basalt, and sandstone geologies in the southwestern United States shows that, in general, limestone is high in calcium (Ca) and potassium (K), while basalt is high in sodium (Na). Magnesium (Mg) is highest in the sand fraction (0.061–2.0 mm) of basalt and the clay-silt fraction (< 0.061 mm) of limestone (Gosz et al. 1980). Often, sandstone has the lowest concentration of these elements, and granite is frequently intermediate. The nutrients adsorbed to sediment particles can be indicative of the type of geologic formation in an area. Vegetation types on a watershed of a given geology primarily affect the organic matter content, total phosphorus, and levels of extractable nutrients of the sediments.

Sediment transport of phosphorus can reduce the chemical quality of surface waters and, as a consequence, result in substantial changes in aquatic ecosystems. Phosphorus is a limited nutrient in many streams and lakes. When phosphorus loading increases in such systems, *eutrophication* (the process of nutrient enrichment leading to dense algae growth) can be accelerated. The resulting increase in algae and biomass in water systems can cause dramatic changes in water quality. For example, small pine-covered watersheds in Mississippi yielded < 60 mg/l of suspended sediment, but the phosphorus (P) concentration averaged from 329 to 515 μg/g of sediment (Duffy et al. 1986). These concentrations of P were from 2 to 3.5 times greater than the concentration found in the soils of the watersheds. Most of the sediment was transported during stormflow events, accounting for 70% or more of total P export and $> 40\%$ of the total nitrogen (N) export, illustrating the importance of maintaining low sediment yields with respect to the chemical quality of water.

In many instances, variations in heavy-metal [zinc (Zn), iron (Fe), copper (Cu), manganese (Mn), lead (Pb), and cadmium (Cd)] levels in a stream are correlated with variations in sediment concentrations. In the southwestern United States, sediment from different geologic strata has different heavy-metal concentrations, increasing in the following order: sandstone, granite, limestone, and basalt (Gosz et al. 1980). From the standpoint of watershed management, land use practices that increase sediment production can increase nutrient and heavy-metal loss via suspended sediment as well.

Thermal Pollution

Water temperature can be a critical water quality characteristic in many streams. The temperature of water, particularly temperature extremes, can control the survival of certain flora and fauna residing in a body of water. The type, quantity, and well-being of flora and fauna will frequently change with a change in water temperature. Of particular concern is a temperature increase due to land use practices. In general, an increase in water temperature causes an increase in the biological activity, which, in turn, places a greater demand on the dissolved oxygen in a stream. That the solubility of oxygen in water is related inversely to temperature has compounded this effect (Table 11.1). Changes in water temperature can result in the replacement of existing species, such as cold-water trout being replaced by warm-water bass or walleye.

TABLE 11.1. Relationship between the saturated solubility of oxygen in water and water temperature

Water Temperature (°C)	Solubility of O_2 (mg/l)
5	12.8
10	11.3
20	9.0
25	8.2

Clearing of riparian vegetation adjacent to a stream channel is one way in which water temperature can be increased. The removal of trees along a streambank increases the exposure to solar radiation, and the rise in water temperature can be predicted if one considers an energy budget for the water in the stream. If trees are removed from the streambanks, the only change in the energy budget is an increase in solar energy entering the system, which will cause a rise in water temperature because there are no new outlets for energy from the system. Increases in stream temperature can range from fractions of a degree centigrade for small openings in a forest overstory to > 10°C for a complete removal of trees along the streambanks. Studies in the northeastern and northwestern United States have reported annual maximum stream temperatures to rise as much as 4°C and 15°C when riparian vegetation was removed from small streams.

Brown (1980) determined that the potential change in daily temperature due to streambank vegetation removal could be estimated from the following:

$$\Delta T = \frac{AR_n}{Q} 0.000267 \tag{11.1}$$

where ΔT = maximum potential daily temperature change due to exposure of a section of stream to direct solar radiation, in °F; A = surface area of stream newly exposed to direct radiation (ft^2); Q = streamflow discharge (ft^3/sec, or cfs); and R_n = net solar radiation received by water surface that is newly exposed (Btu/ft^2/min).

Mathematical models to predict stream temperatures following modifications in the vegetative cover that shade streams have been used on upland watersheds in the western United States. These models generally describe the physical situation of a stream, including the vegetation bordering it. Changes in these variables, as might occur through implementation of land management practices, will often result in corresponding changes in water temperature. Variables in the models can be repeatedly changed to determine the possible effects of different methods of timber harvesting. Predictions of water temperature changes can also be used to estimate corresponding changes in dissolved oxygen and subsequent impacts on aquatic flora and fauna. Doing so can provide a more complete understanding of the water quality consequences of changing riparian vegetation.

Dissolved Oxygen

The *dissolved oxygen content* in water has a pronounced effect on the aquatic organisms and chemical reactions that occur within the water body. The dissolved oxygen concentration of a water body is determined by the solubility of oxygen, which is inversely related to water temperature (Table 11.1), pressure, and biological activity. The solubility of oxygen in water can be estimated from the equation by Churchill et al. (1962):

$$O_s = 14.652 - 0.41022T + 0.0079910T^2 - 0.000077774T^3 \qquad \textbf{(11.2)}$$

where O_s = solubility of oxygen (mg/l) and T = temperature of water (°C).

Dissolved oxygen is a transient property that can fluctuate rapidly in time and space. From a biological perspective, it is one of the most important water quality characteristics in the aquatic environment. However, the dissolved oxygen concentration represents the status of the water system at a particular point and time of sampling. The decomposition of organic debris in water is a slow process; therefore, the resulting changes in oxygen status respond slowly as well. Methods have been developed that estimate the demand or requirement of a given water body for oxygen. In essence, this is an indication of the pollutant load with respect to oxygen requirements and includes measurement of biochemical oxygen demand or chemical oxygen demand.

Biochemical and Chemical Oxygen Demands

The *biochemical oxygen demand* (BOD) is an index of the oxygen-demanding properties of biodegradable material in water. Samples of water are taken from the stream and incubated in the laboratory at 20°C, after which the residual dissolved oxygen is measured. The BOD curve in Figure 11.1 illustrates the two-stage characteristic that is typical—the first stage is related to carbonaceous demand, and the second stage to nitrification. These two stages refer to the oxygen required to oxidize carbon compounds and nitrogen compounds, respectively. Unless specified otherwise, BOD values usually refer to the standard 5 day value, which is the carbonaceous stage. Such values are useful in assessing stream pollution loads and for comparison purposes (Table 11.2).

Chemical oxygen demand (COD) is a measure of the pollutant loading in terms of complete chemical oxidation using strong oxidizing agents. COD can be determined

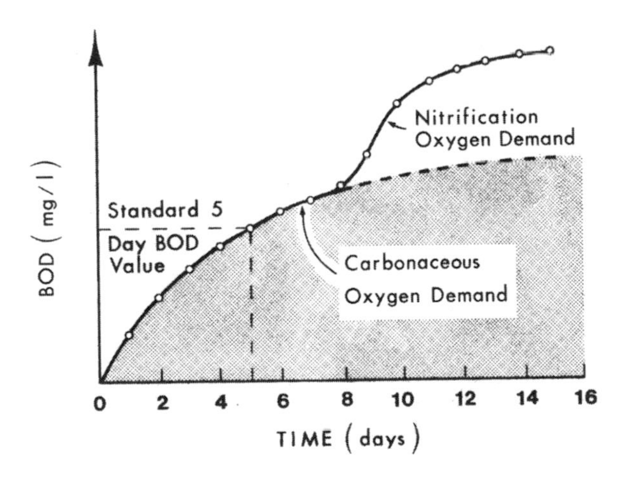

FIGURE 11.1. A biochemical oxygen demand (BOD) curve illustrating the carbonaceous demand phase and nitrification phase.

TABLE 11.2. Examples of biochemical oxygen demand (BOD) values for different conditions

Condition	BOD (mg/l)	
	5-day	90-day
Clean, undisturbed natural stream	< 4	—
Effluent		
Pulp and paper processing	20–20,000	—
Feedlots	400–2000	—
Untreated sewage	100–400	—
Logging residue (needles, twigs, and leaves)	36–80	115–287

Source: Adapted from Ponce 1974, Dunne and Leopold 1978, and others.

TABLE 11.3. Examples of end products from organic loading in water bodies under aerobic and anaerobic conditions

Types of Compounds in Organic Load	End Products	
	Aerobic	Anaerobic
Carbonaceous (cellulose, sugars, etc.)	CO_2, energy, water	Organic acids, ethyl alcohol, methane (CH_4)
Nitrogenous (proteins, amino acids)	NO_3^{--} (in presence of N-bacteria)	NH_4^+, OH^- (NO_2^- temporary)
Sulfurous	SO_4^{--}	H_2S

quickly because, unlike BOD, it does not rely on bacteriological action. However, COD is not necessarily a good index of the oxygen-demanding properties of materials in natural waters. Therefore, BOD is normally used for this purpose instead of COD.

When organic material such as human sewage, livestock waste, or logging debris is added to a water body, bacteria and other organisms begin to break down that material to more stable chemical compounds. If oxygen is readily available and mixed in the water and the organic loading is not too great, oxidation can proceed without any detrimental reduction in dissolved oxygen. When oxygen is limiting or loading is too great, anaerobic processes can occur, which result in a less efficient oxidation process with undesirable by-products in the water (Table 11.3).

A hypothetical sequence of changes that occurs downstream of a heavy pollutant loading of biodegradable material is illustrated in Figures 11.2 through 11.4. These figures show schematically the effects of discharging raw domestic sewage, from a community of about 40,000 people, into a stream with a flow of 100 cfs (2.8 m^3/sec). The BOD increases instantly at the point of discharge, which is followed downstream (or in time) with a reduction in dissolved oxygen (DO) concentrations (Fig. 11.2). The reduction in DO steepens the gradient of oxygen between the atmosphere and the water body,

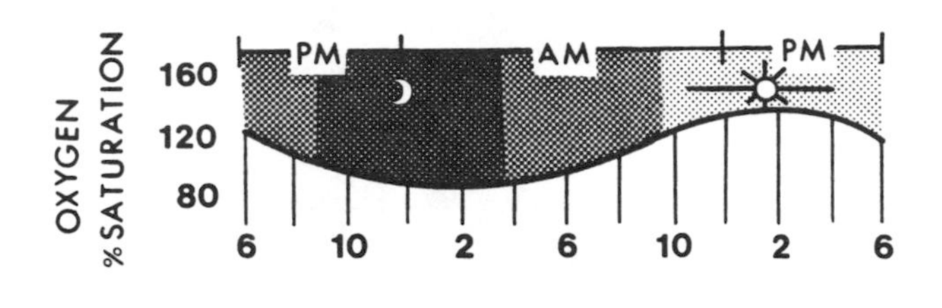

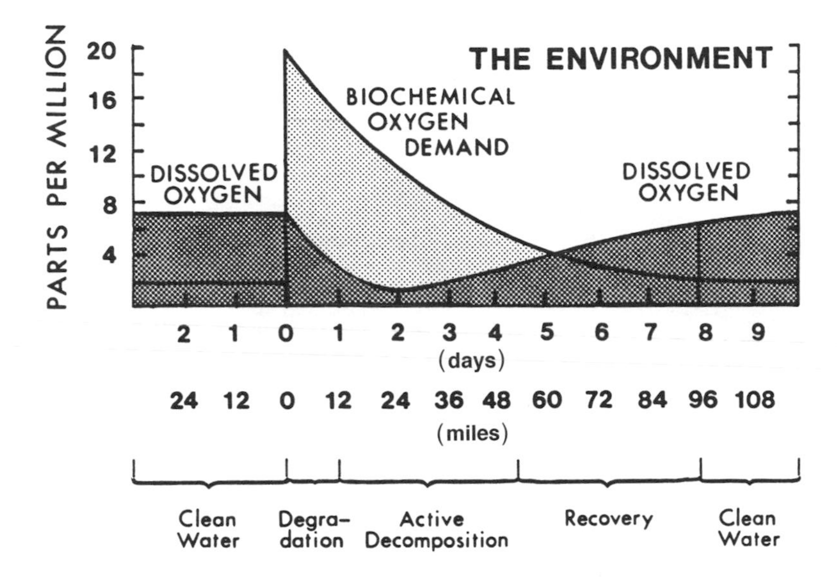

FIGURE 11.2. Effects of disposal of raw sewage in a stream on the dissolved oxygen and biochemical oxygen demand for stream water, either in time or downstream. The effects of reaeration rate and the diurnal characteristics of dissolved oxygen are also shown (from Bartsch and Ingram 1959, © Works J. Corp., by permission).

increasing the reaeration rate. The DO reduction curve is at a minimum in the area undergoing active decomposition. As reaeration takes place, DO concentrations increase and eventually reach DO levels before pollution.

Bacterial growth proceeds exponentially in the degradation and active decomposition zones; the corresponding decomposition of nitrogenous organic matter takes place according to oxygen levels in the stream (Fig. 11.3). Other organisms respond to the modified environment, particularly those organisms adapted to conditions of low oxygen, low levels of light, and high concentrations of organic material. Although many organism populations return to prepollution levels, some organisms do not. Because of the higher nutrient levels in the recovery and downstream clean-water zone, algae populations can flourish. As a result, the habitat for higher organisms is modified to the extent that species diversity does not fully recover to the upstream, prepollution conditions (Fig. 11.4). Notice also that population levels of certain adapted species increase

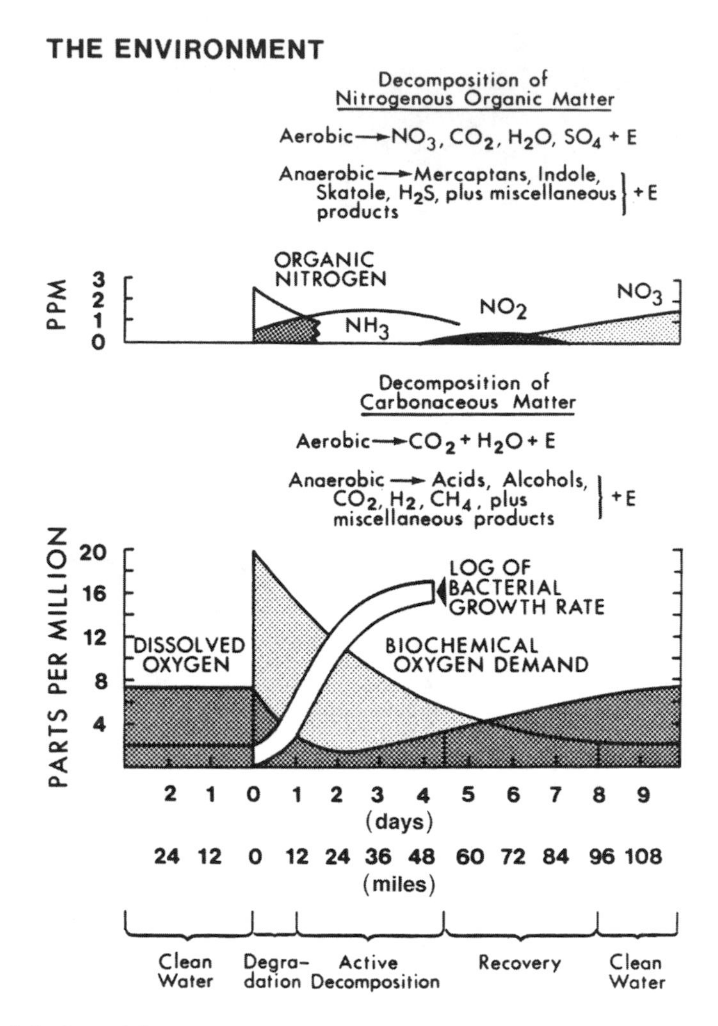

FIGURE 11.3. The relationship of accelerated bacterial growth to changes in dissolved oxygen and biochemical oxygen demand due to disposal of raw sewage in a stream (from Bartsch and Ingram 1959, © Public Works J. Corp., by permission).

in the active decomposition and recovery zones because of limited competition. As water quality conditions improve, the diversity of species recovers, and populations of individuals within species drop to prepollution levels.

Other Characteristics

Several other characteristics, including pH, acidity, alkalinity, specific conductance, and turbidity, describe the physical condition of water. These characteristics can be important indicators of water quality and can directly affect the chemical and biological condition of natural waters.

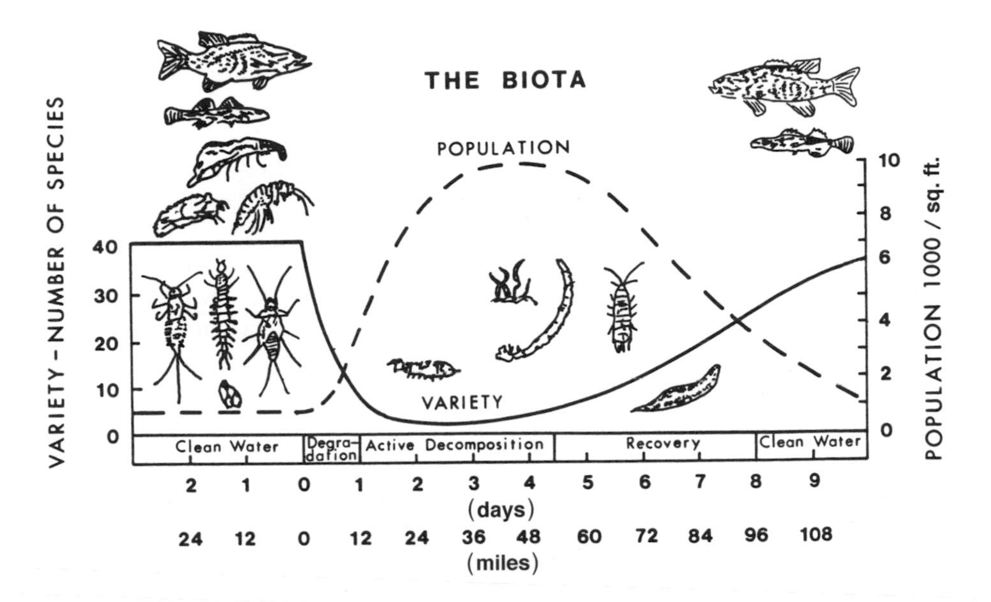

FIGURE 11.4. Effects of sewage disposal on species composition and populations of higher aquatic life-forms in a stream (from Bartsch and Ingram 1959, © Public Works J. Corp., by permission).

pH

The pH of water affects and is affected by chemical reactions in aquatic systems. It also represents thresholds for certain aquatic organisms. When the pH of water is > 7, it is indicative of alkaline water, which normally occurs when carbonate or bicarbonate ions are present. A pH < 7 represents acidic water. In natural waters, carbon dioxide reactions are some of the most important in establishing the pH level. When carbon dioxide (CO_2) enters water either from the atmosphere or by respiration of plants, carbonic acid is formed, which dissociates into bicarbonate; carbonate and H^+ ions are then liberated, influencing pH:

$$CO_2 + H_2O \leftrightarrows H_2CO_3 \leftrightarrows H^+ + HCO_3^- \leftrightarrows 2H^+ + CO_3^{--} \qquad (11.3)$$

The *pH* at any one time is an indication of the balance of chemical equilibria in water and affects the availability of certain chemicals or nutrients in water for uptake by plants. The pH of water also directly affects fish and other aquatic life. Generally, toxic limits are pH values < 4.8 and > 9.2. Most freshwater fish seem to tolerate pH values from 6.5 to 8.4; most algae cannot survive pH values > 8.5.

Acidity

Acidity and *pH* are closely related indicators of H^+ ion activity in water. Linked to pH, acidity is caused by the presence of free H^+ ions from carbonic, organic, sulfuric, nitric, and phosphoric acids. Acidity is important because it affects chemical and biological reactions and can contribute to the corrosiveness of water. The acidity of precipitation that falls on watershed lands emerged as one of the prominent environmental issues concerning water quality and the environment in the 1980s as described earlier.

Alkalinity

Alkalinity, the opposite of acidity, is the capacity of water to neutralize acid. Alkalinity is also linked to pH and is caused by the presence of carbonate, bicarbonate, and hydroxide, which are formed when carbon dioxide is dissolved. A high alkalinity is associated with a high pH and excessive dissolved solids. When water is high in alkalinity, it is considered to be well buffered; that is, large amounts of acid are required to change the pH.

Most streams have alkalinities of < 200 mg/l, although some groundwater aquifers can exceed 1000 mg/l when calcium and magnesium concentrations are high. Ranges of alkalinity of 100–120 mg/l seem to support a well-diversified aquatic life.

Specific Conductance

Specific conductance is the ability of water to conduct electrical current through a cube of water 1 cm on a side, expressed as micromhos per centimeter at 25°C, or as microsiemens per centimeter (older instruments used the former units). By itself, this measure has little meaning in terms of water quality, except that specific conductance increases with dissolved solids. Its measurement is quick and inexpensive and can be used to estimate total dissolved solids (*TDS*) as follows (Hem 1992):

$$TDS\ (\text{ppm}) = A_o \times \text{specific conductance}\ (\mu\text{mhos/cm}) \tag{11.4}$$

where A_o = a conversion factor ranging from 0.54 to 0.96 (normally 0.55 to 0.75), with higher values associated with water high in sulfate concentration.

Specific conductivities in excess of 2000 μmhos/cm indicate a *TDS* level too high for most freshwater fish species. Values of seawater can exceed 50,000 μmhos/cm.

Turbidity

The clarity of water is an indicator of water quality that relates to the ability of light to penetrate. *Turbidity* is an indicator of the property of water that causes light to become scattered or absorbed. The lower the turbidity, the deeper light can penetrate into a body of water and, hence, the greater the opportunity for photosynthesis and higher oxygen levels. Turbidity is caused by suspended clays, silts, organic matter, plankton, and other inorganic and organic particles. Standard instruments, called *turbidimeters*, are used to measure light penetration through a fixed water sample. Turbidity measurements, like specific conductance, can be used to estimate certain other factors affecting water quality. For example, the more easily measured turbidity can sometimes be used to predict suspended-sediment concentrations. A correlation between the two parameters is necessary, however.

Cumulative Effects

The effects of environmental change on sediment have been considered in the previous chapter. However, it is also important to note that many chemicals and compounds are adsorbed and transported with silt or clay particles. As a consequence, land use activities that increase erosion can also increase the transport of nutrients and chemicals from upland watersheds. The long-term productivity of these watershed lands can conceivably be reduced if these erosive processes continue.

The surface water temperature at a stream site is determined by the temperature of inflowing water, radiant energy inputs and outputs, conductive energy inputs and outputs, and heat extraction by evaporation. Riparian (streambank) vegetation influences all of these heat exchanges in a stream (see Chapter 13). Beschta et al. (1987) and others have described relationships between land use activities that reduce riparian vegetation and the resulting increases in stream temperatures. Water transports heat by changes in temperature from one site to another; therefore, the effects of reduction in riparian vegetative covers on water temperature can impact downstream water uses that are dependent on temperature.

Changes in land use activities that reduce dissolved oxygen levels can also decrease habitability for fish. Increased nutrient loads deplete dissolved oxygen by stimulating algal blooms. Detrital organic materials consume oxygen as they decay. The cumulative effects of all such factors on the oxygen and temperature regime of streams and rivers must be recognized throughout the watershed.

Dissolved Chemical Constituents

Natural streams are not separate or distinct from the areas that they drain but instead are an integral part of the watershed and its ecosystems. Water is an effective solvent, and as it comes into contact with each part of the system, the chemical characteristics of the water adjust accordingly. Chemical reactions and physical processes occur, often simultaneously, as the water contacts the atmosphere, soil, and biota. It is these reactions and processes, and the condition of each compartment of the ecosystem, that determine the kind and amount of chemical constituents in solution.

Streams that flow from undisturbed forested watersheds generally exhibit very low concentrations of dissolved nutrients. Because of this, the biological productivity in most of these streams is also low, and the water is generally of sufficient quality to be used for many purposes.

Sources of Nutrients

The major sources of dissolved chemical constituents in water that drains a watershed are the constituents in the precipitation falling on the watershed (see earlier discussion), geologic weathering of parent rock, and biological inputs. The cohesive properties of the bipolar water molecule allow it to wet mineral surfaces and to penetrate into the smallest of openings. Chemical and physical weathering then convert rock minerals into soluble or transportable forms that can be introduced into streams and lakes.

Biological inputs to water systems are primarily the photosynthetic production of organic materials from inorganic substances. Additional inputs are the breakdown of organic into inorganic compounds and the materials gathered elsewhere and subsequently deposited in the ecosystem by humans and their animals. Leaf fall into streams is also an important source of organic matter and can result in periodic changes in nutrient concentrations. Some plants, particularly legumes, can add nitrogen to the soil by fixing free atmospheric nitrogen. Dissolved matter, including organic compounds and mineral ions, are added to the ecosystem by precipitation, dust, and other aerosols. Appreciable quantities of nitrogen, sulfur, and other elements often occur in precipitation and in dust and dry fallout from the atmosphere between precipitation events. However, soils contribute the greatest amounts of dissolved chemical constituents to runoff. Land use activities that affect soil properties and processes, therefore, can affect the chemistry of streamflow.

In undisturbed ecosystems, the rock substrate and soil generally will control the relative concentrations of metallic ions (cations, such as Ca^{++}, Mg^{++}, K^{+}, and Na^{+}). Biological and biochemical processes in watersheds influence, but rarely govern, the anionic (HCO_3^-, NO_3^-, and PO_4^{---}) yield. Anions, such as chloride (Cl^-), nitrate (NO_3^-), and sulfate (SO_4^{--}), originate from the atmosphere, at least in the absence of abundant sulfide minerals. Various soil processes largely regulate the addition of these latter ions to streamflow, however.

Some of the more important dissolved chemicals or nutrients that occur in undisturbed surface waters are described in the following paragraphs to aid in the understanding of the impacts of land use on water quality.

Nitrogen

Sources of nitrogen include the fixation of nitrogen gas by certain bacteria and plants, additions of organic matter to water bodies, and small amounts from the weathering of rocks. Nitrogen occurs in several forms, including ammonia, gaseous N, nitrite-N, and nitrate-N. Organic N breaks down into ammonia, which eventually becomes oxidized to nitrate-N, a form available to plants. In the absence of oxygen, the process of denitrification can convert nitrate back to ammonia and nitrogen gas.

Nitrate-N is generally the primary ion of interest in relation to watershed management practices; all other ions remain at concentrations far below water quality standards in most instances (MacDonald et al. 1991). High concentrations of nitrate-N in water can stimulate growth of algae and other aquatic plants, but if phosphorus (P) is present, only about 0.30 mg/l of nitrate-N is needed for algal blooms. Some fish life can be affected when nitrate-N exceeds 4.2 mg/l. When nitrate-N levels exceed 45 mg/l in drinking water, human health can be affected—in particular, the blue baby disease. Drinking-water standards in the United States limit nitrate-N concentrations to 10 mg/l (Environmental Protection Agency 1976).

Streamflow from undisturbed forestlands usually contains lower concentrations of N and nitrate-N than watersheds with other land uses (Table 11.4). Urban and agricultural development frequently increase nitrate-N and total N concentrations in streamflow.

TABLE 11.4. Mean concentrations of nitrogen and phosphorous for different land uses in the eastern United States

Land Use	Total N	Nitrate-N	Total P	Ortho-P
		(mg/l)		
Forest	0.95	0.23	0.014	0.006
Mostly forest	0.88	0.35	0.035	0.014
Mixed use	1.28	0.68	0.040	0.017
Mostly urban	1.29	1.25	0.066	0.033
Mostly agriculture	1.81	1.05	0.066	0.027
Agriculture	4.17	3.19	0.135	0.058

Source: From Omernik 1976.

Phosphorus

Phosphorus originates from the weathering of igneous rocks, soil leaching, and organic matter. Less is known about the phosphorus cycle than the nitrogen cycle, however. The common forms of phosphorus essentially are defined by the analytical technique used to quantify them rather than by natural processes. In an aquatic environment, phosphorus becomes available to plants by weathering and is taken up and converted into organic phosphorus. Upon decay, this process is reversed.

Phosphorus concentrations in streamflow are affected by land use just as nitrogen concentrations are (Table 11.4). Problems of eutrophication are often associated with accelerated loading of phosphorus to waters that are naturally deficient in phosphorus. Urbanization and agricultural land use result in the greatest problems of phosphorus loading to water bodies.

Calcium

Calcium is abundant in most waters because it is a major constituent of many rock types, especially limestone. Exceptions are acid peat and swamp waters. Soluble calcium occurs as long as CO_2 is present in the water and pH < 7–8. Calcium is one of the major ions contributing to hardness of water, total dissolved solids, and specific conductance. High calcium concentrations do not appear to be harmful to fish and other aquatic life.

Magnesium

Magnesium is abundant in igneous and carbonate rocks, such as limestone and dolomite. Its solubility increases with greater concentrations of CO_2 or lower pH. In concentrations > 100–400 mg/l, magnesium can be toxic to some fish. When concentrations are < 14 mg/l, waters generally support fish fauna.

Sodium

Abundant in both igneous and sedimentary rocks, *sodium* is leached readily into surface and groundwater systems and remains in solution. Sodium does not usually have any adverse impact on fish fauna unless both the sodium concentration and the potassium concentration exceed 85 mg/l. Some levels of sodium have beneficial effects by reducing the toxicity of aluminum and potassium salts to fish.

Potassium

Sources of *potassium* include igneous rocks, clays, and glacial material. Potassium is usually less abundant than sodium, but it is essential to plant growth and is recycled by aquatic vegetation. Pristine waters generally contain < 1.5 mg/l of potassium, but nutrient-enriched or eutrophic waters can contain > 5 mg/l. When potassium levels exceed 400 mg/l, some fish kills occur; levels > 700 mg/l can kill invertebrates as well.

Manganese

Manganese, found in igneous rocks, is leached from the soil. Manganese is essential to plant metabolism and is circulated organically, becoming soluble upon decay. At pH levels ≤ 7, the most dominant form is Mn^{++}. Concentrations of manganese rarely exceed 1 mg/l in undisturbed waters. The drinking-water standard for manganese is 0.05 mg/l.

Sulfur

Sulfur occurs naturally in water from the leaching of gypsum and other common igneous and sedimentary rocks. Weathering processes yield oxidized sulfate (SO_4^{--}) ions that are soluble in water. Sulfate is also found in rainfall at concentrations frequently exceeding 1 mg/l and sometimes > 10 mg/l. The higher concentrations of atmospheric sulfate are largely the result of air pollution and are the main contributors to acid precipitation (Box 11.1).

Under reducing conditions, organic sulfur can be converted to sulfides. Metal sulfides occur, with H_2S being present below pH 7 and HS^- ions occurring in alkaline waters. The sulfide H_2S produces a rotten-egg smell.

Generally, waters with a desirable fish fauna contain < 90 mg/l sulfate; waters with < 0.5 mg/l will not support algal growth. Drinking-water standards are 250 mg/l for sulfate.

Pesticides and Fertilizers

Pesticides and fertilizers that are often used in agriculture and other types of land management have the potential to affect the quality of surface water and groundwater. These are introduced to accomplish management objectives by being applied to a specific *target* organism (pesticides) or location (fertilizers). Of concern to watershed managers are the chemicals that find their way into the water system and become transported to *nontarget* organisms. The risks or hazards to a nontarget organism are determined largely by the likelihood that the organism will come in contact with the chemical (exposure) and the *toxicity* of the chemical to that organism.

Toxicity effects can be either *acute* or *chronic* and are not necessarily lethal. Acute effects are those caused by exposure to large doses of a chemical over a short period, whereas chronic effects are those caused by exposure to relatively small doses of a chemical over a long period. One must realize that the characteristics of the chemical and of the organism affected, in addition to the size of the dose and the frequency and duration of contact, all affect *toxicity*.

Toxicity to nontarget organisms is usually determined with bioassay techniques in which organisms are subjected to increasing concentrations of chemicals and observed over time. The concentration at which 50% of the organisms are killed is the lethal concentration (LC_{50}) or the median tolerance limit (TL_M). Values of TL_M for aquatic organisms and for commonly used chemicals are presented in Table 11.5.

Transport Processes

Dissolved chemical constituents can leave a terrestrial system by subsurface flow through the soil, surface runoff, or groundwater. These processes are part of the nutrient cycling, which consists of inputs, outputs, and movement of dissolved solids and gases within the system (Fig. 11.5). Frequently, outputs (the streamflow from upland watersheds) and the various factors that influence the chemistry of this runoff are a concern. The problem is nutrient losses from forestlands that cause water pollution for downstream uses.

Movement of water through soil, along with the associated biological activity, controls the ionic composition of water leaving upland watersheds as streamflow. The most chemically active components in soil are the clays and organic colloids. Clays have high exchange capacities in comparison to most other minerals in soil because of their large surface areas per unit of volume and their negative electrical charge. Organic colloids also have a large capacity to exchange ions in solutions for those adsorbed on their surfaces.

TABLE 11.5. Values of 48 hr median tolerance level (TL_M) for selected pesticides and aquatic organisms

Pesticide	Range of Concentrations (ppb)		
	Aquatic Insects	Crustacea	Fish
Insecticide			
DDT	10–100	1–10	1–10
Endrin	0.1–1.0	10–100	0.1–10
Aldrin	1–10	10–100	1–100
Malathion	1–10	1–10	100–1000
Dieldrin	1–10	100–1000	1–100
Herbicide			
2,4-D (BEE)	1000–10,000	100–1000	1000–10,000
2,4-D (amine salt)	—	—	100,000–1×10^6
Picloram	10,000–100,000	10,000–100,000	10,000–100,000

Source: Adapted from Brown 1980 after Thut and Haydu 1971.

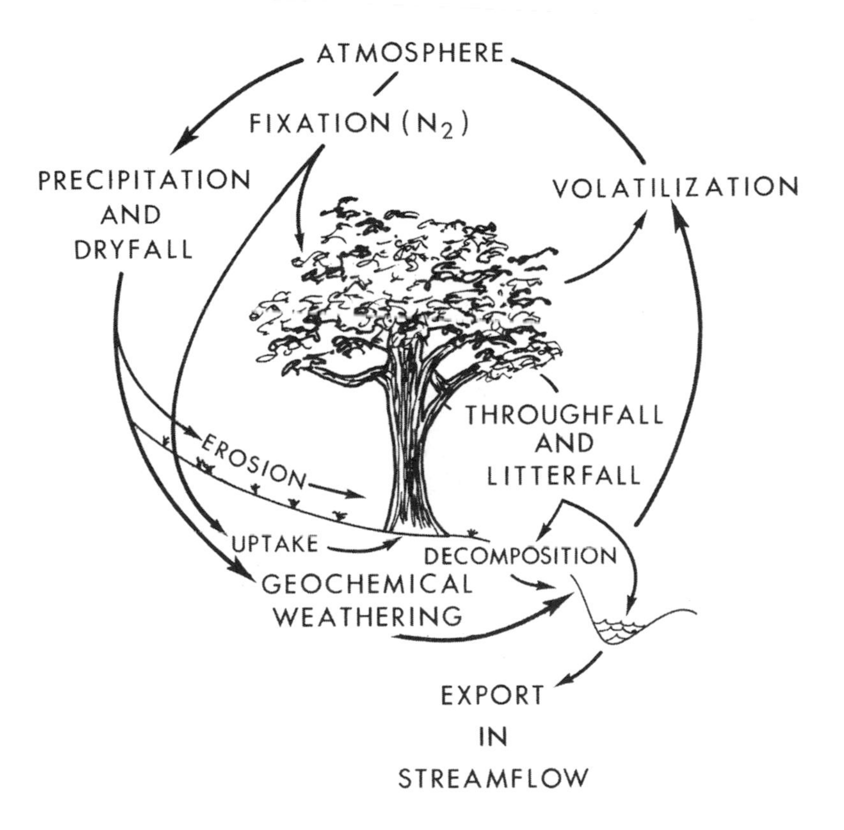

FIGURE 11.5. A nutrient cycle for a forested watershed; some of the gaseous phases are not important for certain nutrients (adapted from Brown 1980).

A simplified concept of the exchange processes of clays and organic colloids is that adsorbed cations are exchanged selectively for hydrogen ions from the soil water. The hydrogen ions, in turn, come principally from the solution of CO_2 in water and the dissociation of the resulting H_2CO_3 molecule into two hydrogen ions and a bicarbonate ion. Dissolved CO_2 originates mainly from the metabolism of microorganisms and plant roots in the soil. Only water that remains for some time in the interstices among soil particles is likely to accumulate an appreciable load of dissolved CO_2 and, consequently, be effective in leaching mineral ions. It is the flushing of these accumulations that creates the initially high levels during a stormflow event.

Ionic concentrations and their relationships to streamflow discharge are variable. For example, a positive relationship exists between H^+ and NO_3^- concentrations and streamflow discharge from forested watersheds in New England, but no relationship exists between discharge and Mg^{++}, Ca^{++}, SO_4^{--}, and K^+. Individual cation concentrations, when considered annually, appear to be independent of discharge rate on watersheds in the Appalachian Mountains of the eastern United States. Inverse relationships between ionic concentrations and discharge have been reported in selected streams of the Rocky Mountains and in California. In large streams, ionic concentrations are commonly lower at times of high discharge than at low discharge, due in large part to the residence time of water in the soil. During periods of dry weather and low streamflow, the slow movement of water through soils enhances the opportunity for chemical reactions. Dilution also occurs when large volumes of water move through soils.

Cumulative Effects

Because the transportation of dissolved chemicals in streams is largely dependent upon water movement, any land use activity that alters the volumes or timing of runoff also affects the rates of chemical and nutrient transport. Timber harvesting often causes a short-term increase in nitrogen concentrations in streams (Mann et al. 1988; Tiedemann et al. 1988). The increases in nitrogen concentrations can result from reduced nutrient uptake, increased subsurface flow, increased alteration of nitrogen to leachable forms, increased volumes of decaying organisms, or combinations thereof. Baker et al. (1989) measured the rates at which nine nutrients were released from decomposing logging residues and found that nitrogen was released most slowly. Other management activities that contribute to these mechanisms, such as herbicide application (Vitousek and Matson 1985) and controlled burning (Sims et al. 1981), can have similar effects.

Some land use activities alter existing chemicals, causing them to interact with the environment in different ways. For example, Helvey and Kochenderfer (1987) found that limestone road gravels can moderate the pH of the runoff from acid rainfall. Birchall et al. (1989) discovered that aluminum released by acid rainfall is less toxic to fish populations if streams are enriched in silica. Bolton et al. (1990) found that vegetative conversions can alter the spatial distribution of the resulting nitrogen mineralization because the plants lend different mineralization potentials to the soil. Other points to consider are:

- Acid rainfall increases the mobility of nutrients and, thus, contributes to nutrient deficiencies (Johnson et al. 1982).
- Chemicals introduced to stream channels can be redeposited downstream. Reservoirs can concentrate dissolved pollutants if evaporation rates are high.

Biological Characteristics

Biological characteristics of a water body include bacteria and protozoa in the water, including some that are health-impairing, and populations of aquatic invertebrates (insects, etc.). In some instances, the nature of the aquatic biota in a water body can be used as indicators of surface water quality and the overall habitat characteristics of aquatic ecosystems.

Bacteria

Waterborne pathogenic bacteria have long been recognized as causes of gastrointestinal illnesses in humans, livestock, and wildlife species. Many of these diseases are transmitted only among members of the same species, while other diseases can be transmitted to hosts of different species. A major concern to managers is that surface water from upland watersheds can become contaminated by health-impairing bacteria and pose hazards to the welfare of humans and other animals.

E. coli (*Escherichia coli*) is an ubiquitous bacterium that is found in the gut of all warm-blooded animals. Long considered to be benign, several pathogenic strains of *E. coli* have developed in recent years and gained attention due to the health risk that they pose. *E. coli* is often used as an indicator species in bacteriological testing to determine if water is suitable for drinking or other forms of human contact (swimming, etc.). An estimate of their number, therefore, represents an index of bacteriological water quality. Other pathogenic organisms are difficult to trap, difficult to analyze, and expensive to process (Buckhouse 2000). As a consequence, testing for fecal coliform has become the accepted surrogate sampling protocol with the understanding that if fecal coliforms are present, pathogens are then potentially present.

Knowledge of the cycles and variability of *E. coli* and other bacteria in natural waters and of the relationships of bacteria to environmental factors on disturbed watershed lands is limited. However, investigations of bacteria-environment relationships in relatively undisturbed areas have provided useful baseline information.

Protozoa

Waterborne protozoal parasites can also cause gastrointestinal illnesses in people, their livestock, and wildlife species. *Giardia* and *Cryptosporidium* have drawn considerable attention recently (Buckhouse 2000). Both parasites can be carried by a variety of warm-blooded animals including rodents, deer and elk, and livestock. One waterborne protozoan, *Cryptosporidium parvum* (*C. parvum*), has been implicated in large-scale outbreaks of gastroenteritis in humans (MacKenzie et al. 1994). Cattle are often perceived to be a leading source of *C. parvum* in surface water, although evidence supporting this claim is incomplete and contradictory in some cases (Atwill 1996).

The possible presence of *Giardia* and *Cryptosporidium* in surface water has been known for decades, but only recently has testing been conducted following gastrointestinal illnesses to determine that these protozoa have been the likely cause. However, managers often have a difficult time in determining if the *Cryptosporidium* found in samples of surface water is *C. parvum* or some other *Cryptosporidium* species not infectious to humans (Atwill 1996).

Box 11.3

Indicating Water Quality by the Presence of Macroinvertebrates: An Index

Macroinvertebrate communities in streams have been found to be sensitive to stream disturbance and can be influenced by stream channel morphology in the Pacific Northwest of the United States (Hershey and Lamberti 1998). Bioassessments of the abundance and diversity of macroinvertebrates in streams and lakes have been successfully used to monitor macroinvertebrate condition or health throughout the world (Levy 1998). The Hilsenhoff Biotic Index (HBI) is one example of a biological approach to monitoring water quality conditions and biological potential that is based upon inventories of macroinvertebrates (Hilsenhoff 1987). This index is based on the tolerance of a wide range of benthic macroinvertebrate species to a diversity of pollutants: some species are intolerant and receive a numerical score different from species that are tolerant.

Aquatic Biota as Indicators of Surface Water Quality

Up to this point, we have discussed water quality characteristics of a physical, chemical, and disease-causing nature. Any body of water can have a range of characteristics that can be difficult to interpret. What has been sought in terms of water quality management, therefore, is some integrative method of characterizing water quality—a synthesis of the collective physical and chemical properties. Some investigators have suggested that the aquatic biota represent such a synthesis.

Any body of water in its habitat can support and maintain certain organisms. As a rule, the better the *aquatic health*, the greater the diversity of species of organisms. As discussed with the earlier BOD loading example, pollution can diminish the variety of organisms, leaving a habitat suitable for only a few, well-adapted organisms. As a result, a section of stream that supports a large variety of species of organisms would be considered in better condition than one supporting a large number of a few species, that is, those adapted to the polluted environment. The advantage of indicating surface water quality through the presence of aquatic biota is that samples of chemical constituents, DO, temperature, and so forth offer a limited picture, at one point in time, of water quality, while the community of aquatic organisms represents more of a long-term indicator of overall impacts on water quality and antecedent flow conditions. It is this concept that underlies biological approaches to water quality classification (Box 11.3).

Groundwater Quality

The usefulness of groundwater for drinking or irrigation depends largely on its quality, which is largely related to the type and location of the aquifer. Water from igneous and metamorphic rocks is generally of excellent quality for drinking. Exceptions occur in arid areas, where recharge water has high concentrations of salts because of high evaporation rates. The quality of water in sedimentary rocks varies; deep marine deposits can yield saline water, but shallow sandstones can have good quality water.

Groundwater is generally of higher quality than surface water. Because it is in direct contact with rocks and soil material longer than most surface water, groundwater is usually higher in dissolved mineral salts such as sodium, calcium, magnesium, and potassium cations with anions of chloride, sulfate, and bicarbonate. If such salts exceed 1000 ppm (mg/1), the water is considered *saline*. High concentrations of dissolved mineral salts can limit the use of groundwater for drinking (laxative effects) and other municipal and industrial uses.

Groundwater that contains high amounts of calcium and magnesium salts is considered *hard water.* The hardness is determined by the concentration of calcium carbonate or its equivalent, as follows: soft, 0–60 mg/l; moderately hard, 61–120 mg/l; hard, 121–180 mg/l; very hard, > 180 mg/l. Although hard water leaves scaly mineral deposits inside pipes, boilers, and tanks, and hampers washing because soap does not lather easily in hard water, it does not represent health hazards. In fact, hard water is considered generally better for human health than soft water.

Naturally occurring concentrations of iron in groundwater can limit water use in some areas. Although not a health problem, iron concentrations in excess of 0.3 mg/l affect the taste and color of water, limiting its use for drinking, cooking, and washing of clothes.

Unconsolidated deposits and other aquifers with high hydraulic conductivities, such as limestone caverns and lava tubes, can become contaminated from biological sources if they are close to surface sources of pollution (Box 11.4). Human garbage, sewage, and livestock wastes can contaminate such aquifers readily. Several factors limit our ability to solve groundwater contamination problems, however. We do not fully understand the hydrology of the vadose zone (see Chapter 5). Furthermore, we often do not have reliable information on the rates and flow paths of groundwater. In addition, the location and extent of recharge areas are not always well known. The vastness of most aquifers, coupled with these deficiencies, presents difficult and long-term problems when groundwater becomes polluted.

SUMMARY

The protection and maintenance of high-quality water are fundamental goals of watershed management. This chapter described water quality characteristics and the impacts of land use activities and disturbances on these characteristics and the aquatic ecosystem. After completing this chapter you should be able to:

- Define and explain the terms *water quality* and *pollution*.
- Discuss precipitation chemistry and its possible impacts on water quality.
- Identify and describe the different physical, chemical, and biological pollutants associated with different types of land use that can occur on upland watersheds.
- Describe how nutrients and heavy-metal concentrations and loadings to a stream can be accelerated above normal background levels, and discuss the implications.
- Describe how thermal pollution affects an aquatic system.
- Explain why dissolved oxygen is an important indicator of water quality, discuss what happens when DO levels are substantially reduced, and define BOD.
- Describe the physical, chemical, and biological effects of introducing a biodegradable pollutant into a flowing stream.

Box 11.4

Effects of Surface Land Use on Water Quality in Karst Aquifers

Many areas with karst topography experience groundwater contamination from surface activities such as farming and livestock grazing. Surface water and water occurring in limestone solution cavities are more directly connected than are most other aquifers; pollutants from the surface can move quickly into groundwater in such cases. Of particular concern in many agricultural and grazing areas in karst topography is nitrate pollution. Boyer and Pasquarell (1995) reported the following levels of nitrate in karst springs associated with different land uses on the watersheds in the Appalachian region of the United States:

Average Nitrate Concentration (mg/l)	Grazing (% of basin)	Cropland	Forest	Urban/ Residential
14	59	16	15	10
10	34	8	40	18
2.7	10	1	80	9
0.4	0	0	100	0

In karst areas, best management practices (described in Chapter 12) that reduce surface water pollution can directly improve groundwater quality as well.

- Discuss how various forest management, livestock, and agricultural practices can affect water quality; consider each major type of pollutant and indicate the kinds of management measures needed to correct or prevent water quality problems.

References

Atwill, E.R. 1996. Assessing the link between rangeland cattle and waterborne *Cryptosporidium parvum* infection in humans. *Rangelands* 18:48–51.

Baker, T.G., G.M. Will, and G.R. Oliver. 1989. Nutrient release from silvicultural slash: Leaching and decomposition of *Pinus radiata* needles. *Forest Ecology and Management* 27:53–60.

Bartsch, A.F., and W.M. Ingram. 1959. Stream life and the pollution environment. *Public Works* 90:104–110.

Beschta, R.L., R.E. Bilby, G.W. Brown, L.B. Holtby, and T.D. Hofstra. 1987. Stream temperature and aquatic habitat: Fisheries and forestry interactions. In *Streamside manage-*

ment: Forestry and fishery interactions, ed. E.O. Salo and T.W. Cundy, 191–232. Seattle: Institute of Forest Resources, University of Washington.

Binkley, D., and T.C. Brown. 1993. *Management impacts on water quality of forests and rangelands*. USDA For. Serv. Gen. Tech. Rep. RM-239.

Birchall, J.D., E. Exley, J.S. Campbell, and M.J. Phillips. 1989. Acute toxicity of aluminum to fish eliminated in silicon-rich acid waters. *Nature* 338:146–147.

Bolton, H., Jr., J.L. Smith, and R.E. Wildung. 1990. Nitrogen mineralization potentials of shrub-steppe soils with different disturbance histories. *Soil Sci. Soc. Amer. J.* 54:887–891.

Boyer, D.G., and G.C. Pasquarell. 1995. Nitrate concentrations in karst springs in an extensively grazed area. *Water Resour. Bull.* 31:729–736.

Brown, G.W. 1980. *Forestry and water quality*. Corvallis: Oregon State Univ. Bookstores.

Buckhouse, J.C. 2000. Domestic grazing. In *Drinking water from forests and grasslands: A synthesis of the scientific literature*, ed. G.E. Dissmeyer, 153–157. USDA For. Serv. Gen. Tech. Rep. SRS-39.

Churchill, M.A., R.A. Buckingham, and H.L. Elmore. 1962. *The prediction of stream reaeration rates*. Chattanooga: Tennessee Valley Authority.

Duffy, P.D., J.D. Schreiber, and S.J. Ursic. 1986. Nutrient transport by sediment from pine forests. In *Proceedings of fourth federal interagency sedimentation conference*, 57–65, Las Vegas, NV.

Dunne, T., and L.B. Leopold. 1978. *Water in environmental planning*. San Francisco: W.H. Freeman and Co.

Environmental Protection Agency (EPA). 1976. *Quality criteria for water.* EPA Rep. 440/976023.

Environmental Protection Agency. 1980. *Acid rain*. Washington, DC: Office of Research and Development, Environmental Protection Agency. EPA-600/979036.

Fahey, T.J. 1979. Changes in nutrient content of snow water during outflow from Rocky Mountain coniferous forest. *Oikos* 32:422–428.

Foster, I., A. Gurnell, and B. Webb, eds. 1995. *Sediment and water quality in river catchments*. Chichester, NY: John Wiley.

Glass, G.E., J.A. Sorensen, and G.R. Rapp, Jr. 1992. *Mercury sources and distribution in Minnesota's aquatic resources: Precipitation, surface water, sediments, plants, plankton and fish.* Minnesota Pollution Control Agency Report to the Legislative Commission on Minnesota Resources. St. Paul, MN.

Gosz, J.R., C.S. White, and P.F. Ffolliott. 1980. Nutrient and heavy metal transport capacities of sediment in the southwestern United States. *Water Resour. Bull.* 16:927–933.

Helvey, J.D., and J.N. Kochenderfer. 1987. Effects of limestone gravel on selected chemical properties of road runoff and streamflow. *North. J. Appl. For.* 4:23–25.

Hem, J.D. 1992. *Study and interpretation of the chemical characteristics of natural water,* 3rd ed. U.S. Geol. Surv. Water Supply Pap. 2254.

Hershey. A.E., and G.A. Lamberti. 1998. Stream macroinvertebrate communities. In *River ecology and management: Lessons from the Pacific Coastal Ecoregion*, eds. R.J. Naiman and R.E. Bilby, 169–199. New York: Springer-Verlag.

Hilsenhoff, W.L. 1987. An improved biotic index of organic stream pollution. *The Great Lakes Entomologist* 20(1): 31–39.

Hutchinson, T.C., and M. Haras, ed. 1980. *Effects of acid precipitation on terrestrial ecosystem.* New York: Plenum Press.

Johannessen, M., and A. Henriksen. 1978. Chemistry of snow meltwater: Changes in concentration during melting. *Water Resour. Res.* 14:615–619.

Johnson, D.W., J. Turner, and J.M. Kelley. 1982. The effects of acid rain on forest nutrient status. *Water Resour. Res.* 18:449–461.

Kapp, W.W., V.C. Bowersox, B.I. Chevone, S.V. Gupta, J.A. Lynch, and W.W. McFee. 1988. *Precipitation chemistry in the United States, Part I: Summary of ion concentration variability 1979–1984,* vol 3. Ithaca, NY: Center For Environmental Research, Cornell University.

Kolka, R.K. 1996. Hydrologic transport of mercury through forested watersheds. Ph.D. Thesis. University of Minnesota, St. Paul.

Levy, S. 1998. Using bugs to bust polluters. *BioScience* 48(5): 342–346.

Likens, G.E., F.H. Bormann, R.S. Pierce, J.S. Eaton, and N.M. Johnson. 1977. *Biogeochemistry of a forested ecosystem.* New York: Springer-Verlag.

MacDonald, L.H., A.W. Smart, and R.C. Wissmar. 1991. *Monitoring guidelines to evaluate effects of forestry activities on streams in the Pacific Northwest and Alaska.* U.S. Enviromental Protection Agency, USEPA/910/9-91-001.

MacGregor, H.G. 1994. *Literature on the effects of environment and forests on water quality and yield.* Fredericton, NB: Canadian Forest Service.

MacKenzie, W.R., N.J. Hoxie, M.E. Proctor, and others. 1994. A massive outbreak in Milwaukee of *Cryptosporidium. N. Engl. J. Med.* 331:161–167.

Mann, L.K., D.W. Johnson, D.C. West, D.W. Cole, J.W. Hornbeck, C.W. Martin, H. Riekerk, C.T. Smith, W.T. Swank, L.M. Tritton, and D.H. Van Lear. 1988. Effects of whole-tree and stem-only clearcutting on postharvest hydrologic losses, nutrient capital, and regrowth. *For. Sci.* 34:412–428.

Norris, L.A., H.W. Lorz, and S.V. Gregory. 1991. Forest chemicals. In *Influences of forest and rangeland management on salmonid fishes and their habitat*, ed. W.R. Meehan, 207–296. Bethesda, MD: American Fisheries Society, Spec. Publ. 19.

Omernik, J.M. 1976. *The influence of land on stream nutrient levels.* EPA Publ. 600/376014.

Pagenkopf, G.K. 1983. *Chemistry of Montana snow precipitation.* Water Resour. Res. Center Rep. 138. Bozeman: Montana State University.

Ponce, S.L. 1974. The biochemical oxygen demand of finely divided logging debris in stream water. *Water Resour. Res.* 10:983–988.

Postel, S. 1984. *Air pollution, acid rain, and the future of forests.* Worldwatch Pap. 58. Washington, DC: Worldwatch Inst.

Reid, L.M. 1993. *Research and cumulative watershed effects.* USDA For. Serv. Gen. Tech. Rep. PSW-GTR-141.

Sims, B.D., G.S. Lehman, and P.F. Ffolliott. 1981. Some effects of controlled burning on surface water quality. *Hydrol. and Water Resour. in Ariz. and the Southwest* 11:87–89.

Stottlemyer, R. 1987. Natural and anthropogenic factors as determinants of long-term streamwater chemistry. In *Management of subalpine forests: Building on 50 years of research,* tech. coord. Troendle, C.A., M.R. Kaufmann, R.H. Hamre, and R.P. Winokur, 86–94. USDA For. Serv. Gen. Tech. Rep. RM-149.

Stumm, W., and J.J. Morgan. 1995. *Aquatic chemistry.* New York: John Wiley and Sons.

Thut, R.N., and E.P. Haydu. 1971. Effects of forest chemicals on aquatic life. In *Forest land uses and stream environment symposium*, 159–171. Corvallis: Oregon State Univ. Press.

Tiedemann, A.R., T.M. Quigley, and T.D. Anderson. 1988. Effects of timber harvest on stream chemistry and dissolved nutrient losses in northeast Oregon. *For. Sci.* 34:344–358.

Vigon, B.W. 1985. The status of nonpoint source pollution: Its nature, extent and control. *Water Res. Bull.* 21:179–184.

Vitousek, P.M., and P.A. Matson. 1985. Intensive harvesting and site preparation decrease soil nitrogen availability in young plantations. *South. J. Appl. For.* 9:120–125.

CHAPTER 12 *Water Quality Management*

INTRODUCTION

Securing clean water will always be a primary goal of watershed management. While management agencies have often assumed a largely regulatory approach to achieving this goal in the past, the trend recently has been to move decision-making and action-taking to the local level. Conservation districts, locally led advocacy organizations, and other local groups have taken on increased roles and responsibilities for ensuring the delivery of high-quality water supplies through watershed management (see Chapter 19). Regardless of how it is accomplished, a major contribution of watershed management is securing flows of high-quality water to all people in the river basin of concern.

Issues of concern to managers of water quality, effects of vegetation management on water quality constituents, and the principles and purposes of water quality monitoring programs are the focus of this chapter. While emphasis has been placed on potential nonpoint water pollution problems, it is important that point sources of pollution also be considered in a general context.

ISSUES

People are increasingly concerned about the quality of water flowing from upland watersheds to downstream users and how this water quality is impacted by land use activities. There is also concern about land use affecting groundwater quality. Part of this concern has evolved from increased public awareness of environmental issues in natural resources management. In large part, this heightened awareness has been manifested in the United States by passage of the National Environmental Policy Act and the Clean Water Act and creation of the Environmental Protection Agency in the early 1970s. From a watershed management perspective, the concern about water quality is exemplified by the need to recognize the impacts of water pollution problems, adhere to the maintenance of water quality standards, and comply with the laws and regulations governing water quality management. Watershed managers in many agricultural regions of the United States must also monitor the hydrologic impacts on water quality levels of removing highly erodible lands from crop production in the Conservation Reserve Program, instituted by Congress in 1985.

The issues of scale and cumulative effects are important both in recognizing water quality problems and in developing management solutions. Water quality problems can be manifested along a reach of stream, for example, where cattle have been concentrated in a riparian area. Intensive agricultural cropping in karst topography can add nutrients and chemicals to groundwater, which can become polluted over a widespread region and can take a long time to show any benefits from changes in management. The hypoxia issue (dead zone due to excessive nutrient loading) in the Gulf of Mexico is largely the response to cumulative effects of agricultural practices and stream channel modifications that have accelerated nutrient discharge from the Ohio and Upper Mississippi River basins. Nutrient loading in the main stem of the Mississippi River has been exacerbated by the loss of wetlands, loss of floodplain areas, and the subsequent increase in streamflow associated with the normal bank-full conditions of streams (see Chapter 10). There are now many entrenched stream channels all along the Mississippi and its tributaries that are no longer in contact with their floodplains on a frequently recurring basis. As a result, the level of denitrification that occurs in wetlands and floodplains (as a result of anaerobic processes) has been drastically reduced. The cumulative effects of such complex interactions over space and time represent some of the greatest challenges to resource managers.

Water Pollution Problems

Point and nonpoint water pollution problems continue to plague the United States and other countries. Deteriorating water quality places in jeopardy the sustainability of goods and services derived from watersheds, threatens the health of people and their livestock, compounds water scarcity issues, adversely affects the functioning of aquatic ecosystems, and ultimately impacts the economy of a region and country. Among the sources of surface and groundwater degradation are increasing discharges of untreated or inadequately treated wastewater; discharges of agrochemicals; discharges of hazardous, toxic, or other industrial wastes into water bodies that serve as a water supply for other users; drainage of saline-water from agricultural lands; and an overdraft of groundwater resources, which results in saline intrusions from the sea.

The emphasis here is on how the management of upland watersheds impacts the quality of water flowing from these watersheds and, more specifically, the quality of naturally occurring surface water. Watershed managers can do much to help alleviate water pollution problems by keeping watershed lands in good condition. Watershed management practices that maintain a watershed in good condition include those that sustain high rates of infiltration into the soil, do not contribute to excessive soil erosion, facilitate a relatively slow streamflow response to precipitation, and sustain baseflow in perennial stream systems between storms.

Maintenance of Water Quality Standards

Another issue of concern to watershed managers is the maintenance of water quality standards. Concerns about point and nonpoint water pollution have been heightened in the United States by the need to adhere to water quality standards specified by the federal and state governments. Water quality standards are established as part of the task of determining the *designated beneficial use* of a water body (Stumm and Morgan 1995). A water body can have a diverse array of designated beneficial uses that are often categorized into water supply and environmental and recreational uses. Designation of beneficial uses into one of these categories is determined by an open process where public comment is solicited to help in establishing these uses (Brown et al. 1993). This process is thought to be the most equitable approach available for establishing a water body's beneficial uses. Once the designated beneficial uses have been determined, the physical, chemical, and biological characteristics of the water in relation to these uses, that is, the established water quality standards, must be maintained.

Regulatory Compliance

Regardless of their goal, watershed management practices must always be planned and implemented in a manner to comply with laws and regulations pertaining to the maintenance of water quality standards. Regulatory compliance is required at both the federal and state levels in the United States.

Federal Regulations

Federal water quality legislation in the United States, with its periodic amendments of increasing specificity, provides the "guidelines" for watershed management on federally controlled lands and, in doing so, water quality protection that "more general" natural resources management legislation often lacks. This legislation has also extended watershed and water quality protection to state and private lands in certain instances. The 1972 amendments to the Federal Water Pollution Control Act, which with its amendments is commonly referred to as the *Clean Water Act*, established a federal presence in controlling water quality.

Clean Water Act. The Clean Water Act with its amendments is the primary law in the United States that addresses the protection of the country's lakes, rivers and streams, aquifers, and coastal areas. A primary goal of the Act is ensuring the implementation of watershed management practices to maintain or, when necessary, improve the physical, chemical, and biological integrity of these bodies of water. For example, the total maximum daily load (TMDL) of total sediments entering into a stream system could be established by the regulations specified in this law (Box 12.1). Because the total sediment load at any point in the stream is a function of everything that influences the erosion and sediment dynamics above that point, the individual and collective influences of these upstream land uses must be addressed to meet the TMDL standards specified to meet the conditions of the Clean Water Act. TMDL regulations only mitigate a problem after it has occurred, however.

The main objective of the Clean Water Act is furnishing mechanisms for maintaining or restoring, when necessary, the integrity of the country's waters. A comprehensive framework of standards, technical tools, and financial assistance is provided by the Clean Water Act to address the causes of water pollution and resulting poor water quality, including municipal and industrial wastewater discharges, polluted runoff from rural and urban areas, and habitat destruction (Brown et al. 1993). The Act requires "major industries" to meet performance standards to ensure pollution control; charges states

Box 12.1

General Procedure for the Preparation of a TMDL Plan

The problem of maintaining water quality can be compounded by the need to develop a TMDL plan for a water body found to have water quality conditions that limit its ability to meet its designated beneficial uses as mandated by the Clean Water Act. TMDLs for a water body are the sum of the waste-load allocations from identified point and nonpoint pollution sources within the water body's watershed. A TMDL plan for a water body should identify current waste loads to the water body and the water-load capacities of the water body. With this information, the waste-load capacity is then allocated to the known point and nonpoint pollution sources within the water body's watershed. However, preparation of a TMDL plan is a difficult task, largely because of the problem of understanding and analyzing the mode of conveyance of diffuse nonpoint sources. More conventional "hydraulic methods" can often be used for monitoring and analyzing the discharge of pollutants from point sources such as through a pipe. Identifying nonpoint pollutants is a more difficult task. Both nonpoint and point sources of pollution must be included in a complete TMDL plan.

with setting specific water quality criteria that are appropriate for their waters and developing pollution control programs to meet them; provides financial assistance to states and local communities to help them meet their clean water requirements; and protects valuable wetlands and other aquatic habitats through a permitting process that ensures that development and other activities are conducted in an environmentally sound manner.

Although the U.S. Environmental Protection Agency, the federal agency responsible for "enforcing" the provisions of the Clean Water Act, initially placed emphasis on the more pressing and traceable problems of point source pollution, the agency now recognizes that water quality degradation from nonpoint source pollution is emerging as the major barrier to achieving the water quality goals. Section 208 specifically addressed nonpoint source pollution and designated silvicultural treatments, such as timber-harvesting, improvement thinning, and road construction activities, and livestock grazing activities as significant sources of nonpoint source pollution. Among other provisions, the amended Clean Water Act of 1977 authorized a program of grants to help in covering the costs to rural land owners of implementing best management practices (BMPs) to control nonpoint source pollution (33 U.S.C. 1289). Best management practices are considered to be the most practical approach to controlling nonpoint source pollution, complying with water quality standards, and securing the flows of clean water.

Best Management Practices. The U.S. Environmental Protection Agency's regulations define *best management practices* (BMPs) as those methods, measures, or practices that prevent or reduce water pollution and include structural and nonstructural controls and operation and maintenance procedures (Brown et al. 1993).

BMPs are applied before, during, and after pollution-producing activities to reduce or eliminate the introduction of pollutants into receiving waters. An iterative process of BMP specification, application, monitoring, and "fine-tuning" for future applications is the key to cost-effective BMP use. A practical approach to this iterative process is to prescribe BMPs that careful study and professional judgment indicate will control nonpoint source pollution to within specified water quality standards in most cases and then reassess the BMPs as new information becomes available.

BMPs to mitigate the erosion-sedimentation processes are already known for many silvicultural treatments, livestock-grazing practices, road-related soil disturbances, and agricultural activities (Moore et al. 1979, Lynch et al. 1985, Chaney et al. 1990, and others). However, BMPs for some types of other pollutants are incomplete or unknown; therefore, research that relates land use to water quality is needed in these instances. Monitoring efforts reveal that the use of BMPs is increasing and that their implementation generally maintains water quality standards. An example of BMPs for forest management aimed at protecting water quality and wetlands is presented in Box 12.2.

The use of BMPs has been justified on the basis as being preventative in contrast to a water quality monitoring program, which would determine problems "after the fact." Furthermore, most believe BMPs to be a cost-effective alternative to intensive water quality monitoring. Information about the overall cost-effectiveness of BMP programs has been lacking, however.

Conservation Reserve Program. The *Conservation Reserve Program* (CRP) is a voluntary agricultural cropland retirement program to reduce soil loss on 18 million ha of highly erodible lands in the United States through their conversion to a cover of trees, shrubs, or grass. In administering the program, the U.S. Department of Agriculture pays enrolled farmers an annual rent and reimburses them for half the costs of establishing the permanent cover in exchange for retiring these lands for 10 yr. The CRP was instituted to decrease soil erosion, increase farm incomes, and improve environmental quality over the life of the program. Major enrollment regions have been the Great Plains (60%), the Corn Belt, and the historic Cotton Belt.

Studies on impacts of the program have shown increases of carbon and nitrogen in soil organic matter pools and potential carbon and nitrogen mineralization in soils on CRP land (Robles and Burke 1997). There have also been the expected reductions in surface erosion and consequent sediment loads in streams. Sediment responses from CRP lands can be delayed, however, as near stream and instream sediments are remobilized (Davie and Lant 1994). Furthermore, CRP lands might not be near instream locations where they would be most effective in trapping upslope sediments. Other water quality services have been nonpoint source pollution abatement; natural purification of water; and dilution of wastewater (Loomis et al. 2000). Often unintentional ecological benefits of CRP have been a reduction in landscape fragmentation and maintenance of biological diversity.

Farmers face a number of options as CRP lands become eligible for release. Among the available options for these lands are returning to agricultural crop production, partially returning to crop production, or retaining forest or grassland cover. Decisions on these options will likely be based in part on people's concerns about possible impacts on local economies, increases in land prices due to decreases in availability, and competition among forest, livestock, and agricultural crop production. Regardless of the decisions made, managers of CRP lands and watersheds will need to carefully monitor these land use decisions and their impacts on hydrology, and especially water quality.

Box 12.2

Best Management Practices in Minnesota: Forest Management for Protecting Water Quality and Wetlands (Minnesota Department of Natural Resources 1995)

A set of Best Management Practices (BMPs) developed in Minnesota for forest resource management targeted the following:

- Managing fuel, lubricants, and equipment—practices include designating the least hazardous areas for fueling and maintaining vehicles and equipment, reporting procedures for spills, etc.
- Providing filter strips of vegetation adjacent to lakes, streams, and open water wetlands to reduce sediment, nutrient, and pesticide runoff from entering the respective water body—practices specify widths of filter strips based on the slope of the land and limitations of soil exposure and disturbance within filter strips.
- Provision of shade strips of vegetation adjacent to lakes, streams, and open water wetlands to moderate water temperature—specific requirements of 100 ft (35 m) are designed for trout streams (cold-water species), required minimum densities of trees for different management objectives, and recommendations for minimizing disturbance.
- Wetland protection measures—stated requirements for planning activities and for minimizing disturbance of wetland sites during forest management activities.
- Building and maintaining forest roads—detailed and comprehensive set of guidelines for planning roads, selecting types of roads, water-crossing specifications, drainage requirements, etc.
- Timber-harvesting guidelines—recommendations for developing a plan, designing the harvesting site to minimize runoff and disturbance to water bodies, and postharvesting recommendations.
- Mechanical site preparation guidelines—planning, design, and operational recommendations.
- Pesticide use guidelines—incorporating integrated peat management strategies, selecting pesticides, measures to minimize contamination potential, and managing all phases of pesticide use.
- Prescribed burning guidelines—planning considerations and practices for eliminating the hazard of wildfires, facilitating revegetation, improving wildlife habitat, controlling insects and diseases, and reducing potential for erosion and sediment deposition into water bodies.

State Regulations

State laws and regulations that encourage the watershed as a planning basis are emerging throughout the United States. These actions are often related to the need to protect an endangered or threatened fish species or meet the needs of those people who rely on "out-of-stream" uses of water (Thorud et al. 2000). Provisions of most of the state laws and regulations are often voluntary and can call for pluralistic representation from state agencies, local government entities, the major economic interests in the area of concern, and the general public. Their main goal is collaboratively developing integrated management plans for land, water, and other natural resources on a watershed basis.

Vegetation Management and Water Quality

The quality of water flowing from upland watersheds is determined largely by precipitation chemistry, climatic factors, soils, geologic strata, vegetation, and land use activities on the watersheds. Therefore, watershed managers need to recognize the linkages among these factors, land use practices, and the resultant water quality. By knowing in particular the effects of vegetation management practices on water quality, insight is gained concerning measures to mitigate the detrimental impacts of land use activities on water quality.

Sediment

Sediment concentrations are often increased following logging operations, livestock grazing, and other land use actions that disturb soil surfaces. Likewise, fire and floods can increase suspended loads in streams—these can be natural occurrences but are sometimes affected by land use. In many cases, the effects of individual land use activities on sedimentation cannot be isolated.

Timber Harvesting

Timber harvesting, silvicultural treatments, and other forestry activities can increase the sediment in streams by increasing surface runoff rates and the risks of soil mass movement. In timber-harvesting operations, the contributions of felling, limbing, and bucking of trees to the sediment levels in streamflow are usually negligible. However, skidding and yarding operations with concentrated vehicular traffic often cause accelerated soil erosion, which can lead to increased downstream sedimentation. Soil disturbances and erosion caused by moving logs from the stump to a landing vary greatly with the type of skidding or yarding equipment (Stednick 2000). Crawler tractors generally cause the greatest disturbance, followed closely by wheeled skidders. One method of mitigating the amount of site disturbance by skidders is through the careful layout of skid trails. Cable logging systems result in less site disturbance because yarding trails are established to the yarding tower machinery, which is restricted to road surfaces. Helicopters and balloons will likely result in the minimal site disturbance. A rule of thumb is that the less compacting and disturbance of the forest floor, the less watershed damage will result from skidding and yarding activities.

Streamside vegetation acting as buffer strips has been used to control the amount of overland flow and eroded soil reaching stream systems. Buffer strips decrease the velocity of flowing water and, in doing so, allow sediment to settle out and permit overland flows to infiltrate into undisturbed soils. The buffer strips can also reduce the amount of road-derived sediment reaching streams. Characteristics that make buffer strips work include their width, vegetative and litter cover, surface roughness, and the local topography (Stednick 2000). Where gully erosion or soil mass movement contribute sediment to the stream, buffer strips are less effective. Furthermore, the effectiveness of buffer strips in controlling soil erosion from most timber harvesting and succeeding site preparation practices has not been rigorously tested.

Unlike many other land uses that disturb soil for long periods, any increases in sediment from timber harvesting or other forestry activities are usually short-lived. While surface soil disturbances can provide an increase in sediment supply, once the finer materials are transported away and other excess materials have moved through the system, sediment yields rapidly return to predisturbance levels (Bunte and MacDonald 1999). Additionally, as long-term plant succession and regrowth of the site occur, the impacted site is less likely to continue eroding.

Fire

Fire that reduces litter depths and exposes mineral soil is also likely to contribute to accelerated erosion and sedimentation because of the loss of protective vegetation and litter and physical changes in the soil surface. On many watersheds, combined timber harvesting and follow-up burning to eliminate the logging residue (slash) have produced increases in mass soil movements, which are generally attributed to the loss of mechanical support by the root systems of trees and herbaceous vegetation. Various studies have demonstrated the effect of fire, timber harvesting, and road building on accelerated mass movements of soil and resulting sedimentation in streams (see Chapter 9).

Annual total sediment yields increased from 100 to 3800 kg/ha following wildfires in ponderosa pine and Douglas-fir forests in the eastern Cascade Mountains of Washington State (Helvey et al. 1985). The increase in sediment increased nutrient losses of N, P, Ca, Mg, K, and Na, largely from the riparian zone where plant growth could be affected. In streamflow, total N increased from 0.004 to 0.16 kg/ha/yr; available P increased from 0.001 to 0.014 kg/ha/yr; Ca, Mg, K, and Na increased from an average of 1.98 to an average of 54.3 kg/ha/yr. When such increases occur, the potential impacts on soil productivity can be of as much concern as the impacts on water quality. The investigators of the above believed that the losses observed would not affect soil productivity.

Firelines created by bulldozers to control the spread of fire can be potential sources of sediment in streams (Landsberg and Tiedemann 2000). Firelines are often constructed in urgent circumstances without adequate time to consider stream protection. Firelines can be difficult to stabilize with vegetation because much of the nutrient-rich top soil is cast aside in their establishment. Application of seed and fertilizer is an effective way to protect firelines from eroding.

Sedimentation following carefully prescribed burning is generally less than that resulting from a high severity wildfire (DeBano et al. 1998). When sediment yields increase as a consequence of burning of any type, however, these yields are often the highest in the first year after the fire, declining in subsequent years as protective vegetation becomes established on the watershed.

Forage Utilization

Utilization of forage plants by livestock under properly specified grazing conditions does not normally increase the amount of suspended sediments in surface water. But intensive and concentrated grazing pressures on steep and unstable terrain and fragile soils can create sedimentation problems. Sediment levels can also increase when live-

Box 12.3

Managing Livestock Grazing in Riparian Areas to Protect Streambanks

Management of livestock in riparian areas has been a concern of managers in recent years. In addressing this concern, efforts have been made to develop livestock grazing prescriptions for these fragile ecosystems that are effective and easily understood. Herbaceous stubble height has been widely used as a proper-use criterion that serves both purposes (Smith 2001). Clary and Webster (1992) recommended that a minimum stubble height of 2-4 cm should remain at the end of the grazing season to maintain plant vigor and provide for streambank protection and sediment entrapment. Clary and Webster recommended that livestock utilization of riparian forage in the spring be limited to 65% of the current growth and that cattle be removed by the middle of July to allow for regrowth. After this time, plant growth appears to slow considerably in many riparian areas. Sheeter and Svejcar (1997) also found little regrowth of riparian forage in southeastern Oregon after this time.

stock are allowed to overgraze riparian plant communities; this activity often leads to streambank erosion and sediment deposition directly into stream channels. It can be possible to manage livestock grazing in ways that enhance riparian vegetation and protect streambanks by timing grazing to accommodate forage plant physiography in some situations (Box 12.3).

Thermal Pollution

The surface water temperature at a stream site is determined by the temperature of inflowing water, radiant energy inputs and outputs, conductive energy inputs and outputs, and heat extraction by evaporation. Riparian vegetation influences all of these heat exchanges (see Chapter 13). Beschta et al. (1987) and others have described relationships between activities that reduce riparian vegetation, such as timber harvesting and fire, and the resulting increases in stream temperatures. Water transports heat by changes in temperature from one site to another; therefore, the effects of reduction in riparian vegetative covers on water temperature can impact downstream water uses that are dependent on temperature. Among the attributes that help to determine the contributions of stream shade in mitigating thermal pollution are stream width and orientation, distance from riparian vegetation to the stream, and canopy characteristics of this vegetation (Quigley 1981).

Changes in land use activities that reduce dissolved oxygen levels can also decrease habitability for fish. Increased nutrient loads deplete dissolved oxygen by stimulating algal blooms. Detrital organic materials consume oxygen as they decay. The cumulative effects of all such factors on the oxygen and temperature regime of streams and rivers must be recognized throughout the watershed.

Dissolved Chemical Constituents

Vegetation management practices can affect the ionic balances in the streamflow originating from upland watersheds. Dissolved chemical loads of streams after forest disturbances such as timber harvesting or fire are largely a function of the biotic and abiotic characteristics that describe the natural ecosystem. For example, leaching rates are influenced by the form, amount, and intensity of precipitation events. Vegetative characteristics, primarily composition and densities of the plant species, influence the rates of nutrient uptake (Stednick 2000). The speed at which regrowth takes place after a disturbance controls, to a large extent, the initiation of nutrient recycling following the disruption. Soil characteristics, including porosity and texture, determine the pathways and rates of water movement in or over the soil matrix.

The intricate relationships among these and other variables largely preclude detailing the effects of vegetation management practices on dissolved chemical loads in water originating from widely differing watershed ecosystems; however, generalizations are often possible. The more common vegetation management practices of interest are timber harvesting, fire, and grazing of forage by livestock.

Timber Harvesting

A number of important changes occur on a watershed that can change the concentrations of dissolved chemical constituents in streamflow when a forest is harvested. Trees are no longer in place to take up nutrients from soil, and noncommercial parts of the trees left as slash accumulations increase the amount of decaying forest litter. The removal of forest canopies also makes the site warmer while reducing evapotranspiration. Less evapotranspiration increases the soil water content, which, in turn, accelerates the activities of microorganisms that break down organic material, including the added slash.

The increased respiration of microorganisms raises the partial pressure of CO_2 in the soil atmosphere, which also increases the bicarbonate anion level and leaches more cations from the system. In addition, nitrogen losses (as nitrate-N) can occur when nitrogen in organic material is oxidized (nitrification) but is not utilized by the forest vegetation that harvesting has removed.

Studies throughout the world show that following intensive timber harvesting on well-drained soils, there is usually an increased loss of nutrients (cations and nitrate-N) from the logged area. For example, clearcutting northern hardwood forests in New Hampshire resulted in increases of 57 kg/ha for inorganic N, 71 kg/ha for Ca, and 15 kg/ha for K for the first 4 yr following cutting (Martin et al. 1986). The largest increases were observed during the second year after clearcutting. Concentrations of most nutrients were back to preharvest conditions by the end of the fourth year. Such increases in nutrient export from clearcut watersheds are often at least partly the result of increases in water yield that usually accompany clearcutting (see Chapter 6). Even if concentrations of nutrients in streamflow do not increase much, the increase in runoff volume can increase the nutrient loading of streams and lakes. This loading is rarely significant when logging operations are properly planned. In general, carefully planned and implemented forestry practices that minimize site disturbances and quickly lead to the establishment of replacement stands do not cause excessive nutrient export from watersheds (Box. 12.4).

Box 12.4

Effects of Clearcutting Trees and Site Preparation on Nutrient Export from Small Watersheds in Southeastern United States

Clearcutting of loblolly pine followed by site preparation with roller-choppers and planting caused nutrient export to increase, but the increase was small and did not last beyond 2 yr (Hewlett et al. 1984). The concentrations of nutrients in streamflow did not increase following clearcutting, but water yield increases of 10 to 20 mm over 2 yr flushed more nutrients from the watershed than a paired (uncut) control. Nitrate-N loading increased 0.3 kg/ha for 2 yr following harvesting. The maximum nutrient flushing following timber harvesting was less than 0.5 kg/ha/mo for P, K, Ca, Mg, and Na; these were short-lived and diminished as regrowth occurred. Such nutrient losses result in neither soil fertility losses nor stream eutrophication.

Experimental watersheds at the Coweeta Hydrologic Laboratory in North Carolina experienced larger nutrient losses following hardwood clearcutting than those reported above. Net losses of 6.2 kg/ha and 3.6 kg/ha of N were measured for the first and second years following clearcutting (Swank and Waide 1979). Again, the losses reported were not considered to be a threat to the quality of streamflow.

Fire

Burning the slash left in a forest after timber harvesting produces an even greater and more rapid release of ions from the forest litter and mineral soil than the harvesting operation itself. The increased release of ions is due largely to the breakdown of organic materials into a soluble form, making them easily removable by leaching (Landsberg and Tiedemann 2000). In general, this process can lead to an increase in the total loss of nutrients with streamflow; however, this increase is only temporary in many instances. In addition, volatilization of nitrogen and sulfur by fire results in losses from the watershed.

It has been reported that moderately severe fires in coniferous watersheds of the western United States have no specific effect on the concentrations of Ca^{++}, Mg^{++}, Na^{+}, and HCO_3^- in streamflow (Helvey et al. 1985). It was postulated that light rainfall dissolved and leached the ash constituents into permeable soil before the first snowfall. Because of the acidic nature of the soils on these watersheds, the dissolved cations were adsorbed by the exchange complex instead of being washed directly into the stream.

Overland flow from rainfall events of high intensities that follow severe fires can move large quantities of soluble ash compounds into streams, especially in the first year or so after the fire. But, as a forest regenerates after the fire, the dissolved chemical load of the stream generally returns to the levels observed before the fire.

Utilization of Forage

Grazing by livestock is a common land use practice on many upland watersheds, especially in the western United States. Except where prolonged overgrazing occurs, this grazing generally does not have a significant impact on the dissolved chemical constituents in streamflow. However, when livestock become concentrated near water bodies (for example, in feedlots), nutrient loading can be high. More often, grazing can adversely affect the biological quality of water, as described in the section on Biological Quality.

Introduced Chemicals

Chemicals are widely used in agricultural, forestry, rangeland, and urban settings to control unwanted pests or to enhance productivity of the land. When such chemicals enter water bodies they can diminish the health and productivity of aquatic ecosystems and can adversely affect humans who consume the water. Preventative and mitigative strategies to prevent such occurrences must be addressed through watershed management.

Pesticides

Management of the vegetation on watersheds can require the use of pesticides to control or kill competing plants or animals; some of these pesticides can have a direct and often detrimental impact on water quality and aquatic ecosystems, however. Pesticide use in general will often vary from year to year depending upon the need. However, pesticide concentrations associated with vegetation management practices are generally many times less than those used on agricultural lands. It has been estimated, for example, that insecticide, herbicide, and fungicide use in national forests in the United States has been only 0.323 $kg/km^2/yr$, compared to 236 $kg/km^2/yr$ for agricultural use (Binkley and Brown 1993).

Pesticides are used widely to control competing vegetation during forest stand establishment. In contrast to mechanical measures of site preparation, using pesticides minimizes off-site soil loss, eliminates on-site soil and organic-matter displacement, and prevents deterioration in soil physical properties (Michael and Neary 1995). Residue concentrations of such pesticides as 2,4-D, hexazinone, picloram, sulfometuron methyl, metsulfuron methyl, triclopyr, and imazapyr tend to be low and do not persist for extended periods of time; the exception is where direct applications are made to ephemeral channels or streams. Pesticide applications that follow regulatory guidelines have not been found to impair water quality (Norris et al. 1991, Michael 2000). The concentrations of herbicides in streams following forest application are generally < 0.1 mg/l; > 2 mg/l would be required to affect stream flora.

We will not attempt a thorough discussion of all possible pesticides used in vegetation management, detailed guidelines for their application, or their effects on water quality constituents. Rather, we briefly discuss some important considerations pertaining to pesticide use from the standpoint of watershed management.

The basic rule of thumb in applying pesticides in vegetation management is to minimize their contact with nontarget organisms and areas and, more importantly, prevent direct application to streams and lakes. Guidelines for application, proper field management, and follow-up monitoring are all needed to minimize the problems in pesticide application (Michael 2000). Wind conditions, temperature, and proximity of target areas to streams, lakes, and other water bodies must all be considered.

Once pesticides are applied properly to forest or rangeland areas, their risk to aquatic systems is dependent upon the persistence characteristics of the pesticide and the hydrologic processes and characteristics of the site. Following their application, pesticides can be either leached into and through the soil or carried off by surface runoff. Therefore, rainfall rates, soil infiltration and hydraulic properties, soil depth, slope of the landscape, and organic matter can all affect transport. A buffer zone in which pesticide application is prohibited is often designated adjacent to water bodies. The size of the buffer zone depends largely on the above factors.

While the nature of the movement of pesticides into streams is relatively well documented, their movement into groundwater aquifers has not been researched as well. However, their movement from streamwater into groundwater should reduce their initial concentrations for several reasons (Michael 2000). Infiltrating pesticides must pass through several physical barriers or layers before reaching the groundwater. As the pesticides pass through each barrier or layer, they are degraded, diluted, and metabolized. Surface water is a medium for dilution, hydrolysis, and photolysis. Aquatic vegetation can metabolize pesticides. Microbes associated with particulate organic matter found naturally in streams also metabolize pesticides.

To reach groundwater, pesticides must also percolate through the soil column, where they are adsorbed to soil particles, reducing the amount reaching the groundwater. Pesticides adsorbed to the soil particles are often irreversibly bound, released slowly, or further metabolized by microbes. Once they eventually reach the groundwater aquifer, pesticides might degrade further.

Fertilizers

Application of nitrogen fertilizers to forests and rangelands to increase plant growth can also increase the nitrogen concentrations in surface water. Urea and ammonia levels appear to remain well below levels of concern, while nitrate-N levels can peak at high concentrations. Studies in the Pacific Northwest have shown that the risk of nitrate-N pollution is small when fertilizing Douglas-fir stands (Bisson et al. 1992).

Biological Quality

Pathogenic bacteria and protozoa in surface water are a concern to watershed managers because of their potential to cause disease. Water containing these pathogens, therefore, must be adequately treated for purification. However, site-specific linkages between the occurrence of these parasites and land use practices are often unclear.

Bacteria

While timber-harvesting activities or fire in themselves do not usually affect the occurrence of pathogenic bacteria in surface water, livestock grazing can cause fecal coliform bacteria densities in streams to rise (Box 12.5), particularly where livestock graze in riparian areas. The importance of this observation in the management of upland watersheds is obvious—to minimize the impacts of grazing by domestic livestock on the bacteriological quality of water, intensive grazing of livestock and concentrations of other ungulates near water bodies and riparian areas must be avoided. Strategically placed watering troughs located away from the streams can be helpful in this regard.

Where they occur, concentrations of bacterial groups in streams are often closely related to physical characteristics of the streams. In the Rocky Mountains, bacterial

Box 12.5

Impacts of Livestock Grazing on Bacteriological Quality of Surface Water: Some Examples

- Doran and Linn (1979) reported bacteriological water quality parameters for grazed and ungrazed pasturelands in eastern Nebraska. Bacterial concentrations generally exceeded drinking water standards in both pastures. However, fecal coliform concentrations in the grazed pasture showed a 5- to 10-fold increase over those in the ungrazed pasture.
- Skinner et al. (1974) examined the concentrations of bacteria in small catchments of the Nash Fork Watersheds in southeastern Wyoming. The concentrations of fecal coliform on a low-use, natural watershed average about 0.2–1.2 colonies/100 ml, compared to concentrations of 20–30 colonies/100 ml for watersheds receiving intensive livestock grazing and recreational use. It was concluded, however, that these levels presented few water quality problems relative to recreational use.
- The impacts of livestock grazing on water quality on sagebrush rangelands in southwestern Idaho were investigated by Stephenson and Street (1978) on the Reynolds Creek Watershed. Fecal coliform concentrations increased significantly when cattle or sheep were introduced to pastures, and they remained elevated up to 3 mo after the grazing stopped. Maximum fecal coliform concentrations were 2500 colonies/100 ml. The authors concluded that livestock grazing is likely to lead to intermittent concentrations of bacteria that exceed the water quality standards.

counts appear to be largely dependent upon the flushing effect of runoff from rainstorms, snowmelt, and irrigation. Seasonal trends for all of the bacteria groups are similar throughout the year. Low counts prevail when the temperature of water approaches freezing, and high counts appear during rising flows and peak flows resulting from late spring snowmelt and rain. A short postflush decrease in bacterial counts takes place as runoff recedes in late spring. High bacterial counts are found again in the summer period of warm temperatures, and finally, counts decline in autumn.

The fate of bacterial groups can be less important in drier environments. Sherer et al. (1992) reported that only 5 days a year experienced enough rainfall to produce overland flow in a Great Basin rangeland of central Oregon. Consequently, the probability of fecal material washing into streams was relatively low.

Accumulations of organic material on the ground can act as a bacterial filter in certain situations. For example, it has been found that rainfall and snowmelt that percolate through a strip of organic material contain fewer bacteria than water that had not passed through the strip.

Protozoa

It has been claimed that *Cryptosporidium parvum* (*C. parvum*), a waterborne protozoan that can be debilitating to humans (MacKenzie et al. 1994), can be carried by livestock grazing on watershed lands (see Chapter 11). Calves up to 4 mo old seem to shed greater numbers of oocyts (eggs) than older animals (Atwell 1996). Where the linkage between cattle and *C. parvum* in surface water is substantiated, restricting livestock—and other ungulate—use in riparian areas is one management practice that can minimize animal contact with streams and other water bodies. Livestock can be provided water in locations away from streams and lakes. Another option is to favor management aimed at grazing only older livestock in contrast to total livestock exclusion from watersheds where *C. parvum* is a concern.

Roads and Water Quality

Roads, trails, and other corridors are common features of many watersheds. Unfortunately, runoff and seepage from roads and other corridors can contain elevated levels of suspended sediments and chemical pollutants that are either adsorbed to the sediment, occur as particles, or are dissolved by the runoff. Even when the corridors are abandoned, they can continue to be a source of sediments and pollutants in streamflow runoff.

Sediments

Roads, their construction, and their maintenance are often the principal sources of erosion and sediment from otherwise undisturbed watersheds (as discussed in chapters 8, 9, and 10). This sediment often originates from surface erosion, which is generally more likely to carry chemical pollutants than other forms of erosion. On steep watersheds, however, more sediments can be derived from gully erosion and soil mass movement, which can bring greater volumes of soil to the stream system than surface erosion. Regardless of its origin, the sediment can adversely impact water quality by increasing turbidity and carrying phosphates, pesticides, and other hydrocarbons into surface water and groundwater resources.

Assuming erosion rates as proxies of the relative differences in sediment originating from roads, immediately after roads are constructed, erosion rates from the bare slopes and road surfaces are relatively high (Fig. 12.1). Erosion rates generally drop rapidly as the exposed slopes revegetate and stabilize. Erosion reductions of 90% or more are common as a road ages (Ketcheson and Megahan 1996). However, road surfaces remain a main source of sediment as long as traffic or maintenance prevents the establishment of vegetation. Applications of high-quality gravel to unpaved roads can decrease erosion rates by 80%.

Burroughs and King (1989) identified the cut slope, the roadway, and the fill slope as specific sources of sediment in the Intermountain Region of the United States and suggested applications of mulch, seed, and sod as mitigation measures for each source. Luce and Black (1999) concluded that the roadway itself and the road ditch were the significant sources of sediment in the Oregon Coastal Range. Slopes and channels downhill from a road can also be the sites of deposition of soil eroded by the influence of the road or a major source of sediment from a segment of the road.

Poor drainage can lead to a saturation of road beds and subsequent soil mass movement. In areas of high rainfall and steep terrain, such as the Coastal Range in Washington and Oregon, more sediment originates from road-caused landslides than from surface erosion from roads. Beschta (1978) reported that roads and timber harvesting caused sediment yields to increase from about 100 Mg/km^2 to 140 Mg/km^2, with much of this increase in sediment attributed to soil mass movement. However, recent studies in Oregon suggest that sediment from landslides in undisturbed areas is similar

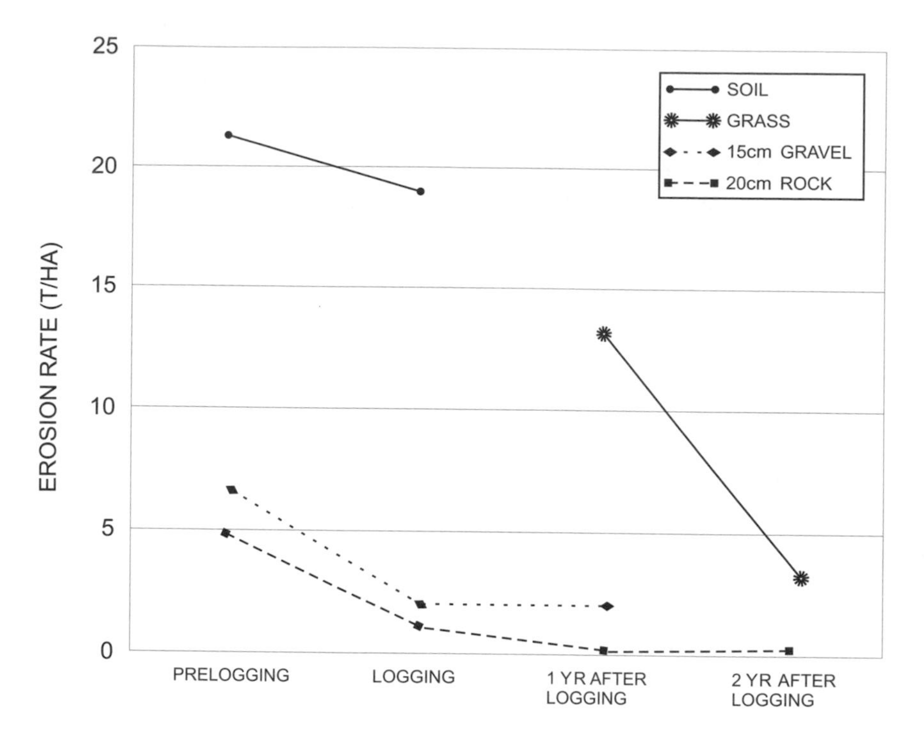

FIGURE 12.1. Erosion rates for different road surfaces before, during, and 2 yr after timber-harvesting operations (from Elliot 2000, based on Swift 1984).

to that in areas with roads (Elliot 2000). While surface erosion is a source of sediment every year, landslides tend to contribute large amounts of sediment during wet years and little or no sediment in dry years.

Common techniques to reduce road erosion include surfacing the road with gravel, decreasing the spacing of cross drainage, locating the road farther from the stream, and limiting road gradients (Burroughs and King 1989). Treatment of cut-and-fill slopes by applying straw and seeding has also been effective in reducing sediment delivery from newly constructed roads. The installation of vegetated filter strips or slash filter windrows below fills and/or sediment basins below culverts are also management practices that have reduced sediment yields. Seasonal closure of roads in wet seasons can be an option to prevent erosion. Erosion rates can drop to background level as the density of vegetation increases on abandoned road surfaces.

Other rights-of-way such as trails for bicycling, walking, or horseback riding are also potential sources of sediment. Any construction that exposes bare soil is likely to increase sedimentation. Much of the sediment eroded from a right-of-way is deposited below the right-of-way and, therefore, never reaches a stream (Rummer et al. 1997). Once installed, rights-of-way can continue to be sources of sediment if revegetation or other erosion control practices are not initiated.

TABLE 12.1. Pollutants observed in runoff from road surfaces

Pollutants	Comments	References
Cadmium, copper, lead, zinc	Treated in wetlands	Mungur et al. 1994
Highway deicing salt	Sodium adsorbed in soil	Shanley 1994
Polycyclic aromatic hydrocarbons	Altered aquatic communities	Boxall and Maltby 1997
Total petroleum hydrocarbons, lead, zinc	Reduced by vegetation	Ellis et al. 1994
Heavy metals, petroleum hydrocarbons, pesticides, sediment, nutrients	Treatment ponds remove up to 95% of pollutants	Karouna-Renier and Sparling 1997

Source: Dissmeyer 2000.

Chemicals

Hydrocarbons and other chemical pollutants in the runoff from roads and other surfaced sites can find their way into surface and subsurface water. Researchers have identified a number of chemical pollutants in runoff from road surfaces (Table 12.1). Some of these pollutants originate from the road material itself, some chemicals occur in the soil and rock on the site and are released during construction, and many chemicals are from vehicles (Elliot 2000). Traffic and road surfacing can contribute undesirable cations, hydrocarbons, and metals to water. Cations released from a road can have a buffering effect on runoff acidity.

Pollutants in runoff from road surfaces can be trapped in natural or constructed wetlands or can be reduced by vegetative buffers between the road and the stream. Another control method is the construction of a partial exfiltration trench that filters out suspended solids carrying undesirable hydrocarbons and other pollutants from road surfaces (Sansalone and Buchberger 1995). Multiple controls can be necessary to decrease chemical loads in runoff from both large and small storms. Detention basins might catch a first flush of polluted runoff, and a filtration system might be used to control chemicals in the runoff from larger storms.

Sediment basins and similar structures built to contain runoff from roads can become sources of pollution through seepage into the soil or other forms of hydraulic or structural failure (Grasso et al. 1997). Sediments with large amounts of adsorbed chemicals can eventually enter a stream system. One of the better controls of such problems is cleaning and maintaining the structures to minimize the risk of failure.

WATER QUALITY MONITORING

Representative water quality measurements are required to evaluate the kinds and amounts of substances present in surface water, and the effects of vegetation and other watershed management activities on the concentrations of these substances. Usually, sampling is necessary because continuous measurements of water quality are difficult and costly. A good sampling and monitoring procedure is based on knowledge of the water system being sampled, of the time and space distribution patterns of the parameters being sampled, and most importantly, of the purpose for which monitoring is being undertaken.

The beginning of the environmental era of the 1970s in the United States led to widespread water quality monitoring programs that ended up with computer data banks filled with many worthless data. Costly physical, chemical, and biological data were collected from streams and lakes with little regard to overall objectives and application to environmental problems. More recently, a greater emphasis has been placed on better design of monitoring programs.

The first and most important step in developing an effective water quality monitoring program is to clearly define the problem, goals, and objectives. The objectives should be stated in terms that are explicit and meaningful. Criteria must be established that can be used to determine whether specific objectives are met. A sampling scheme can then be established in time and space to provide the required information.

Water quality monitoring programs are established to answer specific questions and, therefore, should be designed accordingly. Water quality monitoring programs for natural resource management situations are largely nonpoint in nature and can be classified as one of the following (Ponce 1980):

- Cause-and-effect monitoring is conducted to determine the effects of specific actions on water quality constituents. For example, monitoring can be set up to determine the effect of timber harvesting on suspended-sediment concentrations or, more commonly, on several different water quality constituents.
- Baseline monitoring is conducted to help resource managers determine if there are certain trends in water quality at a particular location. Are certain water quality constituents changing over time?
- Compliance monitoring is carried out to determine if water quality standards are being met. For example, compliance monitoring could be used to determine whether streamflow at a particular location is suitable for drinking water.
- Inventory monitoring is designed to indicate existing water quality conditions. For example, planning may be carried out in which several sites are being considered for recreational development that includes swimming. Sites must be selected that are suitable for water contact recreation.

The location, time, and frequency of water quality sampling are determined by the type of monitoring being done and the normal statistical considerations (variability, cost of samples, accuracy needed, etc.). The water quality characteristics to be measured should also be explicitly stated, and they should relate directly to the objectives.

The design of a monitoring program must locate sampling stations that are appropriate for the objectives and the type of monitoring planned (Kunkle et al. 1987). Cause-and-effect monitoring of streamflow quality can be carried out with the paired-watershed approach (see chapters 3 and 18) or with an above-and-below approach (Fig. 12.2). With the former, water quality measurements are taken at the outflow of a control watershed and at the outflow of one to be treated. Sampling is carried out for a sufficient period before the treatment to establish regression relationships relating the water quality constituents' concentrations between the two watersheds. After the treatment implementation (e.g., timber harvesting), sampling continues, the regression relationships are statistically evaluated for changes, and these changes or lack thereof are used as an indicator of the effect of the treatment. With the second approach, sampling is done at station A, above the treated area, and at station B, downstream of the treated area (Fig. 12.2). The effect of the treatment is determined by statistical methods that either compare station means or compare regression relationships at the two stations.

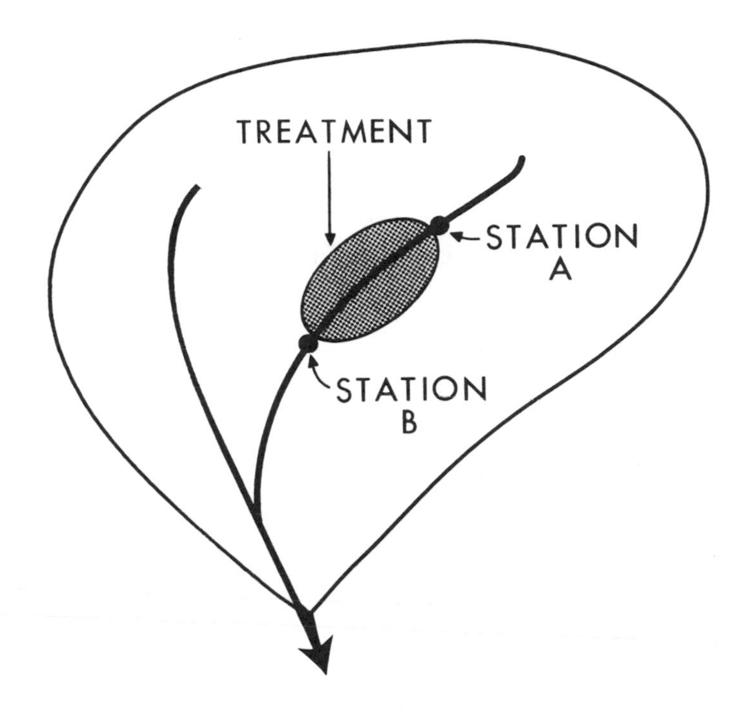

Figure 12.2. Example of station location for cause-and-effect monitoring program where the treatment can be readily isolated (from Ponce 1980).

Compliance monitoring can use the same above-and-below approach, in which sampling stations are located to determine if a particular activity is causing the water quality standards to be exceeded. For example, a forested area might be used as a disposal site for treated sewage from a recreational campsite. The water that is entering the stream channel must be monitored to determine if certain standards are being met.

Baseline and inventory monitoring require that sampling stations be located so that natural stream systems can be characterized. Multiple streams can be sampled concurrently and, over time, the statistical distribution of water quality characteristics compared to determine natural variability and to detect trends.

Much of the sampling done in the above monitoring approaches is based on random "grab" samples, which are samples collected in a container by hand rather than by automated or other mechanical sampling devices. Often, a grab sample obtained in a clean glass or plastic container is satisfactory, at least for preliminary analyses. However, a single grab sample is representative of the stream discharge only at the time of sampling. The magnitude of streamflow discharge affects some constituents, including suspended sediment (see Chapter 9). As a consequence, discharge measurements should accompany all water quality samples.

Once obtained, grab samples might have to be treated to protect against degradation of the contents. If the water samples are to be used later for organic analysis, the sample must be frozen or otherwise preserved. As a result, water quality measurements in remote areas require careful planning and scheduling to ensure that the samples collected are handled correctly and that measurements are made promptly.

Cumulative Effects

Because the transportation of sediments, nutrients, and other chemicals in streams is largely dependent upon water movement, any land use activity that alters the volumes or timing of runoff also affects the rates of sediment and chemical transport (Sidle 2000). Timber harvesting often causes a short-term increase in sediments and chemicals in streams (Mann et al. 1988; Tiedemann et al. 1988). Increases in nitrogen concentrations can result from reduced nutrient uptake, increased subsurface flow, increased alteration of nitrogen to leachable forms, increased volumes of decaying organisms, or combinations thereof. Baker et al. (1989) measured the rates at which nine nutrients were released from decomposing slash and found that nitrogen was released most slowly. Other management activities that contribute to these mechanisms, such as burning (DeBano et al. 1998) and herbicide applications (Vitousek and Matson 1985), can have similar effects.

Some land use activities alter existing chemicals, causing them to interact with the environment in different ways. For example, Helvey and Kochenderfer (1987) found that limestone road gravels can moderate the pH of the runoff from acid rainfall. Birchall et al. (1989) discovered that aluminum released by acid rainfall is less toxic to fish populations if streams are enriched in silica. Bolton et al. (1990) found that vegetative conversions can alter the spatial distribution of the resulting nitrogen mineralization because the plants lend different mineralization potentials to the soil. Chemicals reaching stream channels can be redeposited downstream. Reservoirs can concentrate dissolved pollutants if evaporation rates are high.

Acid rainfall increases the mobility of nutrients and, therefore, contributes to nutrient deficiencies (Johnson et al. 1982).

Summary

The protection and maintenance of high-quality surface water are fundamental goals of watershed management. This chapter considers issues of importance to managers of the quality of water flowing from upland watersheds, describes some of the impacts of land use activities and disturbances on these characteristics and the aquatic ecosystem, and outlines the roles of water quality monitoring programs in land stewardship. After completing this chapter, you should be able to:

- Discuss the more important issues confronting managers concerned about maintaining flows of high-quality water from upland watersheds.
- Describe the key provisions of the Clean Water Act. Outline the preparation of a TMDL plan. Appreciate the role of best management practices in maintaining water quality standards.
- Discuss how various timber harvesting, fire, range management practices, and road-building activities can affect water quality.
- Consider each major type of pollutant and indicate the general kinds of management measures needed to correct or prevent water quality problems.
- Describe the different types of water quality monitoring programs and discuss how to locate sampling stations for each.

References

Atwell, E.R. 1996. Assessing the link between rangeland cattle and waterborne *Cryptosporidium parvum* infection in humans. *Rangelands* 18:48–51.

Baker, T.G., G.M. Will, and G.R. Oliver. 1989. Nutrient release from silvicultural slash: Leaching and decomposition of *Pinus radiata* needles. *For. Ecol. Manage.* 27(1): 53–60.

Beschta, R.L. 1978. Long-term pattern of sediment production following road construction and logging in the Oregon coast range. *Water Resour. Res.* 14:1011–1016.

Beschta, R.L., R.E. Bilby, G.W. Brown, L.B. Holtby, and T.D. Hofstra. 1987. Stream temperature and aquatic habitat: Fisheries and forestry interactions. In *Streamside management: Forestry and fishery interactions*, ed. E.O. Salo and T.W. Cundy, 191–232. Seattle: Institute of Forest Resources, Univ. of Washington.

Binkley, D., and T.C. Brown. 1993. *Management impacts on water quality of forests and rangelands*. USDA For. Serv. Gen. Tech. Rep. RM-239.

Birchall, J.D., E. Exley, J.S. Campbell, and M.J. Phillips. 1989. Acute toxicity of aluminum to fish eliminated in silicon-rich acid waters. *Nature* 338:146–147.

Bisson, P.A., G.G. Ice, C.J. Perrin, and R.E. Bilby. 1992. Effects of forest fertilization on water quality and aquatic resources in the Douglas-fir region. In *Forest fertilization: Sustaining and improving growth of western forests*, ed. H. Chappell, G. Westman, and R. Miller, 173–193. Seattle: Institute of Forest Resources, Contribution 72, Univ. of Washington.

Bolton, H., Jr., J.L. Smith, and R.E. Wildung. 1990. Nitrogen mineralization potentials of shrub-steppe soils with different disturbance histories. *Soil Sci. Soc. Amer. J.* 54:887–891.

Boxall, A.B.A., and L. Maltby. 1997. The effects of motorway runoff on freshwater ecosystems: 3. Toxicant confirmation. *Archives of Environ. Contamination Toxicology* 33:9–16.

Brown, T.C., D. Brown, and D. Brinkley. 1993. Laws and programs for controlling nonpoint source pollution in forest areas. *Water Resour Bull.* 22:1–13.

Bunte, K., and L.H. MacDonald. 1999. *Scale considerations and the detectability of sedimentary cumulative effects*. Research Triangle Park, NC: National Council for Air and Stream Improvement Tech. Bull. 776.

Burroughs, E.R., Jr., and J.G. King. 1989. *Reduction of soil erosion on forest roads*. USDA For. Serv. Gen. Tech. Rep. INT-264.

Chaney, E., W. Elmore, and W.S. Platts. 1990. *Livestock Grazing on Western Riparian Areas*. Washington, DC: U.S. Environ. Protect. Agency.

Clary, W.P., and B.F. Webster. 1992. *Managing grazing of riparian areas in the Intermountain Region*. USDA For. Serv. Gen. Tech. Rep. INT-263.

Davie, D.K., and C.L. Lant. 1994. The effect of CRP enrollment on sediment loads in two southern Illinois streams. *J. Soil Water Cons.* 49:407–412.

DeBano, L.F., D.G. Neary, and P.F. Ffolliott. 1998. *Fire's effects on ecosystems*. New York: John Wiley & Sons, Inc.

Dissmeyer, G.E. 2000. *Drinking water from forests and grasslands: A synthesis of the scientific literature*. USDA For. Serv. Gen. Tech. Rep. SRS-39.

Doran, J.W., and D.M. Linn. 1979. Bacteriological quality of runoff water from pastureland. *Appl. Environ. Microb.* 37:985–991.

Elliot, W.J. 2000. Roads and other corridors. In *Drinking water from forests and grasslands: A synthesis of the scientific literature*, ed. G.E. Dissmeyer, 85–100. USDA For. Serv. Gen. Tech. Rep. SRS-39.

Elliot, W.J., R.B. Foltz, C.H. Luce, and T.E. Koler. 1996. Computer-aided risk analysis in road decommissioning. In *Proceedings of the AWRA annual symposium on watershed restoration management: Physical, chemical, and biological considerations*, eds. J.J. McDonnell, J.B. Stribling, L.R. Neville, and D.J. Leopold, 341–350. Herndon, VA: American Water Resources Association.

Ellis, J.B., D.M. Revitt, R.B.E. Shutes, and J.M. Langley. 1994. The performance of vegetated biofilters for highway runoff control. In *Proceedings of the 4th international symposium on highway pollution*, eds. R.S. Hamilton, D.M. Revitt, R.M. Harrison, and A. Monzon-Caceres, 543–550. London: Middlesex University.

Foster, I., A. Gurnell, and B. Webb, eds. 1995. *Sediment and water quality in river catchments*. Chichester, NY: John Wiley and Sons, Inc.

Grasso, D., M.A. Butkus, D. Sullivan, and N.P. Nikolaidis. 1997. Soil-washing design methodology for a lead-contaminated sandy-soil. *Water Res.* 31:3045–3056.

Helvey, J.D., and J.N. Kochenderfer. 1987. Effects of limestone gravel on selected chemical properties of road runoff and streamflow. *North. J. Appl. For.* 4:23–25.

Helvey, J.D., A.R. Tiedemann, and T.D. Anderson. 1985. Plant nutrient loss by soil erosion and mass movement after wildfire. *J. Soil Water Conserv.* 40:168–173.

Hewlett, J.D., H.E. Post, and R. Doss. 1984. Effect of clearcut silviculture on dissolved ion export and water yield in the Piedmont. *Water Resour. Res.* 20:1030–1038.

Johnson, D.W., J. Turner, and J.M. Kelley. 1982. The effects of acid rain on forest nutrient status. *Water Resour. Res.* 18:449–461.

Karouna-Renier, N.K., and D.W. Sparling. 1997. Toxicity of stormwater treatment pond sediments to *Hyalella aztecia* (*Amphipoda*). *Bull. Environ. Contamination Toxicology* 58:550–557.

Ketcheson, G.L., and W.F. Megahan. 1996. *Sediment production and downslope sediment transport from forest roads in granitic watersheds.* USDA For. Serv. Res. Pap. INT-486.

Kunkle, S., W.S. Johnson, and M. Flora. 1987. *Monitoring stream water quality for land-use impacts: A training manual for natural resource management specialists.* Fort Collins, CO: Water Res. Div., Nat. Park Serv.

Landsberg, J.D., and A.R. Tiedemann. 2000. Fire management. In *Drinking water from forests and grasslands: A synthesis of the scientific literature*, ed. G. E. Dissmeyer, 124–138. USDA For. Serv. Gen. Tech. Rep. SRS-39.

Loomis, J., P. Kent, L. Strange, K. Fausch, and A. Covich. 2000. Measuring the total economic value of restoring ecosystem services in an impaired river basin: Results from a contingent survey. *Ecol. Econ.* 33:103–117.

Luce, C.H., and T.A. Black. 1999. Sediment production from forest roads in western Oregon. *Water Resour. Res.* 35:2561–2570.

Lynch, J.A., E.S. Corbett, and K. Mussallem. 1985. Best Management Practices for controlling nonpoint-source pollution on forested watersheds. *J. Soil Water Conser.* 40:164–167.

MacKenzie, W.R., N.J. Hoxie, M.E. Proctor. 1994. A massive outbreak in Milwaukee of *Cryptosporidium. N. Engl. J. Med.* 311:161–167.

Mann, L.K., D.W. Johnson, D.C. West, D.W. Cole, J.W. Hornbeck, C.W. Martin, H. Riekerk, C.T. Smith, W.T. Swank, L.M. Tritton, and D.H. Van Lear. 1988. Effects of whole-tree and stem-only clearcutting on postharvest hydrologic losses, nutrient capital, and regrowth. *For. Sci.* 34:412–428.

Martin, C.W., R.S. Pierce, G.E. Likens, and F.H. Bormann. 1986. *Clearcutting affects stream chemistry in the White Mountains of New Hampshire.* USDA For. Serv. Res. Pap. NE-579.

Michael, J.L. 2000. Pesticides. In *Drinking water from forests and grasslands: A synthesis of the scientific literature*, ed. G.E. Dissmeyer, 139–150. USDA For. Serv. Gen. Tech. Rep. SRS-39.

Michael, J.L., and D.G. Neary. 1995. Environmental fate and the effects of herbicides in forest, chaparral, and range ecosystems of the Southwest. *Hydrol. Water Resour. Ariz. and Southwest.* 22–25:69–75.

Miner, J.R., J.C. Buckhouse, and J.A. Moore. 1992. Will a water trough reduce the amount of time hay-fed livestock spend in the stream (and therefore improve water quality)? *Rangelands* 14:35–38.

Minnesota Department of Natural Resources. 1995. *Protecting water quality and wetlands in forest management—Best Management Practices in Minnesota.* Division of Forestry, St. Paul, MN.

Moore, E., E. James, F. Kinsinger, K. Pitney, and J. Sainsbury. 1979. *Livestock grazing management and water quality protection.* Washington, DC: U.S. Environ. Protect. Agency EPA 910/9-79-67.

Mungur, A.S., R.B.E. Shutes, D.M. Revitt, and M.A. House. 1994. An assessment of metal removal from highway runoff by a natural wetland. In *Wetland systems for water pollution control*, eds. R.H. Kadlec, and H. Brix, 169–175. London: Middlesex University.

Norris, L.A., H.W. Lorz, and S.V. Gregory. 1991. Forest chemicals. In *Influences of forest and rangeland management on salmonid fishes and their habitat*, ed. W.R. Meehan, 207–296. Spec. Publ. 19. Bethesda, MD: American Fisheries Society.

Perry, J. and E. Vanderklein. 1996. *Water quality—Management of a natural resource.* Cambridge, MA: Blackwell Science.

Ponce, S.L. 1980. *Water quality monitoring programs*. Watershed Systems Dev. Group, Tech. Pap. 00002. Ft. Collins, CO.: USDA Forest Service.

Quigley, T.M. 1981. Estimating contribution of overstory vegetation to stream surface shade. *Wildl. Soc. Bull.* 9:22–27.

Robles, M.D., and I.C. Burke. 1997. Legume, grass and Conservation Reserve Program effects on soil organic matter recovery. *Ecol. Appl.* 7:345–357.

Rummer, B., B. Stokes, and G. Lockaby. 1997. Sedimentation associated with forest road surfacing in a bottomland hardwood ecosystem. *For. Ecol. Manage.* 90:195–200.

Sansalone, J.J., and S.G. Buchberger. 1995. An infiltration device as a best management practice for immobilizing heavy metals in urban highway runoff. *Water Sci. Tech.* 32:119–125.

Shanley, J.B. 1994. Effects of ion exchange on stream solute fluxes in a basin receiving highway deicing salts. *J. Environ. Qual.* 23:977–986.

Sheeter, G., and T. Svejcar. 1997. Streamside vegetation regrowth after clipping. *Rangelands* 19:30–31.

Sherer, B.M., J.R. Miner, J.A. Moore, and J.C. Buckhouse. 1992. Indicator of bacterial survival in streams. *J. Environ. Qual.* 21:591–595.

Sidle, R.C. 2000. Watershed challenges for the 21st century: A global perspective for mountainous terrain. In *Land stewardship in the 21st century: The contributions of watershed management*, tech. coords. P.F. Ffolliott, M.B. Baker, Jr., C.B. Edminster, M.C. Dillon, and K.L. Mora, 45–56. USDA For. Serv. Proceed. RMRS-P-13.

Skinner, Q.D., J.C. Adams, P.A. Rechard, and A.A. Beetle. 1974. Effect of summer use of a mountain watershed on bacterial water quality. *J. Environ. Qual.* 3:329–335.

Smith, S.J. 2001. Rethinking riparian regrowth. *Rangelands* 23:14–16.

Stednick, J.D. 2000. Timber management. In *Drinking water from forests and grasslands: A synthesis of the scientific literature*, ed. G.E. Dissmeyer, 103–119. USDA For. Serv. Gen. Tech. Rep. SRS-39.

Stephenson, G.R., and L.V. Street. 1978. Bacterial variations in a stream from a southwest Idaho rangeland watershed. *J. Environ. Qual.* 7:150–157.

Stumnn, W., and J.J. Morgan. 1995. *Aquatic Chemistry*. New York: John Wiley and Sons.

Swank, W.T., and J.B. Waide. 1979. Interpretation of nutrient cycling research in a management context: Evaluating potential effects of alternative management strategies on site productivity. In *Forest ecosystems: Fresh perspectives from ecosystem analysis*, ed. R.H. Waring and J. Franklin, 137–158. Corvallis: Oregon State Univ. Press.

Swift, L.W., Jr. 1984. Gravel and grass surfacing reduces soil loss from mountain roads. *For. Sci.* 30:657–670.

Thorud, D.B., G.W. Brown, B.J. Boyle, and C.M. Ryan. 2000. Watershed Management in the United States in the 21st Century. In *Land stewardship in the 21st century: The contributions of watershed management,* tech. coords. P.F. Ffolliott, M.B. Baker, Jr., C.B. Edminster, M.C. Dillon, and K.L. Mora, 57–64. USDA For. Serv. Proceed. RMRS-P-13.

Tiedemann, A.R., T.M. Quigley, and T.D. Anderson. 1988. Effects of timber harvest on stream chemistry and dissolved nutrient losses in northeast Oregon. *For. Sci.* 34:344–358.

Vitousek, P.M., and P.A. Matson. 1985. Intensive harvesting and site preparation decrease soil nitrogen availability in young plantations. *South. J. Appl. For.* 9:120–125.

Wetlands take on many forms and functions in watersheds; a forested peatland in northern Minnesota (from Minnesota Department of Natural Resources, Division of Minerals).

PART 4

Riparian and Wetland Management

Special attention is given in this book to the functions and management of riparian systems and wetlands. In many parts of the world, riparian areas have been exploited and wetlands drained, with little recognition of the cumulative effects of such actions. Their value is now recognized in terms of hydrologic function, water quality management, and ecological importance. Riparian areas and wetlands are viewed as integral components of watersheds, influencing water yield and the flow of water and sediment in streams and rivers. Riparian areas directly affect aquatic habitats, and their management is critical to maintaining water quality. Chapters 13 and 14 discuss riparian and wetland systems, respectively, and relate their importance in the planning and management of watersheds. Both chapters build upon topics covered in parts 1, 2, and 3.

CHAPTER 13 *Riparian Management*

INTRODUCTION

Riparian areas are integral and important components of watersheds throughout the world. While they make up only a small percentage of the total land area, their hydrologic and biological functions must be considered in the determination of water and other resource management goals for the entire watershed. Riparian areas represent the interface between aquatic and adjacent terrestrial ecosystems and are made up of unique vegetative and animal communities that require the regular presence of free or unbound water (Fig. 13.1). They occur in both arid and humid regions and include the plant communities along the banks of rivers and streams and around springs, bogs, wet meadows, lakes, and ponds. In this chapter, however, we will discuss only stream and river riparian systems. Riparian areas are found along most major waterways throughout the United States and also are important management areas on small perennial, ephemeral, and intermittent streams.

Riparian areas have high value in terms of diversity because they share features with both adjacent upland and aquatic ecosystems. The kinds of biological and physical diversity vary widely, spatially and temporally, and contribute greatly to the overall diversity of a landscape by providing a wider range in numbers, kinds, and patterns of the land- and waterscape ecosystems along with a multitude of ecological processes that are associated with these patterns (Lapin and Barnes 1995). Also, riparian areas often support a variety of plants and animals that are not found elsewhere. They play less obvious roles relating to enhanced water quality, flood peak attenuation, streambank stability, and reduced erosion and sediment transport.

Although riparian areas vary widely from one region to another, general principles can be established that apply to most riparian areas. The focus of this discussion is riparian areas of the United States, although examples from other parts of the world will be used to illustrate some principles.

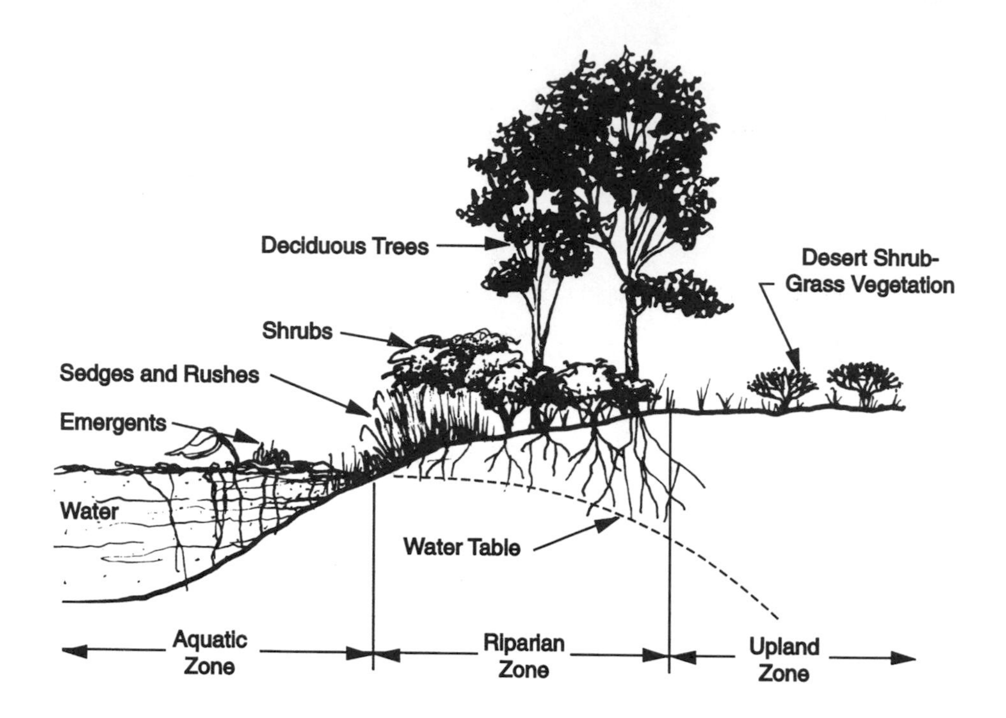

FIGURE 13.1. Riparian systems are most often associated with the presence of vegetation that requires large amounts of free or unbound water; This is an example of a riparian area in a dryland environment (from Thomas et al. 1979).

DESCRIBING RIPARIAN AREAS

The description of riparian areas necessary for management has been facilitated by terminology, definitions, and classification schemes for a range of complex riparian environments. Some of the more common terms used in the following discussion are given in Table 13.1.

Riparian areas are defined in ways that reflect agency concepts, the disciplines involved, or the particular functional role that a riparian area plays in the total ecosystem (Ilhardt et al. 2000). Many agency definitions focus on water quality protection issues along with riparian management practices, identification and mapping, site productivity, wildlife, and cultural concerns. This approach, however, leads to inconsistencies between agencies, depending upon the individual agency regulations and mandates. A technical definition developed by the Society of Range Management and the Bureau of Land Management (Anderson 1987) is:

> A riparian area is a distinct ecological site, or combination of sites, in which soil moisture is sufficiently in excess of that otherwise available locally, due to run-on

Table 13.1. Terminology used in describing riparian systems[a]

Term	Definition
Allochthonous	Food organisms, organic matter, and nutrients originating outside and transported into a system. The term usually refers to organic material of terrestrial origin transported into streams and lakes but can include material deposited by wind, carried by animals, etc.
Autochthonous	Organic substances, materials, or organisms originating within a particular stream or river and remaining in that water course. Also pertains to rocks, mantle rock, peat, soils, or their constituents formed at the site.
Benthic	Plant and animal life whose habitat is the bottom of a sea, lake, or river.
Buffer	A vegetative strip or management zone of varying size, shape, and character maintained along a stream, lake, road, recreation site, or different vegetation zone to mitigate the impacts of actions on adjacent lands, to enhance aesthetic values, or as a best management practice. Synonymous with buffer strip or buffer zone.
Channelization	Mechanical alteration of a stream that can include straightening or dredging of an existing channel or creating a new channel to which the stream is diverted.
Coarse woody debris (CWD)	Any piece(s) of dead woody material (dead boles, limbs, and large root masses) on the ground in forest stands or in streams. Synonyms: Large wood debris (LWD), large organic debris (LOD), downed woody debris (DWD). The type and size of material designated as coarse woody debris varies among classification systems.
Detritus	Small pieces of dead and decomposing plants and animals.
Ephemeral stream	A stream or portion of a stream that flows only in direct response to precipitation, receiving little or no water from springs and no long-continued supply from snow or other sources, and whose channel is at all times above the water table.
Filter strip	A strip or area of vegetation for removing sediment, organic material, organisms, nutrients, and chemicals from runoff or wastewater.
Hyporheic zone	The area under a stream channel or floodplain that contributes water to the stream.
Instream flow	Streamflow regime required to satisfy a mixture of conjunctive demands placed on water while it is in the stream.
Intermittent stream	A stream, or portion of a stream, that does not flow year-round but only when it (a) receives baseflow solely during wet periods, or (b) receives groundwater discharge or protracted contributions from melting snow or other erratic surface and shallow subsurface sources.
Keystone species	A species that increases or decreases the diversity of a system. These species are competitively superior species. A keystone predator is a species that increases the diversity of a system by selective predation.
Lotic	Of or relating to flowing water such as a stream or river. Of or relating to organisms living in a brook, stream, or river.
Perennial stream	A stream that flows continuously throughout the year, usually groundwater fed throughout the year.
Phreatophyte	A plant that derives its water supply from groundwater and is more or less independent of precipitation.
Riffle	A fast section of a stream where shallow water races over stones and gravel. Riffles usually support a wider variety of bottom organisms than other stream sections.
Riverine	Relating to rivers and stream. It is commonly used in classifying and delineating wetlands and deepwater habitat.
Riparian vegetation	Vegetation that grows near a body of water and that is dependent on its roots reaching free, unbound water (e.g., the water table) during a portion of the year.
Scale	In ecology, scale refers to the level of spatial or temporal resolution perceived or considered.

[a]Most definitions from Helms 1998.

> and/or subsurface seepage, so as to result in an existing or potential soil-vegetation complex that depicts the influence of that extra soil moisture. Riparian areas may be associated with lakes; reservoirs; estuaries; potholes; springs; bogs; wet meadows; muskegs; and intermittent and perennial streams. The characteristics of the soil-vegetation complex are the differentiating criteria.

Disciplinary definitions are also highly variable and, in many cases, are oriented to the particular disciplines involved. For example, a soil scientist may base the definition on water availability, and it may include qualifications such as: "... land, inclusive of hydrophytes, and/or with soil that is saturated by groundwater for at least part of the growing season within the rooting depth of potential native vegetation" (Ilhardt et al. 2000). Other disciplines may characterize riparian areas with emphasis on climate, geology, landforms, natural disturbances, vegetation, or other factors reflecting their interests.

Functional definitions focus on the flow of energy and materials as their basis rather than being based on static state variables. The functional definition recognizes that the riparian boundary does not stop at an arbitrary, uniform distance from the stream channel, but can vary in width and shape (Ilhardt et al. 2000). A functional definition that has been proposed is:

> Riparian areas are three-dimensional ecotones of interaction that include terrestrial and aquatic ecosystems, that extend down into the groundwater, up above the canopy, outward across the floodplain, up the near-slopes that drain to the water, laterally into [the] terrestrial ecosystem, and along the water course at a variable width.

Managers are increasingly interested in classifying and delineating riparian areas because of their important and unique values. A basic understanding of the ecology of riparian systems is complicated by extreme variation in geology, climate, terrain, hydrology, and disturbances by humans. As a result, it is often difficult to determine the vegetation potential of riparian sites to develop management options (Kovalchik and Chitwood 1990). Geomorphology is useful on riparian sites where the natural vegetation composition, soils, and/or water regimes have been altered by past natural or human-induced disturbance. The stream classification system developed by Rosgen (1994) is one such classification, as discussed in Chapter 10. When classifying riparian areas, it is important to obtain as complete a documentation as is possible to satisfy evolving uses of the classification system.

The Biophysical Environment and Functional Relationships

Riparian systems are a diverse mosaic of landforms, communities, and environments that serve as a framework for understanding the organization, diversity, and dynamics of plant and animal communities associated with fluvial systems (Naiman et al. 1988). Watersheds surrounding riparian areas vary from a few hectares on first-order streams to millions of hectares on large river systems. The topography of the surrounding watershed can vary from a few meters above the riparian areas in Florida to thousands of meters in the Rocky, Sierra Nevada, and Cascade Mountains of the western United

States. The climate can range from the extremely arid desert environments in Africa, Asia, Australia, and the United States to humid tropical jungles in Asia, Africa, and South America. The only continent not supporting riparian areas is Antarctica.

In each environment, the functioning of riparian ecosystems is controlled by water, vegetation, soils, and biotic organisms extending in size from microorganisms to large mammals. The magnitude and direction of these functional relationships are further influenced by geology, geomorphology, topography, climate, and their inherent interactions.

Hydrology

The presence and persistence of water at the aquatic-terrestrial interface is a key to the formation and maintenance of a riparian system. This water affects the hydrologic processes in the riparian area, which most often differ greatly from those processes both in the adjacent aquatic and terrestrial system. Groundwater can be at or near the soil surface in riparian areas, being fed from upland areas. In drylands, the riparian area can be defined by the shallow saturated zone, whose water table gradient drives water away from the aquatic system (as illustrated in Fig. 3.1). The interplay of unsaturated and saturated zones in riparian areas is complex, as in the hyporheic zones, discussed later. For example, alluvial riparian areas can function as shallow aquifers that recharge at high flows and drain at low flows. The exchange of water between surface flow and groundwater storage tends to moderate high flows and enhance or prolong base flows. Shallow aquifers create a moist soil environment that favors riparian plant growth. Surface processes of watershed runoff, erosion, sediment transport, and streamflow can all occur in this riparian system.

Streambanks occupy a unique position in the riparian system. Bank and channel profiles affect water velocity and sediment flow as discussed in chapters 9 and 10. The morphology of the stream channel, stream bank, and floodplain all affect the types of vegetative cover and their influence on stream temperature, sediment input, and ultimately the habitat of the riparian and the aquatic systems. In addition, the hiding cover and structure provided for fish is important.

Intermittent streamflow and variable annual precipitation are more common in the arid lands than in the more humid areas. As a result, riparian systems in humid regions are less dependent on annual recharge by episodic events to sustain flow and are more dependent on a regular and dependable groundwater flow accompanied by interflow (subsurface flow) and, to a lesser extent, on overland flow. Watersheds in humid regions commonly have lakes, ponds, and other systems that sustain flows over much of the year.

Sediment transport is episodic in the arid West, where pulses of aggradation and degradation are punctuated by periods of inactivity. Unlike on humid watersheds, sideslope erosion in drylands is discontinuous, and there are often long lag periods between watershed events and sediment delivery (Heede et al. 1988). As a result, the macrotopography of riparian areas in arid regions is rearranged mainly during the cycles of aggradation and degradation. Channels in humid regions tend to be more stable, and the macrotopography of these systems, though still responsive to flooding events, tends to evolve much more slowly.

Sediment originates mainly from hillslopes in drylands and is transported downslope from upland areas during major precipitation or wind events. When vegetative cover is limited, natural upland weathering and the formation of rills and gullies contribute sediment to downslope channels during major storm events. Systems in humid areas are better vegetated, and as a result, less erosion occurs from the watershed. Watershed side slopes tend to creep rather than erode, and the sediment is finer. The transport of sediment produced in humid environments depends more on the amounts of energy available than on the rate of weathering (Ritter 1986). As a result, sediment transport down the hillslope into riparian systems in humid regions is less episodic in nature.

Vegetation

Riparian vegetative communities are usually distinct from those in the adjacent uplands. These communities play many important roles in their effect on microclimate, such as providing shade to streams, resulting in lower stream temperatures that are more favorable for fish habitat. Vegetation also provides a critical habitat for many invertebrates, amphibians and reptiles, small and large mammals, and birds. Vegetation acts as a filter, or buffer, because it traps the overland movement of dissolved sediments and nutrients and can take up nutrients from the saturated and unsaturated zones adjacent to water bodies. Vegetation also plays a role in supplying both large and small organic matter to the soil and stream. Litter fall, when covered with sediment, decomposes quickly and releases both nutrients and humus to the soil. Structurally intact living and dead plant material is colonized by microorganisms that provide food for macroinvertebrates, which serve as food sources for fish and wildlife species.

Riparian vegetation consists of trees, shrubs, grasses, rushes, and sedges in varying combinations. Riparian vegetation is adapted to well-drained, periodically flooded soils. Stream riparian systems typically have transported coarse soils and are adapted to high-magnitude, short-duration flood events. The physical structure of riparian vegetation consisting of tall, deep-rooted trees and shrubs enables them to survive flooding. Riparian vegetation, and particularly riparian forests, exhibit high rates of ET that are usually near potential evapotranspiration rates (Tabacchi et al. 2000). Along stream banks this translates into reduced pore pressures that when combined with root strengths of woody vegetation lead to greater streambank stability. Extensive root systems and organic matter help to bind the soil and promote high infiltration rates (Maser and Sedell 1994). Trees and associated vegetation along streams trap sediments, leading to floodplain succession and ultimately community succession.

In many floodplain areas of the western United States, certain plant species have adapted to conditions of shallow water tables or wet environments adjacent to streams and lakes. In dryland regions, extensive communities of these plants can be found along ephemeral stream channels and in expansive floodplains which have shallow water tables. Often, the most prevalent plants are *phreatophytes*, which are plants with deep and extensive rooting systems that allow them to extract water from the water tables or from the capillary fringe (Box 13.1). These water-loving trees include alder, ash, willow, cottonwood, saltcedar, and spruce, and along with certain other species, they are more abundant in riparian areas than in the surrounding watershed.

Soils

Soils provide a storage reservoir for water, which is necessary for the growth of plants and the sustenance of other biological organisms. Soils also provide an essential environment for organic matter and the soil organisms that continuously cycle nutrients.

Box 13.1

Phreatophyte Species in the United States

"Phreatophyte" is a Greek word coined by Meinzer (1923) meaning "well plant"; such plants extract water from the soil and act like a series of miniature water wells. Phreatophytes are native and nonnative trees, shrubs, and grasses than can effectively tap the water table and transpire large amounts of water. Most phreatophytes are deep-rooted, woody perennials, with root systems that can extend to shallow water tables (e.g., 2 m). Native species include Fremont cottonwood (*Populus fremontii*), mesquite (*Prosopis juliflora*), ash (*Fraxinus pennsylvanica*), sycamore (*Platanus wrightii*), walnut (*Juglans major*), alder (*Alnus* spp.), willow (*Salix* spp.), seepwillow (*Baccharis glutinosa*), sacaton (*Sporobolus airoides*), and salt grass (*Distichlis stricta*). These species can transpire up to 2500 m^3 water/ha/yr (Horton and Campbell 1974).

An introduced phreatophyte from the Old World is tamarisk, or saltcedar (*Tamarix chinensis*), which has spread vigorously in the reservoir deltas and along major rivers in the southwestern United States. Floodplain areas provided ideal conditions for a rapid invasion of tamarisk. First introduced into the United States as an ornamental, tamarisk was sold at nurseries in the East as early as 1823. It was being sold in nurseries in California by the middle 1850s. In 1901, it was recorded as a naturalized shrub along the Salt River in Tempe, Arizona. In the 1920s, tamarisk was beginning to spread along the Rio Grande and other rivers. These river floodplains were to a large extent covered with solid, unbroken stands of almost impenetrable saltcedar by the 1940s. It is currently found along many smaller streams, around springs and stock water tanks, by roadsides, and in other areas that have sufficient moisture to germinate and establish the seeds. Saltcedar has physiological characteristics that make it a successful invader of riparian ecosystems containing Gooding willow (*Salix gooddingii*) and Fremont cottonwood (Horton et al. 2001). These characteristics include higher leaf gas exchange rates, faster growth when water is abundant, higher drought tolerance, and the ability to maintain a viable canopy under dry conditions.

Physical, chemical, and biological soil properties vary widely depending on the depositional history of a stream reach and the length of time the soils have been allowed to develop in place. The profile of young riparian soils tends to be stratified, containing discrete layers of fine materials and coarse materials. Because of the physical variability of soils in riparian areas, the drainage can vary over small distances from a saturated condition to being well drained.

Organic matter accumulates rapidly in the riparian environment either from small and large organic debris that is deposited during flooding, or more importantly, by the sloughing off of dead plant material (leaf fall, dead roots, and bark) from riparian vegetation. In warm and moist soil environments, rapid decomposition of woody organic material occurs where adequate soil aeration is present. In these aerobic environments, decomposition occurs via a wide range of organisms ranging from soil microorganisms to large invertebrates. In saturated parts of the soil, organic matter decomposition is much slower because of the anaerobic environment, and organic matter can accumulate into thick layers. It is in such anaerobic zones that denitrification occurs, which constitutes an important role of floodplains that receive excessive nutrient loading from upland watershed practices.

The cycles of aggradation and degradation in drylands often prevent the development of a stable organic soil horizon or a well-developed soil profile. Instead, soils reflect the depositional pattern and history of channel flows. Vegetation that thrives is suited to these coarse-textured soils under either well-drained or well-flooded conditions (Platts et al. 1987). In humid regions, erosional cycles are not as severe, and soils tend to have well-developed organic horizons. In many cases, the soils in humid regions are highly reduced, and vegetation must be able to grow in soils having saturated, low-oxygen conditions.

Wildlife Habitat

The soil-plant-water diversity of riparian areas provides critical habitat for many species of fish, reptiles and amphibians, large and small mammals, and birds. Riparian ecosystems furnish food, water, cover, and a shaded streamside environment for a wide range of wildlife. Some of these animals spend their entire life in riparian areas, while others come and go but still require a riparian habitat during part of their life cycle. As a result, many species of wildlife concentrate in these areas. In California, for example, 25% of the mammals, 80% of the amphibians, and 40% of the reptiles are limited to or depend upon riparian areas, and more than 135 species of birds depend on, or prefer, riparian habitats (Sorenson 1989). In more arid areas, even higher percentages of species are dependent upon riparian habitats. Riparian areas provide critical habitat for threatened and endangered species worldwide.

Fish are obligate aquatic organisms, entirely dependent on water conditions for survival. Fish come in all sizes and shapes with varying water requirements (temperature, rearing habitat, food supply, and migration routes). These include warm-water species (smallmouth bass) and cold-water species (trout). Salmon are an important commercial and game fish both in tributaries to the Great Lakes and in many large rivers and streams in California and the Pacific Northwest. Anadromous fish, including salmon, are recognized as keystone species (Wilson and Halupka 1995). Although many game fish are recognized for the recreational fishing they provide to millions, there are also numerous rare and endangered nongame fish in riparian systems.

Amphibians are indicators of riparian ecosystem health. They are sensitive to changes in the environment because (1) they have the most complex life cycles involving both aquatic and terrestrial stages that expose them to perturbations in both environments; (2) they have permeable skin, gills, and eggs that are susceptible to environmental alterations (e.g., radiation, moisture levels); (3) their dependence on ectothermy for temperature control makes them vulnerable to environmental fluctuations; (4) many of them hibernate or aestivate in soils that can expose them to toxic chemicals; and (5) they are both predators and prey in terrestrial and aquatic food webs (Pauley et al. 2000). Threats to amphibians and reptiles include surface water withdrawal, road construction, bridge crossings, dam construction, flooding, logging, mining, agricultural and urban development, recreation, grazing, exotic species introduction, human-associated debris, and pollution.

Riparian habitats provide food, shelter, cover, and breeding and nesting sites for a variety of bird populations ranging in size from small hummingbirds to eagles. These bird populations are either resident or migratory. Rapidly growing riparian vegetation provides a multistoried habitat, and older trees have holes or cavities that provide homes for many of these birds. The structure of the vegetation and the differences between the riparian vegetation and the vegetation on the surrounding upland watershed affect the concentrations of birds. Greater differences occur in bird communities when the riparian and adjacent upland habitats reflect differences in land use. In New England forests, there are few, if any, differences in breeding bird populations between the streamside area and the surrounding watershed because the vegetation in the two systems does not differ significantly. When riparian areas are adjacent to agricultural land, large differences in bird and mammal populations are present between the two. The arid and semiarid environments of the southwestern United States provide the greatest differences in species richness and abundance between riparian and upland habitats (Szaro 1981). Under conditions in the Southwest, riparian vegetation along streams can interface with desert grassland or shrubs within a few hundred feet, thereby providing an oasis in otherwise hostile environments.

A wide range of large and small mammals is partially or wholly dependent upon healthy riparian ecosystems. In the eastern United States, some of the large mammals that depend upon riparian areas include opossum, racoon, fisher, mink, river otter, and moose. Small mammals include the water shrew, star-nosed mole, and rock vole. Stand type and age are not as important for small mammals as microhabitats that provide food and cover (DeGraaf and Yamasaki 2000). Species of large mammals in the West include the raccoon, coyote, mule and white-tailed deer, elk, and grizzly bear, as well as the river otter, which was once considered extinct in Arizona but was successfully reintroduced in the Verde River. These mammals use riparian habitat for food and water, cover from heat and cold, cover from predators for small mammals, and breeding and rearing areas. The beaver (*Castor canadensis*) is a keystone species in riparian areas because of its historical importance and its capacity to change the stream environment (Box 13.2).

LINKAGES

Physical, biological, and chemical linkages exist among aquatic systems, riparian systems, and the surrounding watershed areas. These linkages involve processes that are responsible for the continual flux of water, air masses, dissolved particulate matter, organisms, and energy both within and among these three components. These processes include runoff, erosion, sedimentation, groundwater recharge and discharge, nutrient cycling, food webs, streamflow, nutrient enrichment of water, and water and nutrient exchanges in the hyporheic zone.

Linkages occur at a multitude of spatial and temporal scales. For example, the spatial scale of the aquatic environment can be divided into a stream system, a segment of the stream system, a pool/riffle system, a microhabitat system, and a particle system. Associated with this spatial scale is a time period, which can extend from local stream surges that last only a few seconds, or fractions thereof, to long-term, wide-scale variations in water levels and temperatures that may extend over seasons from decades to thousands of years (Giller and Malmqvist 1998). These temporal and spatial scales can

Box 13.2

Role of Beaver (*Castor canadensis*) in Riparian Systems of North America, Their Impacts on Stream Channels, and Consequences of Human Development

At the time of European settlement in North America, as many as 400 million beaver occupied the continent (Butler 1995). Beaver dams, felled trees, and other debris impeded the flow of many rivers, resulting in wide meandering streams and extensive floodplains. Riparian wetlands that resulted from beaver activity covered over 200 million acres of land in the 48 contiguous states (Dahl 1990). Many of the once stable rivers are now deeply incised with eroded banks, the result of agricultural development, urbanization, and the many stream channel alterations carried out for navigation and flood control purposes. However, the net cumulative hydrologic effects of human development can be viewed in terms of lost storage and increased conveyance. Storage was lost through beaver dam destruction and drainage of wetlands. Channelization and levee construction removed floodplain storage and increased conveyance of stream channels, which flow at greater velocities. The frequency and severity of flooding and flood damages have increased (Hey and Wickencamp 1998). Although repopulating beavers is not suggested, restoration of riparian floodplain and wetland storage should become important objectives of stream restoration that can lead to reductions in flood damage. Donald Hey of the Wetlands Initiative in Chicago, Illinois, speaks of these important functions of riparian and wetland systems and suggests that "flood-storage credits" could be provided as incentives to landowners for restoring natural streams, floodplains, wetlands, and riparian systems.

extend over about 16 orders of magnitude. The scale is important because it defines the level of generalization that may be generated from the results of a particular study.

Many of the physical linkages involve hydrologic processes when water moves energy and matter usually begins in the upland terrestrial watershed ecosystems, is funneled into and through the riparian plant communities along the streams, and enters the stream channel and the aquatic system. Where riparian vegetation is not present, materials move directly from the watershed into the stream system, lacking the mitigating effect of the buffer action provided by the streamside vegetation. There are also linkages below ground via groundwater and aquifers that move materials from the watershed to the aquatic zones in streams. In many of the arid desert streams in the Southwest water can move down the channel during flow periods and into the adjacent riparian vegetation zone via bank recharge (Martí et al. 2000).

Surface Runoff and Erosion

The most obvious linkage is that of runoff and sediment produced from erosion (see chapters 4 and 7). Runoff and erosion processes are key factors affecting the stability of lands within the stream channel, in the riparian area, and on the surrounding watershed. The loss of soil can lead to a poor-condition watershed that contributes excessive runoff and sediment to downslope riparian and stream systems. These quick responses of streamflow to precipitation must be addressed in any riparian enhancement program.

Groundwater

Streamflow through riparian areas is usually dependent upon continual groundwater inputs (see Chapter 5). Water passing slowly through the soil mantle can sometimes sustain a dependable perennial streamflow necessary for maintaining downslope riparian ecosystems. Also, the chemistry and flow characteristics of streams are dependent upon the processes of infiltration, percolation through the soil profile, and movement by underground flow paths into the riparian and stream areas (Holmes 2000). Different groundwater flow patterns deliver the watershed precipitation to the riparian areas and stream. On watersheds having deep, well-drained soils, water can percolate beneath the rooting depth and never interact with the vegetation. When watersheds have shallow soils, vegetation roots can absorb water, along with dissolved nutrients, before it reaches the riparian area. At the most arid extreme, desert watersheds that have hydrophobic soils can transport precipitation directly to stream channels as overland flow without the water passing through the rooting zone of the riparian vegetation.

Groundwater contains dissolved materials that affect the riparian area and stream composition. The primary nutrients being delivered via groundwater flow are nitrogen and phosphorus. These two nutrients are important to stream ecology because they limit the primary productivity of aquatic ecosystems. Nitrates produced by the leaching of agricultural fertilizers, sewage, and industrial inputs are particularly troublesome. Some riparian areas can reduce nitrate concentrations by the denitrification process and thereby reduce groundwater nitrate concentrations and fluxes from upland to stream ecosystems. Inorganic carbon can also be a major dissolved constituent in areas where the water passes through soils having concentrations of calcite and dolomite.

Food Webs

Food webs include the trophic linkages between consumers and their food source and are based on a series of interlinked food chains (Giller and Malmqvist 1998). In essence, they represent a pattern of energy flow through an ecosystem. Food chains occur in the terrestrial, the riparian, and the aquatic environment.

Important components of the food web include the allochthonous and autochthonous sources of energy. Forest litter composed of deciduous leaves, dissolved organic carbon in groundwater, and photosynthesis by diatoms are the main sources of energy for this food chain. Physical abrasion and decomposition by microbes transform the different forms of organic matter into fine particulate organic matter. This transforms the allochthonous and autochthonous organic substances into a substrate that is consumed by invertebrates (Box 13.3) that become the food source for predator fish and other animals including amphibians, reptiles, birds, and small mammals.

Box 13.3

Invertebrates and Their Role in Aquatic Food Webs

Invertebrates found in aquatic systems vary from unicellular protozoa to large insects. Protozoa act as lower food chain predators and feed primarily on bacteria and algae. They are highly productive and can produce 12 g of dry material per square meter of stream bed per year (Giller and Malmqvist 1998). Other microinvertebrates are flatworms, snails, freshwater mussels, crayfish, hog lice, shrimp, and copepods.

Macroinvertebrates include worms, leeches, mites, and insects. Insects are the most conspicuous of the macroinvertebrates. In most cases, the larvae live in the aquatic environment and the adults are terrestrial. The most important insects in flowing water are stoneflies, caddisflies, true flies, beetles, bugs, alderflies and dobsonflies, and dragonflies and damselflies. These macroinvertebrates feed on algae and fine detritus, while others are predators and feed on other smaller invertebrates. Caddisflies are important as shredders; larval caddisflies often represent the largest biomass of the macroinvertebrate communities in streams.

Shredders are an active component of the aquatic food web (Cummins et al. 1989) and include insects and crustaceans. Important insects are stoneflies, caddisflies, and craneflies. Shredders use coarse particulate organic matter (greater than 1 mm in size), such as leaves, for food. They chew up and ingest this leaf material. The undigested fragmented organic material and the shredder feces form a food source of fine particulate organic matter (< 1 mm but $> 0.5\ \mu m$) for a guild of invertebrates called collectors/filters. Shredders can process more than their own body weight daily and about 60% of this passes through as feces.

Hyporheic Zone

Another important linkage between the stream and the adjacent riparian area is the *hyporheic zone*, a zone of saturated sediments lying below the streambed and extending laterally beneath the stream banks (Boulton 2000). This is the zone where there is a dynamic exchange of water and materials between the groundwater below, the lateral alluvial aquifers, and the river flowing above (Fig. 13.2). The exchange is facilitated by a dynamic hydraulic gradient between the river and the adjacent aquifers that can extend several hundreds of meters in large river systems (Hinkle et al. 2001). Because the water in the hyporheic zone comes from the interaction between the regional groundwater and river water, the composition can vary from a mixture of ground- and river water to 100% river water. Both surface and subsurface flow paths exist. Subsurface flow occurs when river water enters the channel bed and banks and then reemerges downstream (Fernald et al. 2001). The flow of water through the hyporheic zone promotes biochemical processes that are important for water quality and aquatic habitat. Stream riffles exhibit a typical flow pattern that consists of surface water entering the hyporheic zone in a downwelling zone at the head of the riffle and hyporheic water returning to the stream surface in an upwelling zone at the tail of the riffle. The

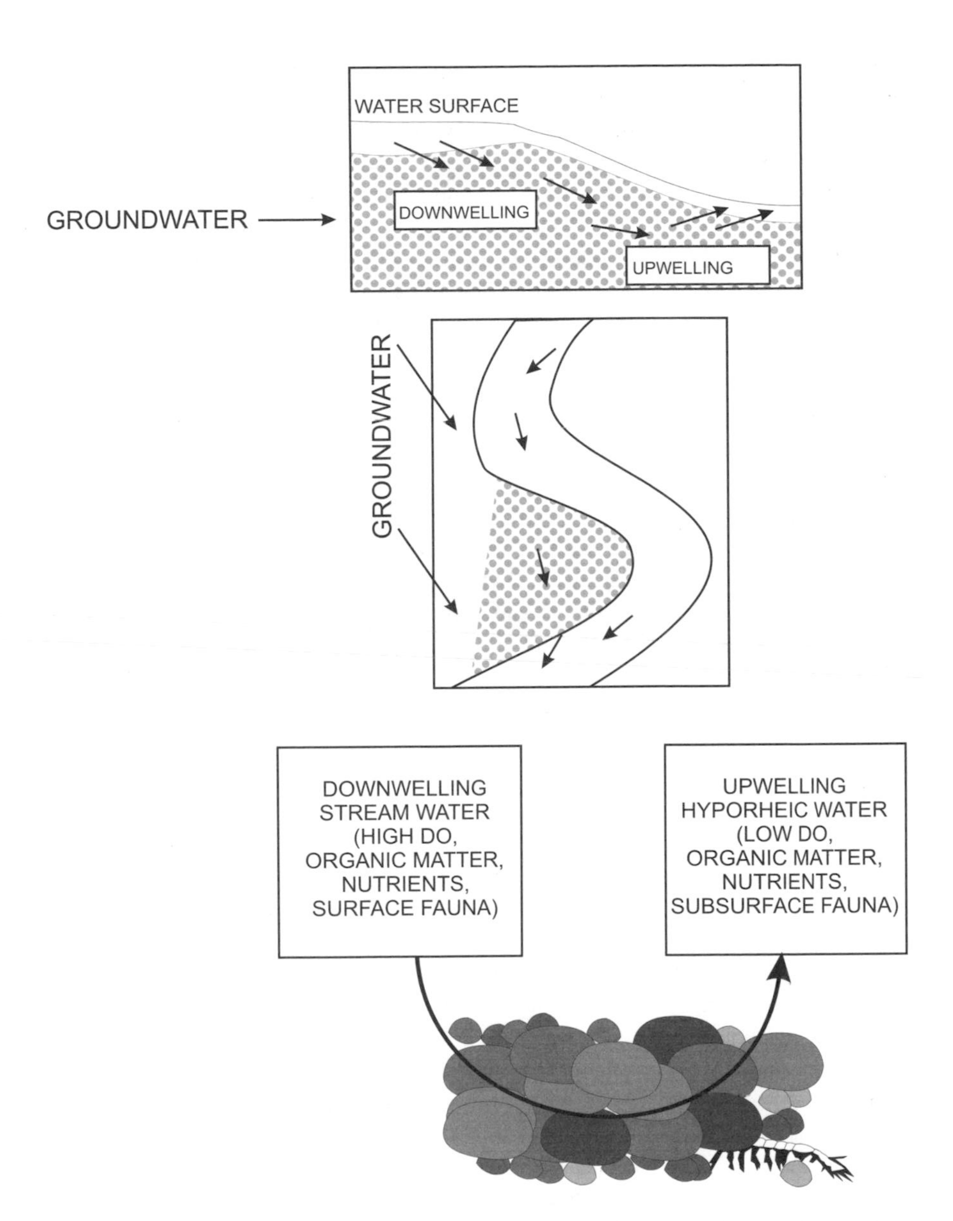

FIGURE 13.2. Water in the hyporheic zone is exchanged above with surface streamwater, below with the groundwater, and laterally with the riparian zone and alluvial aquifers (from Boulton 2000, ©Blackwell Science Ltd., by permission).

downwelling and upwelling zones differ in temperature, pH, redox potential, dissolved oxygen, and nitrates (Franken et al. 2001). The upwelling zones have been reported as preferred spawning areas for bull trout (Baxter and Hauer 2000) and chinook salmon (Geist 2000).

The hyporheic zone is biologically active. From the whole stream perspective, the hyporheic zone can be visualized as an active area in the stream that traps, transforms, and mineralizes organic carbon, and its contained energy, that otherwise would be transported farther downstream (Kaplan and Newbold 2000). As water flows through the hyporheic zone, water chemistry is altered by microbial and chemical processes that include transforming nutrients, consuming oxygen, decomposing organic matter, and other activities (Boulton 2000). Invertebrates and bacterial activities produce a complex series of organic matter and sediment dynamics and interactions. As a result, the water emerging from the hyporheic zone is chemically and biologically different than the water entering this zone. The downwelling water entering the hyporheic zone is usually high in dissolved oxygen, organic matter, nutrients, and surface fauna in contrast to the upwelling water. The abundance and diversity of the invertebrates appear to be positively correlated to bacterial abundance and production (Brunke and Fischer 1999). Microbial biofilms coating sediments act as biological filters, which enhance water quality during the exchange processes involving this zone.

The hyporheic zone of many rivers is threatened by siltation, toxic substances, increased acidity, altered groundwater inputs through pumping, and the physical extraction of gravels (Boulton 2000). Siltation clogs the interstitial spaces, thus preventing the downwelling water from moving into the hyporheic zone. This results in lower levels of dissolved oxygen and nutrients, which lead to the death of aerobic microbes and invertebrates. Heavy metals such as zinc and copper can poison the interstitial vertebrates. Pumping destroys or drastically decreases the groundwater component in the hyporheic zone, which adversely affects all chemical and biological components of the zone.

Nutrient Cycling

Many of the riparian linkages involve the cycling of nutrients. Regular and consistent cycling of nutrients is essential for plant growth and the sustained productivity of natural ecosystems. Nutrients cycle continually within and among the terrestrial, riparian, and aquatic systems. The most important nutrients are carbon, nitrogen, and phosphorus. Carbon is important as both a source of energy for biological processes (e.g., decomposition, respiration) and a reservoir for all the other plant nutrients. Organic matter is created during the fixation of gaseous carbon from the atmosphere during photosynthesis in green plants. Other nutrients enter the system via the soil by natural processes such as precipitation, dry fall, nitrogen fixation, and geochemical weathering. The main source of nitrogen is nitrogen fixation, although some is contributed by precipitation and dry fall. Nitrogen fixation is carried out by symbiotic nitrogen-fixing organisms and by nonsymbiotic free-living microorganisms. Many of the symbiotic microbes are bacteria associated with the roots of leguminous plants. Nutrients can also enter streams as runoff from agricultural crops where fertilizers and soil amendments have been used. Nutrients and toxic materials are also added to rivers and streams by industrial, urban, and animal wastes.

Organic matter and its contained nutrients cycle primarily within the natural systems in which they occur, that is, the terrestrial, the riparian, and the aquatic systems. However, there can be a regular interchange of carbon and nutrients between riparian vegetation and the stream. Furthermore, there can be a significant movement of carbon and nutrients from the watershed slopes, through the riparian vegetation during surface runoff and erosion or in groundwater draining from the watershed.

Terrestrial and Riparian Vegetation

Living trees shed leaves that accumulate on the soil surface and are immediately acted upon by macro- and microorganisms, which decompose them, physically and chemically releasing many of the nutrients. Not only do leaves provide organic material, but large woody materials including stems, twigs, bark, roots, and tree boles also provide storage reservoirs for carbon and nutrients (see section on large woody materials). The readily available nutrients released during decomposition are quickly captured as part of the foraging microbial biomass or are absorbed by plant roots. Small amounts of available nutrients and dissolved carbon can leave the terrestrial ecosystem via percolation into groundwater aquifers or by surface runoff during periods of intense precipitation. Most of the nutrients lost are absorbed by vegetation in the riparian areas or buffer strips before reaching the streamwater. Much the same type of nutrient cycling process occurs within the riparian vegetation along the stream channel, although there can be substantial nutrient additions from these streamside plant communities in the form of leaves and other organic debris as both solid and dissolved carbon.

Aquatic Environment

Carbon transformations and nutrient cycling are complex. In streams, there are two major sources of energy that are utilized by producer and consumer organisms (Giller and Malmqvist 1998). One is solar radiation, which supports instream primary production (autochthonous), and the other is organic matter produced elsewhere (allochthonous). The allochthonous organic matter ranges in size from coarse particulate organic matter (> 1 mm) to dissolved organic matter (< 0.45 μm). Large woody material (see below) can enter the stream as parts of leaves, twigs, flowers, and wood by direct fall or lateral litter blow. The fine particulate organic matter enters the stream through windblow, surface runoff, and bank erosion. Dissolved organic matter enters the stream via groundwater, surface runoff, or canopy drip and stemflow. This autochthonous organic matter is largely passed through the series of food chains described above. The allochthonous organic material is more resistant to breakdown and is broken down more via detritivory and decomposition. Some of the more resistant materials can enter detritus storage pools or be washed downstream from their origins. The amount of autochthonous and allochthonous materials is dependent upon the stream order. Low-order forested streams are subjected to shading and hence have low production of autochthonous material. Riparian forests along high-order streams have less shading effect, and as a result the high light levels increase the amount of autochthonous materials.

Nutrient cycling in stream systems differs from that occurring in terrestrial ecosystems because of the constant passing of water. As a result, only a small portion of the total nutrients is captured and cycled within the system. Those nutrients not immediately captured are washed downstream along with part of the soluble nutrients released during each time the nutrient is cycled. This downstream movement of nutrients is referred to as nutrient *spiraling*. The carbon and nutrients moving downstream may be captured by other food chains in the stream channel, or more importantly, move into the hyporheic zone where they are further transformed by various organisms.

Large Woody Materials

Large woody materials have become recognized as an important component in the productivity of forests and as a source of increased stability in river systems (Maser and Sedell 1994, Hering et al. 2000). This large woody material has been designated as large woody debris (LWD), large organic debris (LOD), and coarse woody debris (CWD). Although the terms are frequently used interchangeably, LWD and LOD are used more

Box 13.4

Specific Types of Large Organic Debris

Affixed logs—single logs or groups of logs that are firmly embedded, lodged, or rooted in a stream channel.

Bole—the stem or trunk of a tree.

Large bole—10 m or more in length; often embedded, remains in the stream for extended periods.

Small bole—less than 10 m in length, usually part of a bole; seldom stable, usually moves downstream on high flows.

Deadhead—log that is not embedded, lodged, or rooted in soil but is submerged and close to the water surface.

Digger log—log anchored to a streambank and/or channel bottom in such a way that a scour pool is formed. A scour pool is a depression in the stream channel caused by concentrated flow that scours out the streambed.

Free log—log (or group of logs) in water bodies that is not embedded, lodged, or rooted in soil or rock.

Root wad—root mass of the tree (also called butt end).

Snag 1—standing dead tree.

Snag 2—fallen tree in large stream; top of the tree is exposed or only slightly submerged.

Sweeper log—fallen tree whose bole or branches form an obstruction to floating objects.

Accumulations of large organic debris include:

Clumps—accumulations of debris at irregularly spaced intervals along the channel margin that do not form major impediments to flow.

Jams—large accumulations of debris partially or completely blocking the stream channel margin, creating major impediments to flow.

Scattered—single pieces of debris at irregularly spaced intervals along the channel.

frequently to describe woody material found in riparian areas, while CWD is used to describe the large woody material in the upland forests of the watershed.

Large woody debris refers to any piece of relatively stable woody material that is at least 1 m long, and 10 cm in diameter (Dolloff 1996). *Large organic debris* (LOD) also describes the large woody component that intrudes into the stream channel and specifically refers to large root wads and tree stems that provide overhead cover and modify flow so as to provide for the effective spawning of anadromous and resident fishes (American Fisheries Society 1985). The classification of LOD is given in Box 13.4.

Large woody materials are produced when trees die and fall to the soil surface. Large woody materials are produced both on the forested watersheds and in the streamside riparian areas. Although there is some exchange of large woody material between the forested watershed area and the riparian area, the larger components of large woody materials are probably not transported great distances except during excessive surface runoff or major floods in the channels. Runoff and erosion can contribute debris to streams and account for water transporting pieces of debris. Large organic debris in channels is frequently added by the undercutting of streamside trees. Downstream movement of LOD within channels is related to the length of individual pieces; most pieces that are moved are shorter than the bank-full width of the stream (Lienkaemper and Swanson 1987).

Effects on Watersheds

Much of the CWD found on forested watersheds is partially or totally covered by soil and humus layers. Large woody debris on the soil surface or covered by soil and humus layers can comprise > 50%, or 37–50 Mg/ha, of the total surface organic matter in old-growth forests in the Inland Northwest (Page-Dumroese et al. 1991). This woody debris and associated smaller organic matter enhance the physical, chemical, and biological properties of the soil and thereby contribute directly to site productivity (Jurgensen et al. 1997). The debris also provides a favorable microenvironment for seedling establishment and growth.

Organic matter and CWD found on forested watersheds improve physical properties of soil by enhancing soil structure and stabilizing the soil surface. The improved physical soil conditions are important for moderating hydrologic responses of watersheds to precipitation. Vegetation, litter, and organic matter provide a soil surface cover that is capable of absorbing precipitation energy, increasing infiltration, and slowly transferring subsurface water flow downslope. Improved soil structure resulting from organic matter provides pore spaces and allows the soil to store and transmit water slowly from the soil surface to underground storage or flow, providing a large temporary storage of water throughout the entire watershed system. The slow subsurface flow decreases the erosive potential produced by overland flow, particularly on steep slopes.

Effects on Streams

In the past, LWD in streams was considered a liability and was removed because engineers considered it a hindrance to the navigation of large rivers and an impediment to drainage of small watersheds. Biologists also worried that the accumulation of LWD would degrade water quality and block fish migration in addition to being unsightly. However, LWD has been found to have many physical and biological benefits ranging from enhancing the production of insects and other invertebrates to forming critical habitat for fish. Benefits associated with large woody debris are given in Figure 13.3.

Woody debris increases the complexity of stream habitats by physically obstructing water flow. Large affixed logs extending partially across the channel deflect the flow laterally, causing it to widen the streambed. Sediment stored by debris also adds to hydraulic complexity in organically rich channels, which are often wide, shallow, low-gradient streams on alluvial valley floors with many riffles and pools. Even when the stream becomes so large that trees cannot span the main channel, debris accumulations along the banks cause meander cutoffs and create well-developed secondary channel

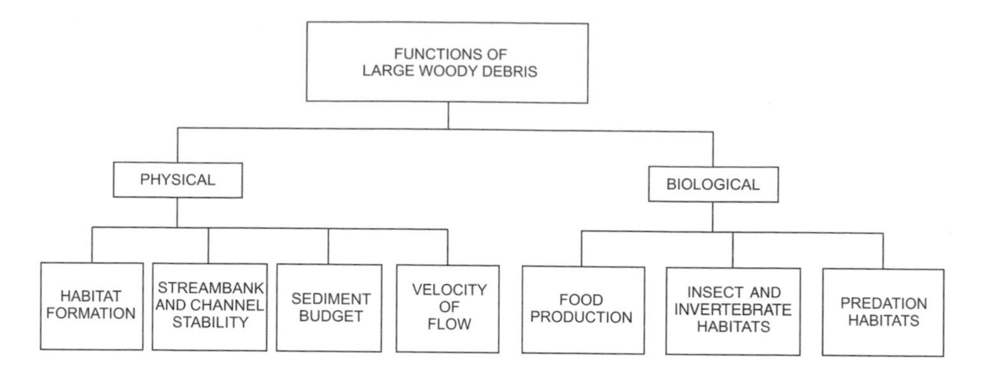

FIGURE 13.3. The various benefits generated by large woody debris in streams (from Dolloff 1996).

systems. Debris creates variation in channel depth by producing scour pools downstream from obstructions. Therefore, wood maintains a diverse physical habitat by (1) anchoring and stabilizing the position of the pools in the direction of streamflow, (2) creating backwaters along the stream margin, (3) causing lateral migration of the channel and forming secondary channel systems in alluvial valley floors, and (4) increasing depth variability.

Fallen trees that protect the outer edge of a thicket and those in the thicket itself create local environments of quiet water where fine sediments and organic debris are deposited during high flows. This debris, coupled with leaf and woody litter from the riparian stand, enhances soil development and vegetative growth. Fallen trees on gravel bars provide sites where some stream-transported species of hardwoods and shrubs can reroot and grow. Also, large fallen trees help a forest stand to reach a stage of structural development that allows it to withstand floods better.

Several factors influence the movement of LOD downstream. The most important are the angle of the piece relative to flow direction, whether or not the log has a root wad, the density of the log, and the piece diameter (Braudrick and Grant 2000). The stability generally increases when the pieces have root wads or are rotated parallel to flow. Piece length does not affect stability unless it is longer than the stream width.

Organic debris in streams increases aquatic habitat diversity by forming pools and protected backwater areas, providing nutrients and substrate for biological activity, dissipating energy of flowing water, and trapping sediment. Large woody debris is important for pool formation, where it develops plunge pools, overhanging logs, and debris jams, particularly with root wads, which protect the fish from predators and excessive competition. Smaller pieces of wood also have a significant role in stream ecology, but they are easily dislodged and moved downstream by flowing water and consequently have less influence on channel morphology and fish habitat.

DISTURBANCES

The functioning of riparian areas is dependent on the condition of both the riparian zones and the surrounding watersheds. As discussed above, there are strong linkages between watershed and riparian areas, and as a result, both natural and human disturbances can influence these interrelationships.

Box 13.5

The Influence of Long-Term Geologic Processes on Stream and Sediment Dynamics: A Case Study (Miller et al. 2001)

The effects of historical climates and landform processes on present-day channel and riparian relationships are illustrated by a study of the Great Basin of central Nevada. Data on stratigraphic, geomorphic, and paleoecological features were used to assess the relationships of late Holocene climate change, hillslope processes, and landforms to present-day channel dynamics. It was concluded that the climatic conditions from approximately 2500 to 1300 YBP shifted to drier and warmer climatic conditions. This shift initially led to massive hillslope erosion and the simultaneous aggradation of both side-valley and alluvial fans and the axial valley system, resulting in fan stabilization and axial channel incision. During the past 1900 years, channel entrenchment was the primary process operating, which led to fan deposits that temporarily blocked the movement of sediment down the main stem of the valley and created a stepped longitudinal valley profile. Stream reaches located immediately up the valley of these fans are characterized by low gradients and alternating episodes of erosion and deposition. Conversely, the reaches coincident with or immediately downstream of the fans exhibit higher gradients and limited valley floor deposition. As a result, the present-day channel dynamics and associated riparian ecosystems are influenced by landforms created by depositional events that occurred about 2000 years ago.

Historical Processes and Disturbances

Natural geological processes and disturbances are sometimes not obvious, but they can affect the current functioning of streams and riparian areas (Box 13.5) Therefore, it is important to consider specific disturbances and their consequences when managing watershed-riparian ecosystems. Also, it is important to be aware of past use and misuse of watershed-riparian ecosystems when considering rehabilitation potentials for a particular situation.

Although geologic history can be useful in interpreting landscapes, it is usually not widely available beyond small study areas. Information on more recent disturbance history (within the last three to four centuries) is more readily accessible to managers and can be useful in interpreting and understanding present-day riparian and watershed responses.

Effects of Natural and Human Disturbances

Natural disturbances that affect watershed-riparian systems include wildfires, unusual hydrologic conditions (e.g., excessive storm precipitation), and natural disasters such as earthquakes, volcanic eruptions, tornados, or hurricanes. Managers have little control over the consequences of natural disasters except to maintain the best possible condition of a watershed-riparian ecosystem so it is better able to withstand these unusual events. Human-related disturbances include prescribed fire, livestock grazing, and timber harvesting. In contrast to natural disturbances, managers have ways to mitigate the impact of human activities on riparian ecosystems and their associated watersheds.

Fire

Both wildfires and prescribed burns affect riparian systems directly and indirectly. If fires burn in the riparian area itself, direct effects such as the consumption of part or all of the vegetation are immediately obvious (DeBano et al. 1998). However, subtle indirect effects such as stream temperature increases, alterations in the quantity and quality of organic matter inputs to streams, aquatic macroinvertebrate population changes, and fish migration can occur long after the fire. When fire burns the surrounding watershed, surface runoff and streamflow can increase, causing higher peak flows and sediment movement into and through riparian ecosystems. Debris flows along with dry ravel and small landslides are common occurrences after fire on steep topography. Nutrient cycling changes, plant and animal community shifts, population declines, and other indirect effects can occur but are not easily visible.

Some riparian areas recover rapidly following wildfires, while others undergo slow recovery. The rate of recovery depends largely on the environment. The recovery of vegetation following fire reflects the magnitude of the combined disturbance of both fire and flooding.

Livestock Grazing

Livestock are attracted to riparian areas and tend to concentrate along stream banks. If not properly managed, livestock grazing can degrade riparian areas, adversely impact the aquatic system, alter stream channel morphology, and influence the balance between surface and subsurface flow from the surrounding watershed. Grazing animals can reduce streambank stability by (1) the reduction of shear strength of stream banks through the removal of riparian vegetation and (2) the added shear stress on the stream banks by animal hooves. Improper livestock grazing can eliminate riparian vegetation through direct consumption and as a result of concentrated animal movement, which can compact the soil, harm plant roots, and break down woody vegetation. Channel aggradation or degradation can result, causing widening or incising of channels, and in dryland environments channel incision can lower the surrounding water tables (Clary and Webster 1990). Cattle, and their influence on riparian systems, can be particularly important agents for changing stream channel morphology (Trimble and Mendel 1995). A sequence of channel degradation that lowers the water table adjacent to a stream is illustrated in Figure 13.4.

The removal of grazing animals from riparian areas in eastern Oregon has been shown to reduce channel width and increase pool areas, but the time for recovery varies with site conditions (Magilligan and McDowell 1997). The process of stream channel recovery is illustrated in Figure 13.5.

Through proper management, livestock grazing can occur without detrimental effects on riparian systems. Stocking the grazing area with the appropriate number and type of grazing animals and preventing livestock from directly accessing the channel can prevent riparian and stream channel degradation. Piping or pumping water from streams to stock tanks located out of the riparian area allows animals to be watered

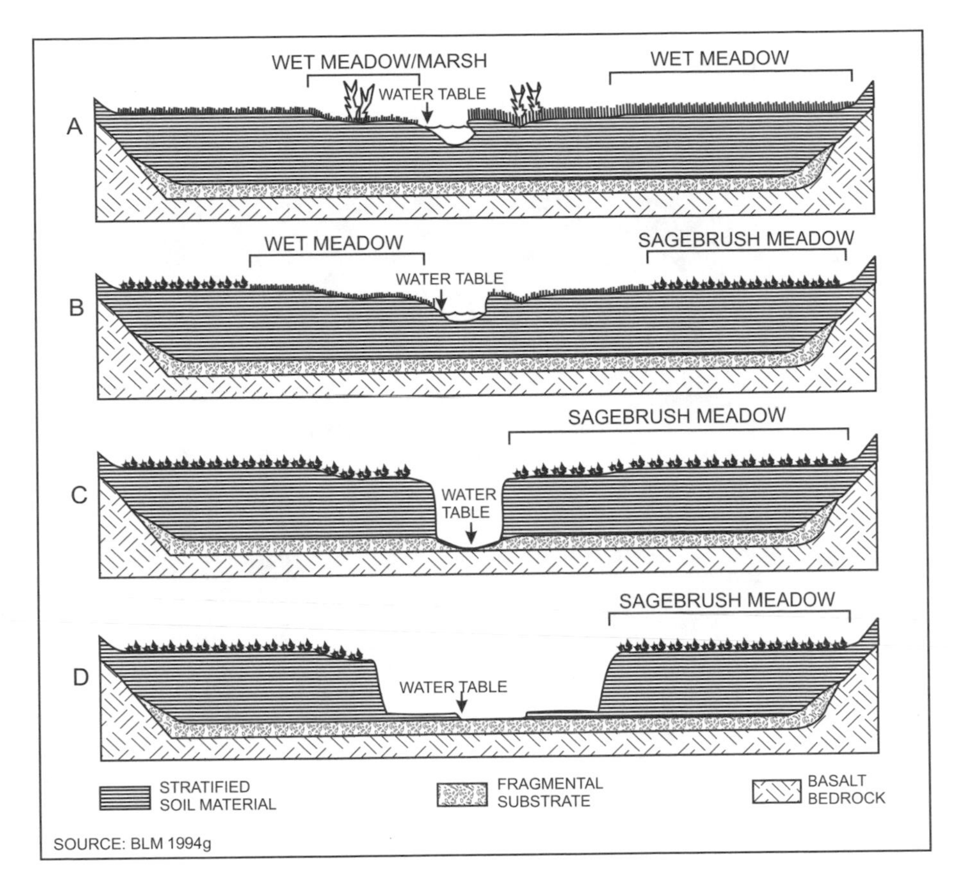

Figure 13.4. Effects of a degrading stream channel on the local water table (from BLM 1994).

without entering the channel. Good range management applies to the surrounding watershed area as well. Through good grazing management, excessive runoff, erosion and sedimentation, and riparian degradation can be minimized.

Timber Harvesting

Timber-harvesting activities include the felling, processing, extracting, sorting, loading, and hauling of timber products. Tree removal is an important activity that affects both the riparian area and the surrounding watershed. When riparian forest cover was removed along the Mississippi River, streambank erosion was increased three-fold (Burckhardt and Todd 1998). The loss of root strength, which adds to the resisting shear strength of the stream banks, is a major reason for the increased streambank erosion. When trees are removed, particularly those with high transpiration rates, the resulting wetter soils will have higher pore pressures, which can further reduce bank stability.

Other aspects of timber harvesting add to the stress on riparian systems. Construction of haul roads, landing areas, and skid trails can have negative impacts on riparian areas unless planned carefully to protect the riparian area (Mattson et al. 2000).

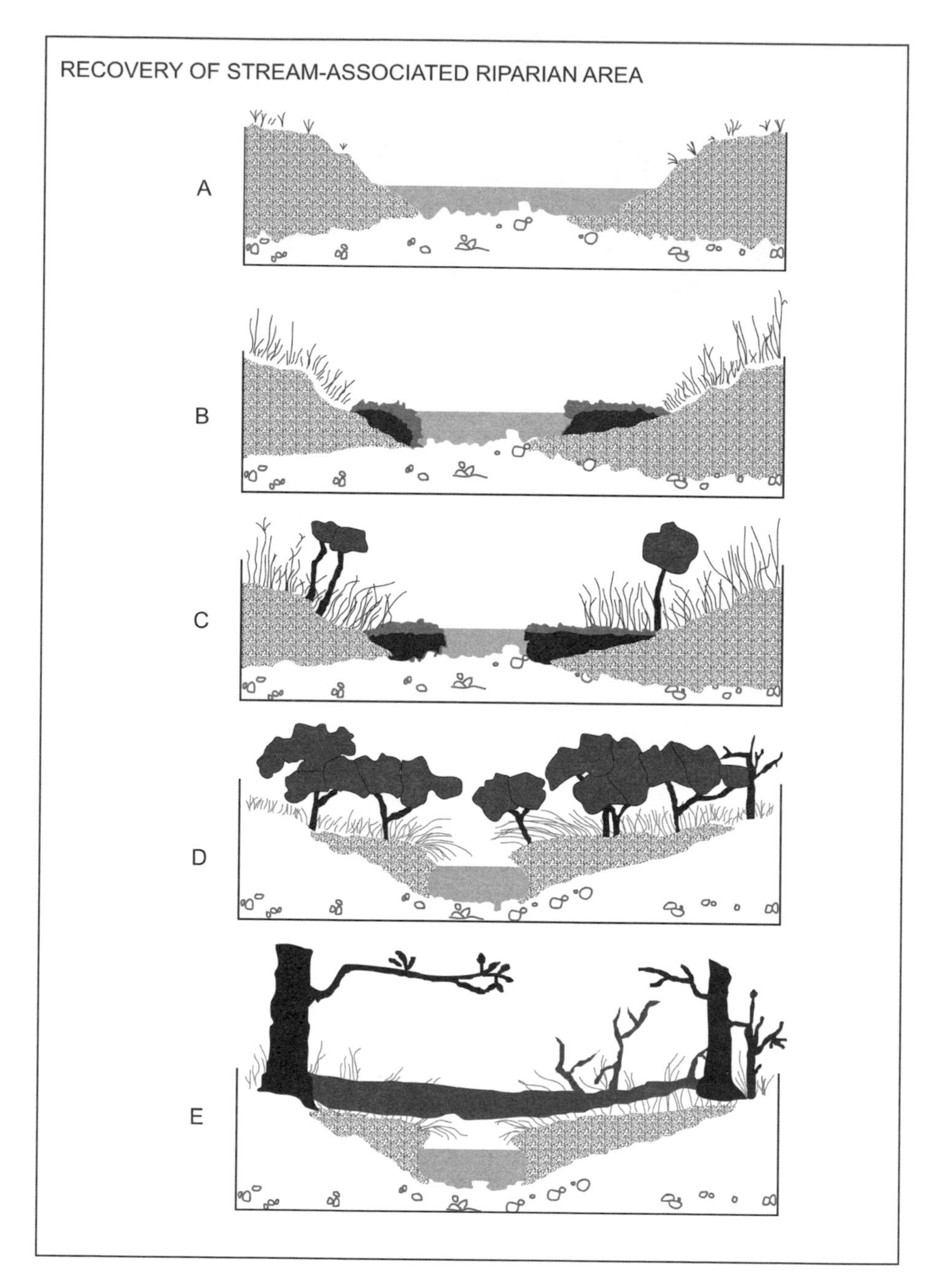

FIGURE 13.5. Recovery of a degraded riparian–stream channel system (from BLM 1994).

Road construction has been notoriously adverse to streams and associated riparian areas. Roads located on steep slopes, erodible soils, or stream crossings have the greatest potential for degrading water quality.

Water Use

The available water provided by streamflow through riparian areas has been in high demand since the settlement of North America. Some of the competing uses of the water include floodplain farming and groundwater pumping. Streamflow has also been altered by the construction of dams and reservoirs for flood protection and water development.

Floodplain farming and the development of large agrarian societies have occurred in the southwestern United States for thousands of years. Agriculture continues to utilize floodplain environments with ever-increasing efficiency, using modern-day technology. Increased floodplain farming, along with rapid urban sprawl, continues to impact the sustainability and health of many historic riparian habitats.

Groundwater is an important source of water in many areas. In early days, water was extracted from shallow depths using windmills. As pumping and irrigation technology improved, it became possible to extract water from ever-increasing depths. However, it was quickly discovered that sources of surface water were also necessary to support growing agricultural and urban needs. As a result, intense competition for groundwater has developed between urban development and the need to sustain perennial streamflow in riparian ecosystems.

Dams produce major environmental effects on streams and associated riparian habitat. The most immediate effects are the obstruction of migration routes for fish, the inundation of nearby upstream habitats, and the creation of reservoirs that trap sediments. The ponding of water by reservoirs inundates upstream riparian communities and leads to permanent loss of habitat. During the initial inundation process, if the trees are not removed, they become partially submerged, and the water-logged soil becomes anaerobic, leading to oxygen stress and death of the trees (Nilsson and Berggren 2000). If the trees remain and are submerged, they release CO_2, CH_4, N, and P. The added nutrients can lead to an increase in aquatic productivity and create the excessive growth of plants, which may reduce water quality. Eventually, new riparian areas develop along the new shorelines.

Less subtle and unpredictable downstream changes following dam construction include the effects on flood dynamics and nutrient transport. The major downstream effect can be a moderated flood peak and a reduced frequency of overbank flooding. Fluctuations in discharge can alter surface and groundwater interactions in the riparian zone. By trapping sediment, dams also change natural geomorphic processes downstream. Water released through gated structures or spillways contains less sediment than the stream carried prior to the dam; thus there is a greater capacity to pick up and transport sediment (see Chapter 9). Many downstream riparian communities no longer receive the watering, fertilizing, cleaning, and scouring of the channel that occurred with an unregulated stream; this all affects the species composition of the downstream riparian forests and can lead to the invasion of exotic species, the extinction of native riparian species, and reduction in biological diversity. Reservoir management (Chapter 16) needs to take these effects into account.

Management and Rehabilitation

The deterioration of riparian areas throughout the world requires that remedial treatments be applied to repair those that have been damaged and, in many cases, to redirect management to maintain and protect existing riparian areas. The management and improvement of riparian areas require considering not only the riparian area and associated stream, but also the surrounding watershed. The following discussion develops a framework to analyze current conditions within the watershed context and arrive at successful management solutions.

Watershed Condition and Riparian Health

The functional dependency and interrelationship of riparian areas with their surrounding watersheds need to be understood when managing riparian areas, particularly if rehabilitation activities are planned. This section provides a framework for discussing and analyzing these functional relationships. The two main factors used to describe the well-being of the watershed and the riparian area are watershed condition and riparian health.

The term *watershed condition* describes the state of a watershed. It comprises such factors as vegetative cover, flow regime, sediment and nutrient output, and site productivity (DeBano and Schmidt 1989). Although dense or sparse vegetation ground cover is sometimes equated with good or poor watershed condition, respectively, water storage on the watershed and the stream channel conditions that affect streamflow conveyance are also important. A watershed in good condition absorbs rainfall energies, maintains high infiltration rates of water into the soil, is managed in a condition that maintains its inherent water storage capacity, and releases storm water slowly into the channels. It also has a minimal density channel network necessary for conveying runoff from the watershed. A watershed in poor condition produces sediment-laden runoff water because a sparse vegetation cover permits detachment of soil particles, sealing of soil pores, and increased erosion. The condition of a watershed is important because it influences the quality, abundance, and stability of downstream riparian resources and habitat. This influence results from controlling production of sediment and nutrients, influencing streamflow, and modifying the distribution of chemicals throughout the environment.

Riparian health refers to the stage of vegetative, geomorphic, and hydrologic development, along with the degree of structural integrity, exhibited by a riparian area. The health of a riparian area also depends upon the condition of the surrounding watersheds (DeBano and Schmidt 1989). A healthy riparian area reflects a dynamic equilibrium between channel erosion and sediment deposition processes. In this condition, the riparian vegetation remains vigorous but does not encroach into the active mean annual flood channel, nor does streamflow rapidly expand stream meander cutting or growth of point bars through the riparian area, or impact it by eroding the channel bed. The equilibrium between channel deposition and downcutting by erosion in riparian areas was illustrated in Figure 9.2. Flows in excess of channel capacity overflow onto floodplains where healthy riparian vegetation and associated debris provide a substantial resistance to flow and act as filters, or traps, for sediment. During these bank overflows, opportunities are available for germination and establishment of certain riparian plant species. The healthy riparian area is also characterized by a shallow water table, which is necessary for sustaining healthy riparian plant communities on nearby floodplains.

Recognizing the interrelationship between watershed condition and riparian health provides a framework for managers to synchronize the improvement of the riparian areas and the surrounding watersheds (LaFayette and DeBano 1990). The surface runoff and erosion relationships between the riparian areas and the surrounding watershed, to a large extent, determine riparian health and watershed condition. The surface runoff and erosion relationships between the riparian areas and the surrounding watershed can be defined in a matrix containing four possible combinations of riparian health and watershed condition, namely: good watershed condition–healthy riparian; good watershed condition–unhealthy riparian; poor watershed condition–healthy riparian;

and poor watershed condition–unhealthy riparian. Some of these combinations are more likely to occur than others. For example, it is likely that healthy and productive riparian areas occur in watersheds that are in good condition. At the other end of the spectrum, it is also likely that unhealthy riparian areas are present when watersheds are in poor condition. It is possible, although less likely, to have other combinations of watershed condition and riparian health because of lag periods between the changes occurring on the watershed slopes and their effect on downstream riparian areas. For example, it is possible to have an unhealthy riparian area while the surrounding watershed is in good condition because concentrated overgrazing or human-induced disturbance has been concentrated in the riparian area. Over long periods of time, misuse of riparian areas can lead to channel incision and the extension of a gully network throughout the surrounding watershed. It is least likely to have a healthy riparian area present when the surrounding watershed is in poor condition, although installation of structures and exclusion of livestock grazing can temporarily improve riparian areas on watersheds that are generally in poor condition.

Rehabilitation

Rehabilitating riparian areas can be complex and requires good knowledge of the watershed and stream channel linkages previously discussed. There is no standardized handbook available to guide the manager when rehabilitating a particular degraded riparian area due to the extreme variation from one site to another. Treatments can be successful in a given situation, but when applied to a nearby area can be totally inappropriate and thereby result in failure. The manager, therefore, must rely on an adaptive approach to riparian rehabilitation. Flexibility of the approach used is necessary; the manager should rely on a basic understanding of the processes involved, make perceptive field observations and interpretations, know as much as possible about the history of the area being treated, and utilize existing expertise. In the end, personal experience is the best attribute one can have in developing rehabilitation programs.

Phases of Rehabilitation

The rehabilitation of riparian areas can be divided into five general phases: (1) overall planning, (2) project design, (3) implementation, (4) monitoring and evaluation, and (5) adaptive management and maintenance (LaFayette et al. 2000).

The planning phase establishes the extent of riparian degradation and the historic and/or present-day reasons for the present concerns. Understanding the historical function of the riparian area cannot be stressed enough, because seemingly unexplainable features in the analysis can become clear when viewed from a historical perspective. In this respect, historical ground level and aerial photographs are useful. The objectives and goals of the project need to be defined at this phase in order to give the project direction. A goal might be to reverse or mitigate the damage that has occurred. For example, the goal may be to establish a new equilibrium condition that supports a viable riparian area regardless of whether such a condition reflects the former riparian situation. Goals are usually broad and not site specific. Conversely, objectives are more site specific, and direct the actions of the project. An example would be to revegetate a disturbed area with native trees and shrubs that also provide wildlife habitat. Justification for the project and discussion of its economic and political dimensions need to be included in the analysis (see Chapter 19).

Project design can be as complex or simple as the situation demands, but the proper scale must be established so that the project is not too ambitious. The design should encompass both the riparian area and the surrounding watershed. Specific treatments should be carefully analyzed. Successful treatments used in the past may not be applicable to a current project. Managers should also be skeptical of "hot" trends because they may be more costly than effective. This does not mean that newer technologies do not have potential application; it means that new technologies should be tested before they are widely applied. When given a choice of alternative methods, use the least intrusive treatment.

Thought and planning should go into the implementation of a desired project. If it is a relatively large project, it might be profitable to utilize a local watershed-level coordinator to make sure all phases are carefully executed. Implementation should be based on a compatible balance between the environmental, economic, and social values and the technical requirements (see Chapter 19). Political, emotional, physical, and financial support can often be gained through the use of interested partnerships.

Monitoring and evaluation are critical components of any riparian-watershed improvement project. Project monitoring includes verifying that the designed features of the project have been adequately incorporated into the plan. Post-treatment monitoring is essential to establish the necessary maintenance activities and schedule to ensure continued success for the project. There are cases of project failures because monitoring and the necessary follow-up maintenance were not performed. Also, monitoring allows an evaluation of the strong points and shortcomings of the treatments used and provides information useful to future projects.

Adaptive management and maintenance should be coordinated closely with post-treatment monitoring. Often, it is more difficult to obtain funds for maintenance than it is to obtain financing for the project. It must be remembered, however, that all projects require regular maintenance to sustain them successfully into the future. It can be less costly to provide small maintenance and improvement to a project on a regular basis than to allow small inadequacies to develop into project-threatening situations.

Techniques for Rehabilitation

It is seldom that riparian systems are managed best for only a single use, for example, livestock production. A compromise form of management usually results in the greatest value to people. In many instances, riparian systems might have to be set aside as natural areas to preserve their values.

A host of physical and biological rehabilitation techniques are available for treating either the watershed or the stream, or both. Some techniques for prevention and control of surface and channel erosion processes are described in earlier chapters. Stream channel restoration can require major engineering work in reshaping the stream channel and reconnecting the channel to its floodplain. Techniques based on modern technology are emerging daily. More than a thousand references concerning wetland and riparian restoration (including planning, establishing goals and objectives, designing, and general techniques, along with specific references dealing with planting, fencing, land forming, installing instream devices, soil treatments, monitoring, evaluating success, and erosion control) have been summarized by Schneller-McDonald et al. (1990).

Buffer Strips

Buffer strips are important components when managing riparian systems. They are sometimes called "filter strips," "riparian zones," "protection strips," or "streamside management zones." One definition of buffer strips is that they are riparian lands maintained immediately adjacent to streams or lakes to protect water quality, fish habitat, and other resources (Belt et al. 1992). The term has also been loosely applied to designated protection zones managed by state and federal agencies. These vegetated buffer areas around water bodies either have been purposely left intact or have been planted to achieve the benefits derived from vegetation.

Functions

Buffer strips maintain the hydrologic, hydraulic, and ecological integrity of stream channels and associated soil and vegetation ecosystems. They furnish large woody debris, which enhances the hydraulic stability of the channel. They also provide a unique habitat for wildlife and fish by supplying food, cover, and thermal protection.

Buffer strips not only can directly enhance stream habitat for fish and other animals but can also produce other benefits such as improved water quality and scenic values. Buffer strips are used to minimize the effect of different activities on the surrounding watershed, including agricultural uses, grazing, use of prescribed fire, urban activity, and timber harvesting.

Designing Buffer Strips

Guidelines for the establishment of buffer strips have been developed to mitigate the effects of logging and agricultural practices. Natural forest buffer strips have high infiltration rates, which when coupled with the roughness of the soil surface and debris, transform overland flow from upslope areas into subsurface flow. Sediment from surface erosion becomes trapped, and waterborne solutes enter the soil, where subsequent uptake by buffer vegetation can occur. Therefore, buffer strips of natural riparian forest are often protected from logging to provide a buffer between upland logging impacts and downstream aquatic ecosystems. Belt et al. (1992) developed a manual for the design of forest riparian buffer strips for the western United States (Table 13.2). Other guidelines specify the width of the buffer strip as a function of the slope of the hillside adjacent to the stream channel (Table 13.3). Riparian buffer strips are not effective in controlling channelized flow that originates outside (upslope of) the buffer; special structural and vegetative treatments can be necessary to control gully erosion in these cases.

General guidelines have been established for designing riparian forest buffers in agricultural areas to reduce nonpoint pollution, as illustrated in Figure 13.6.

If riparian buffer zones are to be instrumental in protecting against nonpoint source pollution, their effectiveness should be related to physical properties and the nature of management on the upland area. Corner and Bassman (1993) suggest that a legal buffer zone width should be calculated as a function of physical parameters (e.g., slope, soil permeability, soil erodibility) and intensity of management practices rather than establishing a fixed distance.

Water Quality

Riparian buffers can reduce the amount of solar radiation reaching the surface of streams, thus allowing cooler water temperatures, which favor higher quality aquatic organisms and maintain higher levels of dissolved oxygen; this is often a major benefit of maintaining streamside vegetation, particularly in small headwater streams. Maintaining streamside buffers of riparian vegetation is also an effective way of reducing the velocity of runoff from upland areas, causing sediment and associated pollutants to be deposited within the buffer area. This filtering action keeps a portion of the sediment from being deposited in the

TABLE 13.2. Guidelines for buffer strips in the western United States

State	Stream Class	Buffer Strip Requirements: Width	Shade or Canopy	Leave Trees
Idaho	Class I*	Fixed minimum (75 ft)	75% current shade[a]	Yes; no./1000 ft dependent on stream width[b]
	Class II**	Fixed minimum (5 ft)	None	None
Washington	Types 1, 2, and 3*	Variable by stream width (5–100 ft)[c]	50%; 75% if temperature > 60°F	Yes; no./1000 ft dependent on stream width and bed material
	Type 4**	None	None	25/1000 ft > 6 in. diameter
California	Class I and Class II*	Variable by slope and stream class (50–200 ft)	50% overstory and/or understory; dependent on slope and stream class	Yes; no. to be determined by canopy density
	Class III**	None[d]	50% understory[e]	None[e]
Oregon	Class I*	Variable; three times stream width (25–100 ft)	50% existing canopy, 75% existing shade	Yes; no./1000 ft and basal area/1000 ft by stream width
	Class II special protection**	None[f]	75% existing shade	None

Source: Belt et al. 1992.
*Human water supply or fisheries use.
**Streams capable of sediment transport (California) or other influence (Idaho and Washington) or significant impact (Oregon) on downstream waters.
[a]In Idaho, the shade requirement is specifically designed to maintain stream temperatures.
[b]In Idaho, the leave tree requirement is specifically designed to provide for the recruitment of large organic debris (LOD).
[c]May range as high as 300 ft for some types of timber harvest.
[d]To be determined by field inspection.
[e]Residual vegetation must be sufficient to prevent degradation of downstream beneficial uses.
[f]In eastern Oregon, operators are required to "leave stabilization strips of undergrowth ... sufficient to prevent washing of sediment into Class I streams below."

stream channel. By reducing sedimentation, the materials attached to sediments, such as phosphorus and heavy metals and pesticides, are kept from entering water bodies.

Welsch (1991) reported that agricultural croplands and pasturelands/rangelands account for 38 and 26% of annual sediment deposition in surface waters, respectively, in the United States each year. Nitrogen inputs to surface waters from croplands and pasturelands/rangelands amount to 43 and 25%, respectively. Multispecies riparian buffer strips, using combinations of grasses, shrubs, and trees, have been developed to reduce agricultural nonpoint pollution in the Great Plains and upper Midwest. Fast-growing species of trees, such as willow (*Salix* spp.) and poplar (*Populus* spp.), are preferred so that functioning buffer strips can be established in a short period of time (Schultz et al. 1995).

TABLE 13.3. Guidelines for filter strip widths as a function of hillslope in Minnesota

Slope of Land between Activity and Water Body (%)	Recommended Width of Filter Strip (Slope Distance)[a] (feet)
0–10	50
11–20	51–70
21–40	71–110
41–70	111–150

Source: Minnesota Department of Natural Resources 1995.
[a]For roads, distance is measured from the edge of soil disturbance; for fills, distance is measured from the bottom of the fill slope.

OTHER RIPARIAN-RELATED TOPICS

Research that is contributing to the management of riparian areas includes the development of stream and watershed models and the evaluation of flood control activities. Additionally, the determination of instream flow requirements necessary to maintain viable riparian areas is becoming of increased interest in areas where streamflow is already fully committed to previous uses.

Stream and Watershed Models

A stream continuum model has been developed to describe food chain resources for organisms ranging from invertebrates to fish (Vannote et al. 1980). This model provides the ecological rationale for maintaining a protective vegetation area along streams and rivers of all sizes. The continuum concept is a framework for studying and managing streams as heterogeneous systems because useful generalizations can be developed.

The stream continuum model depicts a watershed as a landscape that extends from the smallest streams in the headwaters to the mouth of the river. Over this spatial arena, the physical and biological variables represent a continuous gradient in which a continuum of biological, physical, and geomorphological adjustments occur. As key parts of watersheds, riparian corridors possess a diverse array of species and environmental processes. This ecological diversity is related to variable flood regimes, geomorphic channel processes, altitudinal climate shifts, and upland influences on the fluvial corridor. The dynamic environment results in a variety of life cycles of organisms and a diversity of biogeochemical cycles and rates, as organisms adapt to disturbance regimes over broad spatial and temporal scales. Within this framework, riparian corridors play an essential role in water and landscape planning, in the restoration of aquatic systems, and in catalyzing institutional and societal cooperation for these efforts (Naiman et al. 1993).

Watershed models should also reflect the long-term influence of geology, climate, and topography as well as the shorter-term influences of vegetation. Flows resulting from climatic conditions create and maintain streams (Hill et al. 1991). When natural flow patterns are changed, fluvial processes change, and the condition of the valley, the stream, and all other ecological components must change as a consequence (Lotspeich 1980). For understanding stream processes, the watershed has been considered to have

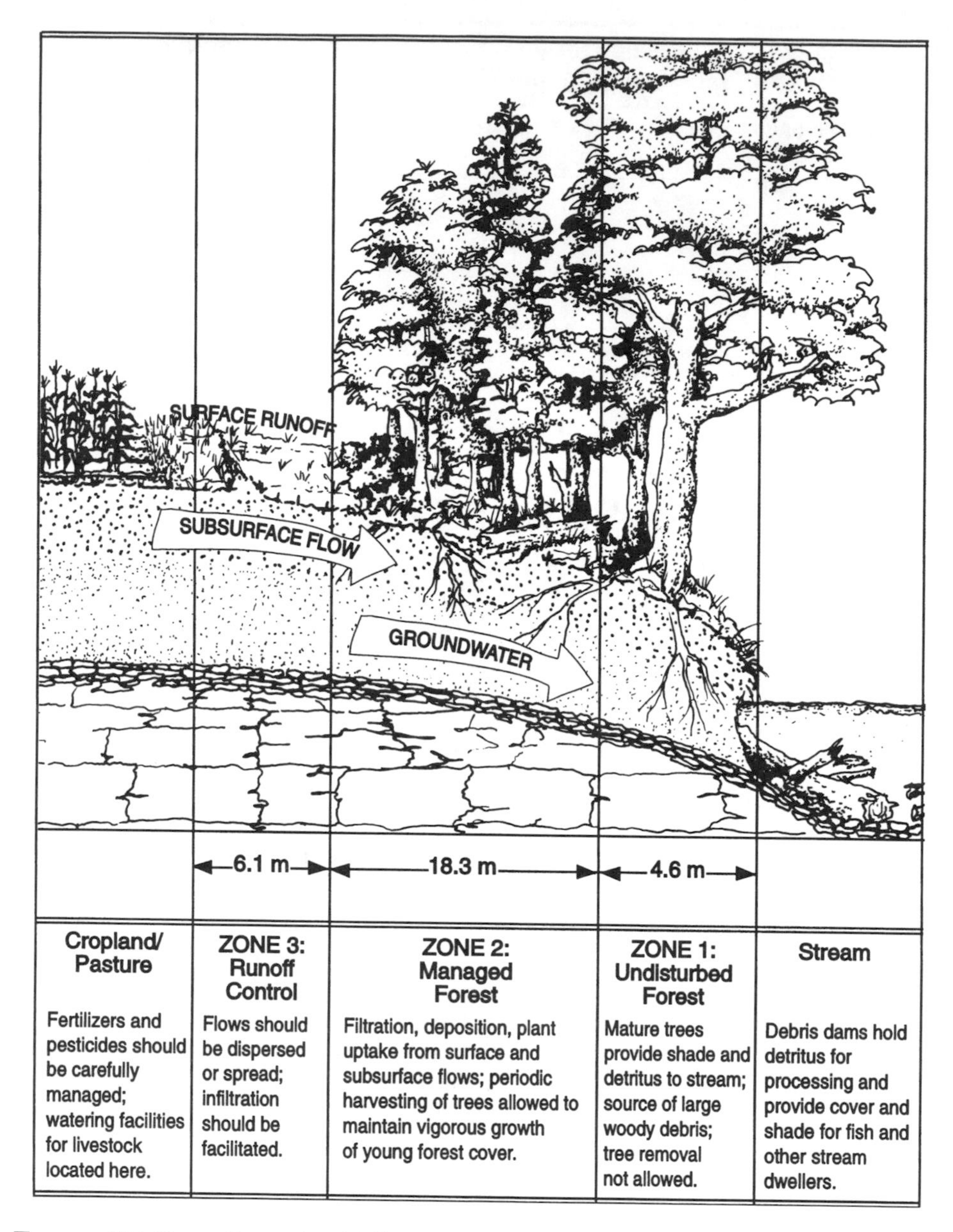

FIGURE 13.6. General guidelines for the management of riparian buffer strips (from Welsch 1991).

the following dimensions (Ward and Sanford 1989): (1) the longitudinal dimension from headwaters to mouth, (2) the lateral dimension extending beyond the channel boundaries, and (3) a vertical dimension resulting from the out-of-channel flows moving downward into the soil and groundwater. Each of these dimensions must be analyzed over time. Furthermore, within an ecosystem-geomorphic context, four types of streamflow regimes are recognized: instream flows, channel maintenance flows, riparian maintenance flows, and valley maintenance flows.

River Health

The term *river health* needs to be mentioned because it is currently attracting worldwide attention and is an important concept relating to riparian ecosystems. This term is similar to the concept of ecosystem health, and it incorporates both the ecological values and the human values into a philosophical framework (Boulton 1999). The ecological values component consists of ecological integrity (capacity to support/maintain natural and balanced, integrative adaptive biological systems) and resilience to stress (ability to recover following disturbance). The human values component includes goods (water supplies for irrigation and industry, clean water for domestic use, and an environment for recreation and spiritual renewal) and services (cleansing/detoxifying water, producing fish and shellfish, providing aesthetic pleasure, maintaining water supply, and storing/regenerating essential elements). The numbers and specific indicators to be used as criteria for river health are currently being discussed and debated.

Flood Control Effects

Different types of riparian systems exert different influences on flooding within a watershed. Forest vegetation along smaller headwater stream channels can provide LWD to channels, which helps reduce flow velocity. Stream stabilization that minimizes sediment delivery into the channel helps maintain stream channel capacity to transmit floodwaters.

In some instances where phreatophyte vegetation occurs extensively along floodplains, there are trade-offs that must be considered in evaluating the role of riparian vegetation. For example, such vegetation increases channel roughness, slowing the velocity of floodwaters and causing higher stages and, where development occurs in the floodplain, more frequent flood damages. However, the benefits of increased storage and reduced flow velocities in upstream areas will reduce flood frequency and damaging flows downstream and can improve downstream water quality as well.

Instream Flow Considerations

The concept of instream flow has become important in watershed management and in the management of riparian areas. *Instream flow* is the streamflow regime required to satisfy a mixture of conjunctive demands on water while it is in a stream channel (American Fisheries Society 1985). Instream flow requirements are discharge rates in a stream course that sustain some predetermined level.

Instream flow rights are legal entitlements to use surface water within a specified area of a stream channel for fish, wildlife, or recreational uses. This use must be nonconsumptive except for the normal needs of wildlife and vegetation. An instream flow right protects a designated flow through a specified reach of a stream from depletion by new water users; it is important where new upstream uses, developments, diversions, or transfers could threaten existing flows. Benefits of instream flow rights include protection of fish and the diversity of riparian plants and animals that live in or along the water (Kulakowski and Tellman 1990).

Most natural resource values in riparian areas depend either directly or indirectly on streamflow (Jackson and Long 1991). Direct requirements are those needed for benefits derived from the existence of surface water in channels, such things as aquatic habitat, wildlife drinking water, recreation water, and aesthetics. Indirect requirements refer to those needed for benefits derived from conditions created by flowing water. Indirect benefits include moist riparian soils, which (in turn) support water-dependent vegetation and habitat benefits associated with the morphology and physical composition of channels and floodplains. Most riparian values are derived from a combination of direct and indirect instream flow benefits.

Urban Riparian Areas

Riparian areas have not fared well during the process of urban development. Since ancient times, rivers and their riparian areas have attracted people and been the basis for the development of communities, villages, and larger cities. Through urban development, surface runoff increases, native vegetation on the watershed and in riparian areas is removed, and construction adds heavy sediment loads to stream channels, all of which can have cumulative, unwanted effects. One of the major problems facing local governments of urban communities involves finding an economically and environmentally acceptable solution to increased flooding of urban streams. Flood control typically includes stream channelization (i.e., widening, straightening, or deepening a stream channel). Channelization tends to adversely affect the physical and biological environment and to reduce the aesthetic quality of the stream.

Much of the physical damage done to streams in urban areas results from the "paving" of the watershed (Finkenbine et al. 2000). Paving reduces infiltration, preventing groundwater recharge, which (in turn) increases surface runoff. Stream sediment is increased in two ways: (1) fine sediment is increased as a result of construction activities, and (2) the impervious surface generates higher peak flows, which cause stream enlargement through bed and bank erosion. In addition, the removal of riparian vegetation eliminates the buffering effect of streamside vegetation and increases stream temperature. An important part of stream rehabilitation includes riparian plantings to increase the future amounts of LWD.

Summary

Riparian areas are the transitional zones between terrestrial and aquatic systems. They are found in nearly every geologic formation, geomorphic setting, and climatic regime. Consequently, the hydrology, vegetation, and soils are spatially and temporally variable. Riparian areas provide food, water, cover, and shade for fish, reptiles, amphibians, large and small mammals, and birds. Physical, biological, and chemical linkages in and among the stream, riparian vegetation, and the surrounding watershed are responsible for the continual flux of water, air masses, dissolved particulate matter, organisms and energy. Healthy riparian systems help maintain a stable stream channel and high quality water. Unfortunately, riparian areas and their surrounding watersheds have been, and currently are, exposed to both natural and human-induced disturbances. As a result, many of the riparian areas throughout the world have been damaged and are in need of remedial treatments.

After reading this chapter, you should be able to:

- Explain the hydrologic and water quality benefits of a healthy riparian system.
- Describe the hyporheic zone and explain how it functions and why is it important to aquatic food webs and fish habitat.

- Explain the role of large organic materials in stream dynamics and know the source of large woody materials in or adjacent to the stream channel.
- Explain why it is important to understand the historical and geologic setting when planning and implementing rehabilitation treatments.
- Discuss the reasons why carbon is an important nutrient in the riparian-aquatic ecosystem.
- Describe the linkages that occur among or within the terrestrial, riparian, and aquatic ecosystems.
- Outline the steps that should be used to initiate, implement, and maintain a riparian-watershed rehabilitation project.

References

American Fisheries Society. 1985. *Aquatic habitat inventory: Glossary and standard methods*, ed. W.T. Helm, 1–34. Washington, DC: Western Division, Habitat Inventory Committee.

Anderson, E.W. 1987. Riparian area definition—A viewpoint. *Rangelands* 9:70.

Baxter, C.V., and F.R. Hauer. 2000. Geomorphology, hyporheic exchange, and selection of spawning habitat by bull trout (*Salvelinus confluentus*). *Can. Jour. Fish. and Aquatic Sci.* 57:1470–1481.

Belt, F.H., J. O'Laughlin, and T. Merrill. 1992. *Design of forest riparian buffer strips for the protection of water quality: Analysis of scientific literature.* Idaho Forest, Wildlife and Range Policy Analysis Group Report 8. Moscow, ID: Univ. of Idaho.

Boulton, A.J. 1999. An overview of river health assessment: Philosophies, practice, problems and prognosis. *Freshwater Biology* 41:469–479.

———. 2000. River ecosystem health down under: Assessing ecological condition in riverine groundwater zones in Australia. *Ecosys. Health* 6:108–118.

Braudrick, C.A., and G.E. Grant. 2000. When do logs move in rivers? *Water Resour. Res.* 36:571–583.

Brunke, M., and J. Fischer. 1999. Hyporheic bacteria—Relationships to environmental gradients and invertebrates in a prealpine stream. *Achiv. fur Hydrogbiologie* 146:189–217.

Burckhardt, J.C., and B.L. Todd. 1998. Riparian forest effect on lateral stream channel migration in the glacial till plains. *J. Amer. Water Resour. Assoc.* 34(1): 179–184.

Bureau of Land Management (BLM). 1994. *Rangeland reform A94: Draft environmental impact statement.* Washington, DC: Department of the Interior.

Butler, D.R. 1995. *Zoogeomorphology: Animals as geomorphic agents.* New York: Cambridge University Press.

Clary, W.P., and B.F. Webster. 1990. Riparian grazing guidelines for the Intermountain Region. *Rangelands* 12:209–212.

Corner, R.A., and J.H. Bassman. 1993. Contribution of legal buffer zones to non-point source pollution abatement following timber harvesting in northeast Washington. In *Riparian management: Common threads and shared interests,* tech. coords. B. Tellman, H.J. Cortner, M.G. Wallace, L.F. DeBano, and R.H. Hamre, 245. USDA For. Serv. Gen. Tech. Rpt. RM-226.

Cummins, K.W., M.A. Wilzbach, D.M. Gates, J.B. Perry, and W.B. Taliaferro. 1989. Shredders and riparian vegetation. *Bioscience* 39:24–30.

Dahl, T.E. 1990. *Wetland losses in the United States: 1780s to 1980s.* Washington, DC: U.S. Department of Interior, Fish and Wildlife Service.

DeBano, L.F., D.G. Neary, and P.F. Ffolliott. 1998. *Fire's Effects on Ecosystems.* New York: John Wiley and Sons.

DeBano, L.F., and L.J. Schmidt. 1989. *Improving southwestern riparian areas through watershed management.* USDA For. Serv. Gen. Tech. Rep. RM-182.

DeGraff, R.M., and M. Yamasaki. 2000. Bird and mammal habitat in riparian areas. In *Riparian management in forests of the continental eastern United States,* ed. E.S. Verry, J.W. Hornbeck, and C.A. Dolloff, 139–156. New York: Lewis Publishers.

Dolloff, A. 1996. Ecological role of large woody debris in forest streams. In *Proceedings of National Hydrology Workshop,* ed. D.G. Neary, K.C. Ross, and S.S. Coleman, eds), 54–57. USDA For. Serv. Gen. Tech. Rep. RM-GTR-279.

Fernald, A.G., P.J. Wigington, and D.H. Landers. 2001. Transient storage and hyporheic flow along the Willamette River, Oregon: Field measurements and model estimates. *Water Resour. Res.* 37:1681–1694.

Finkenbine, J.K., J.W. Atwater, and D.S. Mavinic. 2000. Stream health after urbanization. *Jour. Amer. Water Resour. Assoc.* 36:1149–1660.

Franken, R.J.M., R.G. Storey, and D.D. Williams. 2001. Biological, chemical and physical characteristics of downwelling and upwelling zones in the hyporheic zone of a north-temperate stream. *Hydrobiologia* 444 (1–3): 185–195.

Geist, D.R. 2000. Hyporheic discharge of river water into fall chinook salmon (*Oncorhynchus tshawytscha*) spawning areas in the Hanford Reach, Columbia River. *Can. Jour. Fish. and Aquatic Sci.* 57:1647–1656.

Giller, P.S., and B. Malmqvist. 1998. *The biology of streams and rivers*. New York: Oxford University Press.

Heede, B.H., M.D. Harvey, and J.R. Laird. 1988. Sediment delivery linkages in a chaparral watershed following a wildfire. *Environ. Manage.* 12:349–358.

Helms, J.A., ed. 1998. *The dictionary of forestry*. Bethesda, MD: The Society of American Foresters.

Hering, D., M. Mutz, and M. Reich. 2000. Woody debris research in Germany—An introduction. *Int. Rev. Hydrobiology* 85:1–3.

Hey, D.L. and J.A. Wickencamp. 1998. Effect of wetlands on modulating hydrologic regimes in nine Wisconsin watersheds. In *Water resources and the urban environment*. Weston, VA: The American Society of Civil Engineers.

Hill, M.T., W.S. Platts, and R.L. Beschta. 1991. Ecological and geomorphological concepts for instream and out-of-channel flow requirements. *Rivers* 2:198–210.

Hinkle, S.R., J.H. Duff, F.J. Tiska, A. Laenen, E.B Gates, K.E. Bencala, D.A. Wentz, and S.R. Silva. 2001. Linking hyporheic flow and nitrogen cycling near the Willamette River—A large river in Oregon, USA. *J. Hydro.* 244:157–180.

Holmes, R.M. 2000. The importance of ground water to stream ecosystem function. In *Streams and Ground Waters,* ed. J.B. Jones and P.J. Mulholland, 137–148. New York: Academic Press.

Horton, J.L., T. Kolb, and S.C. Hart. 2001. Responses of riparian trees to interannual variation in ground water depth in a semi-arid river basin. *Plant Cell and Environ.* 24:293–304.

Horton, J.S., and C.J. Campbell. 1974. *Management of phreatophyte and riparian vegetation for maximum multiple use values*. USDA For. Serv. Res. Pap. RM-117.

Ilhardt, B.L., E.S. Verry, and G.J. Palik. 2000. Defining riparian areas. In *Riparian Management in Forests of the Continental Eastern United States,* ed. E.S. Verry, J.W. Hornbeck, and C.A. Dolloff, 23–42. New York: Lewis Publishers.

Jackson, W.L., and B.A. Long. 1991. Southwest riparian instream flow issues and requirements. In *Riparian issues: An interdisciplinary symposium on Arizona's instream flows.* Tucson: Arizona Hydrological Society and Soil and Conservation Society.

Jurgensen, M.F., A.E. Harvey, and T.B. Jain. 1997. Impacts of timber harvesting on soil organic matter, nitrogen, productivity, and health of Inland Northwest Forests. *Forest Science* 43:234–251.

Kaplan, L.A., and J.D. Newbold. 2000. Surface and subsurface dissolved organic carbon. In *Streams and Ground Waters,* ed. J.B. Jones and P.J. Mulholland, 237–258. New York: Academic Press.

Kovalchik, B.L., and L.A. Chitwood. 1990. Use of geomorphology in the classification of riparian plant associations in mountainous landscapes of central Oregon, U.S.A. *For. Ecol. and Manage.* 33/34:405–418.

Kulakowski, L., and B. Tellman. 1990. *Instream flow rights: A strategy to protect Arizona's streams*. Water Resour. Res. Center Issue Paper 6. Tucson: Univ. of Arizona.

LaFayette, R.A., and L.F. DeBano. 1990. Watershed condition and riparian health: Linkages. *In Watershed Planning and Analysis in Action*, 473–484. Proceedings of IR Conference/Watershed Mgt/IR Div/ASE, Durango, CO.

LaFayette, R.A., J. Bernard, and D. Brady. 2000. Riparian Restoration. In *Riparian management in forests of the continental eastern United States,* ed. E.S. Verry, J.W. Hornbeck, and C.A. Dolloff, 303–314. New York: Lewis Publishers.

Lapin, M., and B.V. Barnes. 1995. Using the landscape ecosystem approach to assess species and ecosystem diversity. *Cons. Bio.* 9:1148–1158.

Lienkaemper, F.W., and F.J. Swanson. 1987. Dynamics of large woody debris in streams in old-growth Douglas-fir forests. *Can. J. For. Res.* 17:150–156.

Lotspeich, F.B. 1980. Watersheds as the basic ecosystem: This conceptual framework provides a basis for a natural classification system. *Water Resour. Bull.* 16:581–586.

Magilligan, F.J., and P.F. McDowell. 1997. Stream channel adjustments following elimination of cattle grazing. *J. Amer. Water Resour. Assoc.* 33(4): 867–878.

Martí, E., S.G. Fisher, J.D. Schade, and N.B. Grimm. 2000. Flood frequency and stream-riparian linkages in arid lands. In *Streams and Ground Waters,* ed. J.B. Jones and P.J. Mulholland, 111–136. New York: Academic Press.

Maser, C., and J.R. Sedell. 1994. *From the forest to the sea: The ecology of wood in streams, rivers, estuaries, and oceans.* Delray Beach, FL: St. Lucie Press.

Mattson, J.A., J.E. Baumgras, C.R. Blinn, and M.A. Thompson. 2000. Harvesting options for riparian areas. In *Riparian management in forests of the continental eastern United States,* ed. E.S. Verry, J.W. Hornbeck, and C.A. Dolloff, 255–272. New York: Lewis Publishers.

Meinzer, O.D. 1923. Outline of ground-water hydrology, with definitions. *USDI Geol. Survey Water Supply Pap.* 494:99–114.

Miller, J., D. Germanoski, K. Waltman, R. Tausch, and J. Chambers. 2001. Influence of late Holocene hillslope process and landforms on modern channel dynamics in upland watersheds of central Nevada. *Geomorphology* 38:373–391.

Minnesota Department of Natural Resources. 1995. *Protecting water quality and wetlands in forest management—Best management practices in Minnesota.* St. Paul.

Naiman, R.J., H. Decàmps, J. Pastor, and C.A. Johnston. 1988. The potential importance of boundaries to fluvial ecosystems. *J. North Amer. Benth. Soc.* 7:589–592.

Naiman, R.J., H. Decàmps, and M. Pollock. 1993. The role of riparian corridors in maintaining regional biodiversity. *Ecol. Appl.* 3:209–212.

Nilsson, C., and K. Berggren. 2000. Alterations of riparian ecosystems caused by river regulation. *BioScience* 50:783–792.

Page-Dumroese, D.S., A.E. Harvey, M.F. Jurgensen, and R.T. Graham. 1991. Organic matter function in the Inland Northwest forest soil systems. In: *Management and productivity of western montane forest soils,* 95–100. USDA For. Serv. Gen. Tech. Rep. INT-280.

Pauley, T.K., J.C. Mitchell, R.R. Buech, and J.J. Moriarty. 2000. Ecology and management of riparian habitats for amphibians and reptiles. In *Riparian management in forests of the continental eastern United States,* ed. E.S. Verry, J.W. Hornbeck, and C.A. Dolloff, 169–192. New York: Lewis Publishers.

Platts, W., C. Armour, G.D. Booth, M. Bryant, J.L. Bufford, P. Cuplin, S. Jensen, G.W. Lienkaemper, G.W. Minshall, S.B. Monsen, R. Nelson, J.R. Sedell, and J.S. Tuhy. 1987. *Methods for evaluating riparian habitats with applications to management.* USDA For. Serv. Gen. Tech. Rep. INT-221.

Ritter, D.F. 1986. *Process geomorphology*, 2nd ed. Dubuque, IA: W.C. Brown.

Rosgen, D.L. 1994. A classification of natural rivers. *Catena* 22:169–199.

Schneller-McDonald, K., L.S. Ischinger, and G.T. Auble. 1990. *Wetland creation and management: Description and summary of the literature.* USDI Fish and Wildlife Service. Biological Report 90(3).

Schultz, R.C., T.M. Isenhart, and J.P. Colletti. 1995. Riparian buffer systems in crops and rangelands. *In Agroforestry and sustainable systems: Symposium proceedings,* ed. W.J. Rietveld, 13–27. USDA For. Serv. Gen. Tech. Rep. RM-GTR-261.

Sorenson, J.A. 1989. Managing wildlife associations within riparian systems. In *Proceedings of the California riparian systems conference: Protection, management, and restoration in the 1990's,* tech. coord. Abell, D.L., 305. USDA For. Serv. Gen. Tech. Rep. PSW-110.

Szaro, R.C. 1981. Bird population responses to converting chaparral to grassland and riparian habitats. *Southwest. Nat.* 26:251–256.

Tabacchi, E., L. Lambs, H. Guilloy, A.M. Planty-Tabacchi, E. Muller, and H. Decàmps. 2000. Impacts of riparian vegetation on hydrological processes. *Hydrol. Proc.* 14:2959–2976.

Thomas, J.W., C. Maser, and J.E. Rodiek. 1979. *Wildlife habitats in managed rangelands—The Great Basin: Southeastern Oregon riparian zones.* USDA For. Serv. Gen. Tech. Rep. PNW-80.

Trimble, S.W., and A.C. Mendel. 1995. The cow as a geomorphic agent—A critical review. *Geomorphology* 13:233–253.

Vannote, R.L., G.W. Minshall, K.W. Cummins, J.R. Sedell, and C.E. Cushing. 1980. The river continuum concept. *Can. J. Fish. Aquatic Sci.* 37:130–137.

Ward, J.V., and J.A. Sanford. 1989. Riverine ecosystems: The influence of man on catchment dynamics and fish ecology. In *Proceedings of the international large river symposium,* ed. D.P. Dodge, 56–64. Canadian Special Publication of Fisheries and Aquatic Sciences 106. Ottawa: Department of Fisheries and Oceans.

Welsch, D.J. 1991. *Riparian forest buffers: Function and design for protection and enhancement of water resources.* USDA For. Serv. NA-PR-07-91.

Wilson, M.F., and K.C. Halupka. 1995. Anadromous fish as keystone species in vertebrate communities. *Cons. Biol.* 9:489–497.

CHAPTER 14 Wetland Hydrology and Management

INTRODUCTION

Wetlands, like riparian systems discussed in Chapter 13, are integral and important components of watersheds. Wetlands make up only a small percentage of the total land area in most parts of the world, yet their hydrologic and biological functions must be considered when determining water and other resource management goals for watersheds and entire river basins. Both wetland and riparian systems are basically plant and animal communities affected by the regular presence of water. Although the terms *wetland* and *riparian* are frequently used interchangeably, it is important to distinguish between the two from both a physical-biological and a regulatory standpoint.

Wetlands are areas inundated or saturated by surface or groundwater at a frequency and duration sufficient to support, under normal circumstances, a prevalence of vegetation typically adapted to saturated soil conditions. As a result, wetlands have unique vegetation (hydrophytes) and soils (hydric soils) that distinguish them from adjacent uplands. Riparian areas can be considered as the interface between terrestrial systems and either bodies of water or wetlands. However, some definitions of riparian systems or areas include wetlands.

Often, wetlands form where the groundwater table intersects the surface either permanently or frequently enough that hydric, reduced soils and hydric vegetation, such as sedges and rushes, are common. Wetlands include marshes, swamps, lakeshores, peatlands (bogs, muskegs), wet meadows, cienegas, and estuaries. They can also occur in depressions in the landscape, where surface and subsurface water is collected and where its downward percolation to deeper groundwater is impeded.

Natural or well-managed wetlands provide values and benefits far in excess of the land area that they occupy. Wetlands were once considered by many people to be poorly productive swamplands of little value unless drained. In many parts of the world, policies were established to drain wetlands to increase land area suitable for agricultural crop production and, in some areas, to reduce malaria and other mosquito-borne diseases. Wetlands are now recognized as ecosystems that perform important hydrologic and biological functions and that contain many unique plant and animal communities. Unfortunately, the result of past land use practices has been a net loss of wetlands and their biological attributes and hydrologic functions. Even where wetlands have not been drained, many have become degraded.

Wetlands have value in terms of biodiversity, supporting a variety of plants and animals that are not found elsewhere. They provide valuable habitat for waterfowl species, an attribute that has led to extensive wetland preservation programs through sportsmen groups in North America. Wetlands play other important, but less obvious, hydrologic roles relating to enhanced water quality, flood peak attenuation, and reduced erosion and sediment transport. Since the 1980s, efforts to protect and restore many wetlands, including wetland protection laws in the United States, have focused public attention on the value of these ecosystems. The challenge is how to balance the need for protecting certain wetlands and their many attributes while still producing some of the multiple products from these lands that people desire. The role and importance of wetlands in terms of biodiversity, wildlife habitat, and hydrologic function are related to the types of wetlands that exist on a watershed.

Wetland Types

Wetlands occur in many landforms and climates, as streamside or riverine communities, wet meadows, depressional wetlands, peatlands, playas, saltwater marshes, forested swamplands, and coastal mangrove forests. Although there are several classification schemes for wetlands, none is universally accepted; therefore, these various schemes are not discussed in this chapter. We will discuss some general types of wetlands that are important in the management of watersheds in various parts of the world. First, however, we will describe in general inland and coastal wetland types.

Inland Wetlands

A variety of wetlands exist in watersheds that are not located along coastal areas and, therefore, are not influenced by saltwater:

- Freshwater marshes, also called depressional wetlands, occur in depressions in the landscape or where changes in the surface topography result in groundwater discharge. The vegetation in these areas can be a variety of hydric plants such as cattails, reeds, grasses, sedges, and in some cases trees.
- Peatlands are areas with shallow water tables resulting from abundant moisture, either where annual precipitation exceeds annual potential evapotranspiration (ET) or where depressions concentrate water and where the land is saturated year-round so that organic soils accumulate. Some peatlands are forested and some are not.
- Riverine wetlands occur along streams and rivers and in floodplains that are flooded periodically but can be dry during parts of the year.
- Deep-water swamps occur as woody wetlands in many parts of the world where flooding is more pronounced and water above the soil surface is normal.
- Fringe wetlands occur along lake shores or ponds.
- Playas are wetlands sustained by basin discharge in closed basins.
- Southwestern cienegas and wet meadows.

Coastal Wetlands

Many types of wetlands occur in coastal areas and are influenced by saltwater and, sometimes, freshwater runoff or river discharges. Many coastal wetlands provide valuable food, spawning areas, and general habitat for fish that spend part or all of their lives in saltwater. These are areas that can be affected by changes in discharge, sediment, and pollutants from rivers and coastal areas. They, therefore, can be affected either positively or negatively by upland watershed use and management. Coastal wetlands include:

- Tidal salt marshes, which are often dominated by salt-tolerant grasses and annual and perennial broad-leaved plants and are influenced by tidal action.
- Mangrove wetlands, which occur in subtropical and tropical areas, are productive forested wetlands in tidal zones and at the mouths of rivers and streams.

Coastal wetlands share a common history with many of the inland wetlands—they have historically been viewed as wastelands, of little importance. As a result, many of these wetlands have been exploited and/or destroyed over time. The importance of coastal wetlands cannot be overstated, and most are now recognized as critical systems for fishery production and as important components of ocean ecosystems. A detailed discussion of these systems is beyond the scope of this book, and, therefore, our discussion will concentrate on selected inland wetlands.

Wetland Management Issues

Not all of the wetland types and their management will be discussed in detail here. Rather, we will concentrate our discussion on wetlands of North America, effects of management, and wetland issues that have been of interest in watershed management activities.

Wetlands of the Southwestern United States

Hendrickson and Minckley (1984) recognized three types of wetlands that have received considerable attention from watershed managers in the southwestern United States—wet meadows, cienegas, and riverine wetlands. *Wet meadows* have hydric soils and experience low-velocity surface and subsurface flows; vegetation is dominated by grasses, sedges, and rushes. In higher elevations, they are associated with snowmelt runoff and are characterized by cyclic freezing and thawing. *Cienegas* occur in the middle elevations, trapping organic materials from the adjacent aquatic ecosystems, and are associated with permanently saturated and highly reduced soils (Box 14.1). *Riverine wetlands* occur around low-elevation oxbows, levees, and stream margins and are transitory in nature. These wetlands can form whenever groundwater becomes perched above clays (Mitsch and Gosselink 1993) or where coarse sediments deposited by floods impound a stream. They can also result from springs that occur along geologic fractures or in surface depressions that intercept the water table. Unlike in riparian areas, flow is slow through riverine wetlands, although it is also permanent.

Northern Prairie Wetlands of the Upper Midwestern United States and Canada

The northern prairie wetlands, extending from Alberta, Canada, through southwestern Manitoba, and southeast into Iowa in the United States, have been in the forefront of

Box 14.1

Cienegas in the Western United States

In the western United States, cienegas are important ecosystems. *Cienega* is said to be a contraction of "cien aguas," "a hundred fountains (or springs)." They often occur in headwater situations where water rises to the surface in multiple sites, resulting in a unique type of well-watered flat or valley. These wetlands are located between 1000 and 2000 m in elevation and are characterized by permanently saturated, highly organic, reducing soils and are dominated by sedges. Cienegas and other marshlands in the southwestern United States have decreased greatly during the past century (Hendrickson and Minckley 1984). Cultural impacts have been diverse and are not well documented. While factors such as grazing and streambed modification have contributed substantially to their destruction, climate change can also be involved. Many degraded cienegas could be restored to predisturbance condition by provision of a constant water supply and amelioration of catastrophic flooding events.

wetland issues for decades. Highly valued by waterfowl hunters and wildlife managers, these wetlands have also been subjected to widespread drainage, largely for purposes of expanding cultivated agriculture. Widespread flooding in the Upper Mississippi River basin in 1993 and 2001, and the Red River of the North basin in 1997 raised questions about the extent to which the loss of wetlands in the region may have contributed to these floods. In the Minnesota River basin, > 80% of its presettlement wetlands have been lost to development; the portion of the Upper Red River of the North basin that lies in the state of Minnesota has < 15% of its presettlement wetlands remaining. Therefore, not only has flood storage been lost, but one of the world's most productive habitats for waterfowl has been substantially reduced.

Most wetlands in this region occur in the glaciated prairie of North America and consist largely of the important prairie pothole wetlands. While efforts are underway to restore many of the prairie pothole wetlands for waterfowl production, agricultural drainage programs continue to expand. Benefits of these wetlands for wildlife habitat, maintaining water quality, and flood mitigation are weighed against the benefits of agricultural production, urban development, and other activities on the landscape. The hydrologic role of these prairie pothole wetlands is discussed later in this chapter.

Peatlands

Peatlands are vast areas covering more than 165 million ha globally. Development in many forms is being planned or undertaken in the extensive peatlands of North America and Europe. Peatlands cover approximately 21 and 15 million ha in the United States and Canada, respectively, and are considered important for potential energy production, agriculture, wood products, and other purposes. The former USSR and Finland, with 92 and 14 million ha of peatlands, respectively, have extensively developed their peatlands for peat production, wood products, and agriculture. Peatlands have been drained to enhance the productivity of forests, and peat has been extracted for energy and for horticultural purposes. Most peatlands have limited agricultural value because of low productivity and a short growing season in the northern latitudes.

Peatlands are characterized by deep organic soils, flat topography, shallow water tables, and unique plant communities that reflect the type of water (either regional groundwater or perched water tables derived from rainfall and snowmelt) that sustains these systems. Usually found where annual precipitation exceeds annual potential ET, peatlands develop where water moves very slowly, which results in anaerobic conditions. As a result, organic soils can accumulate to several meters in some areas.

Alaska contains vast areas of peatlands and contains the greatest area of wetlands in the western United States; 74 million ha of wetland area make up 45% of the total land area. A major portion of Alaska's wetlands, however, is underlain by permafrost and thus does not support all the biological or physical functions that wetlands do in nonarctic climates. Wetlands over permafrost have little interaction in terms of recharging or discharging aquifers, and they have little value for flood control or water storage.

Hydrologic Functions of Wetlands

Wetlands are often touted as being important groundwater recharge zones and systems that attenuate flood peaks and sustain streamflow during the dry season. Likewise, wetlands have been represented as functioning like the "kidneys" of a watershed—cleansing or purifying water that enters and then releasing water of a higher quality. As a result of such claims, there has been concern about the loss of wetlands, and active programs have been initiated to restore or create wetlands. To what extent are such claims true? To address such questions, we will first consider the hydrologic conditions under which wetlands form and then review the hydrologic processes and functions of wetlands under different settings. Finally, we will consider the effects of wetland loss on the quantity and quality of groundwater and flow regimes of streams and rivers.

Hydrologic Conditions for Wetland Establishment

Wetlands form where (1) there is a persistent excess of water at the earth's surface and (2) the topography and climatic regime result in slow-moving water (Verry 1997). Excess water occurs at the earth's surface under conditions where annual precipitation exceeds annual potential ET, where the water table of groundwater systems intersects the surface, or where there are depressions in the landscape that collect runoff or subsurface flow. Low topographic relief or depressions result in the slow movement of water, which is conducive to anaerobic conditions in the soils, resulting in the formation of hydric soils and the development of unique plant communities that are adapted to such conditions. Wetlands typically have a poorly defined or nonexistent channel drainage system; drainage density is very low in contrast to upland areas of a watershed. As a result of slow-moving water and saturated soils, dead organic material tends to accumulate; decomposition of organic matter results in even further reduced oxygen levels in the zone of saturation.

The type of wetlands that develop is further determined by climatic regime and the source of water that forms the wetland. The source of water, its quality, and its periodicity all affect wetland development. Although precipitation is the ultimate source of water to all inland wetlands, the residence time of precipitation in the soil and geologic strata prior to entering a wetland site will dictate water chemistry and the types of plant communities that develop. Regional groundwater-fed wetlands tend to be nutrient rich, their development being a function of residence time and the types of geologic strata from which the water originates. Some sources of groundwater are more calcium rich than others, a characteristic that can influence the type of wetland plants that develop. If the water sources are primarily precipitation and surface runoff, the wetland water will generally have lower nutrient content and a lower pH than those that are fed by the more mineral rich regional groundwater. *Depressional wetlands* and those with perched water tables are examples of more nutrient poor wetlands. However, surface runoff from agricultural or urban lands can bring with it excessive nutrients that can affect the productivity of vegetative communities. Furthermore, wetlands occurring near oceans can receive high concentrations of sodium chloride. All these factors affect the types of plants and overall productivity of biomass in wetlands. The vegetation that develops on wetlands can be forests, shrubs, mosses, grasses, sedges, and other hydrophytes.

From the above, it is clear that the role of groundwater is a key factor in determining the type of wetland that develops. Wetlands can develop in dry environments, where water collects in depressions and maintains a persistent, saturated soil condition. Wetlands most often are low-lying areas that are connected with groundwater. In some cases, wetlands are maintained by their own water table, which in essence is a perched groundwater system. In any case, the water table of wetlands is normally above, at, or near the ground surface throughout the year. Examples of the various surface-groundwater relationships for wetland formation are illustrated in Figure 14.1.

The flow-through type wetland (Fig. 14.1A) maintains a fairly stable water table in the wetland throughout the year because of the direct connection with the regional groundwater source. In peatlands, this type would be typical of minerotrophic (mineral rich) fens, which maintain a more diverse plant community than ombrotrophic (mineral poor) bogs where water tables are maintained by precipitation and perched above the regional water table (as in the case of Fig. 14.1D). Raised bogs can also form on top of fen peatlands, a situation in which a perched water table lies immediately above the regional water table. The vegetation in these raised bogs, however, is not in direct contact with the regional groundwater and, as a result, exhibits the same mineral poor characteristics as a perched bog.

Examples of depressional wetland formation are situations where the groundwater flows into a depression, then when the water table in the wetland is sufficiently raised, the wetland discharges water from a channel outlet (Fig. 14.1B). In drier environments, a depressional wetland can form from seasonal precipitation excess and surface runoff and can either discharge to groundwater (from the edges of the wetland) during periods of high water tables, as in Figure 14.1C, or maintain a perched water table above the regional groundwater (Fig. 14.1D). A point to consider with depressional and perched wetlands is the linkage mechanisms between the wetland water table and the regional groundwater table. These wetlands form over time because downward flow of water is impeded. Therefore, one would not expect to see significant exchanges of water vertically through the bottom of these wetlands; most likely the greatest exchange would occur along the edges of these wetlands during wet periods when the water table is raised. If there were an active exchange of water, then one would question how the wetlands were able to form in the first place. For example, depressional wetlands that are elevated above the regional groundwater table would not be able to form under deep sandy soils. Such depressions would represent focused recharge areas for groundwater, but by virtue of the fact that there is an active downward percolation of water, the nec-

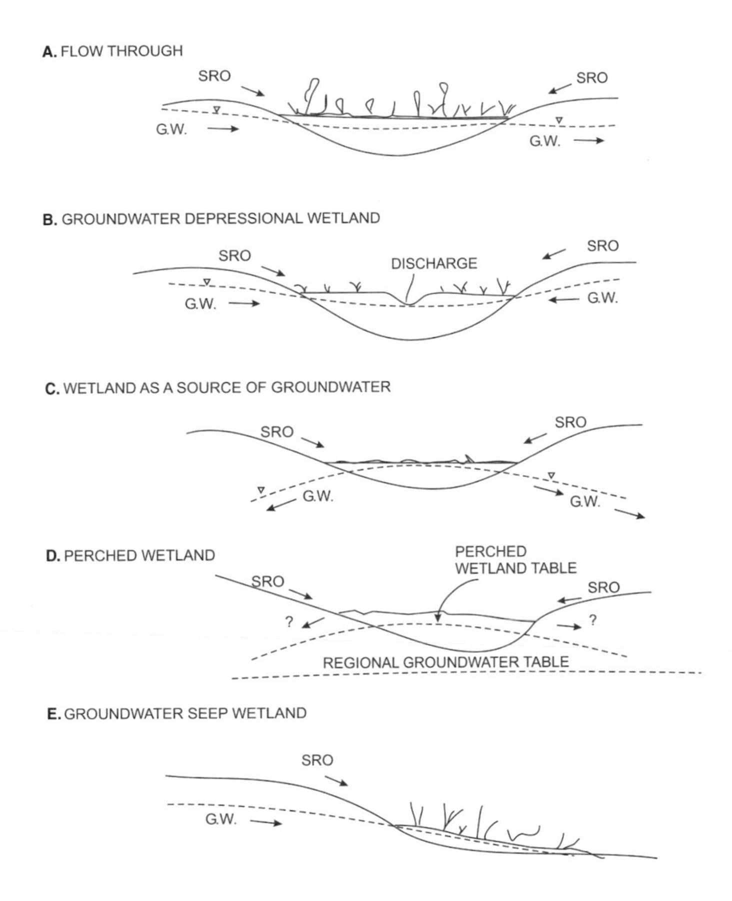

FIGURE 14.1. Surface and groundwater relationships that form wetlands (modified from Mitsch and Gosselink 1993).

essary water table and saturated soil conditions necessary to form a wetland would not likely be maintained. Once such depressions fill in with finer sediments, an impeding layer can form, and a water table could be maintained; but then the active recharge to groundwater is diminished.

Wetlands can also form in areas where springs or seeps occur (Fig. 14.1E). This same process can occur along lakes and coastal areas where the groundwater discharges to the water body through an adjacent, low-lying area.

Water Budget of Wetlands

The linkage and relationships between a wetland and groundwater, as discussed in the previous section, largely determine the water budget and hydrologic behavior of wetlands. Water budget components of perched peatland bogs and groundwater-fed peatland fens differ in their accounting for groundwater input and output, as illustrated in Figure 14.2. Although these examples are of peatlands, the same water budget components would apply to other perched and groundwater-fed wetlands.

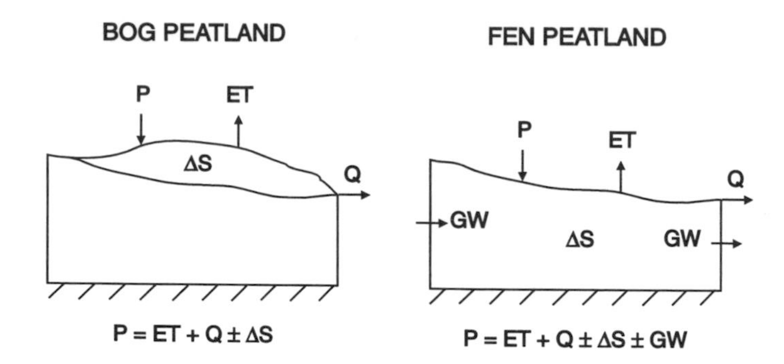

FIGURE 14.2. Water budgets for a perched peatland bog and a groundwater-fed peatland fen. P = precipitation; ET = evapotranspiration; Q = surface discharge; ΔS = change in storage; GW = groundwater.

Wetlands that are fed by regional groundwater exhibit a relatively stable water table. If there is a channel outlet in the wetland, these systems will sustain a relatively even pattern of streamflow throughout wet and dry seasons. Wetlands that have perched water tables, or are otherwise separated from regional groundwater, exhibit greater seasonal fluctuations in both the water table and streamflow discharge (if channel outlets are present). Examples of the seasonal pattern of streamflow from a perched bog and a groundwater fen in northern Minnesota are illustrated in Figure 14.3. Flow duration relationships for these two peatland types are illustrated in Figure 14.4.

Wetland water budget components of ET and outflow, whether as surface discharge to streams or as groundwater discharge, are a function of water table depth. When the water table is located above, at, or close to the surface, both ET and outflow are large. As the water table elevation drops, that is, the water table becomes deeper, ET and discharge from the wetland diminish. The relative magnitude of water budget components for different wetlands varies, but in general, wetlands exhibit high annual ET losses and a corresponding low water yield as discharge to either surface or groundwater systems. In northern latitude wetlands, snowmelt and rain-on-snow events in the spring are normally the periods of greatest discharge to streamflow. During the summer months, flow from wetlands is much less and occurs only from large and intense rainfall events.

Evapotranspiration

Wetlands evaporate and transpire water at or near the potential (PET) rate during the growing season. The readily available water supply for plants and wet surface conditions that characterize most wetlands explain this high rate of ET. When the water table is close to the wetland surface, ET can exceed the evaporation from an open water surface. Although field measurements of ET are sparse, examples of measured rates are listed below:

- Weekly measurements of ET in a northern Minnesota peat bog between July and August averaged between 2.0 and 4.3 mm/day and exceeded pan evaporation rates on three of the six weekly measurements (Bay 1966). At the same location, ET from peat bogs equaled Thornthwaite PET estimates, which aver-

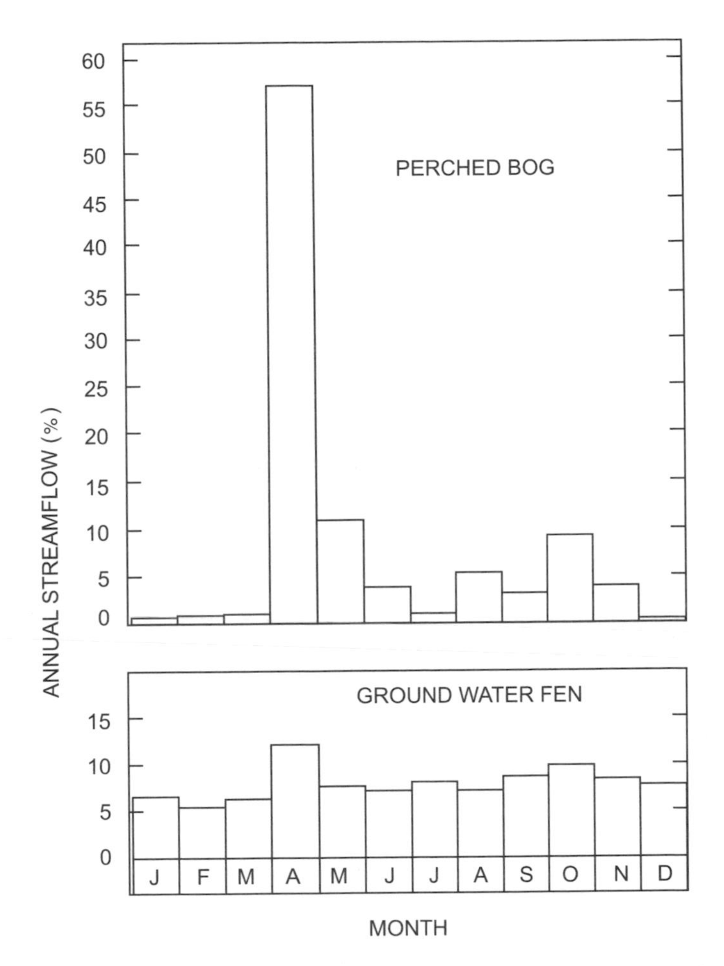

FIGURE 14.3. Seasonal pattern of monthly streamflow from a perched bog and a groundwater-fed fen in northern Minnesota (from Boelter and Verry 1977).

aged between 3.5 and 3.9 mm/day during the growing season (Verry and Timons 1982).

- Eddy correlation methods indicated that from May to early October, ET in northern Minnesota peatlands ranged from 0.9 to 6.0 mm/day and averaged 3.6 mm/day (Verma et al. 1993).
- Daily ET for 6 days in late summer from a mountain bog in Wyoming averaged 3.3 mm/day in contrast to average daily pan evaporation rates of 2.6 mm/day (Sturges 1968).
- Comparisons from a bog in Quebec where the peat was harvested and then allowed to regenerate naturally showed that ET was between 84 and 92% of total outputs and averaged 2.9 mm/day (Van Seters and Price 2001). ET in the harvested and cut-over bogs varied from 1.9 mm/day on raised banks to 3.6 mm/day from moist surfaces with *Sphagnum.*

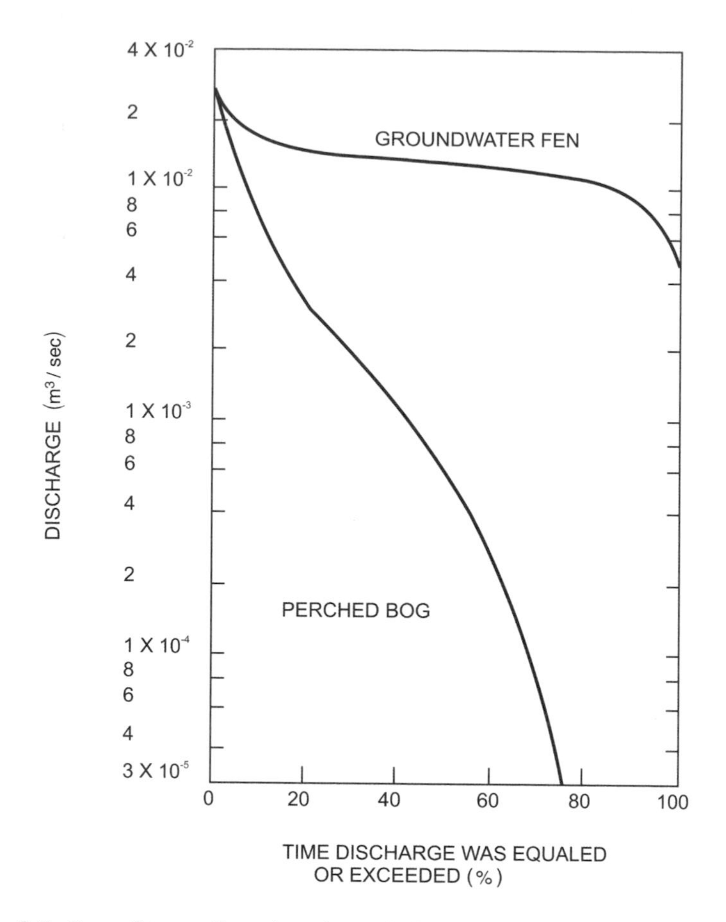

FIGURE 14.4. Streamflow duration relationships for a perched bog and a groundwater-fed fen watershed in northern Minnesota, each of about 53 hectares area (from Boelter and Verry 1977).

- ET rates from three reed (*Phragmites australis*) sites in England were quite variable, ranging from 1.0 mm/day to 6.3 mm/day over a 4 yr period. The maximum mean monthly ET for the three sites was 3.3, 4.4, and 6.3 mm/day (Fermor et al. 2001).

Verry (1997) summarized the relationship between water table depth and the ET/PET ratio for peatlands in Figure 14.5. Note that the ET/PET ratio is close to 1.0 when the water table is between the surface and a depth of about 25 cm. It is somewhat lower when the water table is above the surface and diminishes dramatically when the water table drops below around 30 cm deep. For one of the peat bogs used in this relationship, the water table was found to reside between the surface and 30 cm deep over 85% of the time, based on a 27 yr record (Verry et al. 1988). This tells us that growing season ET of peatlands will occur at the potential rate in all but drought years.

For wetlands such as the prairie pothole wetlands of North America, water tables can fluctuate dramatically from year to year and seasonally and, therefore, would be

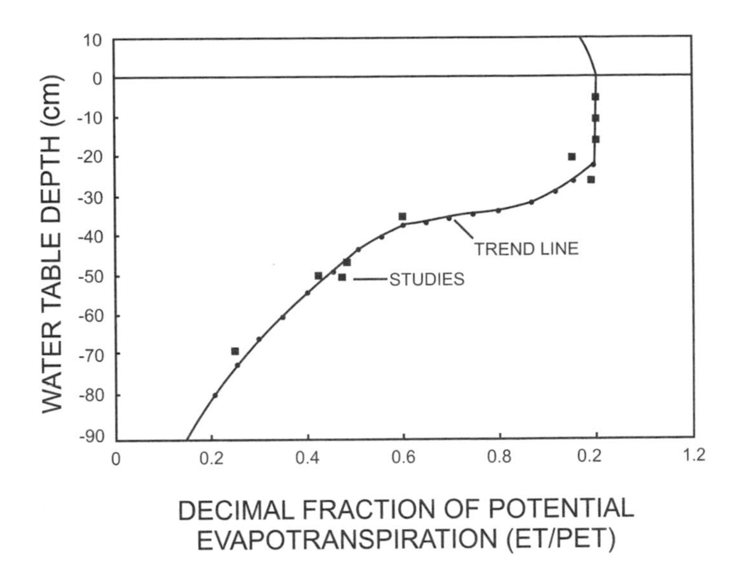

FIGURE 14.5. Relationship between water table depth of peatlands and the ratio of actual evapotranspiration to potential evapotranspiration (adapted from Verry 1997).

expected to have more variable ET. Potholes can become completely dry during periods of drought, greatly reducing ET. Under normal precipitation regimes, ET can be a relatively large component of the water budget. For example, from May through October, pothole wetlands in North Dakota lost 604 and 682 mm from potholes with emergent vegetation and those with little emergent vegetation, respectively (Shjeflo 1968). Average precipitation for the period of study was 418 mm/year. In these cases, the emergent vegetation seemed to limit total ET.

ET measurements from other types of wetlands indicate variable relationships between vegetated wetlands and open bodies of water. For example, ET from a vegetated wetland in New Hampshire was 80% greater than evaporation from an adjacent open body of water (Hall et al. 1972). In contrast, ET from cypress wetlands in Florida were lower than evaporation from open water (Brown 1981).

Water Flow in Wetlands: Surface-Groundwater Relationships

Groundwater recharge from most types of wetlands is limited and generally is not a large component of their water budgets. In northern prairie wetlands of North America, water is primarily lost by means of ET and to a lesser extent by seepage. Isolated wetlands can recharge groundwater by lateral seepage (Kleinberg 1984). However, seepage is often difficult to quantify and can be largely lost through ET from the wet edges of prairie potholes (Su et al. 2000). Recharge to regional groundwater aquifers from these wetlands has been reported to range from 2 to 45 mm/yr when averaged over the area of the catchment (van der Kamp and Hayashi 1998). However, the higher rates only occurred where there were large hydraulic gradients from the wetlands to the groundwater aquifer and where the groundwater is shallow.

Based on how they are formed over time, prairie potholes and similar wetlands would be expected to recharge groundwater more from their edges during periods of high water tables than from percolation through bottom deposits. Even in wetlands where the water table is maintained throughout the year and water is therefore available for recharge, the rates of flow would be expected to be low. This rationale follows when one considers how and why wetlands form in the first place. If they are not groundwater discharge areas, wetlands are the result of an impeding layer that restricts the downward percolation of water, hence the shallow water table. As a result, hydraulic conductivities of wetland soils would be expected to be restrictive.

Organic material in peatlands accumulates over a long period of time (centuries) during which the slow anaerobic decomposition of the lower level peat breaks down the coarser material to finer particles. As the organic matter deposits become thicker, the lower layers become more compressed; consequently, hydraulic conductivities diminish rapidly with depth into the soil, which corresponds to the increased decomposition of peat (Fig. 14.6). The von Post degree of decomposition (von Post and Granlund 1926) is used to characterize peat soils; the higher the value on the von Post scale, the greater the decomposition and the lower the hydraulic conductivities as depicted in Figure 14.6. As a result, vertical, downward flow of water through the peatland is restricted with depth. This forces water to flow in a more horizontal direction in the peatland in response to the hydraulic gradient. Examples of hydraulic gradients and flow rates in three Minnesota peatlands are shown in Table 14.1. For the three peatlands in Table 14.1, the hydraulic gradients remained quite constant throughout the year, even though water levels dropped over the growing season.

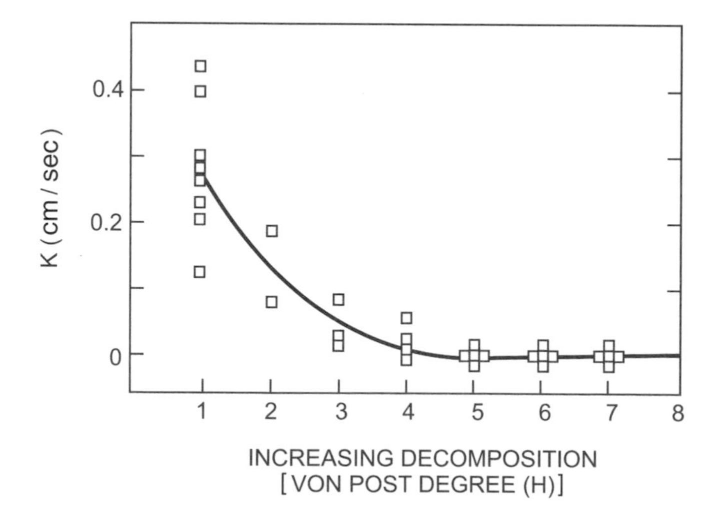

FIGURE 14.6. Relationship between hydraulic conductivity and peat decomposition in peat soils (from Gafni and Brooks 1990).

TABLE 14.1. Hydraulic gradients and measured flow velocities through Minnesota peatlands

Peatland	Hydraulic Gradient (%)	Groundwater Velocity (cm/hour) at Depths of:				
		5 cm	14 cm	25 cm	35 cm	45 cm
Perched bog	0.0534	0.490	0.146	0.103	0.016	0.019
Fen	0.0435	0.475	0.078	0.024		
Raised bog	0.1043		0.122			0.033

Source: Gafni and Brooks 1990.

Flow velocities through peatlands and wetlands in general tend to be very slow because of the low hydraulic gradients. Even where the hydraulic conductivities are relatively high, the low gradient governs flow velocities (refer to Darcy's Law in Chapter 4). In contrast, if one were to impose a greater hydraulic gradient in the soil profile, velocities in the deeper horizons would remain slow because of the lower hydraulic conductivities in the deeper layers. It is not uncommon to see expressions of vertical pressure gradients within wetlands (based on piezometer measurements) that suggest vertical water movement into and out of wetlands. Interpreting such relationships in determining rates of water flow require, again, that hydraulic conductivities of the deeper layers of wetland soils be carefully evaluated.

Runoff and Streamflow from Wetlands

The depth of the water table governs wetland runoff responses to rainfall and snowmelt. Wetlands that have a channel outlet normally yield high amounts of runoff only when the water table is at or above the soil surface. The percentage of rainfall that becomes streamflow is usually small except during the wet season (or snowmelt season in temperate climates) and when plants are relatively dormant. High ET rates during the summer lower the water table; this creates considerable storage that must be satisfied before water tables rise and subsequently discharge increases. In the case of perched wetlands with outlets, it is not uncommon for streamflow to cease late in the growing season or during droughts. An example of the cumulative frequency at which the water table occurs at different depths in a peatland bog is presented in Figure 14.7. The corresponding discharges to these water table elevations indicate that 25% of the time, there is no discharge (corresponding to depths of 14.5 cm or more below the hollow), and significant discharges from the bog occur only 5% of the time when the water table exceeds the hollow elevation. About 70% of the time, the bog discharges water to the stream channel, but at very slow rates.

Depressional wetlands, peatlands, and the like provide storage benefits, much like a shallow reservoir, that help attenuate flood hydrographs. Largely because of their flat topography and lack of well-defined channels, most wetlands function much like simple reservoirs (see Chapter 17); they attenuate flood peaks by temporarily storing or detaining water (Fig. 14.8). There must be well-defined storage–water table elevation and water table elevation–discharge relationships for a wetland for this analogy to hold. The difference between the two systems is that (1) the wetland relationships are based on

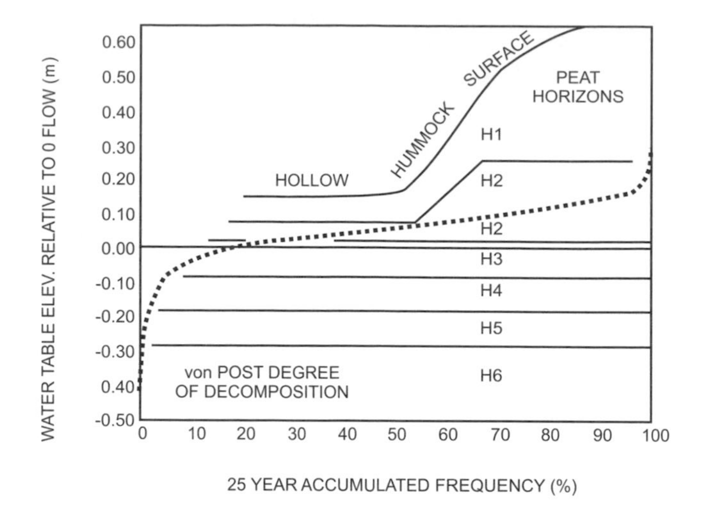

FIGURE 14.7. Cumulative frequency of occurrence of the water table elevation at different peat horizons in a perched bog in northern Minnesota (Verry et al. 1988).

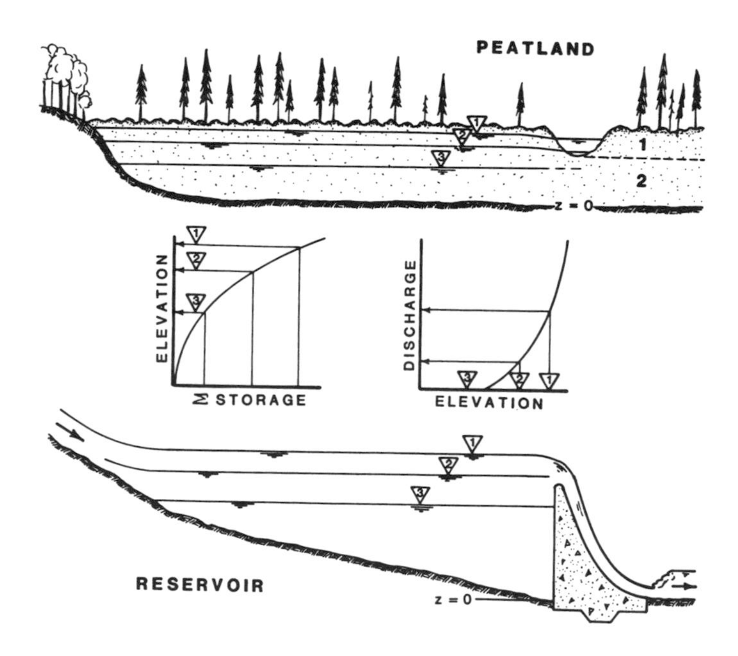

FIGURE 14.8. Streamflow from a peatland is governed by water table elevation similar to the way that discharge from a reservoir is governed by pool elevation (Guertin et al. 1987). Point 1 = high storage and elevation (peatland) correspond to high discharges (reservoir); point 2 = as water table (peatland) or pool (reservoir) drops, discharge decreases; point 3 = when respective elevations drop below the outlet (peatland) or spillway (reservoir), discharge ceases; z = datum.

water table elevations in contrast to reservoir pool elevations and (2) storage in wetlands is a function of wetland soil characteristics, whereas storage in the reservoir is free, unbound water. Unique storage–water table elevation relationships can be determined for any wetland. Developing a water table elevation–discharge relationship for wetlands with an outlet channel is more problematic, but has been accomplished as described in Box 14.2. The development of such relationships facilitates modeling of streamflow from wetlands that have outlet channels. Relationships between the water table elevation, as measured by a well, and outflow from the wetland need to be developed from field data.

The suggested analogy between wetlands and reservoirs suggests that flood storage–flood control benefits can be attributed to wetlands. Although storage is more limited than with lakes or reservoirs, the presence of wetlands can influence peak flow discharge from watersheds. Up to a point, the greater the percentage of watershed area composed of any combination of wetlands, reservoirs, or lakes, the greater the reduction in peak flow discharges for a given storm event. In northern Wisconsin such benefits have been considered significant when up to 20–30% of a watershed has a combination of wetlands, lakes, or reservoirs (Conger 1971). Figure 14.9 illustrates this relationship.

Summary of Hydrologic Function of Wetlands

The hydrologic characteristics of wetlands can be summarized as follows:

- Shallow water tables, flat topography, and low drainage densities are dominant features of most wetlands. As a result, they are areas that lose large quantities of water via ET and produce low levels of discharge to streams.
- The depth of the water table governs ET and streamflow discharge from wetlands; the deeper the water table, the lower the ET and discharge. Wetlands with surface outlets produce high levels of discharge from rainfall and snowmelt only when water tables are at or above the soil surface.
- Annual ET far exceeds annual discharge for most wetlands.
- Wetlands tend to be groundwater discharge areas more often than groundwater recharge areas.
- Largely because of their flat topography, wetlands function much like simple reservoirs: they attenuate flood peaks by temporarily storing or detaining water.
- Wetlands linked to regional groundwater systems exhibit less seasonal fluctuation in water table and streamflow discharge than do wetlands that are perched or otherwise isolated from regional groundwater.

Wetlands and Water Quality

A comprehensive review of wetlands and water quality relationships is beyond the scope of this book. However, a brief and general overview of how wetlands influence certain water quality characteristics is presented below.

The physical location of wetlands in most watersheds suggests that they are largely receivers of water from precipitation, surface runoff, and/or groundwater. The water chemistry of a wetland reflects the source of water. Fen peatlands in northern Minnesota, which are expressions of regional groundwater, have higher mineral content and pH than rainfall-fed bogs. Specific conductance of water discharging from fens is significantly higher than that from bogs and can be used to identify both the source of water and peatland type.

Wetlands that receive overland flow or stream discharge from upland watershed areas are often considered to be sinks in terms of water quality constituents. The velocity of any

Box 14.2

Streamflow Response from a Peatland Similar to That of an Unregulated Reservoir (from Verry et al. 1988)

Water table elevation-discharge relationships of a 3.24 ha peat bog in northern Minnesota were similar to those predicted for a small unregulated reservoir (refer to Fig. 14.8). When the water table elevations in the bog were modified to reflect the specific yield from the peat soils, they matched closely with the same discharges that would occur from a level reservoir (see figure below). Departures from the reservoir elevation-discharge relationship occurred when the water table in the bog raised slightly above the elevation of the hollows of the hummocklike peat surface (the dots in the figure were greater than the solid line of the reservoir relationship). The higher elevation to discharge relationship in this 75–100 m^3/hr range is attributed to the resistance to flow offered by the hummocks and trees in the bog. As the water table elevation was raised further, the discharge rates from the bog were higher than would occur from a level reservoir. These higher flow rates were attributed to channel-like flow in the lagg of the peatland in which a moving wedge of water has a greater velocity than that from a flat reservoir surface.

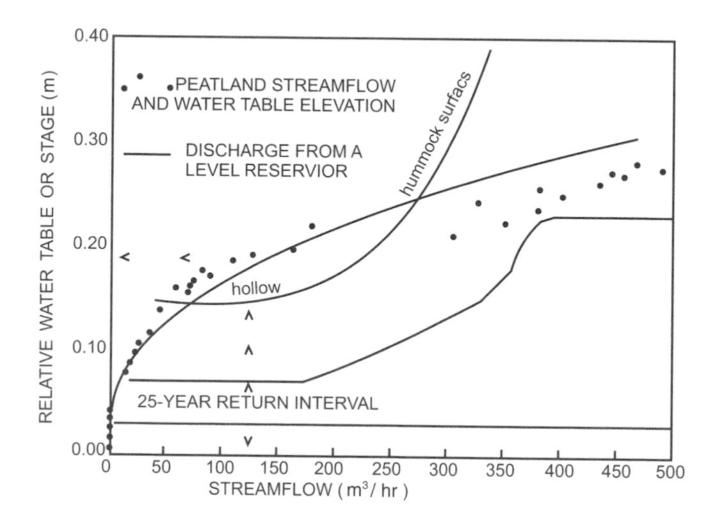

Measured relationships between water table elevation and discharge from a bog and similar relationships for a shallow reservoir with the same dimen-

The above reservoir relationships determined from field measurements were incorporated into a hydrologic model capable of predicting streamflow discharge from peatlands (Guertin et al. 1987). The peatland hydrologic impact model (PHIM) that was developed computes water budget components so that storage conditions in the peatland are determined. Storage-elevation and elevation-discharge relationships are then incorporated into the model to predict streamflow discharge from peatlands.

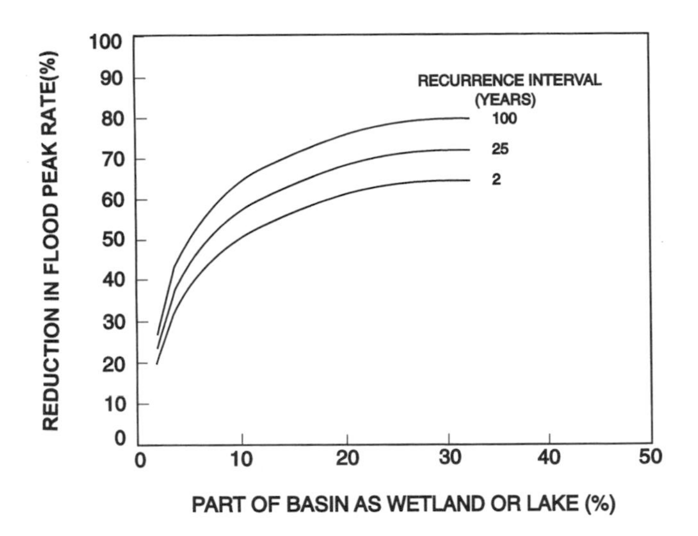

Figure 14.9. Relationship between peak flow discharge (%) and percentage of a watershed in wetland or lakes for different recurrence intervals in northern Wisconsin (adapted from Conger 1971 as presented by Verry 1988).

flow of water into wetlands is reduced, resulting in sediment deposition. Carried along with this sediment are minerals, phosphorus, and sometimes heavy metals. Similarly, many dissolved nutrients that enter a wetland can be taken up by wetland plants, resulting in lower nutrient loads leaving a wetland than entered. It is this "cleansing" function that is often seen as a valuable benefit of wetlands on a watershed. Whether this is true depends on the water quality constituents and the type of wetland. Generalities concerning wetland effects on water quality include:

- Sediment and particulate matter deposited in wetlands can reduce that which would otherwise enter a stream system. The capacity of a wetland for storing particulates is finite. However, in the case of particulate phosphorus (P) that is deposited over time, some P can eventually be converted into soluble forms, which can then leave the wetland and enter streams or groundwater systems in a soluble form.
- Nitrogen entering a wetland can be taken up by plants, and because of anaerobic conditions, denitrification can reduce nitrogen available for export to streams and groundwater systems.
- The role of wetland plants in taking up excess nutrients from upland runoff is a function of the types of plants present, the total biomass, and the fate of these plants. Eventually, plants will die and the nutrients taken up will be recycled. Therefore, some type of management that removes and regenerates vegetation or a natural process such as fire (see below) is needed to sustain a nutrient "sink" function.

Artificial wetlands are being created in urban areas to provide these functions and are designed to treat urban runoff before it enters streams or lakes. Likewise, restored and created wetlands in agricultural areas have been used to reduce nutrient export to streams.

Hydrologic Implications of Wetland Drainage and Loss

Wetland Drainage for Development

For decades, agricultural, urban, and other types of economic development on watersheds have drained, altered and in many cases eliminated wetlands, as discussed earlier. The hydrologic implications of such development have not always been clear, however.

Most development activities on wetlands require drainage to remove excess water. The impacts of this drainage, and associated changes in vegetative cover and other changes, can be assessed in terms of changes in storage and routing within watersheds to understand possible impacts on flooding and pattern of streamflow emanating from wetlands. Water quality changes can be related to these changes in flow regimes as well.

Peat Mining

Peatlands in northern Europe, Canada, and parts of the United States have been cleared and peat soil extracted for energy or horticultural products. They have also been extensively drained to enhance forest growth or allow for crop cultivation in many parts of the world. Development of peatlands usually requires one or more of the following: forest (vegetation) clearing, drainage, and peat extraction. These activities alter the hydrologic response of a watershed in the following ways:

- The initial ditching process and its connection with a surface stream outlet releases water that is otherwise being stored or that would move slowly in response to natural hydraulic gradients. This initial release of water can result in a significant peak discharge, as has been witnessed in northern Minnesota. To mitigate such surges in flow, small storage or holding ponds may be required; such ponds also can serve as sediment basins.
- Ditching and excavation of peat steepens local hydraulic gradients and can alter direction of flow of groundwater adjacent to the altered site. The regional water table adjacent to a 64 ha mined peatland site in northeastern Minnesota was significantly altered following mining of peat (Leibfried and Berglund 1986). The large perimeter ditch around the harvest area changed the direction of regional groundwater flow and, in effect, increased the area contributing to flow at the outlet of the harvested area by almost 85%.
- Water tables are lowered, which, in combination with vegetation removal, reduces annual ET and, as a consequence, increases annual water yield from the harvested area. As a result of ET differences, harvested bogs in Quebec, Canada, yielded 12–24% runoff compared to negligible runoff from a natural bog (Van Seters and Price 2001). Much of this can be attributed to the reduced ET in the harvested bog.
- Runoff from harvested peat fields increases, and as a result of drainage ditches, the water is conveyed quickly to downstream areas.

- Surface peat soils in the summer can become hydrophobic—increasing surface runoff and sediment delivery from the peatland.
- During the winter months, the peat soils in harvested areas freeze as concrete frost, resulting in minimal infiltration for snowmelt or early spring rains.

The net effect of widespread peat harvesting can lead to higher volumes of snowmelt and rainfall runoff, which can aggravate flooding in local areas. In the case of fens, streamflow during the dry season can be augmented by the expanded contributing areas due to changes in hydraulic gradients and direction of regional groundwater flow. For bogs, however, ditching and peat extraction can lower the local water table and result in lower discharges during the dry season. In such cases, there is no regional groundwater system to sustain flow throughout the year. Changes in water quality can take place as well, including increased discharge of suspended solids into receiving waters.

Forested peatlands in northern Europe have been ditched to promote growth of trees through lowered water tables. In such instances, ET can increase as a result of accelerated growth of trees causing water yield to diminish.

Drainage of Wetlands for Agriculture

In contrast to peat harvesting to extract a product from the wetland, there has been widespread drainage of wetlands to increase land area for crop and animal production. There is no question that food production has increased as a result. What has not been as clear are the hydrologic, water quality, and other environmental implications of wetland drainage. This discussion focuses largely on the prairie wetlands of north central North America.

Ditching and the use of drain tiles, some with surface inlets and others as subsurface drain tiles, have effectively drained a large percentage of northern prairie wetlands. In the Minnesota River basin, over 80% of the original prairie wetlands have been eliminated through drainage (Leach and Magner 1992). This process changes the hydrologic regime in several ways (Box 14.3). Water that would otherwise be ponded, temporarily stored on the landscape, and largely lost to the atmosphere via ET is now leaving the watershed as runoff. Lesser amounts of seepage would likewise be converted into quicker flow from the wetland site. The combination of the loss of storage and the improved conveyance systems for water to leave the watershed results in higher flows of water from rainfall and snowmelt. An additional effect is attributed to the fact that many of the original prairie wetlands had no surface outlet. As a consequence of drainage, many areas became connected to streams, which in effect expanded the watershed area for the existing streams. The net effect of wetland losses and expanded drainage of agricultural fields with drain tiles is an increase in the magnitude of annual peak flows up to the 50 yr return period event.

The implications of tile drains on water quality are also significant. Drainage using tile drains essentially bypasses the riparian vegetation in most instances. The outlet tiles route runoff from agricultural fields and discharge directly into downstream ditches and stream channels. There is little or no opportunity for nutrient uptake by riparian plants. In addition, the drainage-induced increase in the 1.5–2 yr recurrence interval peak flows results in stream channel instability and accelerated bank erosion, increasing suspended sediment concentrations (see chapters 9 and 10).

Box 14.3

Effects of Wetland Drainage on Hydrologic Regimes in the Upper Midwest

A modeling study of a tributary watershed of the Minnesota River basin indicated that the net effect of wetland drainage (1) increased annual water yield, (2) increased peak flow discharges associated with the 1.5 yr to 50 yr recurrence interval, (3) and reduced minimum flow rates (Miller 1999). For areas that contained a greater percentage of internally drained basins, that is, those with no outlets, drainage had a more pronounced effect on peak flows. Connecting these previously internally drained systems to a stream channel increased the area of land contributing runoff to stream channels; in essence, the watershed area is thereby increased. The effects of wetland loss on peak flows were shown to diminish for the more extreme floods, such as those with a 100 yr recurrence interval and greater.

A companion study in the Minnesota River basin showed that annual peak flows increased as the percentage of tile drainage on a watershed increased (Mickelson 2001). This study suggested that increasing the amount of tiled wet soil in a watershed from 60 to 100% increased the 10 yr return period peak by three times, and the 100 yr return period peak by five times.

Streamflow changes were determined for nine Wisconsin watersheds in a similar study, demonstrating that the high flows were higher and the low flows lower when wetlands were removed (Hey and Wickencamp 1998).

Special Topics

Role of Fire in Wetlands

Wildfires and prescribed burns occur in many wetland systems. Furthermore, fire has been found to be a valuable tool for managing some wetlands for habitat improvement (Box 14.4). Fire affects wetland systems both directly and indirectly. If fire burns in the wetland system itself, it can consume part or all of the vegetation. When fire also burns the surrounding watershed, excessive runoff is generated, which increases streamflow and sediment movement into the wetland.

There are also some disadvantages to using fire in wetlands. For example, fire in wetlands is the most important single factor influencing forest and tundra in most of Alaska. Fire removes insulation; lowers permafrost depths and, as a result, the surface; affects subsurface drainage; and modifies water-holding capacity of the soil. A general lowering of the water table resulting from fire is also detrimental to waterfowl in Alaska because of the reduction in waterfowl habitat.

During periods of drought, wildfires in peat bogs can be particularly damaging. Under these conditions smoldering fires burn for long periods of time, modifying the water-holding capacity of the peat soils and starting long-term successional stages on the site. These fires usually continue until the water table is restored or the drought ends.

Box 14.4

Benefits of Prescribed Burning in Some Wetlands (Kirby et al. 1988)

Prescribed fire of wetlands has been useful throughout the United States. Benefits associated with fire include (1) removing annual accumulations of dead herbaceous cover, thereby preventing a buildup of debris on the marsh floor; (2) reducing the level of the marsh floor by burning into organic soils; (3) reducing or eliminating woody vegetation that has invaded impoundments; (4) destroying sphagnum moss and bringing about succession to sedge and grasses, thus creating nesting areas for waterfowl; (5) cleaning impoundment basins before flooding; (6) producing open areas that will provide better spring grazing for waterfowl; (7) increasing nutrient mineralization for increased forage quality; and (8) causing a change in albedo to permit earlier spring growth. In many wetland ecosystems, fire was a natural process that helped maintain the structure, species composition and health of natural wetlands.

Mitigation of the Effects of Roads on Wetlands

Watershed managers have long been concerned with the effects of road-generated turbidity on water quality and stream sedimentation. An array of best management practices (BMPs) have evolved to reduce sediments originating from road-related soil disturbances, with attention focused on how road construction and maintenance activities affect the hydrology of wetlands and riparian systems (LaFayette et al. 1992).

Roads, road-related drainage structures, and channel crossings affect wetlands by intercepting, concentrating, accelerating, and diverting surface runoff; intercepting and diverting subsurface flows; initiating and accelerating erosion of wetland soils; causing or contributing to channel incision; reducing overbank flooding and groundwater recharge; and indirectly converting wetland vegetation from wetland to upland types.

In many situations, road construction and maintenance practices can be modified to favor the restoration of sites previously damaged or impaired by roads or other human-induced disturbances such as agriculture, livestock grazing, or logging. BMP techniques for wet-meadow recovery include relocation of site-damaging road segments; modification or replacement of channel-crossing structures by culverts, fords, or bridges installed to reestablish historic stream gradients and channel morphology; and the modification of ditch systems and cross drains to disperse captured flows (LaFayette et al. 1992). The engineering objective, therefore, becomes one of using road drainage structures to conserve available surface and subsurface flows and nurture native wetland vegetation, thereby sustaining ecosystem integrity.

Cumulative Effects

The effects of wetland loss and/or wetland restoration need to be assessed in the watershed landscape and over time to understand the cumulative watershed effects. The cumulative effects of losses of wetland and riparian areas, coupled with other changes in the upper Mississippi River basin, likely increased the severity and impacts of the 1993 flood. The area of wetlands on many watersheds of the basin has been significantly reduced over the years since land was cleared for agricultural cropping. As indicated in Figure 14.9, when the percentage of a watershed that is composed of wetlands is reduced from over 30% to < 10%, flood peaks can be increased significantly. Coincident with the expansion of cropland areas, there has been a conversion from forest cover to pastures and croplands, which has increased annual streamflow discharges (see Chapter 6), soil erosion (see chapters 7 and 8), and sediment delivery to downstream areas (see Chapter 9).

Urbanization also contributes to the problem and has expanded the impervious area of many watersheds. In many instances, riverine wetlands have been replaced with levees, which in effect reduces the flooding locally, only to increase streamflow discharges downstream. At any one point in time and at any one location, any of the above effects may seem trivial. However, the cumulative effects of wetland loss in conjunction with other land use changes can lead to excessive runoff and higher levels of streamflow; stream channels adjust accordingly, and the end result can be unstable channels, excessive bank erosion, and increased sediment loads. Reversing such effects is not easy and involves improved management and mitigation of watersheds, wetland systems, and stream channels. The economics of these interventions can often be prohibitive for local watershed communities.

Summary

Wetlands are important biological and hydrological features of watersheds. In addition to providing many unique plant and animal communities, wetlands reduce streamflow volumes and peak discharges to receiving stream channels. They also can reduce sediment delivery to receiving waters and can improve the quality of water that is discharged to streams and rivers.

After reading this chapter, you should be able to:

- Describe the different surface water and groundwater interactions that are associated with different wetland types.
- Describe the hydrologic and physical landscape conditions under which wetlands are formed.
- Develop a water budget for both a wetland that has a perched groundwater system and one that is an integral part of a regional groundwater system.
- Describe conditions under which wetlands can be groundwater discharge areas and those in which they can be sources of groundwater recharge.
- Explain how wetlands affect flooding.
- Describe the mechanisms by which wetlands affect water quality; indicate how suspended sediments, P, and N from surface runoff into a wetland can be affected.
- Discuss and provide examples of cumulative effects in a watershed that can result from losses and gains in wetlands.

References

Bay, R.R. 1966. Evaluation of an evapotranspirometer for peat bogs. *Water Resour. Res.* 2:437–442.

Boelter, D.H., and E.S. Verry. 1977. *Peatland and water in the northern Lake States*. USDA For. Serv. Gen. Tech. Rep. NC-31.

Brown, S.L. 1981. A comparison of the structure, primary productivity and transpiration of cypress ecosystems in Florida. *Ecol. Monogr.* 51:403–427.

Conger, D.H. 1971. Estimating magnitude and frequency of floods in Wisconsin. Open File Report. Madison, WI: U.S. Geological Survey.

Fermor P.M., P.D. Hedges, J.C. Gilbert, and D.J.G. Gowing. 2001. Reedbed evapotranspiration rates in England. *Hydrol. Proc.* 15:621–631.

Gafni, A., and K.N. Brooks. 1990. Hydraulic characteristics of four peatlands in Minnesota. *Can. J. Soil Sci.* 70:239–253.

Galatowitsch, S.M., and A.G. van der Valk. 1998. *Restoring prairie wetlands: An ecological approach.* Ames: Iowa State University Press.

Guertin, D.P., P.K. Barten, and K.N. Brooks. 1987. The peatland hydrologic impact model: Development and testing. *Nordic Hydrology* 18:79–100.

Hall, T.F., R.S. Rutherford, and G.L. Byers. 1972. *The influence of a New England wetland on water quantity and quality*. New Hampshire Water Resources Center Research Report 4. Durham: University of New Hampshire,

Hammer, D.A. 1992. *Creating Freshwater Wetlands.* Boca Raton, FL: Lewis Publishers.

Hayashi, M.G., G. van der Kamp, and D.L. Rudolph. 1998. Water and solute transfer between a prairie wetland and adjacent uplands. I. Water balance. *J. Hydrol*. 207:42–55.

Hendrickson, D.A., and W.L. Minckley. 1984. Cienegas—Vanishing climax communities of the American Southwest. *Desert Plants* 6:131–175.

Hey, D.L., and J.A. Wickencamp. 1998. Effect of wetlands on modulating hydrologic regimes in nine Wisconsin watersheds. In *Water resources and the urban environment*, ed. E. Loucks. Weston, VA: The American Society of Civil Engineers.

Kent, D.M. 1994. *Applied wetlands science and technology.* Boca Raton, FL: Lewis Publishers.

Kirby, R.E., S.J. Lewis, and T.N. Sexson. 1988. *Fire in North American wetland ecosystems and fire-wildlife relations: An annotated bibliography.* U.S. Fish Wildl. Serv., Biol. Rep. 88(1).

Kleinberg, K. 1984. Hydrology of north central Florida Cypress Downs. In *Cypress swamps*, ed. K.C. Ewel, and H.T. Odum, 72–82. Gainsville: University of Florida Press.

LaFayette, R.A., J.R. Pruitt, and W.D. Zeedyk. 1992. Riparian area enhancement through road management. In *Proceedings of the International Erosion Control Association 24th Conference,* 353–368. Indianapolis, IN.

Leach, J., and J. Magner. 1992. Wetland drainage impacts within the Minnesota River basin. In *Currents—A Minnesota River Valley Review* 2(2): 3–10.

Leibfried, R.T., and E.R. Berglund. 1986. Groundwater hydrology of a fuel peat mining operation. In Proceedings, *Advances in Peatlands Engineering*, 192–199. Ottawa, Canada: Carleton University.

Mickelson, D.L. 2001. The effect of tile drainage on flood events in the Minnesota River basin. M.S. Thesis, University of Minnesota, St. Paul.

Miller, R.C. 1999. Hydrologic effects of wetland drainage and land use changes in a tributary watershed of the Minnesota River Basin: A modeling approach. M.S. Thesis, University of Minnesota, St. Paul.

Mitsch, W.J., and J.G. Gosselink. 1993. *Wetlands*. 2nd ed. New York: Van Nostrand Reinhold.

Price, J.S., and J.M. Waddington. 2000. Advances in Canadian wetland hydrology and biogeochemistry. *Hydrol. Proc*. 14:1579–1589.

Shjeflo, J.B. 1968. *Evapotranspiration and the water budget of prairie potholes in North Dakota*. Professional Paper 585-E. Washington, DC: U.S. Geological Survey.

Sturges, D.A. 1968. Evaporation at a Wyoming mountain bog. *J. Soil Water Conserv.* 23:23-25.

Su, M., W.J. Stolte and G. van der Kamp. 2000. Modelling Canadian prairie wetland hydrology using a semi-distributed streamflow model. *Hydrol. Proc.* 14:2405–2422.

U.S. Environmental Protection Agency. 1993. *Created and natural wetlands for controlling nonpoint source pollution.* Corvallis, OR: U.S. Environmental Protection Agency.

van der Kamp, G., and M. Hayashi. 1998. The groundwater recharge function of small wetlands in the semi-arid Northern Prairies. *Great Plains Res.* 8:39–56.

Van Seters, T.E., and J.S. Price. 2001. The impact of peat harvesting and natural regeneration on the water balance of an abandoned cutover bog, Quebec. *Hydrol. Proc.* 15:233–248.

Verma, S.B., F.G. Ullman, N.J. Shurpali, R.J. Clement, J. Kim, and D.P. Billesbach. 1993. Micrometeorological measurements of methane and energy flux in a Minnesota peatland. *Suo* 43:285–297.

Verry, E.S. 1988. The hydrology of wetlands and man's influence on it. In *Proceedings of the International Symposium on the Hydrology of Wetlands in Temperature and Cold Climates*, 41–61. Helsinki: Academy of Finland.

Verry, E.S. 1997. Hydrological processes of natural, northern forested wetlands. In *Northern Forested Wetlands*, eds. Trettin, C.C., M.F. Jurgensen, D.F. Grigal, M.R. Gale, and J.K. Jeglum, 163–188. Boca Raton: Lewis Publishing.

Verry, E.S., and D.R. Timons. 1982. Waterborne nutrient flow through an upland-peatland watershed in Minnesota. *Ecology* 63:1456–1467.

Verry, E.S., K.N. Brooks, and P.K. Barten. 1988. Streamflow response from an ombrotrophic bog. In *Proceedings of the International Symposium on the Hydrology of Wetlands in Temperate and Cold Regions*, 6–10. Joensuu, Finland.

von Post, L., and E. Granlund. 1926. Sodra Sveriges torvtillganger I. *Sveriges Geologiska Undersokningen Serie C.* No. 335.

Winter, T.C. 1989. Hydrologic studies of wetlands in the northern prairies. In *Northern prairie wetlands*, ed. A. Van der Valk. Ames: Iowa State University Press.

Forest cover affects the accumulation and melt of snow-packs: melting snowpack on a high elevation watershed in Arizona.

PART 5

Special Topics

Part 5 supplements the earlier chapters and topics of this book by substantially revising and consolidating the chapters contained in Part 4 of the second edition. Chapter 15 of this third edition, entitled Snow Hydrology and Management, should be useful to people in geographic areas where snow plays a significant role in the regional water budget. Chapter 16, Watershed Considerations for Water Resource Development and Engineering Applications, describes many structural (engineering) measures that are commonly used in conjunction with nonstructural (vegetative) measures in water resource development. Water harvesting and water spreading are discussed, and a particular emphasis is placed on watershed management–reservoir management relationships.

Chapters 17 and 18, Hydrologic Methods and Tools for Watershed Analysis and Research, respectively, provide the means for quantifying hydrologic relationships and obtaining information needed in watershed management. Standard hydrologic methods are reviewed in Chapter 17, while applications of field and statistical methods, computer simulation models, and geographic information systems are described in Chapter 18.

CHAPTER 15 *Snow Hydrology*

INTRODUCTION

Snowpacks are major sources of freshwater for many regions of the world. Snowpacks that accumulate in mountains—in dryland regions as well as in many humid regions—are often the primary source of freshwater for downstream users. One-third of the water used for irrigation in the world comes from snowpacks. Runoff from snowmelt is not always beneficial to downstream communities, however, and can cause damaging floods when melt occurs too fast or when flood-level flows are sustained over long periods. It is not surprising then that water resource managers and hydrologists are interested in snow hydrology.

The distribution of snowfall and the ripening and melting of snow are particularly affected by forest vegetation. Changes in forest cover, therefore, can result in changes in these processes. Some changes in forest cover occur inadvertently as a result of management activities; however, sometimes forest cover can be manipulated to achieve specific water resource objectives. Natural resource managers in temperate climates, where snow is important, need to be able to recognize how their management affects snowpack accumulation and snowmelt runoff. Definitions of terminology in snow hydrology are given in Table 15.1.

MEASUREMENT OF THE SNOW RESOURCE

Snowfall can be measured with the standard rain gauges (see Chapter 2), but snowfall is more difficult to measure than rainfall. With its low density and high surface area, snow is more susceptible to wind, which results in rain gauges catching less than the true snowfall. In addition, snow tends to cap, or bridge, over the orifices of rain gauges and clings to the sides of containers. Since it usually remains on the ground for some time,

TABLE 15.1. Terminology used in snow hydrology

Term	Definition
Snowpack	Mixture of ice crystals, air, impurities, and liquid water if melting.
Snowpack density	Weight per unit volume; for pure ice = 0.92 g/cm³, for a snowpack, it can vary from < 0.10 g/cm³ to > 0.40 g/cm³.
Snow water equivalent (*WE*)	The weight of snow expressed as the depth of liquid water over a unit of area = (density) × (depth).
Cold snow	Snow with a temperature below 0°C.
Temperature deficit (T_s)	Snowpack temperature below 0°C.
Thermal deficiency	Heat required to raise the temperature of 1 cm *WE* of cold snow by 1°C (the heat capacity of water is 1 cal/g/1°C, for ice it is 0.5 cal/g/1°C, and for air it is 0.24 cal/g/1°C).
Liquid water-holding capacity	Analogous to soil moisture; it is the water held against gravity on snow crystals and in capillary channels in the snowpack (*f*); it varies with density, crystal size and shape, and capillarity; $f = 0.03$ on average, generally less than 0.05; at 0°C, $f = 1 - B$ (*B* is thermal quality defined below).
Ripe snowpack	Snowpack that has reached its maximum liquid water-holding capacity against gravity; the snowpack temperature is isothermal at 0°C; it is primed to transmit liquid water.
Cold content	Heat required to raise the temperature of a cold snow layer of depth *D* to 0°C, $= 0.5\,\rho\,DT_s$.
Latent heat of fusion	Heat required to change 1 g of dry snow at 0°C to a liquid state without changing the temperature, or changing from liquid to solid at 0°C = 80 cal/g = 80 cal/cm³.
Water equivalent of cold content (W_c)	$W_c = 0.5\,(WE)\,T_s/80 = (WE)\,T_s/160$
Thermal quality (*B*)	The ratio of heat required to melt the snow to the heat required to melt an equal mass of pure ice at 0°C (*B*); for a wet, melting snow, $B < 1$" for a dry snow at 0°C, $B = 1$; for a cold dry snow, $B > 1$.

however, snow does not need to be measured as it falls. The most common method of determining the amount of snow is to measure its depth on the ground and, if possible, its weight. Such measurements can be made adjacent to standard rain gauges, or they can be made in surveys over an area in which several measurements are taken over time.

Snow Surveys

Snow surveys are used to estimate the amount of water in the snowpack and the condition of the snowpack during periods of accumulation and melt on courses with a few to 20 or more sampling points along a transect. These surveys are conducted during periods that normally have maximum snow accumulation (early March through late April in the western United States).

Snow depth and snow water equivalent are measured in a snow course using cylindrical tubes with a cutting edge. These tubes are graduated in inches or centimeters on the outside of the tube to measure depth. Large-diameter tubes (3 in.) are used in areas with shallow snowpacks. Smaller diameter tubes, such as the Mount Rose sampler (1.49 in. diameter), are preferable for measuring snowpacks that exceed 3 ft in depth.

The manner in which a snow course is laid out depends largely upon how the survey information obtained is to be used. If snow surveys are to provide an index of snow water equivalent to predict snowmelt-runoff volumes (as in regression methods described below), they do not necessarily need to represent the average of the watershed in question. Instead, a snow course should be located in an accessible area that is flat, is protected from wind, and has a deep snowpack even in the drier years. The volume of snowmelt runoff is then estimated with a regression relationship based on selected snow course data. The regression equations developed are of the general form:

$$\text{Spring Snowmelt Runoff Volume} = PI - LI \tag{15.1}$$

where PI = indices of precipitation inputs, such as snow water equivalent, spring rainfall, or fall rainfall; and LI = indices of losses, for example, evapotranspiration estimates.

Often, multiple regression equations are developed that have the following components:

$$Y = a + b_1X_1 + b_2X_2 + b_3X_3 - b_4X_4 \ldots + b_n X_n \tag{15.2}$$

where Y = volume of snowmelt runoff, X_1 = maximum snow water equivalent (based on snow survey data for April 1, for example), X_2 = fall precipitation (October–November rainfall), X_3 = spring rainfall (April), X_4 = pan evaporation for October–April, and b_i = regression coefficients.

This approach has been used widely by the Natural Resources Conservation Service in the western United States.

If the purpose of a snow survey is to provide an estimate of the mean depth of snow water equivalent over a watershed (e.g., as input data to a hydrologic model), the snow course should be designed differently than explained above. In the latter case, the snow course should represent the spatial distribution of snow over the watershed area. The same principles discussed in estimating the mean depth of rainfall for a watershed (see Chapter 2) should be applied in such instances.

Remote-Sensing Methods

The availability of telemetry and satellite systems has expanded the methods of measuring snowpacks. Telemetry has been particularly useful in transmitting data collected by pressure pillows (snow pillows) in remote sites. These "pillows" are metal plates that measure the weight of snow, usually by means of pressure transducers. The weight of the snowpack is converted into units of snow water equivalent. A network called SNOTEL (Snow Data Telemetry System), operational in the western United States, uses radio telemetry to transmit snow pillow data, temperature, and other climatological data from remote mountain areas to processing centers. These data complement snow course data and have the advantage of representing current conditions on the watershed. Hydrologists can retrieve data from the system any time they wish. For example, conditions that lead to rapid snowmelt can be detected as they occur and allow so-called real-time predictions.

Reliable snow water equivalent measurements are provided in real time to the National Weather Service (NWS) offices upon request from NWS hydrologists through the Airborne Snow Survey Program (Carroll 1987). The airborne measurement technique uses the attenuation of natural terrestrial gamma radiation by the mass of a snow cover to obtain airborne (fixed-wing aircraft) measurements of snow water equivalent. The gamma radiation flux near the ground originates primarily from natural radioisotopes in the soil; in a typical soil, 96% of the gamma radiation is emitted from the upper 20 cm of soil. After a measurement of the background (no snow cover) radiation and soil moisture is made over a specific flight line, the attenuation of the radiation signal due to the snowpack overburden is used to integrate the average areal amount of water in the snow cover over the flight line. As of 2001, flight-line surveys have been established for operational streamflow forecasting by the NWS with a network of over 1900 flight lines in 27 states and seven Canadian provinces (www.omao.noaa.gov/oldies/snowsurv.html).

Satellite imagery can be useful to assess the extent of snow-covered areas, particularly for large river basins. Resolution of the sensor is used to determine the minimum basin sizes for various satellite systems. Hydrologists in the United States generally use NOAA-AVHRR data on basins as small as 200 km^2, Landsat MSS data on basins as small as 10 km^2, and Landsat TM data on basins as small as 2.5 km^2 (Rango 1994). More important than the resolution is the frequency of coverage; for many applications, this is adequate only on the NOAA-AVHRR (one visible overpass per day). Cloud cover is always a potential problem, although estimates of snow cover under a partial cloud cover can often be obtained by extrapolation from the cloud-free portion of the basin. Water resource managers sometimes develop relationships between the percentage of area in a watershed covered with snow and subsequent streamflow due to snowmelt. As applications of satellite technology continue to expand, there will be opportunities for relating many types of snowpack spectral characteristics to snowpack condition, melt, and other related processes.

Snow Accumulation and Melt

Generally, the same factors affecting rainfall over a watershed (see Chapter 2) also affect the deposition of snow on the watershed. On a microscale, snow accumulation can be quite variable because it is influenced by wind, local topography, forest vegetation, and other physical obstructions such as fences. Newly fallen snow usually has a low density, often assumed to be 0.1 g/cm^3, which makes it susceptible to wind action; it also has a high albedo, ranging from 80 to 95%. Between snowfall and snowmelt, several changes take place within the snowpack. These changes, called snowpack metamorphism, are largely the result of energy exchange; are associated with the ripening process; and involve changes in the snow structure, density, temperature, albedo, and liquid-water content.

Snowpack Metamorphism

A snowpack undergoes many changes from the time snow falls until snowmelt occurs. Snow particles, which are initially crystalline, become more granular as wind, solar and sensible-heat energy, and liquid water alter the snowpack. Snow crystals become displaced and the snowpack settles, resulting in an increase in density. Alternate thawing and freezing at the snow surface, followed by periods of snowfall, can cause ice planes or lenses to form within a snowpack. In addition, the temperature of the snowpack changes in response to long periods of either warm or cold weather. As warmer weather dominates, there is a progressive warming of the snowpack. Of course, the snow can never reach a temperature in excess of 0°C. The albedo of snow diminishes over the time that the pack is exposed to atmospheric deposition of forest litter, dust, and rain. As indicated earlier, a new snowpack can have an albedo in excess of 90%; an older, ripe snowpack can have an albedo value of less than 45%.

The above discussion gives a sense of what snowpack metamorphism entails. What is needed, however, is a way to quantify snowpack metamorphism so that we can determine when a snowpack is ready to yield liquid water. A snowpack is considered ripe when it is primed to produce runoff, that is, when the temperature of the snowpack is 0°C and the liquid-water-holding capacity of the snowpack has been reached. Any additional input of either energy or liquid water will result in a corresponding amount of liquid water being released from the bottom of the pack.

Cold Content

The energy needed to raise the temperature of a snowpack to 0°C per unit area is called the *cold content*. It is convenient to express the cold content as the equivalent depth of water entering the snowpack at the surface as rain that upon freezing will raise the temperature of the pack to 0°C by releasing the latent heat of fusion (80 cal/g). Taking the specific heat of ice to be 0.5 cal/g/°C, the following relationship is obtained (Box 15.1):

$$W_c = \frac{\rho D T_s}{160} - \frac{(WE)T_s}{160} \quad \textbf{(15.3)}$$

where W_c = cold content as equivalent depth of liquid water (cm), ρ = density of snow (g/cm^3), D = depth of snow (cm), WE = snow water equivalent (cm), and T_s = average temperature deficit of the snowpack below 0°C.

Liquid-Water-Holding Capacity

Like a soil, a snowpack can retain a certain amount of liquid water. This liquid water occurs as hygroscopic water, capillary water, and gravitational water. The *liquid-water-holding capacity* of a snowpack (W_g) is calculated by:

$$W_g = f(WE + W_c) \quad \textbf{(15.4)}$$

where f = hygroscopic and capillary water held per unit mass of snow after gravity drainage, usually varying from 0.03 to 0.05 (g/g); WE = snow water equivalent (cm); and W_c = cold content (cm).

Total Retention Storage

The total amount of melt or rain that must be added to a snowpack before liquid water is released is the *total retention storage* (S_f):

$$S_f = W_c + f(WE + W_c) \quad \textbf{(15.5)}$$

When the total retention storage is satisfied, the snowpack is said to be ripe.

Box 15.1

Cold Content Calculation for a Snowpack (Snowpack Depth = 100 cm, the Snow Density = 0.1 g/cm³, and T_s = -10°C)

Because the heat of fusion equals 80 cal/g and the specific heat of ice equals 0.5 cal/g/°C, the amount of energy required to bring the temperature of the snowpack up to 0°C can be determined by:

$$WE = (0.1 \text{ g/cm}^3)(100 \text{ cm}) = 10 \text{ g/cm}^2$$

To raise the temperature of the pack (which at –10°C would be solid ice) by just 1°C would require:

$$(10 \text{ g/cm}^2)(0.5 \text{ cal/g/°C}) = 5 \text{ cal/cm}^2\text{/°C}$$

To raise the temperature of the snowpack from –10°C to 0°C then would require:

$$[0\text{°C} - (-10\text{°C})](5 \text{ cal/cm}^2\text{/°C}) = 50 \text{ cal cm}^2$$

To express the energy requirement in terms of equivalent inches of snowmelt:

$$(50 \text{ cal/cm}^2)(1 \text{ g/80 cal})(1 \text{ cm}^3\text{/g}) = 0.63 \text{ cm}$$

Cold content could have been calculated directly by:

$$W_c = \frac{(0.1)(100)(10)}{160} = 0.63 \text{ cm}$$

Snowmelt

Once the snowpack is ripe, any additional input of energy will result in meltwater being released from the bottom of the snowpack. The main sources of energy for snowmelt are essentially the same as those for evapotranspiration. To calculate snowmelt, therefore, requires either an energy budget approach or some empirical approximation of energy available for snowmelt.

Energy Budget of a Snowpack

The energy that is available either to ripen a snowpack or to melt snow can be determined with an energy budget analysis. The energy budget can be used to determine the available energy for ripening or melting, as follows:

$$M = I_s(1 - \alpha) + I_a - I_g + H + G + LE + H_r \qquad \textbf{(15.6)}$$

where M = energy available for snowmelt (cal/cm²); I_s = total incoming shortwave (solar) radiation (cal/cm²); α = albedo of snowpack (fraction); I_a = incoming longwave radiation (cal/cm²); I_g = outgoing longwave radiation (cal/cm²); H = convective transfer of sensible heat at the snowpack surface (cal/cm²), which can be + or − as a function of gradient; G = conduction at the snow-ground interface (cal/cm²); LE = flow of latent heat [condensation (+), evaporation or sublimation (−)] (cal/cm²); and H_r = advected heat from rain or fog (cal/cm²).

If all the above energy components could be measured, snowmelt could be determined directly. The amount of snow that melts from a given quantity of heat energy depends on the condition of the snowpack, which can be expressed in terms of its thermal quality. *Thermal quality* is the ratio of heat energy required to melt 1 g of snow to that required to melt 1 g of pure ice at 0°C, and is expressed as a percentage (Box 15.2). A thermal quality less than 100% indicates the snowpack is at 0°C and contains liquid water; conversely, a thermal quality greater than 100% indicates the snowpack temperature is less than 0°C with no liquid water. Snowmelt, therefore, can be determined by:

$$M = \frac{\text{total energy (cal/cm}^2)}{B\ 80\ \text{cal/g}} \quad \textbf{(15.7)}$$

where M = snowmelt (cm); and B = thermal quality, expressed as a fraction.

Under most conditions, total energy cannot be measured. Therefore, approximations, such as the generalized snowmelt equations or the temperature index (degree-day) method discussed later in this chapter, are normally used to estimate snowmelt.

The generalized snowmelt equations were developed from extensive field measurements at snow research laboratories in the western United States (see below). All major energy components for forest cover conditions are considered. The following discussion indicates how the generalized snowmelt equations were developed by considering each major source of energy separately.

Solar Radiation. The amount of solar radiation reaching a snowpack surface is dependent upon the slope and aspect of the surface, cloud cover, and forest cover. In the Northern Hemisphere, south-facing slopes receive more radiation than north-facing slopes. The more moderate the slope, the more moderate the effect that slope has on solar radiation. By determining the slope and aspect of an area, the corresponding potential solar radiation for a particular latitude can be determined.

The amount of solar radiation striking a snowpack surface in the open is a function of the percentage of cloud cover and the height of the clouds (Fig. 15.1). The corresponding melt that occurs from solar radiation is then defined as:

$$M = \frac{I_s(1 - \alpha)}{B\ 80} \quad \textbf{(15.8)}$$

A ripe snowpack with a 3% liquid-water content reduces the above relationship to:

$$M = 0.0129 I_s\,(1 - \alpha) \quad \textbf{(15.9)}$$

In forests, the percentage of incoming solar radiation that reaches the snow surface depends largely upon the type of cover, density, and condition of the forest canopy. As the density of a forest canopy increases, incoming solar radiation decreases exponentially (Fig. 15.2). As a result, with dense canopies, the effect of forest cover overrides the effects of cloud cover previously discussed. Coniferous forest cover can reduce substantially the amount of solar radiation that reaches a snow surface. In contrast, deciduous forests have less of an effect on solar radiation.

Box 15.2

Determining Snow Thermal Quality by the Calorimeter Method [Snow Depth = 60 cm, Volume of Snow Sample Taken = 15,000 cm³, and Weight of Sample = 2000 g (modified from Hewlett 1982)]

The above sample was placed in a thermos that initially contained 7000 g of water at a temperature of 32°C. After the snow sample was added and all the snow melted, the temperature of the water in the thermos was 8°C. The following calculations were made:

$$\text{snow density} = \frac{2000\ g}{15{,}000\ \text{cm}^3} = 0.133\ \text{g/cm}^3$$

snow water equivalent $(WE) = (0.133)\ (60\ \text{cm}) = 8.0\ \text{cm}$

First, determine the heat energy needed to raise the temperature of the melted snow water from 0 to 8°C:

$$(2000\ \text{g})(8°\text{C})(1\ \text{cal/g/°C}) = 16{,}000\ \text{cal}$$

The heat available energy was:

$$(7000\ \text{g})(32°\text{C} - 8°\text{C})(1\ \text{cal/g/°C}) = 168{,}000\ \text{cal}$$

The heat available to melt ice would equal:

$$168{,}000\ \text{cal} - 16{,}000\ \text{cal} = 152{,}000\ \text{cal}$$

Therefore,

$$\text{ice content} = \frac{152{,}000\ \text{cal}}{80\ \text{cal/g}} = 1900\ \text{g}$$

$$\text{thermal quality}\ (B) = \frac{1900\ \text{g}}{2000\ \text{g}} = 0.95 = 95\%$$

The snowpack contained 5% liquid water and was at a temperature of 0°C.

Longwave Radiation. Snow absorbs and emits nearly all incidental longwave, or terrestrial, radiation. The narrow range of snowpack temperatures limits the amount of radiation emitted from a snowpack. Because a snowpack temperature cannot exceed 0°C, the maximum longwave radiation a snowpack can emit is 0.459 cal/cm²/min.

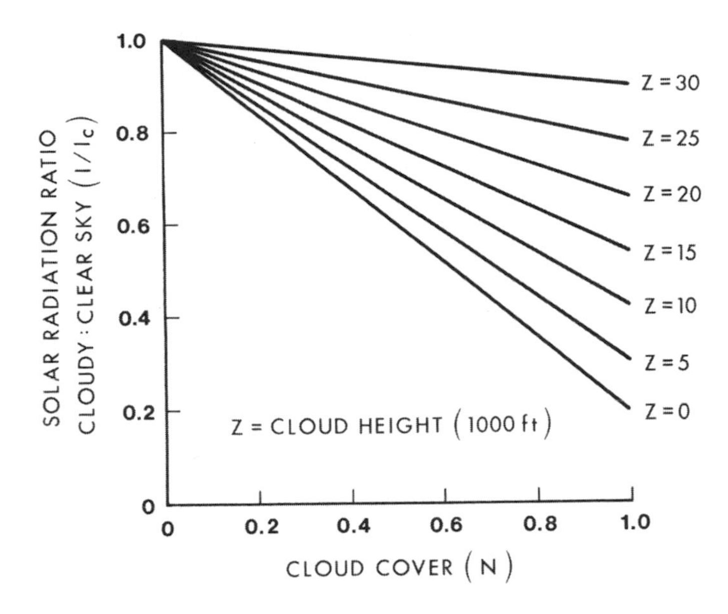

Figure 15.1. Relationship between solar radiation and cloud height and cover (from U.S. Army Corps of Engineers 1956).

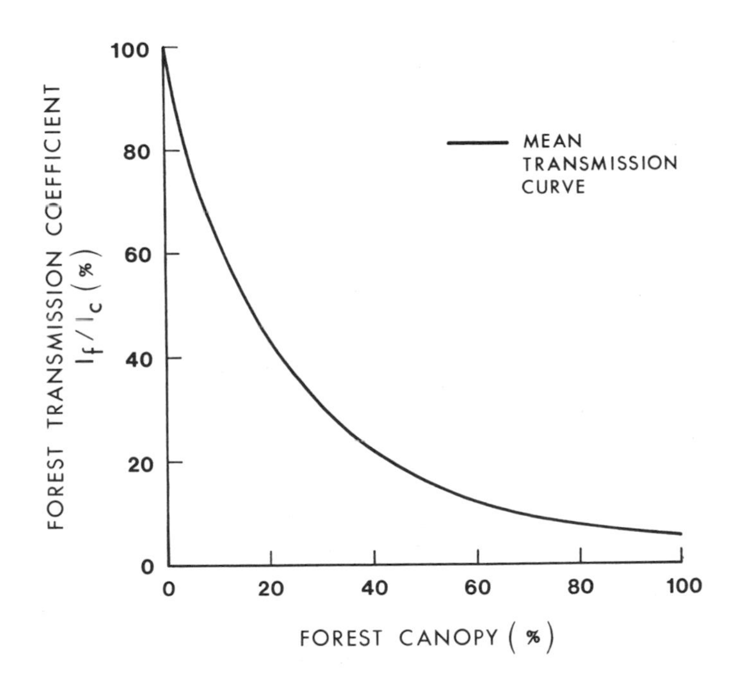

Figure 15.2. Relationship between conifer forest canopy density and transmission of solar radiation (from U.S. Army Corps of Engineers 1956).

Net longwave radiation at a snow surface is determined largely by overhead back radiation from the earth's atmosphere, clouds, and forest canopies. Back radiation from the atmosphere is a function of the water content of air. Over snowpacks, the vapor pressure of air usually varies between 3 and 9 mbar pressure, resulting in a fairly constant rate of longwave reradiation of 0.757 cal/cm²/min. Therefore, under clear skies, the net longwave radiation at a snow surface is approximately:

$$R_c = 0.757\sigma T_a^{\,4} - 0.459 \text{ under clear skies} \qquad \textbf{(15.10)}$$

where Rc = net longwave radiation (cal/cm²/min), σ = Stefan-Boltzmann constant = 0.826×10^{-10} (cal/cm³/min/°K⁴), and T_a = air temperature (°K).

When clouds are present, the temperature at the base of the clouds determines the back radiation to the snow surface, and the net longwave radiation becomes:

$$R_{cl} = \sigma\,(T_c^{\,4} - T_s^{\,4}) \qquad \textbf{(15.11)}$$

where R_{cl} = net longwave radiation (under cloudy skies), T_c = temperature at the base of clouds (°K), and T_s = snowpack temperature (°K).

Under partial cloud cover, the loss of longwave radiation from the snowpack is:

$$R_c\,(1 - K_c N) \qquad \textbf{(15.12)}$$

where $K_c = f$ (cloud type and ceiling) and N = portion of the sky covered by clouds (fraction).

Similarly, for a snowpack beneath a dense conifer canopy, the net longwave radiation (R_f) is:

$$R_f = \sigma\,(T_f^{\,4} - T_s^{\,4}) \qquad \textbf{(15.13)}$$

where T_f = temperature of the underside of the forest canopy (°K).

Because temperatures of the trees are rarely measured, T_f is often estimated from air temperature (T_a). Under clear-sky conditions and a ripe snowpack, the net longwave radiation under a conifer canopy is:

$$R_{lw} = \sigma T_a^{\,4}[F + 0.757\,(1 - F)] - 0.459 \qquad \textbf{(15.14)}$$

where F = canopy density (decimal fraction).

Snowmelt that is caused by net longwave radiation under conditions of dense forest cover or low clouds can be approximated by:

$$M_{lw} = 0.142T_a \qquad \textbf{(15.15)}$$

where M_{lw} = snowmelt caused by longwave radiation (cm) and T_a = air temperature measured at 2 m above the snowpack (°C).

Net Radiation. The net radiation (all waves) that is available for snowmelt is governed largely by forest cover conditions and cloud conditions. Because watershed management activities can affect forest cover density directly, we will focus on this aspect.

There is a trade-off between shortwave and longwave radiation at a snowpack surface as the forest cover changes. As forest cover increases, the solar radiation at the snowpack

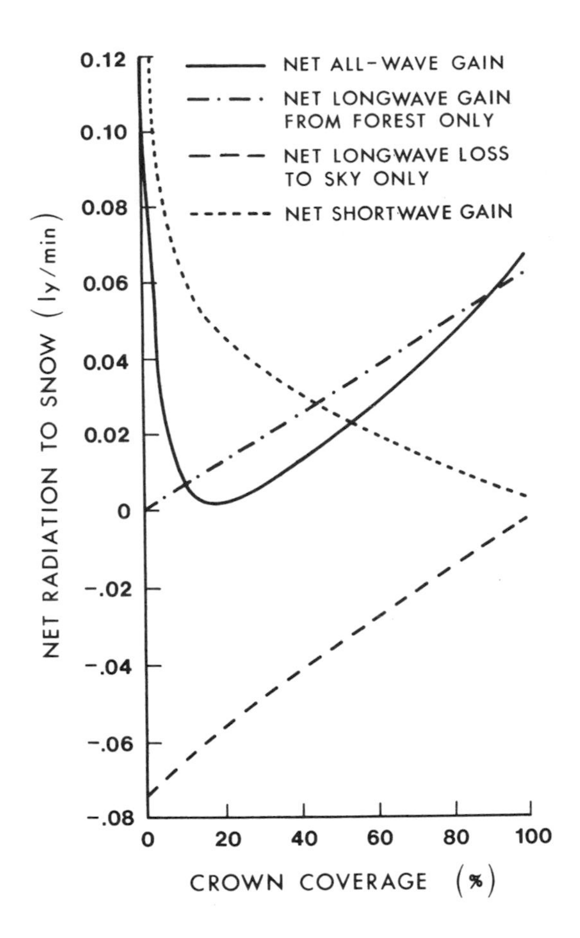

FIGURE 15.3. Net radiation of snowpack as related to forest canopy coverage (from Reifsnyder and Lull 1965).

surface is reduced greatly; the longwave radiation loss from the snowpack is reduced; and the longwave gain component from the canopy increases (Fig. 15.3). Between 15 and 30% canopy cover, net radiation at the snowpack surface is at a minimum; net radiation is highest at 0% cover, but it is also relatively high at dense forest canopy conditions because of the much higher net longwave component. These relationships have significant implications for forest and snowpack management, which will be discussed later.

Convection-Condensation Melt. Heat energy can be added to snowpack by the turbulent exchange of sensible heat from the overlying air and by direct condensation on the snow surface. In general terms, this energy exchange can be approximated by:

$$q_e = A_e \frac{dq}{dz} \tag{15.16}$$

where q_e = energy exchange (cal/cm²/time), A_e = exchange coefficient, and dq/dz = vertical gradient of temperature or water vapor.

For a ripe snowpack, the snow surface would have a temperature of 0°C, with a corresponding vapor pressure of 6.11 mbar. Using these values for the snow surface and determining air temperature and relative humidity measurements in the air above the snow surface allow us to estimate the energy exchange.

Snowmelt from convection can be estimated from:

$$M_c = 0.0005 \frac{P}{P_o} (z_a z_b)^{-0.167} T_a v \qquad \textbf{(15.17)}$$

where M_c = convective snowmelt (cm); P, P_0 = air pressures at the snow surface and at sea level (mbar), respectively; z_a, z_b = heights (m) above the snow surface at which air temperature and wind speed are measured, respectively; and v = wind speed (km/day).

Snowmelt from condensation can be approximated from a similar relationship:

$$M_e = 0.0024 (z_a z_b)^{-0.167} (e_a - e_s) v \qquad \textbf{(15.18)}$$

where M_e = condensation snowmelt (cm), e_a = vapor pressure of the air at height z_a (mbar), and e_s = vapor pressure of the snow surface (6.11 mbar for a melting snow surface).

The above equations can be combined and simplified to provide an estimate of melt due to convection-condensation. In doing so, the P/P_o ratio is assumed to be a constant that can have values from 0.7 to 1.0 (at sea level); 0.8 is assumed for mountainous areas. Also, dew point temperatures can be used to represent vapor pressures. Under these assumptions, and assuming a ripe snowpack (T_s = 0°C), the combined equation becomes:

$$M_{ce} = 0.00078v (0.42T_a + 1.51T_d) \qquad \textbf{(15.19)}$$

where M_{ce} = convection-condensation melt (cm/day), v = mean wind speed (km/day), T_a = mean air temperature (°C at 2 m above the snowpack), and T_d = mean dew point temperature (°C at 2 m above the snowpack).

Rain Melt. Snowmelt caused by the addition of sensible heat from rainfall is relatively small. It can be estimated from air temperature and rainfall measurements for a snowpack with a thermal quality of 100% as:

$$M_p = \frac{P_r T_a}{80 \text{ cal/g}} = 0.012 P_r T_a \qquad \textbf{(15.20)}$$

where M_p = daily snowmelt (cm); P_r = daily rainfall (cm); and T_a = average daily air temperature (°C).

Again, the above relationship holds for a ripe snowpack. If the snowpack has a temperature below 0°C, additional energy can be released to the pack by virtue of the release of the heat of fusion (80 cal/cm of rain that freezes).

Conduction Melt. Heat energy can be added to the base of a snowpack by conduction from the underlying ground. For any given day, the amount of energy available by conduction and therefore the amount of melt are relatively small. Daily values of 0.5 mm are frequently assumed, and sometimes this component is simply ignored if snowmelt is being estimated over a short period. For seasonal estimates of snowmelt, however, conduction melt (ground melt) can be significant. Monthly totals in the central Sierra Nevada in California ranged from 0.3 cm in January to 2.4 cm in May (U.S. Army Corps of Engineers 1956).

Combined Generalized Basin Snowmelt Equations

The relationships discussed in the preceding sections provide the foundation for a set of generalized equations that can be used to estimate snowmelt for a small watershed (U.S. Army Corps of Engineers 1960). Because of the influence of forest cover on energy exchange, the forest cover condition of a watershed determines which equation to use. For rain-free periods, the daily melt from a ripe snowpack (isothermal at 0°C, with a 3% liquid-water content) is calculated by one of the following equations, which are differentiated on the basis of forest canopy cover (F):

Heavily Forested Area ($F > 0.80$):

$$M = 0.19T_a + 0.17T_d \quad \textbf{(15.21)}$$

Forested Area ($F = 0.60 - 0.80$):

$$M = k\,(0.00078v)\,(0.42T_a + 1.51T_d) + 0.14T_a \quad \textbf{(15.22)}$$

Partly Forested Area ($F = 0.10 - 0.60$):

$$M = k'\,(1 - F)\,[0.01I_s\,(1 - \alpha)] + k\,(0.00078v)\,(0.42\,T_a + 1.51\,T_d) + F\,(0.14T_a) \quad \textbf{(15.23)}$$

Open Area ($F < 0.10$):

$$M = k'\,(0.0125I_s)\,(1 - \alpha) + (1 - N)\,(0.014T_a - 2.13) + N\,(0.013T_c) + k\,(0.00078v)\,(0.42T_a + 1.51T_d) \quad \textbf{(15.24)}$$

where M = daily melt (cm); T_a = air temperature (°C) at 2 m above the snow surface; T_d = dew point temperature (°C) at 2 m above the snow surface; v = wind speed (km/day); I_s = observed or estimated solar radiation (cal/cm²/day); α = snow surface albedo (decimal fraction); k' = basin shortwave radiation melt factor, which depends on the average exposure of the open areas to solar radiation compared to an unshielded horizontal surface; F = average forest canopy cover for watershed (expressed as a decimal fraction); T_c = cloud-base temperature (°C); N = cloud cover (expressed as a decimal fraction); and k = basin convective-condensation melt factor, which depends on the relative exposure of the watershed to wind.

The only empirical fitted parameters in the above equations are k and k'. The basin shortwave radiation melt factor (k') can be estimated from solar radiation data for a given latitude, slope, and aspect. Because one has to average several slopes and aspects for a given watershed, the k' value represents an average for the watershed. In general, watersheds with a southern exposure would tend to have a $k' > 1$; those with northern exposures would have a $k' < 1$. A watershed that has a balance between north- and south-facing slopes would have a $k' = 1.0$. The basin convective-condensation melt factor (k) is similarly an average value for a particular watershed that indicates the exposure of the snowpack to wind; values range from $k = 1.0$ for open areas to $k = 0.8$ for dense forest cover.

Temperature Index Method

The data requirements for the generalized snowmelt equations restrict their use in many situations, particularly for remote areas and for day-to-day operational conditions. As a result, simplified methods that depend only on air temperature data have been developed to estimate snowmelt.

The temperature index method is an empirically derived equation of the form:

$$M = MR\,(T_a - T_b) \tag{15.25}$$

where M = daily snowmelt (cm); MR = melt-rate index or degree-day factor (cm/°C-day); T_a = daily air temperature value, usually either the average daily or maximum daily temperature (°C); and T_b = base temperature, at which no snowmelt is observed (°C).

The difference $T_a - T_b$ yields the degree-days of heat energy available for melt. The melt-rate index relates this heat energy to snowmelt that occurs on the watershed. Although we know that the air temperature is but one source of energy for snowmelt, it correlates well with radiation inputs, particularly during the snowmelt season, and it is also a reasonable index for forested conditions. The equation for the temperature index method (Eq. 15.25) is derived from regression analysis of daily melt versus air temperature (Fig. 15.4). The melt-rate index is the slope of the regression line, and the base temperature (T_b) is that air temperature at which no melt is observed. Examples of temperature index coefficients for three watersheds are presented in Table 15.2.

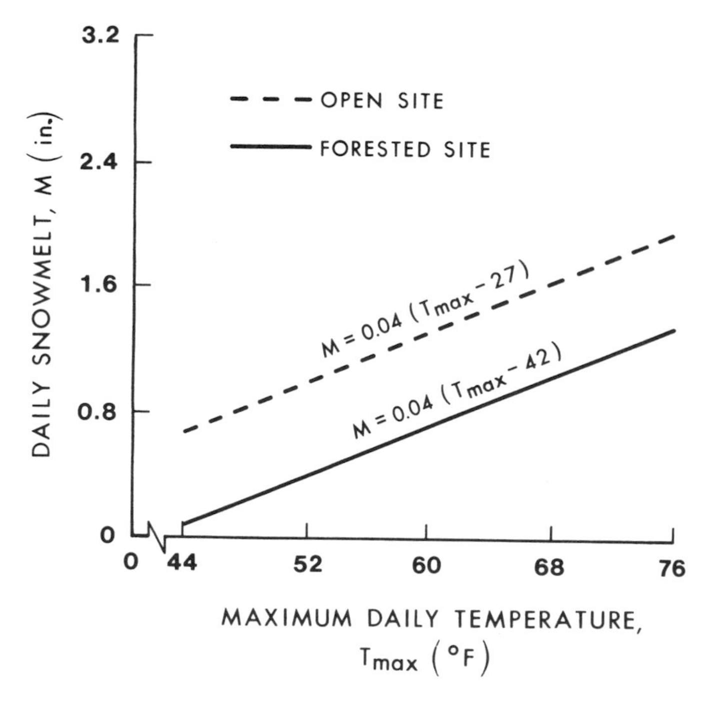

FIGURE 15.4. Temperature index relationships for maximum daily air temperatures (from U.S. Army Corps of Engineers 1960).

TABLE 15.2. Base temperature and melt-rate indices for three watersheds in the western United States

	Mean Temperature			Maximum Temperature		
Watershed	Base T °F	Melt-Rate Index		Base T °F	Melt-Rate Index	
		Apr	May		Apr	May
Central Sierra, CA	26	0.036	0.062	29	0.020	0.038
Upper Columbia, MT	32	0.037	0.072	42	0.109	0.064
Willamette, OR	32	0.039	0.042	42	0.046	0.046

Source: From U.S. Army Corps of Engineers 1956.

Because the temperature index method is developed for a particular watershed, both the melt-rate index and the base-temperature values can differ from one area to the next. Furthermore, the degree-day factor or melt-rate index can vary seasonally rather than, as is often assumed, remain constant (Rango and Martinec 1995). The elevation difference and proximity of the air temperature station to the watershed will affect the values derived.

The temperature index method has been used successfully for over 60 yr in mountainous watersheds. It has been shown to be a reliable alternative to the more physically based, and data-demanding, energy budget methods in predicting snowmelt for computer simulation models (Rango and Martinec 1995).

FOREST MANAGEMENT—SNOWPACK MANAGEMENT RELATIONSHIPS

Because trees affect snow accumulation and melt, it follows that snow accumulation and melt patterns can be changed by manipulating forest density. Much of the snowmelt runoff in temperate regions of the world derives from high-elevation forested zones, which suggests even further the possibilities of using forest management practices to enhance snowmelt water yield. Two questions must be asked in considering these possibilties, however. Can water yield be enhanced without increasing the potential for flooding from snowmelt runoff? Also, are snowpack management practices compatible with other forest uses?

Many studies have indicated that forest management affects snowpacks in ways that can produce greater snowmelt runoff, but with the potential to also increase flooding. Furthermore, thinning and clearing of forest overstories to increase water yield can frequently be made compatible with the demands for wood, forage, wildlife, and recreational use of forest lands.

The possibility of increasing the water yield from forested watersheds appears to be greater for snow than for rainfall in many temperate areas. Snow generally accumulates on forested sites throughout winter, providing a large reservoir of stored water potentially available for use in the spring. For example, the Salt-Verde basin in central Arizona often stores between 3 and 6 billion m^3 of water as snow before the beginning of snowmelt in the spring. If snowmelt water yields were increased by 10%, an additional 300–600 million m^3 of water could be captured annually.

Enhancing Water Yield

Forest cover can be altered in basically two ways to increase recoverable water yield: reduce forest densities by thinning and remove forest overstories by clearcutting in different spatial arrangements. More snow accumulates in sparsely stocked forest stands and in small clearings in forest stands than in dense conifer stands (Table 15.3). These greater accumulations can contribute to increased runoff, particularly when such increases occur in areas that already have wetter soils. The problem is one of determining the most efficient method of cutting the forest that is also compatible with other forest management objectives.

Thinning forest stands has been shown to increase snowpack accumulations, although the increases are often more variable than those observed after clearcutting (Meiman 1987). The stand structure following thinning is often an irregular pattern of small, randomly spaced openings. Although snow water equivalents are often increased because of thinning, the resultant effects on water yield are usually slight. Creating small clearcuts in forest stands has also been shown to result in increased snowpack accumulations. Depending on the size, shape, and orientation of the openings, such clearings can also result in increased water yield.

Deposition and redistribution processes of snowpacks in and adjacent to forest openings involve at least three factors: (1) more snow can accumulate in openings than under forest overstories during snowfall events due to wind eddies in the openings; (2) snowpacks in openings can be augmented during and after storms by snow blown from surrounding forest canopies; and (3) the snowpack can be greater in openings than under forest canopies because there are no interception losses, although in some regions snow interception losses may be minimal.

Ablation (i.e., melt and evaporation) factors affecting snowpack profiles involve variations in shortwave and longwave radiation fluxes. Shaded sides of openings receive less solar radiation than exposed sites. Observations suggest that the highest rate of ablation occurs along the north side of east-west clearcut strips. Exposure to more solar radiation explains such observations.

Rates of ablation are also influenced by spatial variations in longwave radiation emission from trees adjacent to an opening. Trees exposed to solar radiation are warmed and emit more longwave radiation than trees not warmed by solar radiation. A snowpack absorbs almost all of the longwave radiation striking it. Therefore, snow on the side of a forest opening exposed to solar radiation is likely to be exposed to more longwave radiation (emitted from adjacent, heated trees) than snow on the shaded side. Consequently, a combination of shortwave and longwave radiation causes, at least in part, the relatively high rates of ablation along exposed sides of forest openings.

It is difficult to isolate the effects that all the processes of deposition, redistribution, and ablation have on snowpack profiles extending through openings and into adjacent forests. The profiles represent the net result of all these processes prior to the time of observation. However, several studies have shown some consistent responses that have led to guidelines for enhancing water yield (Box 15.3).

Snowmelt-Runoff Efficiency

The amount of water derived from snowmelt is determined largely by the dynamics of snowpack accumulation and melt and the snowmelt-runoff efficiency. *Snowmelt-runoff*

TABLE 15.3. Increase in maximum snow accumulation (water equivalent) after cutting in western conifer forest

Location and Forest Cover	Treatment	Maximum Increase in Snow Accumulation (in.)	(%)
Fraser Experimental Forest, CO	Uncut: 11,900 fbm*	0.0	
Mature lodgepole pine	Cut (residual volume):		
	6000 fbm	0.81	12
	4000 fbm	1.01	15
	2000 fbm	1.49	21
	0 fbm	1.99	29
Young lodgepole pine	Heavy thinning (reduction from 4400 to 630 trees per acre)	2.3	23
	Light thinning (reduction from 4400 to 2000 trees per acre)	1.7	17
Mature Englemann spruce-subalpine fir	Removal of 60% of volume by strip cutting and group and single-tree selection	2.8	22
Front Range, Rocky Mountains, CO			
Ponderosa pine and Douglas-fir	Selection cut	0.45	6
	Commercial clearcut	1.21	29
North central Wyoming			
Lodgepole pine	Clearcut blocks	2.5	40
Willamette Pass, OR			
Mountain hemlock and true fir	Strip-cut, strips two chains (about 40 m) wide	5	15
Central Sierra Snow Laboratory, CA (NE)			
Red fir	Clearcut	11	23
	East-west strip, one tree-height wide	12	26
	Block cutting, one tree-height wide	15	34
	Selection cutting; crown cover reduced from 90% to 50%	2	5
	90% to 35%	9	19
	Commerical selection cut	7	14
	Wall-and-step forest	19	25
Minnesota	Single-tree selection	0.1	4
Black spruce	Shelterwood	0.7	28
	Clearcut strip	1.5	60
	Clearcut patch	2.0	80

Source: From Anderson et al. 1976.
*fbm = foot board measure (board feet per acre).

Box 15.3

Effects of Strip Cutting Forests on Snowpacks and Water Yield

Strip cutting lodgepole pine forests in Colorado has been shown to affect wind patterns, snow accumulation, and melt (Gary 1975). When strips were cut at a width equal to from one to five times the height of surrounding trees, snow water equivalents were from 15 to 35% higher in the strips than in the adjacent forest. Nearly 30 yr after harvest, average peak snow water equivalent in the cut watershed still remained 9% above that of an uncut watershed (Troendle and King 1985). Increases in annual flows above predicted levels are decreasing slowly in response to the regrowth of forests. Apparently, one-third of the increase in annual flows is attributed to this net increase in snow water equivalent; two-thirds of the increase is largely due to reductions in evapotranspiration.

Similar results have been observed in ponderosa pine forests in Arizona. The snow water equivalent within the cut portion of a strip that was cut at a width equivalent to the height of the surrounding forest was increased 60%, or about 3 cm, compared to the uncut forest (Ffolliott and Thorud 1974). In the figure below, the snowpack in the forest adjacent to the strip cut was reduced, resulting in a zone of influence that was much wider than the cut itself. Brown et al. (1974) reported on the cut of a mature stand of ponderosa pine on a watershed in the same area; strips equal to the height of the trees were cut in between strips two times the height of the trees. They found that this treatment increased water yield by about 3 cm, and the effect was sustained for at least 5 yr.

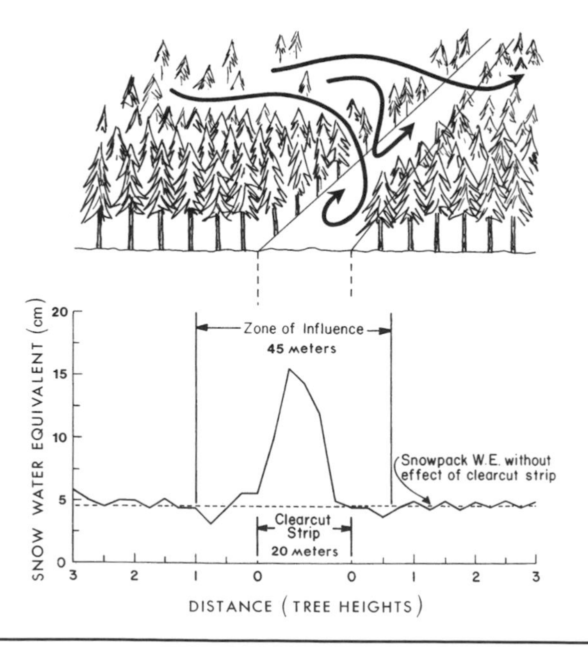

Effects of strip cutting a ponderosa pine forest in Arizona on wind patterns and snow deposition (adapted from Ffolliott and Thorud 1974).

efficiency, defined as the portion of the snowpack on-site that subsequently is converted into runoff, can be derived as:

$$SRE = \frac{Q}{P - SWE} 100 \tag{15.26}$$

where SRE = snowmelt-runoff efficiency (%), Q = runoff (cm), P = precipitation input (cm), and SWE = change in snowpack water equivalent (cm) (a decrease is negative).

Snowmelt-runoff efficiency can vary considerably from one watershed to the next or from year to year on a given watershed, due to climatic patterns and changes in watershed condition. Snowmelt-runoff efficiencies for watersheds in the southwestern United States have varied from 20 to 45% on a given watershed from peak snowpack accumulation to the end of snowmelt runoff. The amount of snowfall and the timing of snowmelt account for much of the variability. Within a given region, snowmelt-runoff efficiencies on individual watersheds have ranged from 25 to 85%, with much of this difference being attributed to physiographic features.

In general, snowmelt-runoff efficiency depends largely upon watershed slope, soil depth, soil texture and structure, vegetative characteristics, and climatic variables. The presence and type of soil frost that can occur have a pronounced effect on runoff efficiency in the western mountains and higher latitudes of North America. Other variables of importance include moisture conditions before the snow season, peak snowpack accumulation, and duration of snowmelt runoff. As a rule, high efficiencies occur on watersheds that experience concrete soil frost, heavy precipitation inputs before the start of snowpack accumulation, deep snowpacks at the peak of accumulation, or large amounts of rainfall (rain on snow) during the snowmelt season.

Effects of Forest Management on Flooding from Snowmelt

The earlier discussion focused on the effects of using different clearcutting configurations to augment water yield. In the process of increasing water yield from snowmelt, a concern for water resource managers is the potential for increased snowmelt flooding.

The results of controlled watershed experiments indicate that the effect of forest cutting on snowmelt streamflow peaks and volumes varies from one area to the next, and even from year to year. When small catchments are 100% clearcut, the odds are that snowmelt runoff will increase. Whether such increased runoff results in an increase in flooding downstream depends on many factors, including the condition of other parts of the basin that contribute flow to flood-prone areas and the basin characteristics that affect timing and routing of flow.

In general, peak discharges from snowmelt have increased after forest removal. Exceptions have been observed in the coastal range of the Pacific Northwest. Harr and McCorison (1979) reported a 32% reduction in snowmelt peak size following clearcutting in Oregon and suggested that condensation-convection melt from snow on tree crowns occurred at a higher rate than snowmelt at ground level (where most of the snow would be after clearcutting). More recent work has clarified the situation further (Box 15.4).

Changes in snowmelt runoff are often attributed to changes in the timing of the melt, which can be managed by manipulating forest cover to alter (or desynchronize) the timing from different parts of a watershed or river basin. This effect has been observed in

Box 15.4

Rain-on-Snow Events and the Effect of Forest Canopy

Some of the most severe flooding in areas that receive snowfall is attributed to rain-on-snow events. The combination of rainfall and the large influxes of sensible heat that are associated with rainstorms can add large amounts of liquid water to the soil surface. Rain-on-snow floods have been particularly frequent in areas that have a transient snowpack: areas where the air temperature frequently hovers around 0°C during the winter. Such is the case in the Pacific Northwest of the United States, where influxes of warm, moist air from the Pacific Ocean can result in high streamflows and in saturated soils that lead to frequent landslides.

Berris and Harr (1987) studied the influence of forest cover on snow accumulation and melt during rain-on-snow events in the western Cascades of Oregon. Their work suggests:

- If air temperature is near 0°C when snow falls or if snow is present on the forest canopy when rain occurs, higher outputs of water occur from forested areas than from cleared areas. The snow-covered canopy offers a greater surface area exposed to convection-condensation processes than the snowpack surface in a cleared area does; more rapid melt occurs from the snow on the canopy.
- If no snow is present on the forest canopy when rain occurs and rainfall rates exceed 5 mm/hr, clearcut areas yield more water than forested areas once the snowpack is ripe. Wind accentuates these differences.

These apparently conflicting results pose a dilemma for the forest manager trying to prevent high streamflows and landslides. If snowfalls are interspersed with rainstorms, situation 1, above, could lead to high streamflows and sufficient water input to saturate soils and promote landslides in forested areas. If snow accumulates over time without appreciable canopy-interception melt, then situation 2 could lead to similar problems from clearcut areas. Although management guidelines may be difficult to establish in such an instance, the most reasonable approach might be to maintain a diversity of cover conditions on a watershed and restrict the percentage of a watershed that can be clearcut at any one point in time.

Minnesota where a mature mixed aspen stand on a watershed was cleared only partially in two successive years (Verry et al. 1983). When forest cover from one-half of the watershed was cleared, the snowmelt-runoff peak was reduced (1971 in Figure 15.5); when more than 70% of forest cover was cleared the next year, snowmelt peak discharge was nearly doubled (1972 in Figure 15.5). Open and forested sites experience snowmelt at such different rates and times that they contribute to streamflow at very different times. As the percentage of a watershed that is clearcut exceeds 60%, snowmelt peak discharge will be expected to increase in Minnesota. Furthermore, the effects can persist for several years (through 1979 in Figure 15.5). Other areas with similar climate, vegetation, and topography might experience a similar response to forest clearing.

Clearing patches of forest cover in mountainous areas would not be as effective as in the above example in reducing peak discharges. Runoff from mountainous water-

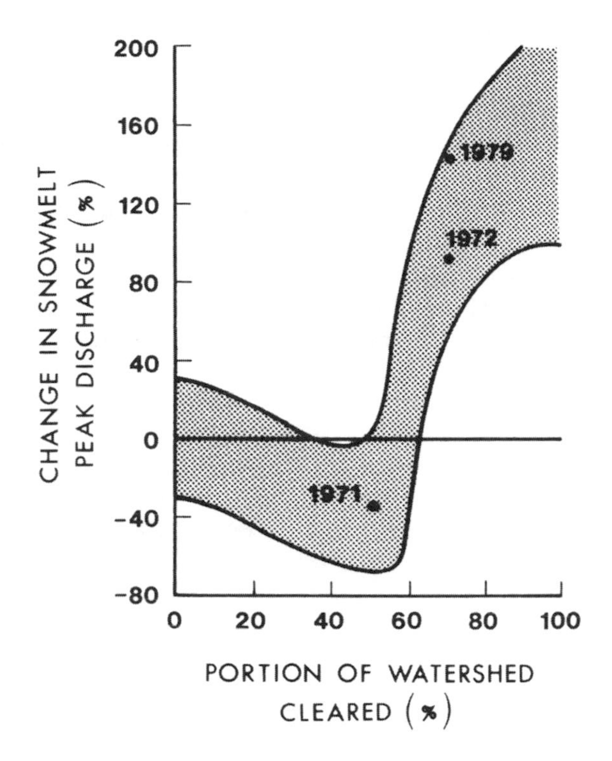

FIGURE 15.5. Relation between the portion of a watershed clearcut and the change in annual snowmelt peak discharge compared to a mature stand of hardwoods in Minnesota (from Verry et al. 1983).

sheds is already more desynchronized than it is on flat watersheds because of differences in elevation, slope, and aspect.

In cold continental climates, changes in forest cover can also affect snowmelt runoff by altering soil frost conditions and, thereby, runoff efficiency. Snowmelt on frozen soil runs off the soil surface rapidly. Soils are usually wetter in the fall under clearcut conditions, and they tend to freeze deeper and experience concrete frost more often. If large areas within a watershed are converted from forest cover to agricultural croplands or are under clearcut conditions, the combination of more rapid snowmelt and more rapid (and more efficient) runoff can lead to more frequent downstream flooding.

Cumulative Watershed Effects

Land use activities and environmental change on the watershed can affect the dynamics of snowpack accumulation and its subsequent melt. These effects can vary spatially and temporally and will influence overall resource management on a watershed.

Snow accumulation and its water equivalent are generally higher in sparsely stocked forest stands and small openings than in adjacent highly stocked forest covers (Table 15.3). Therefore, increased snowpack accumulations are a common result of thinning forest stands or clearing forest overstories; these activities, either singly or together, are components of silvicultural prescriptions for many forest communities.

Snowpacks melt more quickly in sparsely stocked stands and small clearings than under dense forest canopies because of the greater air circulation and increased solar

Box 15.5

Constraining Timber Harvesting to Reduce Excessive Snowmelt Runoff

A generic environmental impact statement was prepared in Minnesota to address the statewide effects of expanded timber harvesting (Jaakko Pöyry Consulting, Inc. 1994). The effects on snowmelt flooding were considered in this analysis. In addressing the effects of proposed scenarios of different levels of timber harvesting, the effects of changing percentages of forest cover on watersheds were analyzed, using the relationship in Figure 15.5. The net effect of changing forest cover on snowmelt peaks had to be considered along with existing percentages of open area and the percentage of forest cover that was 12 yr or younger. Clearcut areas, areas with young stands, and other open areas had to be considered in aggregate to determine how changes in percentage of forest cover impacted snowmelt peak flows.

radiation. In combination with the often higher snowpack accumulations on these sites, the increased melt rates can generate higher peak flows from a watershed than otherwise expected (Box 15.5). The resulting higher peak flows affect both the rate of water delivery and the transport of sediment, chemical, and other watershed products to points downstream.

Increased flood peaks can occur during rain-on-snow events after clearcutting on watersheds with transitional snowpacks (Harr 1986). Urbanization causes earlier melting of snow and increased peak flows, especially from rain-on-snow events (Buttle and Xu 1988). Snowmelt is also increased along roadways, both surfaced and unsurfaced. It is likely that changes in snowmelt-runoff patterns will continue to accelerate with the increasing fragmentation of natural ecosystems due to expanding urbanization. The consequences of these changes will likely confront watershed managers in the years to come (Reid 1993).

Methods for Predicting Snowpack-Snowmelt Relationships

The basic information and empirical studies discussed have been incorporated into methods for predicting snowpack-snowmelt relationships; these methods range from simple regression equations to complex computer simulation models. Although a detailed discussion of all such models is beyond the scope of this book, examples are presented to illustrate the conceptual nature of existing methods and models.

Prediction of Snowmelt

As discussed earlier, several methods are available to estimate snowmelt, ranging from the simple temperature index method, to the generalized snowmelt equations, to detailed energy budget models. The first two methods have been used most frequently in practice [e.g., the Hydrologic Engineering Center's HEC-1 and streamflow simulation and reservoir routing (SSARR) models include both methods]. These methods are used mostly for engineering design and streamflow forecasting; because of their empirical nature, they have limitations in evaluating the effects of forest cover change on snowmelt runoff. More theoretical or physically based models are needed to more accurately predict the effects of forest cover on snowpack accumulation, melt, and the resulting streamflow discharge. The principal components of a snow accumulation and melt model are illustrated in Figure 15.6.

Prediction of Streamflow from Snowmelt

Modeling snowpack dynamics is but one part of predicting snowmelt runoff, as discussed in the section on snowmelt-runoff efficiency. The entire process, from snowfall to streamflow, requires that all of the components illustrated in Figure 15.7 be considered. All the processes need to be considered either explicitly or implicitly in a snowmelt-runoff model. The degree to which processes are simulated within a given model varies with the purpose of the model (Box 15.6). The model presented by Anderson (1978) is a point energy and mass balance model of snow cover that is theoretically complete but also complex. In contrast, the SSARR model was developed for routine operational streamflow forecasting and therefore has minimal data and initialization requirements (U.S. Army Corps of Engineers 1972). Many of the energy budget relationships and snowpack condition characteristics must be simplified for operational use.

Some simplified relationships have been developed to estimate streamflow from snowmelt. For example, if watersheds have been studied during previous snow accumulation-melt seasons, one might be able to estimate the water available for streamflow runoff (Q_m) from:

$$Q_m = M_o A_s (SRE) \tag{15.27}$$

where M_o = daily snowmelt (cm/day), and A_s = snow-covered area on the watershed (area units).

Snowmelt-runoff efficiency can be entered directly to solve the above if the appropriate values are known for the watershed. If not known, *SRE* values can be calculated for specific conditions from known relationships for the area.

Equation 15.27 could be used with the temperature index method to predict the fraction of the meltwater that appears as streamflow on a given day. Because snowmelt water yield is often discharged from a watershed primarily as movement through the soil to the stream channel, all of the snowmelt from a single day does not necessarily appear as streamflow on that same day. Instead, only a fraction can appear on that day, and the rest can be spread out over several days.

Approaches have been outlined to compute streamflow for a given day based on snowmelt from previous days and any existing channel flow (Rango and Martinec 1979). In essence, these methods compute a recession coefficient of streamflow from a watershed. This coefficient is calculated as the ratio of one day's streamflow divided by that of the previous day. It is used in relationships such as the following:

$$Q_n = J_n (1 - k_b) + Q_{n-1} k_b \tag{15.28}$$

where Q_n = streamflow on day n (m^3), J_n = water released from the snowpack available for streamflow on day n (m^3), Q_{n-1} = streamflow on previous day (m^3), and k_b = recession coefficient.

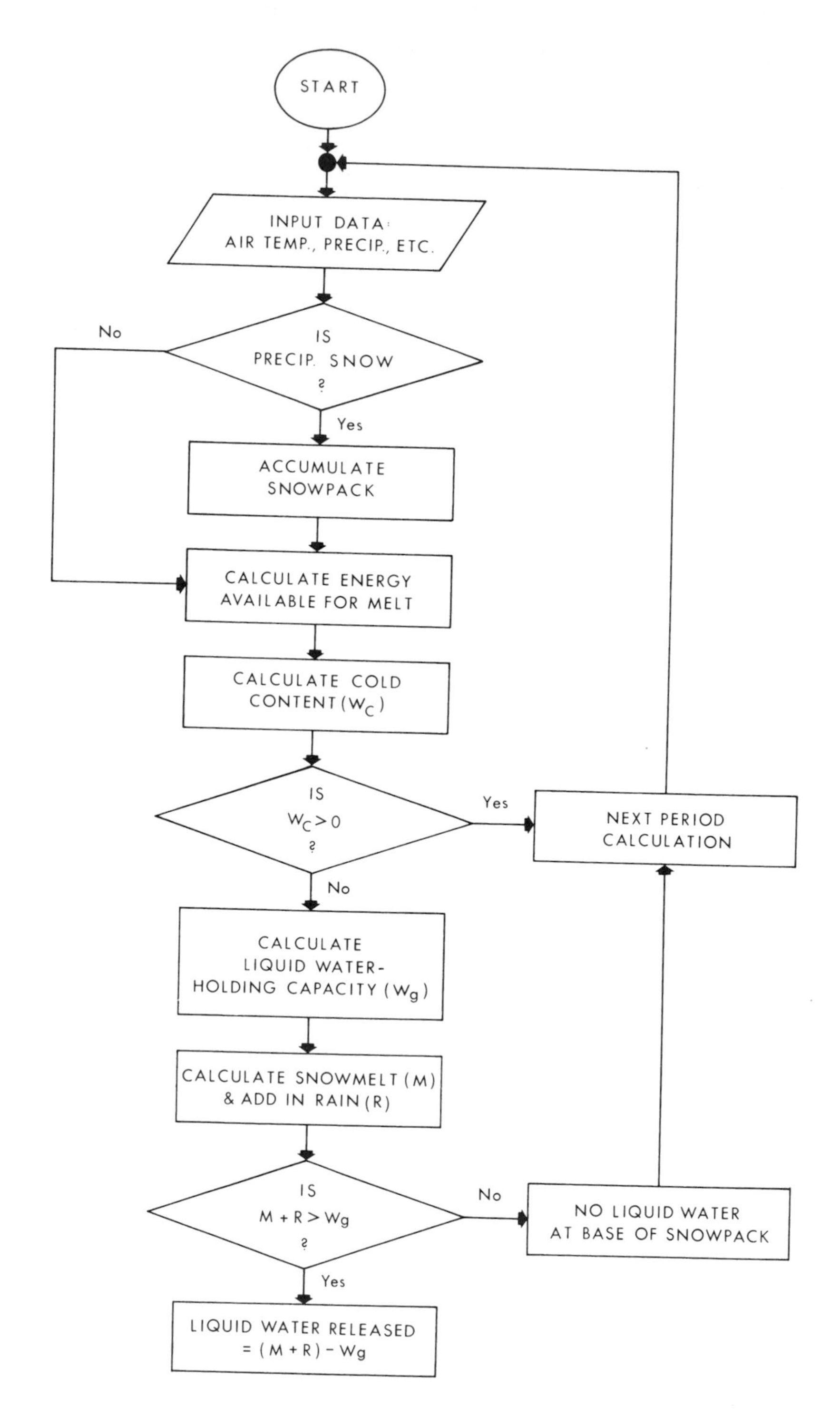

FIGURE 15.6. Flowchart illustrating the important relationships and components in a "typical" snow accumulation and melt model.

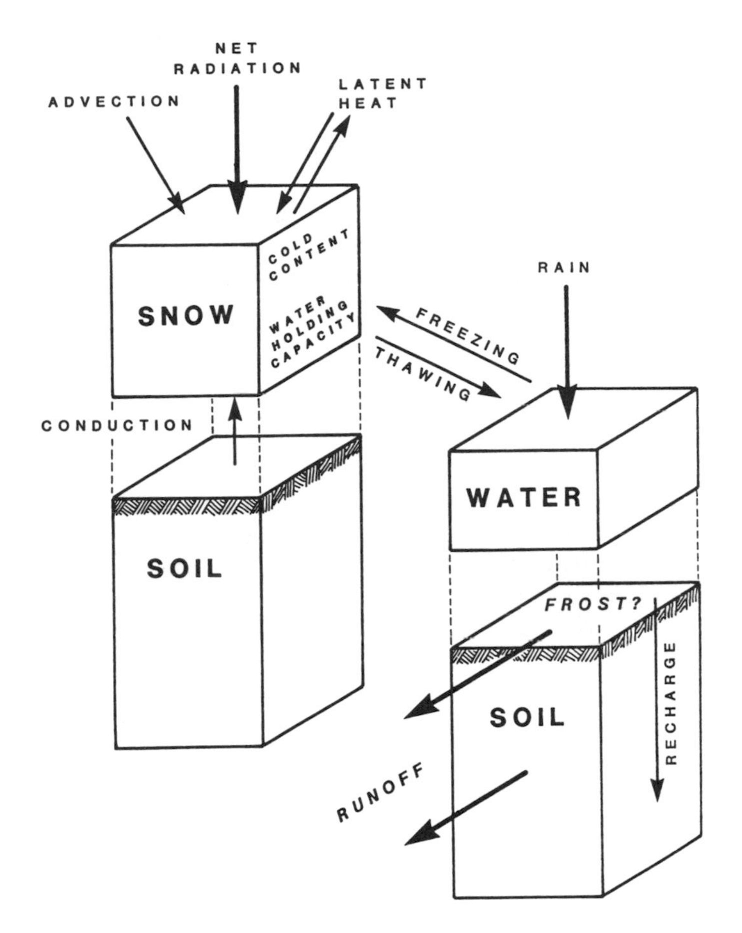

FIGURE 15.7. Factors that affect snowmelt and runoff from a snowpack.

A relationship of average daily discharge of streamflow can be derived from Equation 15.28 by dividing the daily volumes of streamflow by 86,400 (the number of seconds in a day). If the volumes of streamflow are expressed in cubic meters, average daily discharge can be given in cubic meters per second as follows:

$$Q_n = I_n (1 - k_b) + Q_{n-1} k_b \qquad \textbf{(15.29)}$$

where Q_n = average daily discharge on day n (m³/sec), I_n = average discharge of snowmelt on day n (m³/sec), and Q_{n-1} = average daily discharge on previous day (m³/sec).

The recession coefficient (k_b) is assumed to be a function of discharge. In general, it is set to represent specific watershed conditions.

Models developed from these methods are used to predict snowpack dynamics, to describe these dynamics in relation to forest management activities, and to estimate the magnitude of the resulting snowmelt runoff. These analytical tools are valuable to watershed managers concerned with snow hydrology.

Box 15.6

Snowmelt-Runoff Models: Two Examples

A large number of snowmelt-runoff models have been developed by researchers and managers to meet specific needs. It is beyond the scope of this book to consider all of these models. However, two physically based, spatially distributed, snowmelt-runoff models that are widely used by hydrologists and watershed managers in the United States are discussed briefly below.

Subalpine Water Balance Model

The subalpine water balance model (WATBAL) was developed originally by the USDA Forest Service to simulate the hydrologic impacts of watershed management practices on the snowmelt process (Troendle and Leaf 1980). WATBAL was subsequently linked to a model that simulates the total water balance of a watershed on a continuous year-round basis, including the contributions of snowmelt-runoff to annual water yields of a basin. Data requirements for WATBAL are (1) watershed characteristics (topography, vegetation, and soils), (2) snowpack conditions and cover, and (3) climatological information (daily maximum and minimum temperatures and precipitation amounts). Information routinely obtained by the USDA Forest Service, the USDA Natural Resources Conservation Service, NASA, the U.S. Geological Survey, and the National Weather Service is adequate for most applications. Simulation results that are obtained for individual "hydrologic response units" are compiled into a "composite overview" of an entire basin.

Snowmelt Runoff Model

The snowmelt runoff model (SRM) is designed to simulate and forecast daily streamflow in mountain basins where snowmelt-runoff is a major factor (Rango 1995). While originally formulated for applications on smaller basins, SRM has been used successfully on basins up to 122,000 km^2. Knowledge of daily temperatures is used to calculate the degree-days above a threshold value, which (in turn) are used to index the energy balance leading to snowmelt. Snowpack cover obtained by remote sensing is used in applying a snowmelt algorithm to that portion of a basin with a snowpack to the time of simulation. Daily precipitation is required to calculate the amount of rainfall-derived runoff (if any occurs) to be added to the daily snowmelt-runoff component. Information needed for model application includes a degree-day factor, a runoff coefficient, a recession stage coefficient, the temperature lapse rate, and streamflow discharge time lag. SRM can simulate the effects of climate change on the runoff regime for an entire hydrological year.

SUMMARY

Knowledge of snow hydrology is essential for the wise management of watersheds and water resources in many parts of the world. Snow accumulates, ripens, melts, and contributes to streamflow in response to climatological factors and the soil-vegetation system present on a watershed. Not only do we need to understand the hydrologic processes and factors involved in snow hydrology, but we also need methods to predict the streamflow response that results from a given set of climatic variables and vegetative cover conditions. After reading this chapter, you should be able to:

- Explain the factors affecting snow accumulation, snow ripening, and snowmelt.
- Explain and be able to quantify the snow-ripening process of a snowpack; explain the effects of snowpack condition on spring snowmelt floods.
- Describe the differences between snowmelt under a conifer forest and that in an open field using an energy budget approach.
- Calculate snowmelt using either the temperature index method or the appropriate generalized snowmelt equation; discuss the advantages and disadvantages of the two methods.
- Describe the conditions under which forest management can increase water yield due to changes in snowmelt runoff. What types of management activities are most effective? How can snowmelt flooding be moderated?

REFERENCES

Anderson, E.A. 1978. Streamflow simulation models for use on snow covered watersheds. In *Modeling of snow cover runoff*, ed. S.C. Colbeck and M. Ray, 336–350. Hanover, NH: U.S. Army Cold Regions Research and Engineering Laboratory.

Anderson, H.W., M.D. Hoover, and K.G. Reinhart. 1976. *Forests and water: Effects of forest management on floods, sedimentation, and water supply*. USDA For. Serv. Gen. Tech. Rep. PSW-18.

Berris, S.N., and R.D. Harr. 1987. Comparative snow accumulation and melt during rainfall in forested and clearcut plots in the western Cascades of Oregon. *Water Resour. Res.* 23:135–142.

Brown, H.E., M.B. Baker, Jr., J.J. Rogers, W.P. Clary, J.L. Kovner, F.R. Larson, C.C. Avery, and R.E. Campbell. 1974. *Opportunities for increasing water yields and other multiple use values on ponderosa pine forest lands.* USDA For. Serv. Res. Pap. RM-129.

Buttle, J.M., and F. Xu. 1988. Snowmelt runoff in suburban environments. *Nordic Hydrol.* 19:19–40.

Carroll, T.R. 1987. Operational airborne measurements of snow water equivalent and soil moisture using terrestrial gamma radiation in the United States. In *Proceedings of symposium on large-scale effects of seasonal snow cover*, 213–223. Int. Assoc. Hydrol. Sci. Publ. 166. Vancouver, BC: IAHS.

Ffolliott, P.F., and D.B. Thorud. 1974. A technique to evaluate snowpack profiles in and adjacent to forest openings. *Hydrol. Water Resour. Arizona and the Southwest* 4:10–17.

Gary, H.C. 1975. Airflow patterns and snow accumulation in a forest clearing. In *Proceedings of the 43rd Western Snow Conference*, 106–113. Coronado, CA: Western Snow Conference.

Gray, D.M., and D.H. Male. eds. 1981. *Handbook of snow*. New York: Pergamon Press.

Harr, R.D. 1986. Effects of clearcutting on rain-on-snow runoff in western Oregon. *Water Resour. Bull.* 22:1095–1100.

Harr, R.D., and F.M. McCorison. 1979. Initial effects of clearcut logging on size and timing of peak flows in a small watershed in western Oregon. *Water Resour. Res.* 15:90–94.

Hewlett, J.D. 1982. *Principles of forest hydrology*. Athens: Univ. of Georgia Press.

Jaakko Pöyry Consulting, Inc. 1994. *Final generic environmental impact statement on timber harvesting and forest management in Minnesota*. Tarrytown, NY: Jaakko Pöyry Consulting, Inc.

Meiman, J.R. 1987. Influence of forests on snowpacks. In *Management of subalpine forests: Building on 50 years of research*, tech. coords. C.A. Troendle, M.R. Kaufmann, R.H. Hamre, and R.P. Winokur, 61–67. USDA For. Serv. Gen. Tech. Rep. RM-149.

Rango, A. 1994. Application of remote sensing methods to hydrology and water resources. *Journal des Sciences Hydrologiques* 39:309–320.

———. 1995. Snowmelt Runoff Model. In *Computer models in watershed hydrology*, ed. V.P. Singh, 477–520. Highlands Ranch, CO: Water Resources Publications.

Rango, A., and J. Martinec. 1979. Application of a snowmelt-runoff model using LANDSAT data. *Nordic Hydrol.* 10:225–238.

———. 1995. Revisiting the degree-day method for snowmelt computations. *Water Resour. Bull.* 31:657B669.

Reid, L.M. 1993. *Research and cumulative watershed effects*. USDA For. Serv. Gen. Tech. Rep. PSW-GTR-141.

Reifsnyder, W.E., and H.W. Lull. 1965. *Radiant energy in relation to forests*. USDA For. Serv. Tech. Bull. 1344.

Troendle, C.A., and R.M. King. 1985. The effect of timber harvest on the Fool Creek watershed 30 years later. *Water Resour. Res.* 21:1915–1922.

Troendle, C.A., and C.F. Leaf. 1980. Hydrology. In *An approach to water resources evaluation of nonpoint silvicultural sources*, III-1–III-173. Environ. Res. Lab., EPA-600/88012, Athens, GA: Environmental Protection Agency.

U.S. Army Corps of Engineers. 1956. *Snow hydrology*. Portland, OR: U.S. Army Corps of Engineers, North Pacific Division. (Available from U.S. Dept. Commerce, Clearinghouse for Federal Scientific and Technical Information, as PB 151660.)

———. 1960. *Runoff from snowmelt*. Eng. Man. 111021406.

———. 1972. *Program description and users manual for SSARR model, streamflow synthesis and reservoir regulation* (revised 1975). Portland, OR: U.S. Army Corps of Engineers, North Pacific Division.

U.S. Soil Conservation Service (U.S. Natural Resources Conservation Service). 1972. *Snow survey and water supply forecasting*. SCS Natl. Eng. Handb., Sect. 22.

Verry, E.S., J.R. Lewis, and K.N. Brooks. 1983. Aspen clearcutting increases snowmelt and storm flow peaks in north central Minnesota. *Water Resour. Bull.* 19:59–67.

CHAPTER 16 *Watershed Considerations for Water Resource Development and Engineering Applications*

INTRODUCTION

Upland watersheds form the headwater areas of streams and rivers and are important recharge areas for many groundwater aquifers. As such, their management is an essential and integral part of any water resource development project or management plan. Rarely are watersheds managed solely for water resource purposes, an exception being municipal watersheds that are managed to provide municipal and industrial water supplies. Most watersheds are managed not only for water, but also for the production of food, wood, fiber, minerals, and energy. As has been demonstrated in earlier chapters, land use activities on watersheds affect the flow and quality of water; conversely, the development of water resources affects land use. As a result, it is essential that watersheds and water resource engineering projects be managed in concert with one another.

Watershed management should be an essential ingredient of water resource development projects, but should not be viewed as a panacea for all water resource problems. Likewise, engineering solutions, such as reservoirs, levees, water diversion schemes, hydropower plants, drainage projects, water harvesting, and the like, cannot by themselves solve water resource problems completely. Floods, droughts, and landslides are natural phenomena that occur no matter what types of management or engineering measures are instituted. Nevertheless, water resource engineering practices are often essential to meet objectives of flood control, enhanced water supply and energy production of a growing global population. What we want to avoid are the unwanted and often unintended effects that engineering practices can have on watersheds, streams, rivers, and groundwater systems. Likewise, we need to manage watersheds in a manner that complements and enhances the effectiveness and the life of water resource development projects.

The purpose of this chapter is to demonstrate the importance of including watershed management considerations in water resource development programs and engineering solutions. Specifically, we intend to provide an understanding of the linkages between watershed management, land use practices, and some of the most common engineering works used in water resource development.

Management of Surface Water

Issues of surface water can be usually characterized as either not enough usable water or too much water at one time, which results in hazards to human activity. Water supply shortages arise because of the variability in the timing, location, and quality of water yield. As indicated in Chapter 1, there is no shortage of fresh liquid water on the earth as a whole. Because this water is dispersed spatially and is in constant flux, many locations on the earth's surface do not have adequate water supplies to meet human demands on a dependable basis. Even in the humid Tropics, periodic droughts or annual dry spells can cause crop damage and losses of livestock and human life. Conversely, damaging floods can occur in even the most arid regions. It is not surprising, then, that nations and local communities make large investments in engineering projects to gain some measure of control over water resources.

Most water resource development and management programs have multiple objectives, including (1) providing dependable water supplies, (2) reducing the adverse effects of hydrometeorological extremes, and (3) maintaining and protecting water quality and natural aquatic ecosystems. Surface water development can involve a variety of structures and practices designed to alter the amount, timing, and quality of water yield. Benefits derived from such development include increased water supplies for municipal-industrial needs, irrigation, and livestock. Other benefits include flood control and sustained streamflow during dry periods for navigation, hydroelectric power production, or downstream fisheries.

Increasing Water Supplies

Water yield from a watershed is the residual of precipitation minus evapotranspiration (ET) and deep-seepage loss (see chapters 4 and 6). Therefore, methods of increasing water supplies on-site logically would involve either increasing precipitation or reducing ET. However, as streams and rivers deliver water downstream and eventually to the ocean, little of the water yield from a watershed is actually captured for human use. Furthermore, the abundant water supplies that exist as saltwater, polar ice and snow, and groundwater suggest that these are where we should concentrate efforts to increase water supplies. Many schemes have been devised over the years to alleviate water shortages, including:

- Reservoir storage and interbasin transfers of water—Water is stored during wet periods and released during times of need, or it is transferred from areas with excess water to locations with water deficiencies.
- Weather modification—The seeding of clouds has been tested and applied successfully in only a few instances to increase localized precipitation or to dissipate thunderstorms or flash-flood-producing clouds.
- Desalinization of seawater—The technology is now available to take ocean water from along coastal areas and remove the salt so that water can be used for human and crop consumption.

- Evaporation or transpiration suppression—Techniques that reduce evaporation from small water bodies and decrease transpiration losses have been tested but are likely to be useful only on a limited scale.
- Manipulation of vegetation to reduce annual consumptive use—Water yield can be increased in many areas by converting from one vegetative type to another or by implementing forest-cutting practices (see chapters 6 and 15).
- Towing icebergs—The transport of icebergs from polar regions to arid coastal areas has been suggested as a means of increasing water supplies; this has been suggested for coastal areas like southern California and countries in the Middle East.
- Water harvesting—Developing small impervious catchments that efficiently produce runoff from rainfall has been widely used to increase water available for drinking water for humans and domestic livestock. Storage of the harvested rainfall is required, and the small scale of such practices suggests a means of increasing household or farm-level water supplies.
- Pumping of groundwater—Extraction of groundwater from shallow aquifers has provided water supplies for municipal, industrial and agricultural uses for centuries, leading to an overexploitation of groundwater in many parts of the world. Pumping of deeper, previously untapped ancient groundwater raises many environmental and economic questions.

Each of the above methods has some limitations, but they represent alternatives for increasing water yield at certain locations. For example, the development of a desalinization operation along the coast of an arid country might preclude the need for reservoir construction in upland areas. The agricultural and industrial growth that would likely follow in the coastal area could provide jobs for rural watershed inhabitants, which can be important in developing countries. Benefits to watershed lands could include reduced intensity of land use, such as grazing and cropping practices in fragile areas, and consequently, improved health of upstream watersheds.

Other alternatives for increasing water yield include water reuse and conservation measures in households, industrial plants, and farms that reduce the demand for water. Activities that promote groundwater recharge during periods of abundant rainfall can be encouraged; the subsequent increase in groundwater supplies can be used during periods of high demand. The conjunctive use of surface and groundwater supplies represents a viable option for many locations. All options should be considered when water shortages are anticipated; by studying the alternatives and adopting an integrated approach, less costly and more environmentally sound solutions may be gained.

Even though there can be many alternative means of increasing water yield, reservoirs and associated transport systems are by far the most commonly implemented for surface water. The flow of water and sediment from watersheds to reservoirs and canals affects their functioning and economic viability. Therefore, the linkages between reservoirs, conveyance systems, and upstream watersheds must be well understood.

Coping with Floods and Related Hazards

Flooding, soil mass movements, and related debris flows result in widespread damage and loss of life on a global scale. Such phenomena are often called *extreme events*, but they are not necessarily rare.

Coping with floods and related hazards has always been a challenge to human development activities. General options for mitigating natural disasters caused by such

TABLE 16.1. Examples of measures that are commonly used to mitigate problems associated with excesses of water resulting from hydrometeorological extreme events

Water Resource Problems	Structural Measures	Nonstructural Measures
Floods	Dams/reservoirs	Floodplain management
	Levees	Maintain watershed vegetative cover; watershed management
	Channelization	Riparian management
	Channel diversions	Flood warning/forecasting
Landslides	Retaining walls	Zone hazardous areas
	Pilings Buttresses	Maintain vegetative cover with appropriate species; forest watershed management
Debris flows	Structures for energy dissipation	Hazard management and risk mapping
	Debris flow traps	Forest watershed management
	Dam systems	Warning system
	Training systems (diversions)	

Source: From Brooks 1998.

extremes include (1) modifying the natural system, (2) modifying human behavior to minimize risk or to avoid the hazard, or (3) some combination of (1) and (2). The first option commonly involves some type of engineering activity or structure or some modification of the watershed or stream channel. The second option is based on the fact that natural phenomena become disasters only when they impact humans; therefore, a viable option is to remove humans from hazardous areas, or restrict human activities to minimize risk. There are several measures that can be implemented under these options, as indicated in Table 16.1. These measures can be either structural (engineering) or nonstructural (zoning of hazardous areas, managing vegetative cover on watersheds and along riparian areas, protecting wetlands, etc.).

We will focus our attention on some of the most common measures used in surface water resource management today.

RESERVOIRS

From a historical perspective, reservoirs have been the main tool by which water resources are managed. Reservoirs represent large investments of capital, and they are frequently justified on the basis that they provide multiple benefits, such as increased water supplies, hydropower production, flood control, and maintenance of streamflow levels needed for aquatic ecosystems, dilution of pollutants, and navigation in downstream channels. Water-based outdoor recreation and fisheries industries can also benefit. Local and regional economic growth is often stimulated from the development of large reservoir projects. Nevertheless, critics of reservoirs point to their high economic costs and to the fact that many have been constructed regardless of adverse environmental impacts.

The construction and operation of large reservoirs or reservoir systems can cause dramatic changes in the types and patterns of land use, in both upstream and downstream areas. Reservoir operations change downstream flow patterns and amounts, thereby altering stream channel morphology and aquatic habitat (see Chapter 10). In addition, rural inhabitants must often be relocated, and productive bottomlands and valuable wildlife habitat can be lost. Altered land use can also change the hydrologic characteristics and erosion-sedimentation processes of upstream watersheds.

Without careful planning, design, and operation, the economic life of reservoirs may be shortened. The goods and services for which the project was constructed may not be sustained over the period for which the project was designed. The result of such a failure can impact local and regional economies and lead to social disruption. Even when reservoir projects function according to plan, some unanticipated problems can arise. For example, people might encroach into flood-prone areas below reservoirs because of an unwarranted feeling of security. Floods, which inevitably occur, sometimes result in much greater economic loss than would have occurred before the construction of the project. Therefore, floodplain management and zoning (i.e., keeping people and buildings out of flood hazard areas) and upstream watershed protection measures should be integral parts of any reservoir project.

Management of Reservoirs

Reservoirs designed, constructed, and operated in concert with upstream watershed management should be capable of achieving the goals for which they were built (Table 16.2). To a limited extent, the upstream vegetative cover can be manipulated to complement reservoir operations (see chapters 6 and 15). The type and extent of vegetative cover can influence the amount and timing of streamflow from a watershed. For example, converting from deep-rooted (trees) to shallow-rooted (grass) vegetation can increase water yield in many areas. Without a reservoir, much of the increased water yield likely would flow from the area during high-runoff periods and would be unavailable during drier periods when irrigation is needed. When flood control is an objective, watershed management practices designed to maintain high infiltration rates and low surface runoff are needed. Wetland areas in a watershed attenuate stormflow peaks, and if flooding is a concern, their preservation or protection should become part of the management program. Such practices can affect the magnitude and frequency of flooding, except flooding associated with extremely large storms that occur infrequently.

Vegetation influences are minimal for major hydrologic events such as > 100 yr return period floods or for severe droughts. On the other hand, reservoirs are usually designed to function during such extreme events. Whether or not a reservoir remains functional over time, however, depends largely on vegetation characteristics and the overall condition of upstream watersheds. Accelerated rates of sedimentation can cause a premature loss of reservoir storage and an inability to meet demands.

Determining Storage Requirements

Determining the storage capacity of a reservoir requires that water yield and rates of sediment deposition be estimated and balanced against projected demands for water. The total storage capacity of a reservoir is the sum of the active storage and dead storage. *Active*

Table 16.2. Watershed and reservoir management as they relate to several water resource management goals

Water Resource Management Goals	Watershed Management Objectives/Activities[a]	Reservoir Management Objectives/Activities
Increase available water supplies for irrigation, municipal, supplies, etc.	Manipulate vegetative cover to increase water yield (convert from deep-rooted to shallow-rooted species on watershed).	Provide storage to accommodate increased water yield, to provide assured yield for demands.
	Minimize or control erosion and sedimentation to maintain storage space in reservoir for the duration of design life; rehabilitate watersheds in poorly managed areas.	
Reduce flood hazard.	Maintain vegetation and soils in good condition through proper land management; maximize infiltration, and minimize surface runoff; rehabilitate where necessary.	Provide storage space for runoff during flooding; provide storage for high runoff to be released during dry periods.
	Minimize sedimention of channels to maintain channel capacity.	
	Preserve and protect wetlands; minimize wetland drainage and development.	
	Construct contour trenches and furrows to increase detention storage or lengthen infiltration time.	
	Enhance infiltration capacity of disturbed lands with mechanical treatments such as soil ripping or chisel plowing, followed by revegetation.	
Maintain surface water supplies with a high quality for human consumption or fisheries resources.	Restrict or carefully manage land use activities conducive to production of sediments, disease-carrying organisms, and excessive nutrient loading; maintain vegetation and soils in good condition; rehabilitate watersheds where needed.	Regulate outflow to enhance desirable downstream water quality characteristics (dissolved oxygen and temperature).
Provide hydroelectric power.	Manage watersheds to sustain life of reservoir as above.	Maintain a dependable storage with sufficient head to generate enough electricity to meet demands.

[a]Objectives and activities may not apply in every situation.

storage is that which is needed to meet all demands, that is, to prevent shortages and in some cases to provide flood control benefits. *Dead storage* is that part of the reservoir allocated to trap and hold sediment; it should be of sufficient volume to prevent sedimentation from affecting the active storage for a period equal to the economic design life of the reservoir. The gates in a dam used to release water in reservoir operations are immediately above the upper level of the dead storage zone. Detailed procedures for determining reservoir storage needs can be found in most hydrologic engineering texts and therefore will only be briefly discussed here.

Active Storage

Two general approaches exist to estimate active-storage requirements: simplified methods for quick estimation or first approximations and a detailed sequential analysis.

Simplified Methods. Simplified methods include the sequential mass curve, or ripple, method and the nonsequential mass curve (U.S. Army Corps of Engineers 1977). Only the ripple method, which involves the construction of a mass curve of accumulated streamflow volumes over a critical period, is discussed here. This period corresponds to the duration of a drought of a magnitude for which the reservoir is designed (Fig. 16.1). The expected total demand for water (municipal, irrigation, hydroelectric power, etc.) is expressed as a constant rate over time and then plotted as a straight line with the slope equal to the demand rate. This straight line (*DEF*) is constructed tangent to the mass curve at the beginning of the drought period. A second, parallel line (*ABC*) is constructed tangent to the mass curve at the lowest point during the drought period. The vertical distance between these two lines (*BE*) represents the storage needed to meet the demand during the drought period.

Detailed Sequential Method. The detailed sequential method uses a water budget approach that involves tabulating the demands for water and all the inputs and outputs in the sequence in which they occur at the reservoir site. Tabulations can be made on a daily, weekly, or monthly basis and represent seasonal variability in inputs, outputs, and demands. Demands must represent *all* demands for water, both at the reservoir site and downstream. For example, municipal water supplies can be withdrawn directly from the reservoir, but irrigation water can be withdrawn downstream from the dam. There can also be *minimum flow requirements* immediately below the dam; such a demand must be added to the irrigation requirement but is complementary to releases from the reservoir for generating hydroelectric power. Therefore, demands can be considered as competing or complementary. All competing demands must be totaled for each tabulation period (often on a monthly basis). The storage required to meet all demands is determined by:

$$S_t = Q - E + P - D_m \quad \textbf{(16.1)}$$

where S_t = storage required (in units of volume), Q = inflow from streams and local areas (in units of volume), E = evaporation from the reservoir pool (in units of volume), P = precipitation on the reservoir pool (in units of volume), and D_m = demands for water (in units of volume).

The detailed sequential approach has several advantages over simplified methods: (1) demands can vary over time, (2) either observed or simulated water yield (inflow)

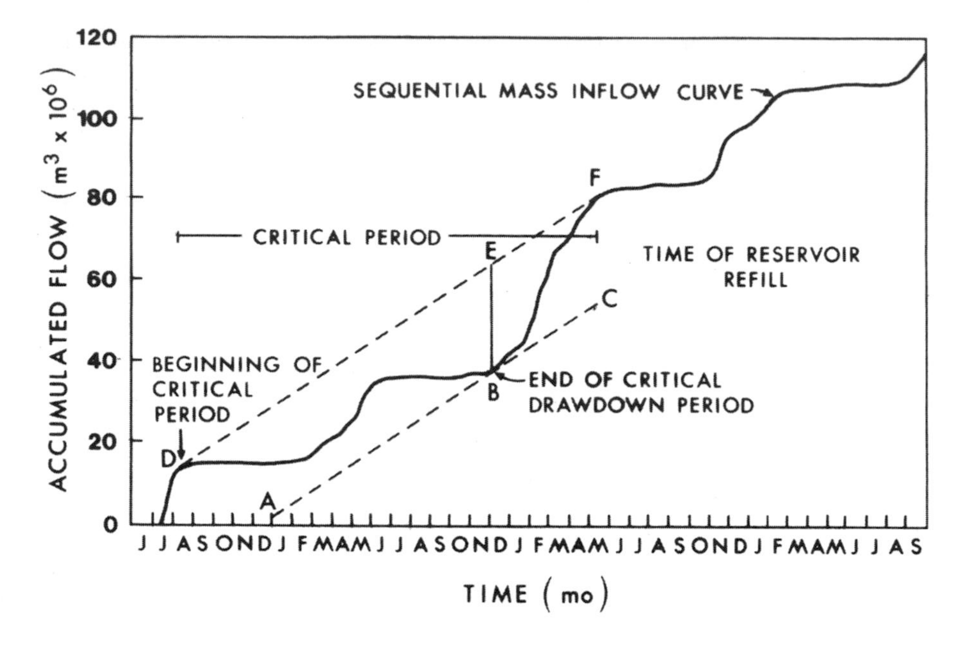

FIGURE 16.1. The sequential mass curve, or ripple, method of determining active reservoir storage (from U.S. Army Corps of Engineers 1977). Given a desired yield of 38 × 10^6 m^3/yr, construct line *ABC* with slope = 38 × 10^6 m^3/yr tangent to mass curve at *B* (lowest possible point of tangency). Construct line *DEF* parallel to *ABC* and tangent to mass curve at *D* (highest point of tangency prior to *B*). Line *BE* represents the maximum storage required to produce the desired yield (about 26 × 10^6 m^3/yr is required).

data can be examined, (3) changes in land use that affect water yield can be considered in the analysis, and (4) the analysis is more amenable to a systems approach and computer simulation. The last advantage is important when more than one water resource development activity is being considered within a large watershed.

Historical streamflow records of a severe drought are usually used in determining *conservation storage requirements* for a reservoir. An alternative to using historical streamflow data as input (Q) in Equation 16.1 is to use a *stochastic model* to generate streamflow. Such models predict streamflow events based upon the statistical characteristics of streamflow records, including the randomness of hydrologic events. A stochastic model generates streamflow records that follow a certain statistical pattern but that have a random component. As a result, each generated streamflow record is unique. For reservoir studies, one can generate several hundred years of records and then examine low-flow sequences to estimate conservation storage requirements. These sequences have a chance of occurring and therefore help to overcome the limitation of basing storage requirements only on observed historical records.

Any large-scale changes in the type or extent of vegetative cover on a watershed can lead to changes in inflow to the reservoir. If such changes occur after a reservoir has been designed, problems can result in meeting the projected demands (see Box 16.1).

Box 16.1

Effects of Land Use on Sedimentation Levels in the Panama Canal Watershed (from Larson and Albertin 1984)

The passing of each ship through the Panama Canal requires 197,000 m^3 (52 million gal) of water to operate the locks. This freshwater is supplied from a 3339 km^2 watershed; water storage for operation is supplied by Gatun Lake, through which the canal passes, and by the Alhajuela Reservoir, which is located at a higher elevation. More than 92% of the Alhajuela watershed has a slope exceeding 45%; 13,400 ha of this area were cleared of forest and cultivated or pastured during 1972–1973. Surface, gully, and mass soil erosion resulting from these land use practices have contributed sediment to the 800 million m^3 of reservoir storage space. Sedimentation yields from the watershed before and after clearing were 10.5 and 29.1 t/ha/yr, respectively. Sediment soundings made in Alhajuela Reservoir showed that deltas were filling the narrow and shallow parts near the mouths of tributaries. Sedimentation depths by 1978 averaged 1.7 m, ranging from 1 m in deep areas to 3–6 m in the delta areas. Sedimentation rates prior to land clearing and cultivation averaged 3.32 cm/yr, or 937,000 m^3/yr. After clearing, sedimentation averaged 9.6 cm/yr, or 2,595,000 m^3/yr. This almost three-fold increase in sedimentation was attributed to the clearing of only 18.2% of the watershed.

Because of the potential impacts of accelerated reservoir sedimentation on the operation of the canal, a watershed management plan was prepared in 1978. The watershed program consisted of institution building, education and research, and implementation of watershed practices, including agroforestry, plantation forestry, forest reserves, pasture management, land use zoning, and soil conservation. More intensive sediment sampling of streams and the reservoir was accompanied by the establishment of three experimental watersheds to quantify the effects of forest-clearing practices on runoff, erosion, and sediment yield.

Dead Storage

The magnitude of dead-storage space required for a reservoir depends upon the quantity of sediment flowing into the site, the percentage of inflowing sediment trapped, and the density of the deposited sediment. The quantity of sediment flowing into the site depends on local and upstream surface erosion, mass soil erosion, and streamflow-channel erosion processes. The *trap efficiency* of a reservoir is a function of the ratio of reservoir storage capacity to annual inflow; as this ratio increases, the amount of sediment that becomes trapped increases. The density of sediment in the reservoir depends on the type of material, its specific weight, and the amount of consolidation that takes place over time.

The quantity of sediment flowing into a reservoir can be estimated by several methods, including the detailed methods presented in chapters 7, 8, and 9 and those following:

- Sediment surveys from nearby reservoirs or ponds, which provide the best information for existing land use conditions.
- Erosion and sediment delivery ratio data for landscapes and channels similar to the watersheds in question and in the same climatic regime.
- Application of locally derived equations or simulation models verified with data from similar watersheds.

Changes in land use, particularly those that affect the vegetative cover, litter, and soil surface of watersheds, can change sedimentation rates. The development of roads and the construction activities associated with the engineering projects themselves can lead to accelerated surface erosion, mass soil erosion, and channel erosion. Activities that result in higher stormflow peaks and volumes can accelerate channel scour, streambank erosion, and sedimentation. It is imperative, therefore, that anticipated changes in land use be evaluated in designing reservoirs.

Generalized equations or models are useful in examining the effects of several different land use activities on erosion and sedimentation. The results of modeling work can be helpful in pointing out needed land management constraints and erosion-sediment control measures. Options can then be developed to correct or reduce problems of sediment deposition at the dam site.

If erosion control practices are deemed necessary before or during dam construction, the effectiveness of alternatives must be evaluated carefully. Excessive levels of sedimentation can add significantly to the cost of a project (Box 16.2). Structural and nonstructural erosion control measures are costly and should be aimed at the most critical sources of sediment. For example, if the main source of sediment at the site is derived from within the channel system, surface erosion control measures are not the answer. If surface and gully erosion are the main problems, structural and nonstructural solutions must usually be considered together. Structural solutions alone are usually not economically feasible, unless accompanied by land use practices that are compatible with soil conservation principles. In addition, all activities taking place on the watershed, and those anticipated, should be considered collectively. In other words, sedimentation from road construction and maintenance activities should be considered along with other sources such as surface, gully, and streambank erosion. Even if erosion can be reduced effectively, there may already be sufficient sediment within the channel system to continue sediment delivery at high levels over a long period (a method of evaluating changes in sedimentation at a reservoir site is outlined in Box 16.3).

Once a dam has been constructed and it is discovered that rates of sedimentation have been underestimated, the large investment of capital and human and land resources practically dictates some type of sediment control. Unfortunately, it can be too late to carry out effective erosion-sedimentation reduction after the dam has been constructed. Watershed management should be considered in the early planning stages of reservoir design.

Reservoir Operation

The effects of watershed use and management on reservoir operation can best be understood by considering an actual reservoir project, as illustrated in Figure 16.2. This multipurpose reservoir has storage space allocated for both conservation (water supply) purposes and flood control. Within the conservation storage is a buffer zone, the top of which is used as a threshold to allocate water releases from the reservoir for different

Box 16.2

Sediment and Reservoir Design (from Gregersen et al. 1987)

Sedimentation of reservoirs poses a serious threat to India's extensive irrigation system. Reducing sedimentation rates can significantly reduce the cost of dam construction (Sinha 1984). A 25% reduction in sediment export from a 629 km^2 watershed to one reservoir would allow the dam height to be constructed 0.53 m lower than the originally designed 36.6 m height. This translates into a savings of 4.5% in dam construction, not to mention the reduction in the number of hectares flooded by the reservoir pool. At another project, sedimentation from a 114 km^2 watershed would have to be reduced by 75% to realize a 6% reduction in cost from reducing the height of the dam by 0.61 m. In the first project, a combination of watershed management and engineering measures may be capable of reducing sedimentation significantly. However, a reduction of 75%, as required in the latter project, would be difficult to achieve. In either example, economic analysis with sound technical data (Chapter 19) would be needed.

Box 16.3

Method of Evaluating Current and Projected Land Use Impacts on Reservoir Life (Adapted from Brooks et al. 1982)

Situation

A multipurpose reservoir will be constructed with a planned completion date 5 yr from the present. The design life of the project is 40 yr. The reservoir has the capacity to store 14 million m^3 of sediment (once this storage is exceeded, the reservoir may be unable to meet all demands). The 18,200 ha watershed for the reservoir is in poor hydrologic condition; overgrazing and poor cultivation practices have resulted in high rates of soil erosion and downstream sedimentation. A watershed rehabilitation project, in which 50% of the uplands will be reforested and the remaining 50% will be reseeded with perennial grasses and forbs, is planned. For the project to be effective, only limited grazing will be allowed on the nonforested area; the forested area will be protected. This work, coupled with gully control structures in critical areas, is deemed necessary to reduce sedimentation rates at the reservoir site. The rehabilitation efforts should be in full effect in 8 yr.

(continues)

(continued)

Data

- Existing Rates of Erosion (by modified universal soil loss equation defined in Chapter 7):

$$A = (R)\,(K)\,(LS)\,(VW) = (80)\,(0.3)\,(10)\,(0.17)$$
$$= 40.8 \text{ t/acre/yr} \times 2.23 = 91 \text{ t/ha/yr}$$

- Sediment density = 1.5 t/m3
- Sediment delivery ratio = 0.39
- Total sediment deposited at reservoir site is estimated to be about 50% from surface erosion and 50% from channel erosion.

Assumptions Made

The watershed project will likely take 8 yr before it is in place and effective; after 8 yr, the soil condition should be improved (the soil erodibility factor K = 0.28) and the watershed vegetative cover will be:

50% watershed area = 50% forest canopy with a 60% ground cover (grass)

50% watershed area = 40% grass cover with 25% short shrub canopy

Gully control structures will be in place in headwater areas by the end of 3 yr following reservoir construction; assume all sediment delivered from upstream areas will be negligible for 1 yr (i.e., the structures are effective for only 1 yr); after that, consider the upstream erosion-sedimentation process to continue, but at the new rate.

Questions Posed

- Under existing conditions, how many years will it take to exceed the designed reservoir sediment storage capacity?
- If the watershed project is implemented, how many years will it take before the sediment storage capacity is exceeded?

Solution

1. Existing rates of erosion:

$$\frac{(91 \text{ t/ha/yr})}{(1.5 \text{ t/m}^3)} = 60.7 \text{ or } 61 \text{ m}^3\text{/ha/yr}$$

For entire watershed:

$$(61 \text{ m}^3\text{/ha/yr})\,(18{,}200 \text{ ha}) = 1{,}110{,}200 \text{ m}^3\text{/yr}$$

Using a delivery ratio of 0.39, the amount of sediment delivered from upstream surface erosion is:

$$0.39\,(1{,}110{,}200 \text{ m}^3\text{/yr}) = 432{,}978 \text{ m}^3\text{/yr or } 433{,}000 \text{ m}^3\text{/yr}$$

If this is half the total sediment delivered, total delivery is approximately:

$$(2)\,(433{,}000\ \mathrm{m^3/yr}) = 866{,}000\ \mathrm{m^3/yr}$$

At this rate, the storage capacity will be exceeded:

$$\frac{14 \times 10^6\ \mathrm{m^3}}{866{,}000\ \mathrm{m^3/yr}} = 16.2\ \text{yrs after construction}$$

2. Since it will take 8 yr before the watershed practices are effective, the first 3 yr after the reservoir is constructed will have approximately the same rates of sedimentation.

In year 4, the sedimentation will be effectively controlled from upstream areas; the channel erosion will contribute 433,000 m^3. By year 5, K will change from 0.3 to 0.28, and VM will change for the reforested and grassland areas, and the new rates of surface erosion are:

Forested Area

$$VM = 0.06$$

$$\begin{aligned} A &= (80)\,(0.28)\,(10)\,(.06) \\ &= (13.4\ \mathrm{t/acre/yr})\,(2.23) - 29.97\ \mathrm{t/ha/yr}\ \text{or}\ 30\ \mathrm{t/ha/yr} \end{aligned}$$

This corresponds to 20 m^3/ha/yr from half of the area or 9100 ha, resulting in a total erosion of 182,000 m^3/yr. With 0.39 sediment delivery ratio, this becomes 70,980 m^3/yr sediment.

Grassland Area

$$VM = 0.09$$

$$A = (80)\,(0.28)\,(10)\,(.09) = (20.2\ \mathrm{t/acre/yr})\,(2.23)$$

= 45 t/ha/yr = 30 m^3/ha/yr from the other half of the area, or 9100 ha, with a total erosion of 273,000 m^3/yr

With 0.39 sediment delivery ratio, this becomes 106,470 m^3/yr.

The total annual sediment delivery after watershed management practices are in place is:

$$70{,}980\ \mathrm{m^3} + 106{,}470\ \mathrm{m^3} = 177{,}450\ \mathrm{m^3}$$

Sedimentation after the Reservoir Project Is Completed

Years 1–3 at existing sediment rate = (3)(866,000 m^3/yr) = 2,598,000 m^3
Year 4 only channel erosion = 433,000 m^3
Year 5 and on 177,450 m^3 + 433,000 m^3 = 610,450 m^3/yr

(continues)

(continued)

Years After Reservoir Sediment Completed	Sedimentation (m³/yr)	Remain Storage (m³)
1	866,000	13,134,000
2	866,000	12,268,000
3	866,000	11,402,000
4	433,000	10,969,000
5	610,450	10,358,440
.	.	.
.	.	.
.	.	.
22	610,450	0

Therefore, with the project, the sediment storage capacity will be exceeded after 22 yr, in contrast to 16.2 yr without the project. To determine if the watershed project is feasible, an economic analysis would need to be performed.

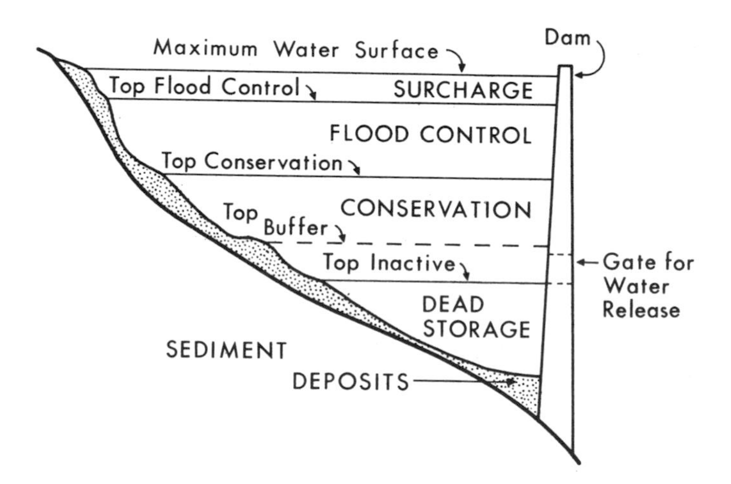

FIGURE 16.2. Storage allocation for a multipurpose reservoir with associated sedimentation.

demands during critical dry periods. Once the reservoir pool elevation drops to the top of the buffer zone, water can be released only for those purposes predetermined to be most important, such as providing municipal water supplies. Other, less essential needs will not be met until the pool elevation is higher than the top of the buffer pool. Several buffer zones can be used as a mechanism to allocate water on a priority basis during periods of shortages. If the project has a hydroelectric-power-generating capacity, the amount of head required to drive the turbines and the corresponding storage would be an added operational zone in the reservoir. Once sediment begins to encroach into the various storage spaces, operating the reservoir to meet the respective demands becomes more difficult. As shown in Figure 16.2, sediment typically does not settle out only in the dead-storage space. Coarse materials are deposited at various inflow points in the upper reaches of the reservoir pool, whereas the finer sediments tend to settle out near the dam.

As sediment fills the active-storage space, the capability of a reservoir to meet all demands becomes limited (see boxes 16.3 and 16.4). Storage volumes in the conservation

Box 16.4

Influence of Land Use on Reservoir Sedimentation (McIntyre 1993)

Long-term sedimentation data at the 55 ha Tecumseh Reservoir in Oklahoma were compared with land use changes on the 1189 ha watershed by examining aerial photographs from 1934 to 1987. The age of sediment cores was determined using ^{137}Cs. Land use was related to sedimentation rates through estimates of surface erosion using the USLE and gully erosion estimates provided by the U.S. Soil Conservation Service (now the U.S. Natural Resources Conservation Service). Land use changed markedly over the period of 1934 to 1987, with a major shift from abandoned gullied fields and cultivated fields to permanent pasture. Over the study period, sediment deposition in the reservoir was not affected by roadways, streambank erosion, stored channel sediment, and long-term changes in precipitation. Land use was the dominant factor affecting reservoir sedimentation.

Period (yr)	Change in Land Use	(%)	Rate of Sediment Deposition (m^3/yr)
1934–54	Abandoned gullied fields	22 to 7	
	Cultivated fields	38 to 17	
	Perennial pasture	4 to 41	
	Forest	34 to 33	
	Other	2 to 2	
			7,417
1954–87	Abandoned gullied fields	7 to 1	
	Cultivated fields	17 to 1	
	Perennial pasture	41 to 66	
	Forest	33 to 27	
	Other	2 to 5	
			1,926

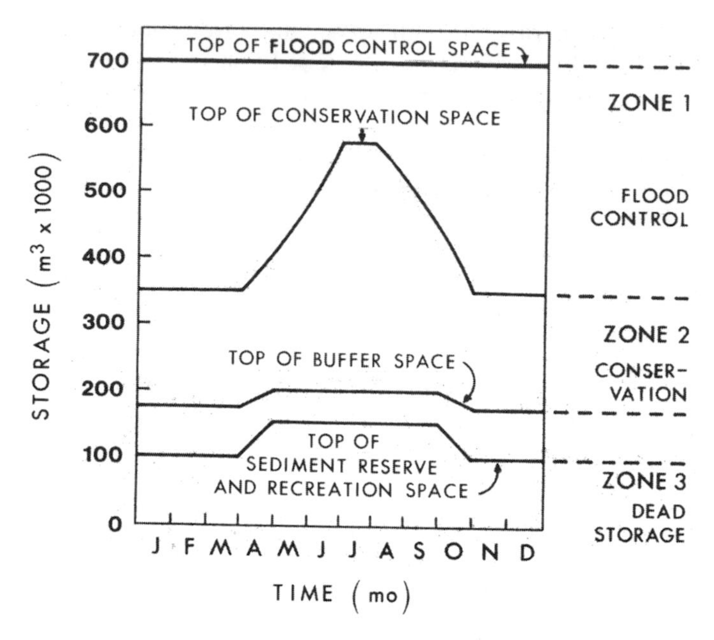

FIGURE 16.3. Operational rule curves illustrating seasonally varying storage requirements of a multipurpose reservoir (from U.S. Army Corps of Engineers 1977).

pool can be inadequate to meet demands during periods of drought. Likewise, there might not be adequate storage space in the flood control pool to control major flood events. Some of these problems can be overcome, at least partly, by using an operational rule curve approach to reallocate storage space for different purposes seasonally, as illustrated in Figure 16.3. *Rule curves* provide target pool elevations that vary with the season; their purpose is to provide operational guidelines that allow the most efficient use of reservoir storage. Typically, the season in which irrigation supplies are needed is a period in which the threat of large floods is slight; therefore, the elevation of the conservation pool can be raised into the normal flood control pool during that time. When the flood season approaches, the conservation pool is then lowered to provide flood control space. Rule curves allow demands to be met with a smaller total storage capacity in the reservoir. But even rule curve operations may not suffice when sedimentation becomes excessive.

Development of rule curves involves considerable knowledge of the variability of inflow to the reservoir and a clear understanding of seasonal water demands, priorities, and the downstream channel capacity. High- and low-streamflow sequences observed in the past are used to develop and test rule curves to ensure that water requirements and all constraints are met. When streamflow records are limited at the site, regional data can be used or streamflow records can be simulated with either deterministic or stochastic models.

Rule curves serve as guides in the operation of a reservoir. The ability to adhere to these guides is improved by providing the reservoir manager with accurate, real-time forecasts of inflow. Many operating constraints must be considered along with the rule curves. For example, most reservoirs have restricted rates of change of outflow to prevent rapid surges of water downstream; in some cases, this can cause the reservoir pool elevation to deviate from the rule curve. Similarly, rule curves may occasionally be violated to accommodate temporary anomalies in the system. Examples would include changes in operation to accommodate the passage of migrating fish or to assist downstream efforts to remove a barge that has gone aground. However, such deviations in rule curve operations should not be allowed to affect some other part of the system adversely.

Although the focus of this discussion has been on reservoirs, it should be recognized that erosion and sedimentation can reduce the capacity of water conveyance systems needed to transport water from the reservoir to the locations where it is needed. Erosion control may be needed, along with channel stabilization, to keep conveyance systems operational.

Design Floods

Design floods (or design storms) are selected for reservoir studies as standards against which the performance of the facility can be evaluated. The design flood is simply the streamflow from a large storm and is defined by some selected criteria.

Design floods typically are spillway, reservoir, or project design floods. The spillway design flood is selected to determine the size of the spillway of the reservoir. The reservoir design flood, usually of a lesser magnitude, is used for sizing the flood control storage of a reservoir—the amount of space to be reserved for flood control.

There are several approaches for determining design floods. The *probable maximum flood* (PMF) is the flood that can be expected from the most severe combination of hydrometeorological conditions that are reasonably possible for an area. Typically, such a large event is used in the design of a spillway, particularly where failure would be disastrous in terms of loss of life. The *standard project flood* (SPF) is used as a design flood for major reservoirs or local protection projects. Often, 40–60% of the PMF rainfall is used to generate the SPF.

Estimated flood events are based on knowledge of the meteorological conditions in the region. A certain amount of judgment is used in determining both of these floods. An alternative approach is to determine floods with specific return intervals, such as the 20 yr, 50 yr, or 100 yr event. A risk is associated with these design floods. Reservoir spillways probably should not be designed based on recurrence intervals except where loss of life is not a factor or where the economic impacts of failure are insignificant. It must be realized, for example, that a reservoir designed on the basis of a 100 yr recurrence interval flood has associated with it a risk of failure of 67% over a 100 yr period (see Chapter 17). In the case of multiple projects, the design based on the 100 yr event could be very misleading; if 100 reservoirs are designed in this manner, there is a 67% chance that one of these reservoirs will fail in any given year.

As with low-flow analyses used in conservation-storage studies, stochastic streamflow data can be generated for flood analyses. The concept of the design storm or design flood simply allows us to preselect the performance criteria that we wish to use for a particular water resource facility. Again, these performance criteria can be based on many factors other than hydrologic considerations. The economical, institutional, political, environmental, and social aspects can also be determining factors in the selection of performance criteria.

Selection criteria for design floods are usually established by an agency responsible for designing and operating the reservoir. If a project has a high dam with a large volume and it is determined that the structure cannot be overtopped without disastrous consequences, a PMF could be chosen for spillway design. On the other hand, run-of-river hydroelectric power plants or diversion dams can be designed by means of an SPF or some historical flood event, in which case the structure could be overtopped without serious damage resulting. Small dams that impound 12 million m^3 of water or less or small recreational lakes and farm ponds can be designed on the basis of floods with a specified recurrence interval. When a dam is low enough and the storage small enough that no serious hazard exists to downstream inhabitants if a failure occurs, an event such as the 50 yr recurrence interval flood may be selected.

Methods of simulating stormflow or flooding events are discussed in Chapter 17. For an SPF or PMF generation, one likely would select a model and use meteorological (rainfall) data to simulate streamflow. In the case of recurrence interval design floods, the study would involve a frequency analysis of existing streamflow records or the use of generalized criteria established by a governmental agency.

Small Hydropower Development

As hydraulic turboelectric technology progressed, efforts were mostly directed toward large-scale hydropower installations on large river systems and water bodies. However, there can also be opportunities to generate smaller amounts of hydropower by utilizing small flows of water. *Small hydropower plants* are those with a capacity of 5000 W or less. Other definitions mention installation sites with flows less than a minimum discharge, for example, 1 m^3/sec, regardless of the head. Being of small capacity, the plant machinery and auxiliaries generally are simple and inexpensive in design. In upland watersheds, the potential number of sites for such development is much greater than for large-scale hydropower dams.

Hydrologic Requirements

The basic requirements for developing a small hydropower facility include knowledge of the total head, net head, and quantity of water at the proposed site. The *total head* (H) of a stream is the difference in elevation between the turbine of the hydropower system and the water level above the turbine, or the vertical distance between the turbine and the water level at the intake. Frequently, a hydropower station will require a dam. With high-head turbines, the dam can be located at a considerable distance upstream and the water conducted to the turbine through a pipe. When selecting a dam site, remember that the greater the vertical distance between the turbine and the water surface behind the dam, the greater the head will be. The greater the head, the more power a given amount of water will produce. Therefore, dam sites should be selected to obtain the highest possible head. For possible sites, the head can be estimated from topographic maps.

The *net head* (h) is the actual head or pressure available to drive the turbine when friction and other losses have been deducted from the total head. Friction losses vary with the type of pipe used, its diameter, and the length of pipe. Concrete and tile pipes have the highest losses, while PVC pipe has some of the lowest. The larger the diameter of pipe, the less the friction loss will be. Friction losses also increase with increasing length of pipe and with the number of bends or curves in the pipe. A hydropower installation site should be where the highest head can be obtained in the straightest line and the shortest distance.

There are other losses in the total head at the turbine. In large part, these losses vary with the type of turbine utilized. Impulse turbines are commonly used in small hydropower installations that operate on a high head. The head loss for these turbines is across the gap between the nozzle and the tail water. Therefore, the net head (h) is the total head (H) less the friction losses in the pipe and the loss at the turbine:

$$h = H - (\text{pipe friction loss} + \text{loss at turbine}) \quad \textbf{(16.2)}$$

Friction losses in pipes can be read from standard tables obtained from the pipe manufacturers; head losses in turbines can be obtained from the manufacturer of the turbine.

The quantity, or volume, of water flowing in the stream determines the amount of hydropower that can be developed at a site. In most instances, streamflow varies with the season of the year; the minimum flow is the amount of water that can be used to drive the turbines continuously. As the streamflow increases, the amount of power that can be developed also increases. Consequently, it is necessary that a stream be measured (see Chapter 4) at various times of the year or that streamflow sequences be simulated with models (see Chapter 17).

Determining the power output requires knowledge of the net head (h) and streamflow (Q); the theoretical power (TkW) that a stream can produce is calculated by:

$$TkW = \frac{hQ}{102} \quad \textbf{(16.3)}$$

This equation gives the potential power that a stream can produce when the efficiencies of the turbines and generators in the system are 100%; however, this is seldom the case. In small hydropower plants, turbines usually drive the generators directly, either through a gearbox or with belts, which reduces efficiency normally to about 60%.

Types of Hydropower Systems

There are two basic types of hydropower systems. *High-head systems* depend largely on the head of a stream rather than on the quantity of water available. These systems are best suited to locations where the power demand is not great. Many of these systems, with power outputs of 510 kW, have been installed throughout the world. *Low-head systems* depend on the quantity of water, not the head of the stream. These systems frequently are installed on relatively level terrain but in streams and rivers with relatively high volumes of flow.

Capital expenditures of small hydropower installations are two to five times those of an equivalently sized diesel or steam power plant, depending upon the size and layout. However, annual operating costs (labor, maintenance, repairs, etc.) of small stations are low. An important cost consideration, however, is the cost of the transmission lines when the power output has to be transported to distant users. Small hydropower developments commonly are most economical for remote and isolated areas where the transmission investments are small and fuel transport is a problem. The relative simplicity of small hydropower equipment also makes it suitable for more isolated locations.

Water Harvesting

Water harvesting is a technique (1) for developing surface water resources to augment the quantity and quality of existing water supplies or (2) to provide water where other sources are either not available or too costly. *Water-harvesting systems* can be defined as artificial methods for collecting and storing precipitation until it can be used for watering livestock, small-scale subsistence farming, and/or local domestic use. These systems require a *catchment area,* usually prepared to improve runoff efficiency, and a *storage facility* for the harvested water, unless the water is to be immediately concentrated in the soil profile of a smaller area for growing plants. A water distribution scheme is also required for those systems devoted to irrigation. The technology of water harvesting can be applied in almost any dryland region of the world with at least 80–100 mm of annual rainfall (Box 16.5). If the annual rainfall is no more than 150 mm, the planting of indigenous plant species is suitable (National Academy of Sciences 1974). Systems for livestock water can also be built, but they will require storage tanks protected from evaporation losses. Farming systems are possible where annual rainfall is 250 mm or more if adequate storage facilities exist.

Configuration and Use

The geometric configuration of water-harvesting systems depends largely on the topography, the type of catchment treatment, the intended use, and the personal preference of the designer. Microcatchments, strip harvesting, roaded catchments, and harvesting aprons are some of the more common types.

Microcatchments and strip harvesting can be successful in years of normal or above-normal rainfall and are best suited for situations in which drought-resistant trees or other drought-hardy perennial species are grown (Box 16.6). The microcatchment technique can be used in complex terrain and on steep slopes, where other water-harvesting techniques can be difficult to install. The collection area can range from 10 to 1000 m^2, depending upon the precipitation in the area and plant requirements; one or several plants are usually grown on the downslope side.

Strip farming is a modification of the microcatchment method. Berms are erected on the contour, and the area between them is prepared to serve as the collection area. Runoff between the berms is then concentrated above the downslope berm to irrigate the vegetation planted there. Only drought-hardy plants should be grown with this type of system.

Apron-type water-harvesting systems, used primarily for livestock, wildlife, and domestic water supplies, are designed for minimum maintenance and must be fenced (Fig. 16.4). The catchment area (apron) is treated to obtain a high runoff efficiency, unless an existing impermeable surface is in place. Gravel-covered asphalt-impregnated fiberglass is a common treatment. A storage tank with evaporation control is required, along with the necessary pipes and valves to conduct the water to drinking troughs or households.

The apron-type water-harvesting system is the simplest to design. As a first approximation for the size of apron required, the following equation is helpful:

$$A = b\frac{U}{P} \tag{16.4}$$

where A = catchment area (m^2); b = 1.13, a constant; U = annual water requirement (l); and P = average annual precipitation (mm).

Box 16.5

Water-Harvesting Techniques Applied in Dry Regions

Australia

Australia was among the first of the Western countries to install operational water-harvesting systems. The systems were designed to provide water for livestock and domestic needs. In 1948, the Public Works Department, Western Australia, initiated a program of construction of roaded catchments. These catchments were made by clearing, shaping, and contouring to control length and degree of slope and by compacting with the aid of pneumatic rollers. About 2500 roaded catchments supply water principally for livestock use. These average approximately 1 ha in size. Also, there are 21 roaded catchments totaling 706 ha and ranging in size from 12.1 to 70.8 ha presently being used to furnish domestic water for small towns in Western Australia (Burdass 1975).

Israel

Researchers in Israel were the first to experiment with new techniques of water harvesting. They have found various methods effective in increasing runoff from land surfaces, including land smoothing and compaction and the formation of sodic crusts by spraying applications of various asphaltic materials. Among the asphaltic formulations, heavy fuel oil diluted with kerosene proved to be both effective and economical. Ratios between contributing and receiving areas on the order of 3:1 to 6:1 were found to be effective in rainfall zones of 200–250 mm. Unirrigated orchards provide a livelihood for an appreciable number of farmers (Hillel 1967). Water received in the planting zones of experimental plots provided soil moisture that was equivalent to the entire normal winter rainfall in the Mediterranean climatic zone of the country.

United States

The village of Shungopovi was located on the Hopi reservation in northeastern Arizona. The village was built on top of a sandstone rock mesa and had no source of water. From the time of first establishment, the villagers had carried water up from the valley, initially on foot and later on the backs of burros. In the early 1930s, a small water-harvesting system was installed to partially relieve the water shortage of the village. About one-third of a hectare was cleared, and the loose soil was removed to expose the sandstone bedrock. Below the area, a deep cistern was hewed into the rock and a concrete roof constructed. This system was a functional part of the village water supply for about 30 yr, until a community well and pump on the valley floor and an uphill water distribution system were installed (Chiarella and Beck 1975).

Box 16.6

Experimental Water-Harvesting System in Southern Arizona (Karpiscak et al. 1984)

A water-harvesting system consisting of a gravity-fed sump, a storage reservoir, 16 catchments, and an irrigation system occupies nearly 2 ha of retired farmland near Tucson, Arizona (see the figure below). The combined designed capacity of the gravity-fed sump and the storage reservoir is approximately 2400 m^3 of water. The sump and the storage reservoir were treated with NaCl to decrease infiltration, and the main reservoir was covered with 250,000 empty plastic film cans to decrease evaporation.

The 16 catchments, also treated with NaCl to decrease infiltration, have been used to concentrate rainfall runoff around planted agricultural crops and tree species in untreated planting areas at the base of the catchments. Excessive runoff flows directly into a collecting channel and then into the sump. Each catchment, about 1.5 ha in size, is approximately 90 m in length, varies from 6 to 18 m in width, and slopes about 0.5%.

The irrigation system consists of a 6000 W centrifugal pump, an 8 cm pipeline connecting the sump and the storage reservoir to the pump, two 5 cm PVC pipelines connecting the pump to the field plots, and 2 cm polyethylene driplines equipped with 0.01 m^3/hr drip emitters. The valving system permits the movement of water from the sump to the storage reservoir, from the storage reservoir to the sump, and from either the sump or the storage reservoir to the field. A water meter records the amount of water applied to the plants.

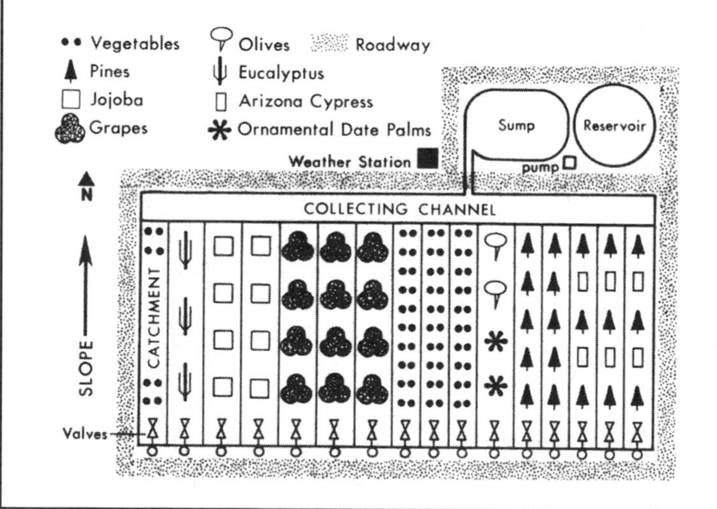

An example of a water-harvesting system as part of an agroforestry project in Avra Valley, near Tucson, Arizona (from Karpiscak et al. 1984).

Roaded catchments are well suited to growing high-value horticultural crops such as fruit trees, nut trees, and grapes and for providing water for livestock. These catchments are best adapted to very gently sloping ground. A roaded catchment consists of

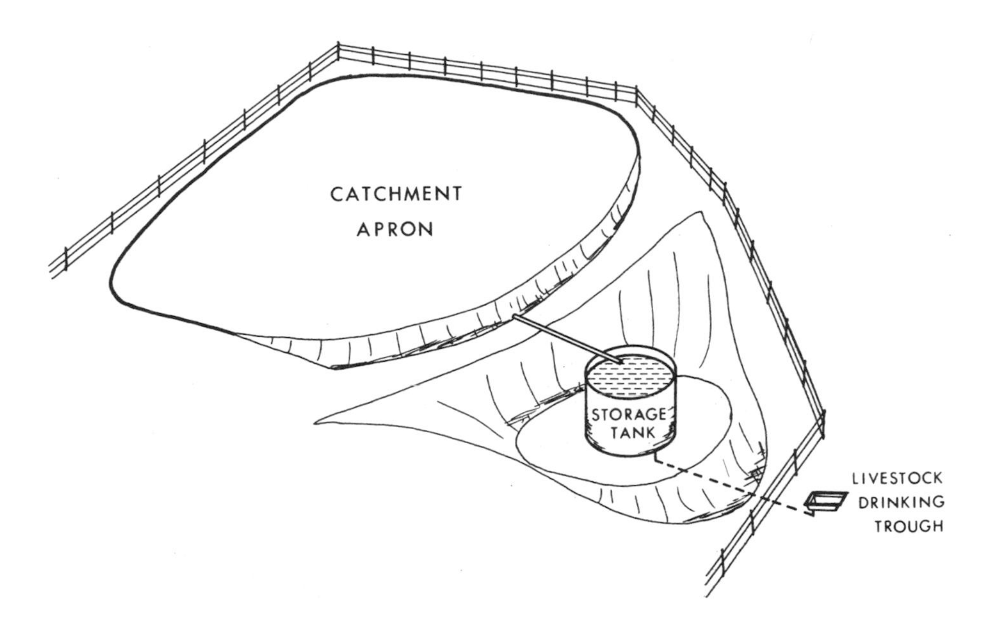

FIGURE 16.4. An example of a simple water-harvesting system (from Frasier and Myers 1983).

parallel rows of drainages 100 m or less long and spaced 15–18 m apart. Trees or horticultural species are planted in the drainages. The areas between drainages are shaped much like high-crowned roads to serve as catchments. Side slopes of the catchment roads and longitudinal slopes of the drainages should be no more than about 2% to prevent erosion. The catchments are cleared of vegetation, smoothed, and treated to reduce infiltration; treatments with NaCl has been effective on expanding-clay soils. If high-value horticultural crops are grown, water storage is necessary to provide supplemental irrigation water; this is accomplished easily by diverting excess water from the drainages into a storage facility.

Water harvesting for agriculture requires a more complex system. The size of the catchment area in relation to that of the agricultural area must be balanced against crop demands and water storage capability. However, the system can be readily adapted to existing topography, provided there is a level area to farm and care is taken with catchment construction to provide gradual slopes or, in steep terrain, short slopes broken by diversions. Compartmental reservoirs consisting of three storage areas are recommended. Electric engine or wind-driven pumps are usually necessary to transfer water between ponds and to drive the irrigation system, although gravity systems might be possible in steep terrain. Since relatively large quantities of irrigation water are usually required, applications to catchments are necessary, but treatments can be expensive. NaCl, one of the least expensive treatments, is effective in locations where the soil has a sufficient quantity (about 10% or more) of expanding clays.

Water harvesting for subsistence farming has been successful in alleviating the food and nutrition problems of people living in drylands. However, training in water-harvesting techniques is necessary; more information is needed from around the world on crop phenology and water requirements; and additional demonstration projects under varying economic, social, and climatological conditions are needed to develop universal prescriptions.

Catchment Areas

The catchment area of a water-harvesting system is impermeable to water and can be used to produce runoff. Examples of different catchment surfaces are:

- Natural surfaces, such as rock outcrops.
- Surfaces developed for other purposes, including paved highways, aircraft runways, or rooftops.
- Surfaces prepared with minimal cost and effort, such as those cleared of vegetation or rocks and smoothed, or both smoothed and compacted.
- Surfaces treated chemically with sodium salts, silicones, latex, or oils.
- Surfaces covered with asphalt, concrete, butyl rubber, metal foil, plastic, tar paper, or sheet metal.

The surface treatment selected depends largely on the cost and availability of materials and labor. In general, the greater the runoff efficiency and life of a treatment, the greater the cost. At one end of the scale, simple smoothing and compaction of nonporous soils is effective but requires annual maintenance. On the other end of the scale is asphalt-impregnated fiberglass covered with gravel, which can last 20 yr or more.

Desirable characteristics of catchment treatments include:

- Runoff from the surface must be nontoxic to humans, plants, and animals.
- The surface should be smooth and impermeable to water.
- The surface material should have high resistance to weathering and should not deteriorate because of chemical or physical treatments.
- The surface material need not have great mechanical strength, but it should be able to resist damage by hail or intense rainfall, wind, occasional animal traffic, moderate water flow, plant growth, insects, birds, and burrowing animals.
- The surface material should be inexpensive and require minimum site preparation.
- Maintenance should be simple.

No single treatment will have all these characteristics. Some trade-off is necessary, but lowest cost over the long term is often the overriding objective. Estimates of the costs of water harvesting using various catchment treatments in the United States are given in Table 16.3.

Selection based only on cost can be a mistake, however. For example, simple, smoothed catchments produce water at low cost, but they do not provide runoff from small storms that characterize rainfall periods in many arid zones. A large, expensive structure might have to be built to store water for use during the period when no runoff occurs. The cost of a simple, smoothed catchment plus the large storage required could be greater than the cost of a more expensive catchment that provides runoff from small storms.

A catchment surface has no standard shape. Flexibility is encouraged to utilize the natural topography to minimize the construction costs. The slope of the surface should be only as steep as necessary to cause runoff; ideally, the slopes should be less than 5%. Slopes that are too steep can erode and produce high amounts of sediment in the runoff water. The catchment surface must be cleared of vegetation, rocks, and other debris that might reduce the durability of a treated surface or retain water on the surface.

TABLE 16.3. Water costs for various water-harvesting treatments

Treatment	Runoff (%)	Estimated Life of Treatment (yr)	Initial Treatment Cost ($US/ha)	Annual Amortized Cost[a] ($US/ha)	Water Cost in a 500 mm Rainfall Zone ($US/1000 m³)
Rock outcropping	20–40	20–30	<120	<240	58–119
Land clearing	20–30	5–10	120–230	<120	79–119
Soil smoothing	25–35	5–10	600–840	120–240	66–188
Sodium dispersant	40–70	3–5	840–1440	120–240	34–119
Silicone water repellents	50–80	3–5	1440–2160	240–480	58–188
Paraffin wax	60–90	5–8	3600–4800	600–1200	132–394
Concrete	60–80	20	24,000–60,000	2040–5280	499–1725
Gravel covered membranes	70–80	10–20	6000–8400	480–1200	119–335
Asphalt fiberglass	85–95	5–10	12,000–24,000	1680–5760	346–1321
Artificial rubber	90–100	10–15	24,000–36,000	2520–4920	494–1056
Sheet metal	90–100	20	24,000–36,000	2040–3120	399–679

Source: Modified from Frazier 1975.
[a]Based on the life of the treatment at 6% interest.

Water Storage

There are three basic types of storage: the soil profile, excavated ponds, and tank or cistern containers. Ancient water-harvesting systems were simple arrangements where water was directed from hillsides onto cultivated areas to immediately store the water in the soil for plant use. The problem with this arrangement was whether or not sufficient water could be stored to offset a prolonged drought. However, the method is still in use today and can be used to grow drought-resistant varieties of trees and other economic plants.

Excavated ponds are often the only economical means of storing the large quantities of water needed for farming, but evaporation and seepage are serious problems. Evaporation suppression on water impoundments is still in the experimental stage; surface area reduction, reflective methods, surface films, mechanical covers, floating Styrofoam balls, and empty plastic film canisters have all been used. Most have been somewhat effective, but a simple economical method has yet to be developed. Surface films are generally not economical on small impoundments. Reflective methods (beads or dyes floating on the surface) are ineffective in windy conditions. Some types of floating mechanical covers and floats have worked well in experimental situations, but most are short-lived and all are expensive.

Seepage from ponds can account for 60 to 85% of the total annual water loss. Seepage control is simpler and less expensive than evaporation control. Chemical dispersing agents, bentonite, soil cement, membrane liners, asphalt, salt, and simple compaction have all been used successfully to seal impoundments. A treatment method and application rate are determined by soil type, purpose of the impoundment, severity of wetting and drying cycles, and economics.

Tanks or cisterns can be used effectively for livestock watering and domestic supplies. Seepage and evaporation are less difficult and less expensive to control. Any container capable of holding water is a potential water storage facility. External water storage, a necessary component of a drinking-water supply system, can also be a part of a runoff-farming system, where some form of irrigation system applies the water to the cropped area. In many water-harvesting systems, the storage and water distribution facility is the most expensive single item, representing up to 50% of the total cost.

A large number of types, shapes, and sizes of wooden and reinforced-plastic storage containers exist. Costs and availability are primary factors in determining suitability. One common type of storage container is a steel tank with vertical walls and a concrete or other type of impermeable bottom. Containers constructed from concrete and plaster are relatively inexpensive but require considerable hand labor. Roofs over the containers to suppress evaporation are common, although they are usually expensive. Floating covers of low-density synthetic foam rubber are an effective means of controlling evaporation from vertical-walled, open-topped storage containers, and they are not expensive.

Constraints and Strategies for Water Harvesting

The strategy to be taken in developing a water-harvesting system depends on a number of constraints, including:

- The need for acceptance by the local community, whether the system is to be used for livestock, domestic purposes, agroforestry, or farming.
- The quantity and quality of water required to meet the demand.
- The availability of alternative, less expensive sources that could be developed.
- The amount, seasonal distribution, and variability of rainfall.
- The materials, labor, and machinery available and suitable for installing a water-harvesting system within budgetary limitations.
- The provisions for maintenance.

With some exceptions, such as microcatchments and strip catchments, most water-harvesting systems must have storage facilities to supply the quantity of water needed at the time it is needed. For livestock, the need depends largely on the type of livestock, the grazing systems employed, and the monthly distribution of rainfall. Many combinations of catchment and storage sizes will provide the desired quantities of water, but the problem is to find the most economical combination. For domestic supplies, people in the United States require from 20 to 40 l/day for cooking, drinking, and washing. The system should be designed to account for this minimum requirement, in addition to any losses that would occur by evaporation or seepage from storage.

Water-harvesting systems for agriculture are more difficult to design. There is frequently little information on the minimum total water requirements of agricultural crops, although consumptive use data are available for crops abundantly supplied with water. Equally important to the total water requirement is the timing of water needs.

The design of the water-harvesting system must satisfy the seasonal pattern of use from initial establishment of the crops to harvest. This type of information has been developed for many crops under intensive irrigation, although these values can be higher than needed for many runoff-farming applications. Relationships of this nature must be developed or estimated for proper design of agricultural water-harvesting systems, and matched with the water supply to determine frequency and amount of irrigation.

Water collected from a catchment can contain organisms and water-soluble impurities from windblown dust deposited on the surface, chemical pollutants directly from the treatment (salt, silicone, tars, or oils), and weathering by-products created by deterioration of the treatment materials (asphalt and certain plastics, for example, deteriorate in sunlight and heat into water-soluble products). Animal feces can be a source of bacterial and viral contamination if the area is not fenced and properly graded. The quality of water from most surface treatments is usually adequate for livestock, but filters are needed in most cases if the water is for human consumption. None of the surface treatments, even with sodium, appear to affect plant production.

The feasibility of developing water-harvesting systems should be evaluated in the context of other alternatives to increase water supply. For instance, untapped springs, a shallow groundwater table that may receive reliable recharge along a mountain front, and perched water that might be tapped with horizontal wells offer possibilities. All sources of water should be evaluated with respect to costs, location, yield, dependability, and quality before embarking on a project. If other convenient sources can be developed economically but are deficient in yield or dependability, they may be used to supplement a system. When groundwater quality is poor (high salt content, for example), harvested rainwater might provide sufficient dilution for the intended use.

Incorporating intermittent water sources into the total water supply system can permit the installation of a smaller water-harvesting facility in some cases. The harvested water can be saved for periods when the ephemeral sources are insufficient or dry up entirely. This combination not only saves time and money but can be important during extended drought periods.

The amount, distribution, and variability of seasonal or annual rainfall are key factors that must be evaluated in designing a water-harvesting system. The frequency of rain and probability of specified rainfall intensities is often more important than the annual quantity (Renner and Frasier 1995a). Long-term daily records of precipitation are the most desirable; in arid lands, at least 15–20 yr of records are usually needed. If large variations exist among years, data from the two wettest years should be eliminated. If sufficient long-term data are available, stochastic methods can be used to determine the probabilities of extreme periods. Mean annual rainfall is not a good indicator of available water, because there will be more years with rainfall less than the mean than years with rainfall greater than the mean. A conservative approach would be to design water-harvesting systems based on the lowest rainfall years.

To compensate for dry years, the size and efficiency of the catchment areas and storage facilities can be increased. Regardless of the design, risk will be involved because of the uncertainty of rainfall. The user must decide the amount of risk that can be accepted.

The ultimate size of the catchment area should be determined by computing a weekly or monthly water budget of collected water and then comparing these values with the water requirement to help ensure that no critical periods exist when there will be insufficient water. Smaller systems can frequently be used when the periods of maximum rainfall

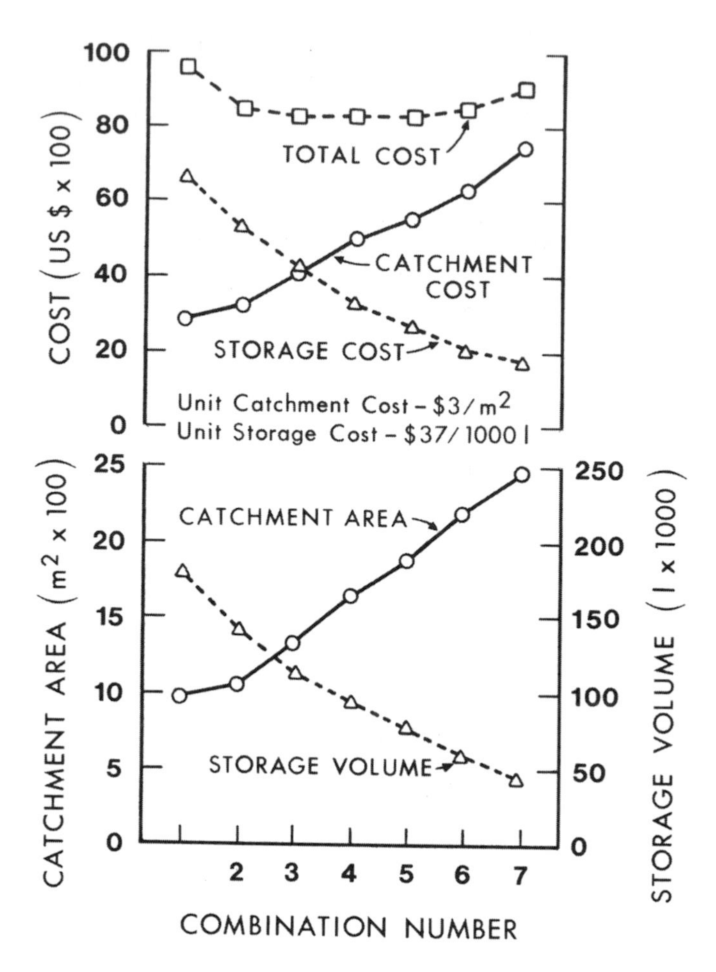

FIGURE 16.5. Cost estimates for different catchment sizes and storage volumes for water-harvesting systems (from Karpiscak et al. 1984).

coincide with periods of maximum use. Larger systems with adequate storage capacity are necessary when the periods of greatest precipitation occur after the periods of greatest water needs. Here, it may be necessary to store water for 6–9 mo.

Economic Considerations

A proposed water-harvesting system should be analyzed to determine economic feasibility. People and communities who are the beneficiaries of water harvesting may see advantages to a system that are different from those recognized by outside technical advisors (Renner and Frasier 1995b). The system must provide both economic and noneconomic benefits to make water harvesting successful (see Chapter 19).

There is no best material for catchment and storage facilities. The cost of alternative water sources and the importance of the water supply determine the costs that can be justified in a system (Fig. 16.5). Systems that supply drinking water are constructed from materials that, in general, are more costly than can be justified economically for runoff-farming applications. One must also balance the cost of materials with the cost of labor. Some materials and installation techniques are labor intensive but have a relatively low capital cost. Other materials can be higher in initial cost but require minimum labor for proper construction. The cost of maintenance must be considered, as the failure to provide for maintenance will result in early failure of any water-harvesting system.

Financial benefits can be profits from the sale of agricultural crops, livestock, or other production items (Oron et al. 1983). There can be environmental benefits as well, such as the reduced grazing of riparian areas, and lower erosion and sedimentation rates downstream. Other advantages of water-harvesting systems include more sustainable agriculture, heightened self-reliance, and reduction of future food shortages.

Water Spreading

Water spreading is a method of using ephemeral flows of streams to produce agricultural crops or to improve rangeland production. Water-spreading systems are best suited for gently sloping alluvial valleys with upstream watersheds that produce at least three or four runoff events before or during the growing season. The method entails diverting stormflow from an ephemeral stream channel; the stormflow is then physically spread onto a gently sloping land surface. Often, a complex system of long, low earth dams, or dikes, is arranged in a zigzag pattern to conduct water away from the stream channel and achieve greater coverage. The series of cascading dikes are situated so that each splits the water it receives so that water is spread over an increasing area in the downslope direction (Fig. 16.6). The diverter can conduct all or part of the streamflow directly into the dikes, which fill and spill into the other dikes that are downslope. In some cases, excess water that reaches the lowest dike can be stored for later use. Grain crops are often planted in rows directly in the wash behind the dams.

Water-spreading systems should be designed to utilize existing floodplains and adjacent terrace features on the landscape. Steep channels and flashy streams are not conducive to water-spreading systems. Slopes should not be too steep to minimize flow velocity and, by doing so, reduce soil erosion and sediment movement. Ideally, broad, flat valleys that provide low velocities and greater resistance to flow and allow a longer time for spreading are best suited. All systems are designed to return runoff to the channel from large runoff events.

The distribution of rainfall and the frequency of high intensity storms are important factors in deciding upon a water-spreading system. A minimum rainfall of 80 mm during the rainy season is generally required to justify a water-spreading system. More rainfall will be needed if the rainfall occurs during warm temperatures. The pattern of rainfall is also important. If the average rainfall is 200 mm and one-half of it occurs in a 2 mo period, for example, it is important to know whether there are 5 or 15 days with rainfall during this period in the average year. Such information gives us a better idea of the frequency of streamflow in the channel.

Streamflow data are often lacking for rural, ephemeral streams. When there are no data available, high-water marks can be used to estimate peak flows (see Chapter 4). Some estimate of streamflow is needed to determine if there is sufficient volume and frequency of flow to produce a crop. A small flow volume might not be worth the cost of diversion, especially if combined with infrequent flow. In contrast, a large volume of flow can be too difficult to manage for the size of the available water-spreading area.

A. PLAN VIEW

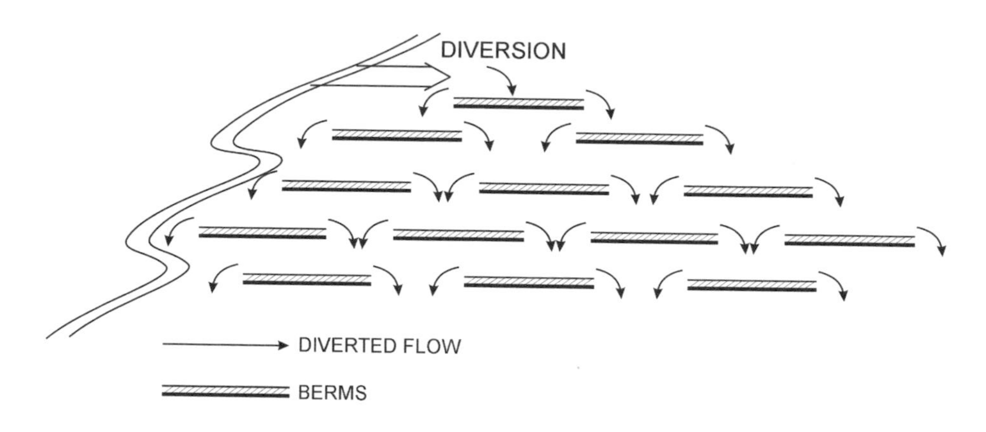

B. IMPLEMENTED WATER - SPREADING PROJECT

FIGURE 16.6. A water-spreading system.

The desired rate of water application to agricultural crops or forage depends on the infiltration characteristics of the soils, the tolerance of the crops to inundation, the soil water-holding capacity, and the plant water requirements during (or immediately following) the season of flow. The distribution of rainfall and runoff events during the growing season is crucial for a successful harvest of agricultural crops, particularly with shallow-rooted crops. For forage production, applications of up to 300 mm in depth over the area are appropriate, particularly if flows are expected only once or twice during the growing season. If the spreading area is large and the water supply limited, 75 to 150 mm of water can produce more forage than concentrating the same amount of water over a smaller area.

Deposits of low volumes of sediment of a high silt content can be beneficial in water-spreading areas. The spreading system that tends to deposit sediments in a gently sloping plane that fills the low spots can add to the soil depth and improve growing conditions. Dikes may need to be heightened periodically to maintain a functioning system.

Excessive sediment deposits result in more costly maintenance and are indicative of serious soil erosion problems in the watershed.

While well-designed and functioning water-spreading systems can improve land productivity at a location, downstream interests must also be taken into account. If there are successive water-spreading systems along the course of a stream, those occupying the most downstream areas may not receive sufficient water. Some type of institutional agreements may be needed to be made to prevent conflicts over water in such situations.

Flood Control

Flood control has been a major objective of many reservoirs and is the principal reason for constructing levees and river channelization projects. Such approaches attempt to keep the floods and related hazards away from humans. Similarly, energy-dissipating devices and related structural development of channels have been developed to reduce debris flow damage. The resulting changes in river response and morphology to changing flow regimes and channel systems downstream can be detrimental to the flood control purposes for which the projects are intended in the first place (see chapters 9 and 10). It is well recognized that alternatives to such structural measures can have financial and environmental benefits. Following the 1993 flooding on the Mississippi River, which resulted in over $12 billion in damages, there was a call for investigating alternatives to structural solutions (Galloway 1995). In this regard, some of the problems with the use of levees in flood control are described in Box 16.7.

Box 16.7

Levees and Flood Control—A Solution or Part of the Problem (from Tobin 1995)

Levees have long been an integral part of flood management in the United States, beginning as early as 1717 along the Mississippi River in New Orleans. Although levees have been breached over the years, they have remained a key component of flood control efforts; such structures occur on over 25,000 miles of waterways in the United States. In spite of billions of dollars of investments, flood damages have continued to rise. Over the years, some rivers have become elevated above their floodplains because of levee construction. In the flood of 1993 on the Mississippi River, about 70% of the levees failed, which exacerbated flooding in many urban areas and farmlands. Even where levees did not fail, flooding occurred upstream and downstream of levees in some instances. The loss of floodplain storage and function adds to stage and peak discharges that levees must be designed to accommodate. Levees can and will continue to play a role in flood protection. However, experience suggests that to achieve effective flood control requires that watersheds and river basins be managed as integrated systems, recognizing linkages among various structural and nonstructural measures. Changing land use on watersheds and the hydrologic role of wetlands, floodplains, lakes, reservoirs, and structures such as levees need to be considered together in formulating flood control programs.

A more comprehensive approach to flood control has been advocated in recent years. This approach takes the position of formulating strategies that keep people away from floods, rather than the converse. *Floodplain zoning* has been widely used in the United States over the past several decades and, in many instances, is linked directly to the issuance of flood insurance. Flood insurance rates are adjusted to the likelihood of flood occurrence adjacent to rivers and streams. Therefore, people who wish to inhabit or construct industries within flood-prone areas must pay higher insurance premiums than those living outside flood-prone areas. Similar approaches are needed in mountainous areas where people occupy areas that are vulnerable to debris flows during high intensity rainfall.

An alternative to floodplain zoning is the *flood-proofing* of structures, so that the floodwaters expected to occur will cause minimal damage. Many civilizations that have developed along river channels have used such approaches for centuries. For example, constructing houses on stilts that elevate the living area several meters above the average high-water elevation is a common practice.

Designation of floodplains and flood-prone zones involves the use of hydrologic methods (see Chapter 17) in conjunction with methods of establishing water surface profiles that are associated with floods of different recurrence intervals. Unfortunately, growing populations of people in many parts of the world have encroached on floodplains as areas to live and grow food, with serious economic consequences and loss of life.

Summary

Water resource projects and programs are developed for multiple purposes, including enhancement of water supplies, flood and related hazard control, and water quality improvement. A combination of structural (engineering works) and nonstructural measures are used in water resource development. To be successful and sustainable, existing and future land use activities and the management of upland watersheds must be considered in the planning, design, and implementation of such projects. Reservoirs, small hydropower systems, water harvesting, and spreading systems are examples of structural modifications that can become integral components of water resource development. After reading this chapter, you should have some insight into how such components of water resource development function and how watershed management relates to their viability. Specifically, you should be able to:

- Explain how different vegetation on a watershed can affect reservoir design and operation.
- Answer the following questions:
  - What are the most critical factors that determine whether a reservoir will function as designed?
  - What can happen to the design life of a reservoir when land use changes after the construction of a reservoir?
  - What role can small hydropower plants play in overall watershed development?
- Discuss why and suggest how watershed management should become an integral part of reservoir, hydropower, and other water resource management and development programs.

- Determine the water-harvesting catchment area necessary to meet a specified annual water requirement, knowing the annual precipitation amount.
- Present examples of different water-harvesting catchment surfaces and describe the desirable characteristics of catchment surfaces in general.
- Describe alternative methods of reducing evaporation from a storage pond or tank.
- Explain the constraints and strategies to be considered in developing a water-harvesting system.
- Identify the benefits and costs associated with a water-harvesting system in the context of economic feasibility.
- Describe how water-spreading systems function and the site conditions necessary for a viable system.
- Discuss the benefits of nonstructural alternatives over engineering methods for reducing damages associated with floods and debris flows.

References

Brooks, K.N. 1998. Coping with hydro-meteorological disasters: The role of watershed management. *J. Chinese Soil and Water Conservation* (Taiwan) 29(3): 219–231.

Brooks, K.N., H.M. Gregersen, E.R. Berglund and M. Tayaa. 1982. Economic evaluation of watershed projects—An overview methodology and application. *Water Resour. Bull.* 18:245–250.

Burdass, W.J. 1975. Water harvesting for livestock in western Australia. In *Proceedings of the water harvesting symposium,* ed. G.W. Frasier, 826. USDA Agric. Res. Serv. ARS-W-22.

Chiarella, J.V., and W.H. Beck. 1975. Water harvesting catchments of Indian lands in the southwest. In *Proceedings of the water harvesting symposium,* ed. G.W. Frasier, 104–114. USDA Agric. Res. Serv. ARS-W-22.

Frasier, G.W. 1975. Water harvesting: A source of livestock water. *J. Range Manage.* 28:429–433.

———. 1988. Technical, economic and social considerations of water harvesting and runoff farming. In *Arid lands: Today and tomorrow,* eds. E.E. Whitehead, C.F. Hutchinson, B.N. Timmermann, and R.G. Varady, 905–918. Boulder, CO: Westview Press.

Frasier, G.W. , and L. Myers. 1983. *Handbook of water harvesting.* USDA Agric. Res. Serv. Handb.

Fritz, J.J. 1984. *Small and mini hydropower systems.* New York: McGraw-Hill.

Galloway, G.E. Jr., 1995. New directions in floodplain management. *Water Resour. Bull.* 31:351–357.

Goodman, L.J., J.N. Hawkins, and R.N. Love, eds. 1981. *Small hydroelectric projects for rural development, planning and management.* New York: Pergamon Press.

Gregersen, H.M., K.N. Brooks, J.A. Dixon, and L.S. Hamilton. 1987. *Guidelines for economic appraisal of watershed management projects.* FAO Conserv. Guide 16, Rome.

Hillel, D. 1967. *Runoff inducement in arid lands.* USDA Final Tech. Report, Proj. A 10-SWC-36.

Jackson, W.L., ed. 1986. Engineering considerations in small stream management. *Water Resour. Bull.* 22:351–415.

Karpiscak, M.M., K.E. Foster, R.L. Rawles, N.G. Wright, and P. Hataway. 1984. *Water harvesting agrisystem: An alternative to groundwater use in Avra Valley area, Arizona.* Tucson: Office of Arid Land Studies, Univ. Arizona.

Larson, C.T. and W. Albertin. 1984. Controlling erosion and sedimentation in the Panama Canal watershed. *Water Int.* 9:161-164.

Linsley, R.K., and J.B. Franzini. 1969. *Water resources engineering.* New York: McGraw-Hill.

McIntyre, S.C. 1993. Reservoir sedimentation rates linked to long-term changes in agricultural land use. *Water Resour. Bull.* 29:487–495.

Mehdizadeh, P. Howsar, A. Vaziri, and L. Boersma. 1978. Water harvesting for afforestation, efficiency and life span of asphalt cover 11, survival and growth of trees. *Soil Sci. Soc. Am. J.* 42: 644–657.

Myers, L.E. 1975. Water harvesting 2000 B.C. to 1974 A.D. In *Proceedings of a water harvesting symposium,* ed. G.W. Frasier, 17. USDA Agric. Res. Serv. ARS-W-22.

National Academy of Sciences. 1974. *More water for arid lands: Promising technologies and research opportunities.* Washington, DC: National Academy of Sciences.

Oron, G., J. Ben-Asher, A. Issar, and T. Boers et al. 1983. Economic evaluation of water harvesting in microcatchments. *Water Resour. Res.* 19:1099–1105.

Reig, C., P. Mulder, and L. Begemann. 1988. *Water harvesting for plant production.* Washington, DC: The World Bank.

Renner, H.F., and G. Frasier. 1995a. Microcatchment water harvesting for agricultural production: Part I: Physical and technical considerations. *Rangelands* 17:72–78.

———. 1995b. Microcatchment water harvesting for agricultural production: Part II: Socioeconomic considerations. *Rangelands* 17:79–82.

Sinha, B. 1984. Role of watershed management in water resource development planning: Need for integrated approach to development of catchment and command of irrigation projects. *Water Int.* 9:158–160.

Thames, J.L., and J.N. Fischer. 1981. Management of water resources in arid lands. In *Arid land ecosystems: Structure functioning and management,* vol. 2, eds. D.W. Goodall, R.A. Perry, and K.M.W. Howes, 519–547. New York: Cambridge Univ. Press.

Tillman, G. 1981. *Environmentally sound small-scale water projects: Guidelines for planning.* Arlington, VA: Volunteers in Technical Assistance (VITA) Publ.

Tobin, G.A. 1995. The levee love affair: A stormy relationship. *Water Resour. Bull.* 31:359–367.

U.S. Army Corps of Engineers. 1977. *Reservoir storage-yield procedures, methods systemization manual.* Davis, CA: Hydrology Engineering Center.

CHAPTER 17 *Hydrologic Methods*

INTRODUCTION

Hydrologic information is needed for watershed management planning, for performing economic and other types of analyses needed for project design, and for making land use impact assessments. Adequate hydrologic, meteorological, and biophysical data are not always available at locations of interest, and even when data are available, deciding which is the most appropriate method to use can be difficult. Hydrologists must apply the tool or method appropriate for a given situation. When hydrologic information cannot be obtained from existing analytical methods, some type of field research or monitoring can be necessary. This chapter presents an overview of analytical methods commonly used to obtain hydrologic information for watershed management and discusses guidelines for their application.

CRITERIA FOR SELECTING METHODS

Numerous methods exist for estimating streamflow characteristics such as peak flow, stormflow volume, annual water yield, and low-flow sequences. These methods range in complexity from simple equations to comprehensive computer simulation methods. Selecting the appropriate hydrologic method requires the careful consideration of:

- The type and accuracy of information required.
- All available data.
- The physical and biological characteristics of the watershed.
- The technical capabilities of the individual performing the study.
- The time and economic constraints.

Study objectives usually dictate the type of information required (Table 17.1) and, in some cases, indicate which method should be used. For example, sizing culverts for

Table 17.1. Examples of project objectives and the corresponding hydrologic information needed

Objective	Hydrologic Information Needed
Culvert design for storm drainage	Peak flow rates for small contributing area corresponding to a particular return interval.
Flood plain delineation	Peak flow rates and associated stages (elevations of associated water levels).
Spillway design for a dam	Hydrographs for extreme meteorological and hydrological events.
Conservation storage requirements for a reservoir	Streamflow rates or volumes during critical drought periods.
Water quality assessment	Volumes and discharge rates for selected events and during dry season periods.
Feasibility for minihydro project	Low streamflow sequences during dry season flow and drought periods.
Determine land use impacts on water yield	Runoff volumes over time for respective land uses.

rural roads requires peak discharge estimates associated with some predetermined risk. Runoff is usually from small, simple drainages, and errors in the estimate of peak flow do not result in great economic loss. Therefore, simple, quick methods are applicable. Conversely, peak discharge estimates for floodplain mapping can require a more complex, but reliable, method. More time and effort can be justified when floodplain mapping is involved, particularly if errors could result in large economic losses or loss of life.

After determining which hydrologic characteristics are of interest, the extent of the data available for the study should be examined. Sometimes, the data are insufficient for all but the simplest of the hydrologic methods.

Watershed characteristics (including size, shape, vegetative cover, topography, soils, and geology) and availability of climatic data and localized hydrologic methods are used to select an appropriate hydrologic method. Some methods are applicable only for small, homogeneous drainages, while others can be applied to large, complex watersheds.

The capacity or capability of the individual or organization performing the study is another important consideration. One should be knowledgeable about the methods used and their limitations. A common constraint with current technology is the availability of appropriate software. Suitable computer programs are invaluable for hydrologic investigations. Time and resources often constrain the methods that can be used (discussed in Chapter 19).

Hydrologic Models

Hydrologic models are simplified representations of actual hydrologic systems that allow us to study the functioning of watersheds and their response to various inputs. Furthermore, they can allow us to predict the hydrologic response of watersheds. For the most part, hydrologic models are based on the systems approach, and differ in terms of how and to what extent each component of the hydrologic cycle is considered.

Models in general can be classified as *material* or *mathematical* (Woolhiser and Brakensiek 1982). Material models can be either *physical,* scaled-down versions of a real system, or *analog models*, which use substances other than (but analogous to) those in a real system. An example of a physical model would be a miniature, scaled version of a particular watershed and channel system; simulated rainfall can be applied in different patterns or quantities to evaluate differences in streamflow response. Physical models are sometimes used to analyze channel structure and engineering constructions (usually hydraulic), but they are expensive and not for general application. They are also not practical for most watershed hydrologic analyses because the physical and biological system cannot be reproduced exactly. An example of an analog model is an electric analog in which the flow of electricity represents the flow of water. Analog models have been used with success in modeling groundwater flow. Because material models have limitations for most watershed applications, except engineering purposes, we will confine our discussion to mathematical models.

Mathematical models can be either *empirical* or *theoretical* and can be further classified as *deterministic* or *stochastic*. Theoretical, or physically based, models rely on physical laws and theoretical principles. It is assumed that the hydrologic functions or relationships in a system are well understood and can be mathematically approximated directly from system characteristics. In contrast, empirical models are based on observed input-output relationships and do not necessarily simulate the actual processes involved. Empirical models rely on data and observations and simply relate output response to a given input.

Deterministic models mathematically characterize a system and give the same response or results for the same input data. For example, a given rainstorm with a particular intensity and duration will yield the same hydrograph response, once the model parameters and initial conditions are fixed. Conversely, stochastic models use the statistical characteristics of hydrologic phenomena to predict possible outcomes and have a random element that results in a different outcome for each execution of the model. Therefore, one can examine an array of possible outcomes that might never have been observed in the past but have some likelihood of occurring in the future. Stochastic models can be applied to specific problems, such as the estimation of drought sequences for the design of reservoir storage for irrigation (see Chapter 16). Several hundred years of simulated streamflow can be generated to investigate potential low-flow periods.

Deterministic models can be grouped according to how and to what extent hydrologic processes of a watershed are represented. The simplest are regression models that statistically relate one or more measurable watershed or climatological characteristics to some hydrologic response of interest, such as annual water yield. Regression models are empirical and useful for the watershed or the region from which they were developed, but they should not be applied elsewhere. In many instances, regression models have led to the development of simple formulas or equations that have more widespread application.

The terms *lumped* and *distributed* refer to the nature of the data representation of a computer simulation model. A lumped model considers a watershed (or land units within a watershed) as an individual unit, and therefore, simulations within the watershed will likely use an average value for the watershed—or an average value for each land unit within a watershed. For example, the rainfall on a watershed varies with time

Box 17.1

Digital Terrain Modeling as Applied in Hydrologic Models (from Moore et al. 1991)

Explicitly including topographic information in hydrologic models is now facilitated with the use of geographic information systems (GISs) (see Chapter 18). The inclusion of topographic features, using digital elevation model (DEM) structures, allows more physically realistic models. Elevation, slope, slope position, aspect, and proximity to stream channels are examples of topographic features that can influence hydrologic response. More physically based models that take into account the variable source area functions, or those that predict erosion from the landscape, can better account for spatial relationships that affect response. Relationships can be developed that allow the simulation of saturated-area expansion or contraction as a function of antecedent moisture and topographic features. The application of DEM and GIS tools allows for more realistic models that account for the role of landscape features in simulating hydrologic response.

and space, but a lumped model considers only the average rainfall on the watershed. All of the processes simulated in a distributed model are calculated from point to point by using a distribution function or a spatially derived algorithm (Box 17.1). Coupled with geographic information systems (see Chapter 18), distributed models are becoming more easily applied.

Although deterministic models can be empirical or theoretical, most hydrologic models are a composite of mathematical relationships, some empirical, some based on theory. As one attempts to explain or predict events from complex systems, more detail and complexity are needed in model formulation; however, the reality is that our understanding of hydrology is not sufficient to represent every process mathematically. This problem leads to the development of models, or parts of models, that need to be calibrated; relationships and parameters must then be fitted for a given watershed. Such a process involves adjusting parameters until the computed response approximates the observed response. Once calibrated, models can then be used to estimate the hydrologic response of the watershed to new, independent input data.

The remainder of this section discusses hydrologic models from the simplest, single-equation models to complex computer simulation models that employ state-of-the-art technology.

Simplified Methods and Models

The need for hydrologic information from ungauged watersheds has led to the development of a wide array of analytical methods, empirical formulas, and models to relate some hydrograph characteristic to measurable watershed characteristics. Most are empirical models developed from observation or experimentation, without necessarily identifying or simulating the processes involved. Sometimes, these are called *black-box models*; they relate hydrologic output directly to input variables using one or more simple (regression) equations.

Direct Transfer of Hydrologic Information

Occasionally, hydrologic information can be transferred from a gauged to an ungauged watershed if the two watersheds are hydrologically similar. The hydrologic similarity depends largely on the following:

- Both watersheds should be within the same meteorological regime.
- Physical and biological characteristics (such as soils, geology, topographic relief, watershed shape, drainage density, and type and extent of vegetative cover) and land use should be similar.
- Drainage areas of the watersheds should be about the same size, preferably within an order of magnitude.

The *direct-transfer method* is quick and easy to use and is normally employed for rough approximations; however, it is applicable only if the assumption of hydrologic similarity is met. If a significant amount of adjustment is needed to transfer hydrologic information from a gauged to an ungauged watershed, there can be little confidence in the result.

Once it is determined that the involved watersheds are hydrologically similar, direct transfer can be applied in two ways. First, the entire historical record from a gauged watershed can be transferred, with minor adjustments for differences in the size of drainage areas. For example, the observed streamflow record could be multiplied by the ratio of ungauged to gauged watershed areas. The second approach is to transfer only certain hydrologic characteristics (such as the 100 yr return period flood peak), again adjusting for drainage area differences. The entire flood frequency curve for an ungauged location can be estimated by adjusting the curve developed for a similar but gauged watershed.

Rational Method

The *rational method* is perhaps the most commonly used simplified formula for estimating peak discharge from rainfall. The basis of the rational method is that the maximum rate of runoff occurs when the entire watershed area contributes to flow at the outlet. Therefore, peak discharge estimates are valid only for storms in which the rainfall period is at least as long as the watershed *time of concentration*. Time of concentration (*Tc*) is the time required for the entire watershed to contribute runoff at the outlet, or specifically, the time it takes for water to travel from the most distant point on the watershed to the watershed outlet.

Peak discharge from small (less than 1000 ha), relatively homogeneous watersheds is estimated from:

$$Q_p = \frac{CP_g A}{K_m} \quad \textbf{(17.1)}$$

where Q_p = peak discharge (m^3/sec); C = runoff constant (Table 17.2); P_g = rainfall intensity (mm/hr) of a storm with a duration at least equal to the time of concentration of the watershed; A = area of the watershed (ha); and K_m = constant, 360 for metric units (1 for English units).

TABLE 17.2. Examples of runoff coefficients (*C*) for the rational method

Soil Type	Cultivated	Pasture	Woodlands
With above-average infiltration rates; usually sandy or gravelly.	0.20	0.15	0.10
With average infiltration rates; no clay pans; loams and similar soils.	0.40	0.35	0.30
With below-average infiltration rates; heavy clay soils or soils with a clay pan near the surface; shallow soils above impervious rock.	0.50	0.45	0.40

Source: Adapted from American Society of Civil Engineers 1969 and Dunne and Leopold 1978.

The assumption that rainfall intensity is uniform over the entire watershed for a period equal to the time of concentration is seldom met under natural conditions. To apply this method, therefore, rainfall intensity and duration values associated with an acceptable risk are used.

Models much like the rational method have been developed throughout the world, and many express regionalized relationships (see Gray 1973 for examples). Such models can be applied when peak discharge estimates are needed quickly or when few data are available. One application would be sizing a road culvert using regionalized rainfall data of some specified design criteria or recurrence interval (Box 17.2).

Hydrologic Response Factor

The stormflow response of a watershed can be characterized by calculating the average ratio of *stormflow volume* (quickflow volume) to precipitation volume for several periods of stormflow. Precipitation can be rainfall, snowmelt, or both. Stormflow volume, that portion of the hydrograph that responds quickly to a rainfall or snowmelt event, is determined by separating *baseflow* from total streamflow during the storm event (Box 17.3). By comparing response factors for watersheds with different vegetative cover and land uses within the same climatic region, the flood-producing potential for different areas can be estimated. In addition, the response factor can be used to develop quick estimates of peak discharge where a consistent relationship exists between stormflow volume and peak. A constraint with this method is that precipitation and streamflow data must be measured to determine the response factor. Also, the response factor is dependent on the antecedent moisture condition of the watershed. However, this does not preclude a regional analysis if regression models can be used to predict response factors based upon measurable watershed characteristics such as size, slope, vegetative cover, and land use.

Stormflow analyses of small watersheds in Georgia by Hewlett and Moore (1976) yielded prediction equations of the form:

$$Q_s = 0.22R\,(SD)\,P^2 \qquad \textbf{(17.2)}$$

where Q_s = predicted stormflow in the area (in.); SD = sine-day factor = sin[360(day no./365)] + 2, where day 0 = November 21; P = total storm precipitation in the area (in.); and

Box 17.2

Using the Rational Method to Size a Culvert

A culvert system is to be designed with a risk of 10% for a design life of 5 yr for a 145 acre forested watershed, based on the following information:

- Soils are medium-heavy clays with good structure.
- Maximum distance (measured from a map) along the stream to the most distant ridgetop is 2430 ft.
- Maximum elevation difference along the maximum stream pathway is 75 ft.

First, the time of concentration must be calculated to determine the appropriate rainfall intensity to use. Using the Kirpich formula as reported by Gray (1973):

$$T_c = \frac{0.0078L^{0.77}}{\left(\frac{H}{L}\right)^{0.385}}$$

where T_c = time of concentration (min), L = distance from main stream outlet to the most distant ridgetop (ft), and H = difference in elevation between main stream outlet and the most distant ridgetop (ft).

In this case:

$$T_c = \frac{0.0078(2430)^{0.77}}{(0.031)^{0.385}} = 12 \text{ min}$$

The probability that a peak will be equaled or exceeded in the next 5 yr is $P_n = 1 - q^5$ where q is the probability of nonoccurrence (Eq. 2.11). Since we want the probability to equal the risk (0.10), the culvert must be designed to convey a peak with $(1 - q)$ probability or a $1/(1 - q)$ return period:

$$0.10 = 1 - q^5$$

$$5 \log q = \log 0.90$$

$$q = 0.98$$

$$\frac{1}{1 - q} = 50 \text{ yr}$$

From a rainfall intensity-duration frequency curve, the 50 yr rainfall event for 12 min is 2.90 in./hr; therefore:

$$Q = CP_gA = (0.30)(2.90)(145) = 126.2 \text{ cfs}$$

$$R = \sum_{1}^{n} (Q_s / P) / n$$

for n observations and $P \geq 1$ in. The sine-day factor approximates antecedent moisture conditions as a seasonal coefficient for the region.

Box 17.3

Baseflow Separation

Baseflow (delayed flow) must be separated from total streamflow for several single-event methods of stormflow analysis. The stormflow volume is that portion of the hydrograph above baseflow and is sometimes called direct runoff or quickflow. There is no universal standard of separating baseflow, because flow pathways through a watershed cannot be directly related to the hydrograph, which represents the integrated response of all flow pathways. This does not present a problem for most flood analyses, however, because the baseflow contribution is typically a small fraction of stormflow (10% or less). Therefore, efforts to devise elaborate baseflow separation routines are usually not warranted. The following is recommended:

- Graphically separate baseflow from stormflow for several storm hydrographs, as with method I or II in the figure below. Once one method has been adopted, it should be used for all analyses.
- After examining several storms, determine if there is a consistent relationship that can be expressed as follows:
  - Draw a straight line from the beginning point of hydrograph rise to a point on the recession limb defined by N days after the peak, where $N = A^c$, A = watershed area in square miles, and c = coefficient (typically a value of 0.2 is used; Linsley et al. 1982).
  - Determine if the separation line I yields a consistent rate in terms of cubic feet per second per square mile per hour. Hewlett and Hibbert (1967) found that 0.05 cfs/mi^2/hr was satisfactory for watersheds in the southeastern United States.

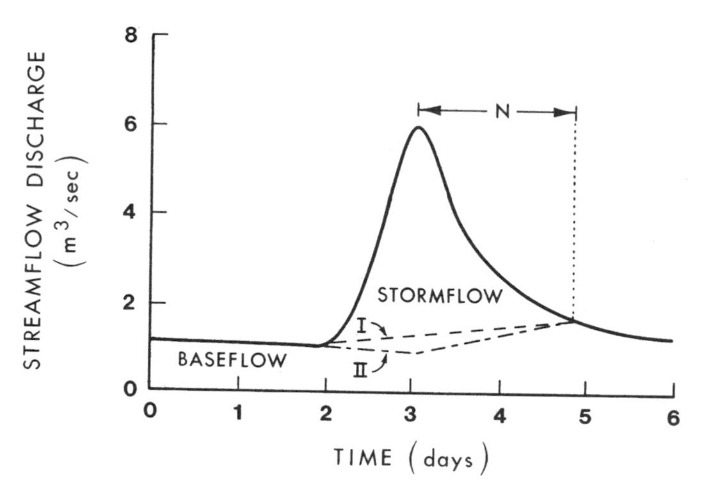

Methods of separating baseflow from stormflow; I and II are different approaches.

Comparisons of stormflow volumes and peaks (Q_p) yielded:

$$Q_p = 25R\,(SD)^{0.5}\,P^{\,2.5} \qquad \textbf{(17.3)}$$

where Q_p = predicted peak discharge above baseflow (cfs/mi^2).

Applications of the response factor method are given in Boxes 17.4 and 17.5. Simple relationships (like the response factors above) are useful, but they should not be applied to watersheds outside the region in which they were developed.

GENERALIZED MODELS

Generalized models are mathematical methods developed and tested on more than one watershed or stream system that can be applied directly to similar systems elsewhere. Implicit with generalized models is that they require some level of input data to describe a given hydrologic system so that one system can be distinguished from another. Generalized models include simplified formulas (e.g., the rational method) that predict some hydrograph characteristic, techniques of hydrograph analysis, dynamic hydraulic routing, and continuous streamflow simulation. No matter how complex, each model is an abstraction of the physical system and uses generalized mathematical functions to estimate hydrologic relationships.

Studies and projects involved with flooding, structural design, and reservoir management require detailed and complex stormflow hydrographs that cannot be obtained with the simplified methods previously discussed. In addition, when the hydrologic response of more than one watershed within a larger basin is desired, hydrographs must be routed and combined to obtain an integrated response. The following paragraphs describe methods of developing stormflow hydrographs, estimating low-flow sequences, and predicting streamflow sequences over time.

BOX 17.4

Using the Stormflow Response Factor Method to Estimate Culvert Size

To select the appropriate size of culvert needed, the peak discharge for a 300 acre forested watershed with R = 0.14 for a rainfall of 3.5 in. is to be determined using the response factors developed by Hewlett and Moore (1976). The wettest month, February, will be used to obtain the maximum response (SD = 2.99).

$$Q_p = (25)\,(0.14)\,(2.99)^{0.5}\,(3.5)^{2.5} = 138.70\ \text{cfs/mi}^2$$

For the 300 acre watershed:

$$\left(\frac{138.70\ \text{cfs}}{\text{mi}^2}\right)\left(\frac{300\ \text{acres}}{640\ \text{acres/mi}^2}\right) = 65.02\ \text{cfs}$$

If the average baseflow for the stream is 4 cfs, then the culvert (culverts) must be capable of conveying a discharge of 69 cfs.

Box 17.5

Applying the Response Factor Method to Estimate Storage Requirements

Determine the amount of storage required to hold the total discharge from a 24 hr, 50 yr return period rainstorm (7.5 in.) on a 3000 acre watershed. The watershed has a 70% cover of old forest (R = 0.10), and 30% of the area is pasture and cultivated land (R = 0.18). Assume SD = 2.99 (February storm).

Forested

$$Q_s = (0.22)\,(0.10)\,(2.99)\,(7.5)^2 = 3.70 \text{ area in.}$$

$$\text{Volume} = (3.7 \text{ area in.})\,(2100 \text{ acres})\,(1\text{ft}/12 \text{ in.}) = 647.5 \text{ acre-ft}$$

Pasture and Cultivated

$$Q_s = (0.22)\,(0.18)\,(2.99)\,(7.5)^2 = 6.66 \text{ area in.}$$

$$\text{Volume} = (6.66)\,(900 \text{ acres})\,(1\text{ft}/12 \text{ in.}) = 499.50 \text{ acre-ft}$$

Total storage required is 1147 acre-ft.

Unit Hydrograph

One of the more widely used methods of stormflow analysis is the *unit hydrograph* (UHG), which is the hydrograph of stormflow resulting from 1 unit (1 mm is used for metric units; 1 inch is used in English units) of effective precipitation occurring at a uniform rate over some period and some specific areal distribution over the watershed. It uniquely represents stormflow response (hydrograph shape) for a given watershed. The *effective precipitation* is the amount of rainfall or snowmelt that is in excess of watershed storage requirements, groundwater contributions, and evaporative losses. It is the portion of total precipitation that ends up as stormflow; therefore, the volume of effective rainfall equals the volume of stormflow.

The UHG method is a black-box model that empirically relates stormflow output to a given duration of precipitation input. No attempt is made to simulate the various hydrologic processes involved in the flow of water through the watershed. The UHG concept provides the basis for several hydrologic models of greater complexity and wider application than simplified formulas such as the rational method.

Development of a Unit Hydrograph

The UHG concept can be best understood by examining the method for developing a UHG from an isolated storm (Fig. 17.1). Records of the watershed are examined first for single-peaked, isolated streamflow hydrographs, which result from short-duration, rainfall or snowmelt hyetographs of relatively uniform intensity.

Once a hyetograph-hydrograph pair is selected, dividing streamflow by the watershed area converts the scale of the hydrograph to millimeters or inches of depth.

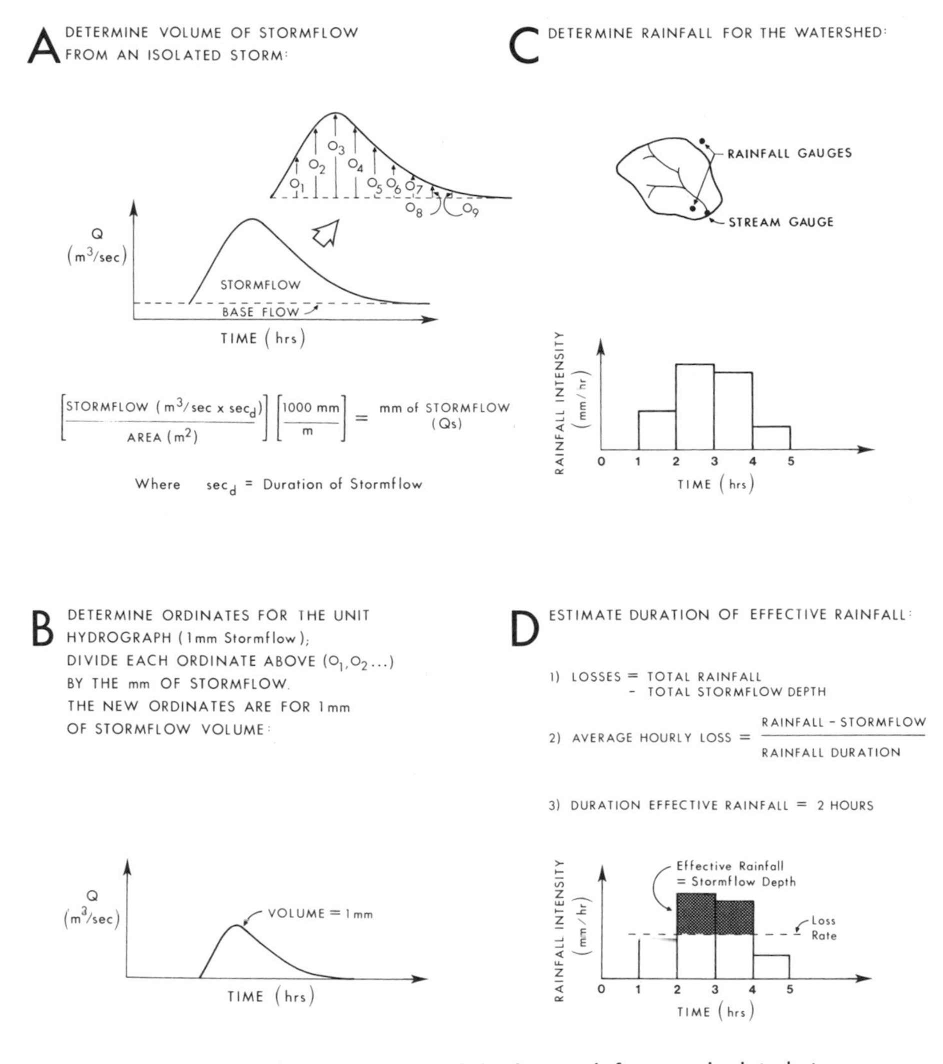

FIGURE 17.1. Development of a unit hydrograph from an isolated storm.

Separating the more uniform baseflow from the rapidly changing stormflow component determines the total stormflow depth from the hydrograph (Fig. 17.1A). Each ordinate of the stormflow hydrograph is then divided by the total stormflow depth, resulting in a normalized hydrograph of unit value under the curve (Fig. 17.1B).

The area-weighted hourly precipitation distribution that caused the stormflow hydrograph is the next factor to be analyzed (Fig. 17.1C). Effective rainfall is defined as being equal to stormflow. Therefore, the total loss (interception, storage, and deep seepage) is determined as the difference between total rainfall or snowmelt and total stormflow (Fig. 17.1D). The magnitude of these losses is largely a function of antecedent moisture conditions. In Figure 17.1D, a uniform-loss rate was assumed, but a diminishing-loss curve can be used if available. The effective rainfall in Figure 17.1D was uniform during the 2 hr period bracketed by the loss curve. This duration of effective rainfall characterizes the UHG, not the duration of the UHG itself. For example, a UHG developed from an effective rainfall of 3/4 hr duration is a 3/4 hr UHG, that of 1 hr duration is a 1 hr UHG, and that of 6 hr duration is a 6 hr UHG. Unit hydrographs can also be developed from multipeaked stormflow hydrographs by means of successive approximations. Such complex storms are best analyzed with computer assistance.

The normalized shape of the UHG is characteristic of the watershed for a given intensity of effective rainfall and is actually an index of stormflow for a particular watershed; it represents the integrated response of that area to a given rainfall input. The UHG method assumes that the effective rainfall and loss rates are relatively uniform over the entire watershed area. Also, the watershed characteristics that affect stormflow response must remain constant from the time that the UHG is developed until the time it is applied.

Application of a Unit Hydrograph

To apply a UHG, the rainfall quantity and distribution over time from a design storm must first be obtained from the watershed (see Chapter 2). Estimated loss rates are then subtracted from total rainfall to obtain the quantity, distribution, and duration of effective rainfall. The UHG must be of the same duration as the duration of effective rainfall, or of increments of effective rainfall. Often a 1 hr UHG is developed and applied to 1 hr increments of effective rainfall. The ordinates of the selected UHG are then multiplied by the quantity of effective rainfall for each period. For example, in Table 17.3, 20 mm of effective rainfall over the first hour yields a stormflow hydrograph with ordinates 20 times those of the corresponding UHG for the first period, and 30 mm of effective rainfall over the second hour yields 30 times the UHG, but delayed, for the second period. The calculated stormflows, plus any baseflow, are then added up for each period to obtain the total stormflow hydrograph.

The assumption of linearity is not always valid, however. As effective rainfall increases, the magnitude of the peak actually can increase more than the proportional increase in the rainfall amount. The consequence of ignoring such a nonlinear response can be to underestimate the magnitude of peak discharge for large storm events. If a nonlinear response is suspected, two or more UHGs should be developed from observed hydrographs resulting from substantially different rainfall amounts. The appropriate UHG would then be applied only to precipitation amounts similar to those used to develop the UHG in the first place.

It is emphasized that the UHG developed from a specific duration of effective rainfall can be applied only to effective rainfall that fell over the same duration. For example, a 6 hr UHG cannot be applied directly to analyze a storm with an effective rainfall that occurred over 2 hr. The duration of a UHG must be changed in such cases. The most common method of converting a UHG from one duration to another is the *S-curve method* (Fig. 17.2). Once a UHG of a specified duration (say, 4 hr) is developed, an S-curve can be constructed by adding the UHGs together, each lagged by its duration. The summation of the lagged UHGs is a curve that is S-shaped and represents the stormflow response expected for a rainstorm of infinite duration but at the same intensity of precipitation as the component UHGs (for a 4 hr UHG, the intensity would be 0.25 mm/hr). Once the S-curve is constructed, a UHG of any duration can be obtained by lagging the S-curve by the desired UHG duration, subtracting the ordinates, and correcting for precipitation-intensity differences, as illustrated in Figure 17.2.

TABLE 17.3. Application of a 1 hr unit hydrograph to a storm of 2 hr of effective precipitation

Time (hr)	1 hr UHG Ordinates (m^3/sec)	Effective Rainfall (mm)	Stormflow: Time 1 (m^3/sec)	Stormflow: Time 2 (m^3/sec)	Stormflow: Subtotal (m^3/sec)	Base Flow (m^3/sec)	Total Discharge (m^3/sec)
0	0.00	0	0.00	0.00	0.00	1.2	1.2
1	0.05	20	1.00	0.00	1.00	1.2	2.2
2	0.50	30	10.00	1.50	11.50	1.2	12.7
3	1.00	0	20.00	15.00	35.00	1.2	36.2
4	0.75	0	15.00	30.00	45.00	1.2	46.2
5	0.50	0	10.00	22.50	32.50	1.2	33.7
6	0.25	0	5.00	15.00	20.00	1.2	21.2
7	0.00	0	0.00	7.50	7.50	1.2	8.7
8	0.00	0	0.00	0.00	0.00	1.2	1.2

Before any empirical model such as a UHG is used for study or design purposes, the model should be tested with observed data. If historical streamflow responses can be reconstructed with a UHG model, then the model can be assumed to be valid. One of the most difficult problems with the testing (verification) and application of UHGs is determining the appropriate loss rates. Such losses are largely dependent on watershed characteristics and antecedent moisture conditions.

Loss Rate Analysis

Loss rates, as used in the UHG method, represent the rainfall or snowmelt that does not contribute to stormflow. Early engineering hydrology textbooks used the terms *loss rates* and *infiltration rates* interchangeably, which led to the idea that stormflow occurs only when infiltration capacities are exceeded and, therefore, results entirely from surface runoff. This is not the case with forested watersheds, where subsurface flow often predominates. As applied in the UHG, it is neither necessary nor desirable to equate loss rates with infiltration rates. Perhaps a more appropriate definition of "losses" would include precipitation stored on vegetative surfaces (interception), in the soils (where soil moisture deficits occur), as detention storage, or as water that percolates to groundwater or is otherwise delayed.

Therefore, a loss rate function need not approximate infiltration curves and can be approximated in most instances with either a constant loss rate (ϕ index) or a constantly diminishing loss rate. Loss rates have to be determined empirically for the entire watershed.

Synthetic Unit Hydrographs

The UHG method can be of limited value for many watershed studies because both rainfall and streamflow data must be available. Because streamflow data are seldom available at locations of interest, synthetic UHG models have been developed. These models

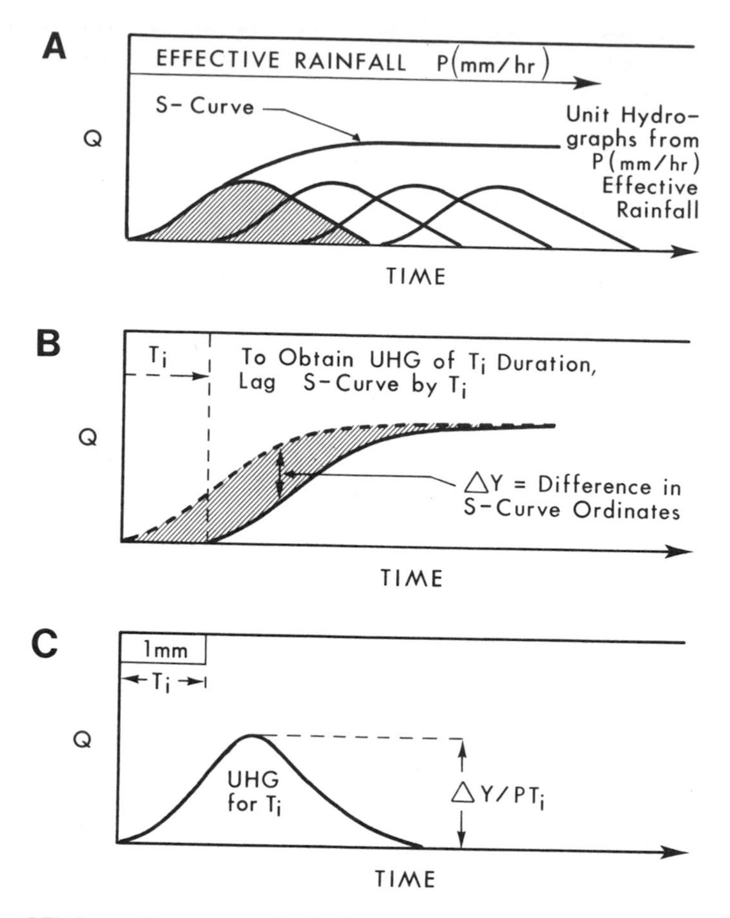

FIGURE 17.2. S-curve method of converting a unit hydrograph of one duration of *P* (mm/hr) effective rainfall to another duration (T_i). A: S-curve is the sum of UHGs that result from *P* (mm/hr) of effective rainfall; B: S-curve is lagged by T_i; difference in ordinates = ΔY; C: the resulting UHG of T_i duration. *Q* axis = streamflow discharge.

consist of mathematical expressions that relate measurable watershed characteristics to UHG characteristics. Runoff hydrographs for ungauged watersheds can be estimated with synthetic UHG models if loss rates can be approximated.

The Soil Conservation Service (SCS) method (U.S. Soil Conservation Service 1972) incorporates generalized loss rate and runoff relationships developed from watershed studies in the United States. The following equation (developed in English units) is used to estimate the stormflow volume from a given storm:

$$Q = \frac{(P - 0.2\,S_t)^2}{P + 0.8\,S_t} \tag{17.4}$$

where Q = stormflow (in.), P = rainfall (in.), S_t = watershed storage factor (in.), and $0.2S_t = I_a$, an initial loss that was a consensus value determined in the original development of the method (Box 17.6).

Box 17.6

Initial Abstraction and Loss in the Curve Number Method

Demonstration of the $I_a = 0.2S_t$ assertion in the original formulation of the curve number method has been recently challenged. Using an event analysis of 52,910 rainfall-runoff events and general curve-fitting methods, and rainfall-runoff data from 114 mostly agricultural watersheds at 26 sites in the United States, Hawkins and Khojeini (2000) found that the value of I_a/S_t is consistently less than 0.2. Values in the vicinity of 0.05 were more in keeping with their data analysis. The differences implied with the current practice were also more pronounced in situations of small rainfalls (that is, low P/S_t values) and lower curve numbers. Relative to using a I_a/S_t value of 0.05, the assumption of $I_a/S_t = 0.2$ calculates lower runoffs. Differences in equation behavior due to variations in I_a/S_t became smaller at higher rainfalls and higher curve numbers. While the researchers indicated that the applied-hydrology community should be aware of these findings, they also stated that these effects must be investigated further.

As seen in Box 17.6, the initial loss, $I_a = 0.2\,S_t$ may be an overestimate. Values in the vicinity of 0.05 S_t may be more accurate. The differences implied with the current practice were more profound in situations of smaller rainfalls and lower curve numbers. These researchers emphasized that the effects that they observed should be investigated further to determine the scope of their results.

A watershed index, or curve number (CN), is related to S_t as follows:

$$CN = \frac{1000}{10 + S_t} \tag{17.5}$$

Soil, vegetation, and land use characteristics are related to curve numbers that indicate the runoff potential for a given rainfall (Table 17.4). Soils are classified hydrologically into four groups:

A = high infiltration rates; usually deep, well-drained sands and gravels with little silt or clay.
B = moderate infiltration rates; fine- to moderate-textured, well-structured soils, e.g., light sandy loams, silty loams.
C = below-average infiltration rates; moderate- to fine-textured, shallow soils, e.g., clay loams.
D = very slow infiltration rates; usually clay soils or shallow soils with a hardpan near the surface.

Three antecedent moisture conditions (AMC) are considered, depending on the amount of rainfall received 5 days before the storm of interest:

AMC I = dry, < 0.6 in.
AMC II = near field capacity, 0.6–1.57 in.
AMC III = near saturation, > 1.57 in.

TABLE 17.4. Runoff curve numbers for hydrologic soil cover complexes for antecedent moisture condition II

Cover			Hydrologic Soil Group			
Land Use	Treatment or Practice	Hydrologic Condition	A	B	C	D
Fallow	Straight row	—	77	86	91	94
Row crops	Straight row	Poor	72	81	88	91
	Straight row	Good	67	78	85	89
	Contoured	Poor	70	79	84	88
	Contoured	Good	65	75	82	86
	Contoured and terraced	Poor	66	74	80	82
	Contoured and terraced	Good	62	71	78	81
Close-seeded legumes[a] or rotation meadow	Straight row	Poor	66	77	85	89
	Straight row	Good	58	72	81	85
	Contoured	Poor	64	75	83	85
	Contoured	Good	55	69	78	83
	Contoured and terraced	Poor	63	73	80	83
	Contoured and terraced	Good	51	67	76	80
Pasture or range		Poor	68	79	86	89
		Fair	49	69	79	84
		Good	39	61	74	80
	Contoured	Poor	47	67	81	88
	Contoured	Fair	25	59	75	83
	Contoured	Good	6	35	70	79
Meadow		Good	30	58	71	78
Woods		Poor	45	66	77	83
		Fair	36	60	73	79
		Good	25	55	70	77
Farmsteads		—	59	74	82	86
Roads (dirt)[b]		—	72	82	87	89
(hard surface)[b]		—	74	84	90	92

Source: U.S. Soil Conservation Service 1972.
[a]Close-drilled or broadcast.
[b]Including right-of-way.

After the *CN* is determined, rainfall is converted to stormflow graphically (Fig. 17.3).

For the SCS model to be valid, *CN* relationships should be determined for each hydrographically different region. The relationships developed in the United States are generally applicable to small watersheds (less than 13 km^2) with average slopes less than 30%.

Once effective rainfall is determined, a hydrograph can be produced. Peak discharge (Q_p) can be approximated with the triangular hydrograph method (Box 17.7):

$$Q_p = \frac{K_o A Q}{T_p} \tag{17.6}$$

where K_o = constant (a value of 484 means that 3/8 of the UHG volume is under the rising limb; for mountainous watersheds a value near 600 might be appropriate; wetlands

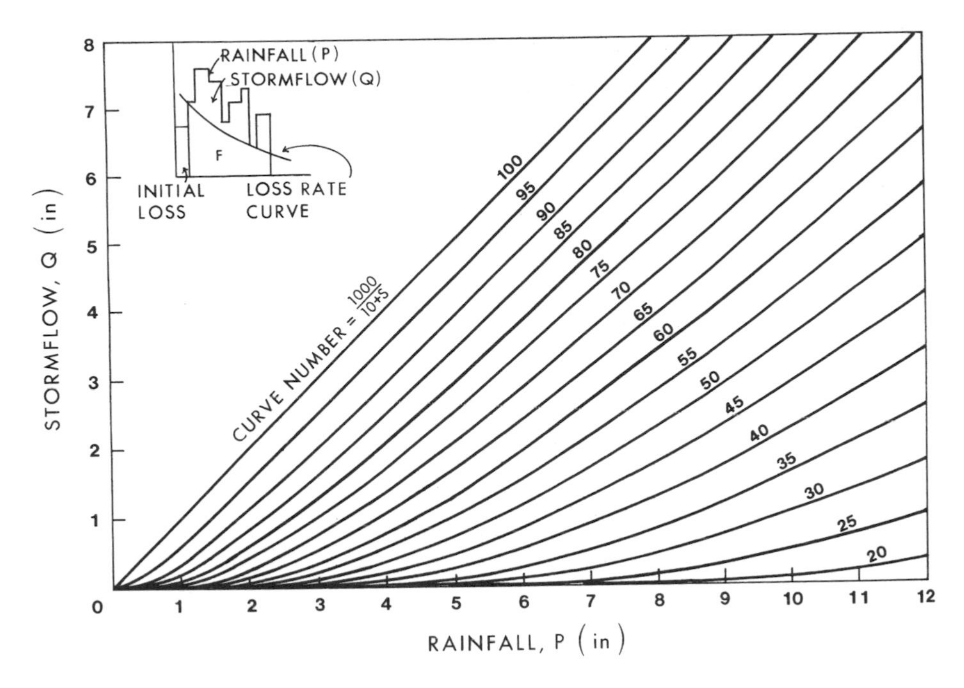

Figure 17.3. Rainfall-stormflow relationships for curve numbers (from U.S. Soil Conservation Service 1972).

can be closer to 300; this is not a universal constant); A = area (mi^2); Q = stormflow volume (in.); T_p = time to peak (hr), where $T_p = \Delta D/2 + 0.6T_c$; ΔD = duration of unit excess rainfall (hr)(= $0.133T_c$); and T_c = estimated time of concentration (hr).

Changes in stormflow associated with changes in the soil-vegetation complex are determined largely by the *CN* relationships. For example, changing from a good pasture condition, within hydrologic soil group C, with AMC = II, to a poor pasture condition changes the *CN* from 74 to 86. In turn, this would result in an increase in stormflow volume from 3.2 in. to 4.6 in. for a 6 in. rain. In some instances, the time of concentration could also be altered, which would affect the peak and general shape of the stormflow hydrograph.

Regression Models

Regression models are mathematical functions that statistically describe relationships between a dependent variable (hydrologic response) and one or more independent variables (watershed and climatological characteristics). Regression models are approximations based on sampling and are therefore subject to sampling variation (see Chapter 18).

Many of the early approaches to predicting hydrologic response involved applications of regression models. Multiple-regression models have been useful for predicting streamflow characteristics for watersheds over a large area or region. Examples include regional relationships between peak discharge or low-flow discharge of a specified return period and other characteristics, such as watershed area, annual rainfall, percentage of area in lakes or wetlands, and watershed or channel slope. Models can be developed for an array of flooding or low-flow events with different recurrence intervals.

Box 17.7

Application of the SCS Method.

A 4 in. rain fell over a 4.6 mi² watershed in a good pasture condition, soils in group D, and an AMC II. The duration of rainfall was 6 hr. Time of concentration was estimated to be 2.3 hr. The stormflow and the corresponding peak flow are estimated as follows:

1. From Table 17.4, $CN = 80$.
2. Enter Figure 17.3 with 4 in. of rain and $CN = 80$. The stormflow is approximately 2.0 in.
3. Peak flow discharge for the watershed with a time of concentration (T_c) of 2.3 hrs is determined from Equation 17.6 as follows:

$$T_p = \frac{\Delta D}{2} + 0.6T_c$$

$$= \frac{0.31}{2} + (0.6)(2.3) = 1.53 \text{ hr}$$

$$Q_p = \frac{(484)(4.6)(2)}{1.53} = 2910 \text{ cfs}$$

Regional analyses of this type are useful for estimating streamflow characteristics for ungauged watersheds.

Multiple-regression models can also be used to estimate hydrologic processes from readily obtained data. For example, an equation for a multiple regression using air temperature, wind velocity, and relative humidity as independent variables might be acceptable in estimating potential evapotranspiration, the dependent variable. Such estimates can then be used as input to a water budget or other simulation model.

Analysis of Recession Flows

At the beginning of a dry season or in periods between isolated storms, the flow of water from a watershed usually diminishes with time. Without further replenishment, much of the water that reaches downstream users during such periods is *recession flow.* These flows typically originate from the upland watersheds, which have different vegetative covers and land use patterns. Therefore, an analysis of recession flows can provide a basis for comparing different watershed management activities and their effects on streamflow between storm events.

While considerable information exists on the influence of vegetative cover types and land use patterns on water yield (see Chapter 6), knowledge of recession flow response to land management activities is less complete. It is commonly assumed that recession flows are largely dependent upon the storage characteristics of a watershed. The variability in recession flows is apparently due to the loss of water by evapotranspiration. The magnitude and duration of recession flows are both related to the rate of water movement through the soil, which is affected by vegetation and land use characteristics. Vegetative cover types, land use patterns, and the season determine the variability in soil moisture conditions and hence affect the rate of recession flow.

Hydrograph Analysis

A number of regression models are available to evaluate recession flow characteristics (Table 17.5). By consolidating variables such as evapotranspiration, soil moisture, and land use patterns into one constant term for a given recession flow event, recession flows from different watersheds can be characterized and compared.

Of the regression models available, the equation reported by Barnes (1939) is used widely throughout the United States. This equation is:

$$Q_t = Q_0 k^t \tag{17.7}$$

where Q_t = streamflow discharge after a time period t (cfs), Q_0 = the initial discharge (cfs), k = the recession constant per unit of time, and t = the time interval between Q_0 and Q_t in hours or days.

A linear form of the equation can be obtained by a transformation onto semilogarithmic paper, with discharge on the logarithmic scale and time on the arithmetic scale:

$$\log k = \frac{\log Q_t - \log Q_0}{t} \tag{17.8}$$

The slope k evaluated in this equation is considered indicative of differences in recession flows, attributed collectively to the effects of vegetative cover, land use patterns, season of the year, and soil moisture.

Streamflow data are frequently reported as daily averages; however, analysis of these averages can prevent the identification of individual hydrographs resulting from small storms not shown by the average streamflow for a day. Furthermore, the peaks identified by data points might not represent the actual peaks of the storms that caused the recession flows.

For prediction purposes, it is a common practice to arrange the recession flows chosen for analysis in a manner such that a composite recession flow is formed to represent

Table 17.5. Recession flow equations

Type of Function	Reported or Reproduced by
$Q_t = Q_o k^t$	Barnes 1939
$Q_t = Q_o e^{-bt^n}$	Horton 1933
$Q_t = Q_o^{-at}$	Hall 1968
$Q_t = Q_o/(1 + at)^2$	Hall 1968
$Q_t = 1/(at) + Q_0$	Indri 1960
$Q_t = a/(t^n) + b$	Toebes and Strang 1964

Q_t = streamflow discharge after time t (cfs).
Q_o = initial streamflow discharge (cfs).
k = recession constant.
b, n, a = constants defined by respective authors.

the complete recession for a watershed (Singh 1992). However, such a procedure can be subject to judgmental errors and errors attributed to factors that contribute to the differences between short-event recessions (e.g., evapotranspiration losses), which vary greatly from time to time. Therefore, evaluation of the constant k for a recession based on a composite curve might not be justified in a study. Instead, the average recession flow might be evaluated, with the variability measured as differences between slope constants of individual recession limbs.

From an analysis of recession flows of all major streams in Iowa, Howe (1966) found that there were no major differences in the values of the recession constants for watersheds that were less than 100 mi^2. He concluded that the variability in recession flows due to area differences is small and therefore might not be important in most evaluations. However, in describing the recession flow characteristics of watersheds in the forests and woodlands of Arizona, Brown (1965) found that the average slope of the recession flows differed according to vegetative cover.

Generalized Continuous Simulation Models

The previous sections have dealt with single-event methods and models that allow estimates of hydrograph characteristics such as peaks, stormflow volumes, or recession flows with limited data. However, an estimate of streamflow response over an extended period is needed for many hydrologic investigations. Such information is useful for determining reservoir conservation storage requirements (see Chapter 16), for investigating water quality, and for estimating differences in annual streamflow or streamflow patterns before and after watershed modifications. More complex models that require more extensive information are usually needed for such applications.

Continuous simulation models compute streamflow discharge over periods that include more than one storm event and intermittent low-flow sequences. Such models use a water budget approach and simulate hydrologic processes such as interception, evaporation, transpiration, detention storage, and infiltration to varying degrees. Processes must be linked mathematically so that the conservation of mass principle is not violated. Many of these models are *lumped models*, which means that the spatial variability of processes and characteristics over the watershed unit is largely ignored; therefore, land units that are being modeled should be relatively homogeneous in terms of hydrologic characteristics. However, for most watersheds of any size, the area must usually be subdivided into many relatively homogeneous units modeled separately, and then streamflow values are combined and routed to obtain the overall system response. *Distributed models* attempt to characterize the spatial variability of a hydrologic property such as infiltration capacity.

Continuous simulation models should be conceptually sound, flexible in design, and physically relevant. Input data requirements should not be too detailed or complex, and the output should be presented in such a way that it can be used conveniently. The utility of such models depends largely upon input-output capabilities and on the applicability of their constants or parameters to ungauged or altered watersheds.

The utility of continuous simulation models is that they can be used for a variety of hydrologic investigations. For example, models like the Hydrocomp simulation program (HSP) have been used as transport models for water quality components. However, the temporal and spatial variability of most water quality constituents limits the application of these models to studies requiring only approximate answers.

Generalized continuous simulation watershed models evolved primarily as a result of engineering and flood-forecasting needs; most have the general form illustrated in Figure 17.4. The HSPF model, the streamflow synthesis and reservoir regulation model (SSARR), the Sacramento model, the National Weather Service streamflow forecasting system models, and the U.S. Department of Agriculture Hydrograph Laboratory (USDAHL)-74 watershed model are examples. Hydrologic processes, represented mathematically, are calculated as flows and storages. Model parameters (or variables) of an *empirical model* cannot usually be measured exactly and therefore require fitting, which is accomplished by comparing observed and simulated streamflow. Parameter optimization routines are available sometimes, but even so they are trial-and-error procedures, which match computed results with those observed. The greater the number of fitted parameters in a model, the more difficult the application to ungauged or altered areas. A regional analysis using regression techniques (described later in this chapter) is often useful to estimate parameters for ungauged areas.

Soil and vegetation parameters in some of the models are not of sufficient detail to directly simulate the hydrologic effects of land use activities such as forest clearcutting

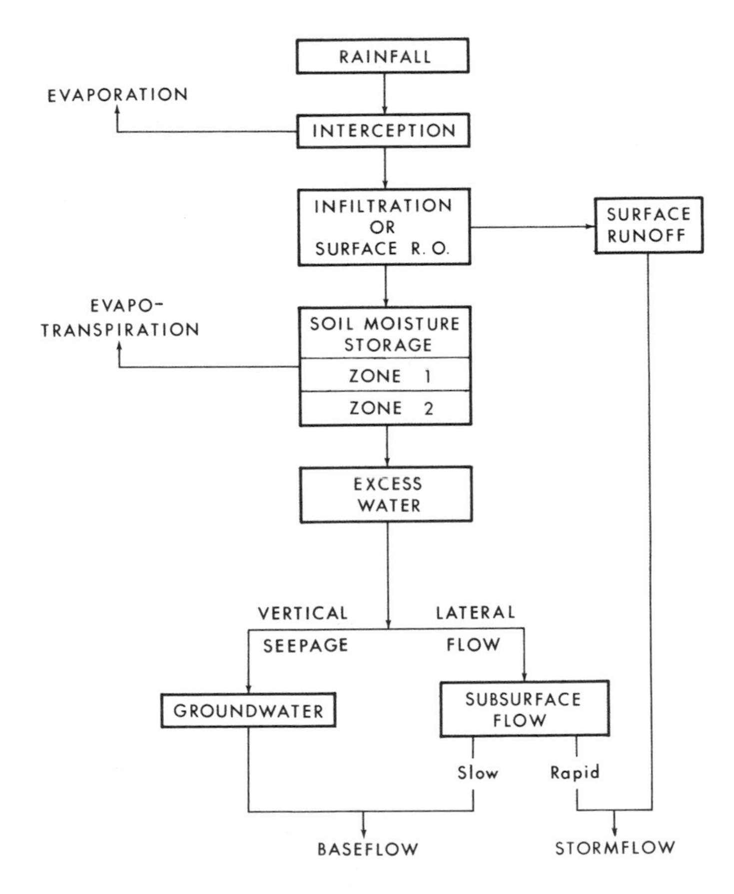

FIGURE 17.4. Hydrologic processes and runoff relationships commonly found in generalized continuous simulation models.

or converting from rural agriculture to urban areas. As a result, detailed, more theoretical, *physically based* models have been developed to estimate the hydrologic effects of land management activities.

STREAMFLOW ROUTING

Stormflow hydrographs for a particular watershed become modified as the streamflow moves downstream through a channel. For example, the net effect of a flood wave moving through a channel is an attenuation of the peak flow, which is the result of the temporary water storage in the channel. Furthermore, the shape, speed, and magnitude of a flood wave change as it moves downstream. *Streamflow routing* is a mathematical procedure for predicting these changes. Routing is essential to predict the changes in the magnitude of the peak and the corresponding stage of flow as a flood wave moves through the channel. Furthermore, routing is required if we attempt to determine how changes in the streamflow hydrograph upstream become transferred to downstream locations. As streamflow enters larger order streams, the routing and combining of flows of all tributary streams need to be taken into account to determine the total hydrograph downstream.

Several methods are available for routing flow. Hydraulic routing takes into account the channel features and computes flow along several cross sections of a channel over time. Hydrologic routing is an empirical method of routing flow over time through a channel using a generalized relationship for the entire channel reach. It is beyond the scope of this book to discuss the array of routing procedures used in hydrology, but a generalized hydrologic routing method is described for illustrative purposes. The generalized hydrologic routing method described is referred to as a lumped flow, reservoir routing as described by Fread (1993). This method is based on the conservation of mass principle, which states:

$$I(t) - O(t) = \frac{dS}{dt} \tag{17.9}$$

where $I(t)$ is inflow at the channel reach as a function of time, $O(t)$ is the outflow from the channel reach as a function of time, and dS/dt is the time rate of change of storage in the channel reach.

The storage S is a function of both inflow and outflow and can be related to either or both by an empirical storage function. If it is assumed that the water surface is level throughout the reach (as in a lake or reservoir), storage can be empirically related to outflow or to the surface elevation (h), that is, $S = f(h)$. A graphical depiction of this routing procedure is shown in Figure 17.5.

The generalized method described above is the basis for several mathematical routing methods, including the streamflow simulation and reservoir regulation (SSARR) model, Puls, modified Puls, Muskingham, and linear reservoir routing. The reader is referred to Singh (1992) and Fread (1993) for a more detailed discussion of these routing applications.

MODELING HYDROLOGIC EFFECTS OF WATERSHED MODIFICATIONS

The need to evaluate the hydrologic consequences of a specific watershed modification, such as clearcutting forest cover, or the implications of a comprehensive management

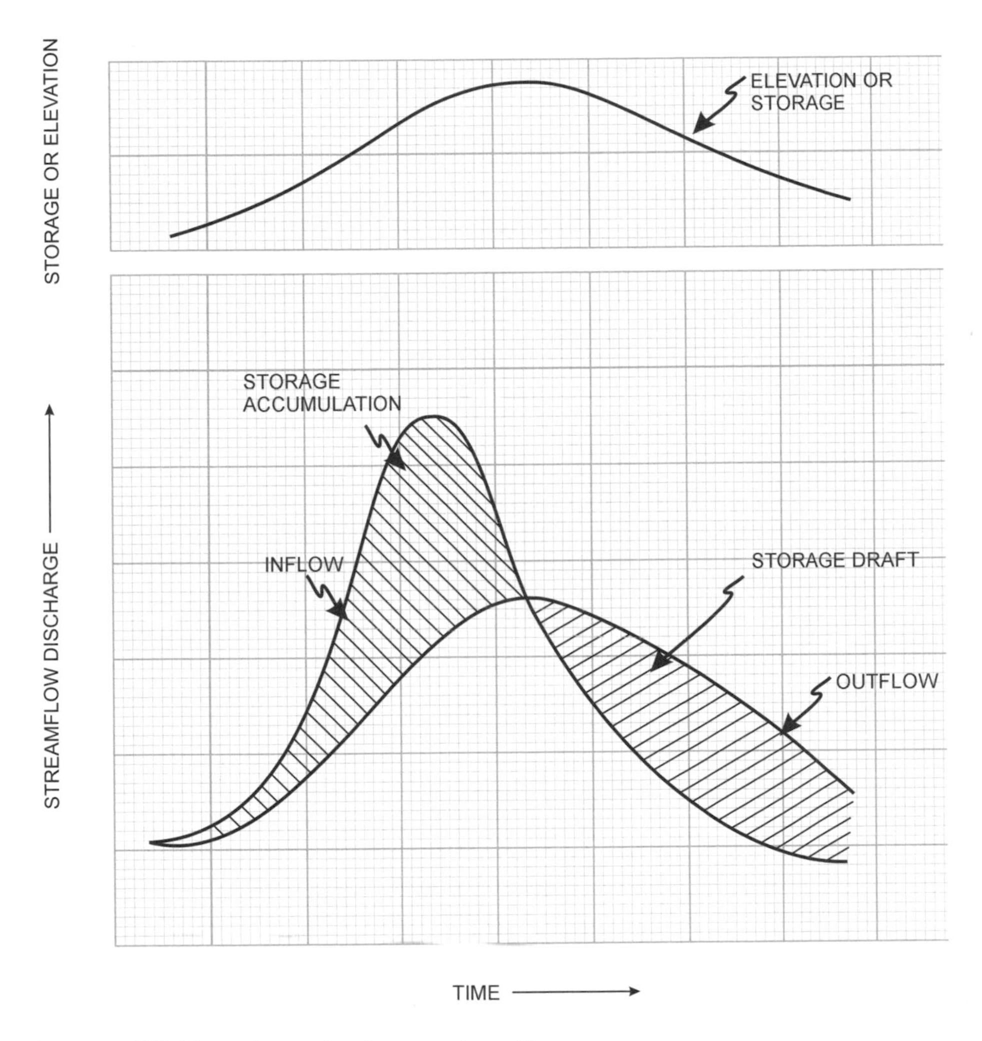

FIGURE 17.5. Example of reservoir routing.

plan that involves several different land use changes led to the development of models oriented to watershed modifications. The hydrologic response of altered watersheds can be modeled in essentially two ways. The first and perhaps most appealing approach is the application of a physically based model in which all hydrologic processes changed by the modification must be represented and determined from measured characteristics (Larson 1973). The empirical approach, which offers a practical alternative to physically based models, links regional relationships (usually determined from experimental watershed studies) with readily available data, and empirically predicts an outcome. The latter approach relies on extensive data from experimental studies.

Few physically based watershed models have been used in practice, although many have been developed and tested as part of research projects. Some physically based hydraulic models have been developed and used to estimate the effects of channel modification on streamflow. Extensive data are needed for even the simpler hydraulic models, but data requirements and the degree of complexity increase exponentially when water flow and storage processes are modeled for the entire soil-plant-atmosphere system. Examples of some continuous simulation models that have a more physical basis are listed in Table 17.6.

Models capable of predicting hydrologic effects due to changes in vegetative cover or land use practices have not been developed for global application. Rather, models have been developed to simulate streamflow response to specific land use practices for a particular region and/or ecosystem, such as the subalpine water balance model (WBMODEL) (Table 17.6), which is designed to simulate streamflow response due to timber management activities in the subalpine elevations of the Rocky Mountains. The development of such regional models is continuing, and with the use of long-term watershed research data for testing and verification, useful models should be forthcoming for more areas of the world.

Application of Isotopes

Isotopic tracers are useful in studying hydrologic processes. Where surface water and groundwater interact, as in the case in riparian areas and wetlands, it is often of interest to be able to determine the flow pathways and contributions of groundwater and surface water to a body of water. If one wanted to restore or create a wetland, the contributions of groundwater and surface water would be of importance in determining the type of wetland that could be restored or constructed. Stable isotopes of oxygen and hydrogen can be used to determine the contributions of water from rainfall runoff (new water) in contrast to the contributions of water that has had significant residence time in the soil and/or groundwater aquifers (old water).

An *isotope* is an element with a specific number of neutrons and protons in its nucleus. The isotopes ^{18}O, ^{2}H, and ^{3}H are components of natural water molecules that fall to earth as precipitation and therefore are useful tracers for watershed analyses. Hydrogen has two stable isotopes, a lighter isotope, protium or ^{1}H, and a heavier isotope, deuterium (D) or ^{2}H. There is also a third isotope, tritium or ^{3}H, that is radioactive and has a half-life of 12.3 years. About 99.985% of the hydrogen atoms in the hydrosphere are comprised of ^{1}H, while ^{2}H only accounts for 0.015%. Oxygen has three stable isotopes: ^{16}O comprises 99.859%, ^{17}O 0.038%, and ^{18}O 0.2% of the oxygen atoms in the hydrosphere. The heavier isotopes represent a small fraction of the total, but this difference can be measured, and the ratio between the lighter and heavier isotopes determined. An agreed upon standard that has been set by the International Atomic Energy Agency, known as the Vienna standard mean ocean water (V-SMOW), is used to report the differences between the ratio of a heavy isotope to a lighter isotope. Values of δ are reported in units of parts per thousand (reported as parts per million or $^{o}/_{oo}$) relative to a standard with a known composition (Kendall and Caldwell 1998) as:

$$\delta(^{o}/_{oo}) = (R_x/R_s - 1) \times 1000 \tag{17.10}$$

where R is the ratio of the heavy isotope to light isotope, R_x is the ratio of the sample, and R_s is the ratio of the standard.

Processes of evaporation and condensation can produce variations in the ratio between lighter and heavier isotopes. For example, when water evaporates from surface water, molecules containing heavier isotopes evaporate at a slightly slower rate than the molecules containing lighter isotopes. During a thunderstorm, molecules containing heavier isotopes will tend to fall more readily in the precipitation than molecules con-

TABLE 17.6. Examples of continuous simulation models that have been used to examine the hydrologic effects of land use changes

Model	Processes Simulated				Flow		Application	Reference
	Rainfall	Snowfall	Infiltration	ET	Surface	Subsurface		
HSP	X	X	X	X	X	X	Simulates streamflow records for forested, rangeland, agricultural, and urban watersheds; engineering applications.	Hydrocomp 1976
PROSPER			X	X	X	X	Simulates water flow through soil-plant-atmosphere system, computes water yield for different soil-plant systems.	Goldstein et al. 1974
USDAHL-74	X	X	X	X	X	X	Simulates streamflow records for agricultural watersheds; considers zones of infiltration and exfiltration; similar to variable source area approach.	Holtan et al. 1975
WBMODEL	X	X	X	X	X	X	Simulates hydrologic changes resulting from watershed management in Colorado subalpine zone; produces year-round water budget.	Leaf and Brink 1973
PHIM	X	X	X	X	X	X	Simulates streamflow records for forested upland and peatland watersheds in upper Midwest; computes effects of timber harvest and peat mining activities.	Barten and Brooks 1988, Guertin et al. 1987
SWRRB	X	?	X	X	X	?	Simulates hydrologic and related processes in ungauged rural basins; computes effects of watershed management changes on outputs.	Arnold et al. 1990

taining lighter isotopes. These processes redistribute the relative abundance of δD and $\delta^{18}O$ as water moves through the hydrologic cycle. The term used to describe this redistribution is known as *fractionation*.

Isotopes have been used in watershed studies to estimate evaporation, groundwater recharge areas, groundwater age, relative hydraulic residence time, hydrograph separation in runoff studies, and characterization of interactions between surface water and groundwater. The isotopic composition of water in deep and shallow aquifers suggests different recharge source areas. Elevation differences between plains and mountains produce isotopic variation because mean annual temperature changes with elevation. Magner et al. (2001) observed isotopic compositions that were similar to snow for the region in a confined outwash aquifer below the Red River near Moorhead, Minnesota. Additional carbon dating confirmed that the aquifer water was of glacial origin. These applications of stable isotopes were successful largely because of unique hydrologic, geologic, and landscape features of the study area. Stable isotopes are commonly used in hydrology for estimating source waters and hydrologic pathways.

Burns and McDonnell (1998) estimated hydraulic residence time by comparing the amplitude of a best-fit isotopic sine curve for precipitation to the amplitude of a similar curve for water that is of interest. Seasonal changes in the $\delta^{18}O$ composition of precipitation at temperate latitudes tend to follow a sinusoidal pattern. The pattern occurs over a period of 1 yr, reflecting the seasonal changes in tropospheric temperature. For a given location, measured changes in $\delta^{18}O$ composition for the water body of interest (e.g., lake, pond, stream, soil water, or groundwater) are made during different seasons. Mean hydraulic residence can be calculated, if the seasonal waters are considered in a steady-state, well-mixed reservoir with an exponential distribution of residence times as follows:

$$T = \grave{u}^{-1}\,[(A/B)^2 - 1]^{1/2} \tag{17.11}$$

where T = the hydraulic residence time (days), $\grave{u}$ = the angular frequency of variation ($2\pi/365$ days), A = the input amplitude, and B = the output amplitude.

Streamflow Frequency Analysis

Frequency analysis is performed for purposes such as the design of water control works, determining conservation storage requirements, and delineating a floodplain according to the flood risk. Frequency curves, the product of such analyses, are simply an expression of hydrologic data on a probability basis. The frequency with which some magnitude of a selected variable is equaled or exceeded can be determined from a frequency curve. Frequency curves can be developed for hydrologic variables such as annual flood peaks, rainfall amounts (see Chapter 2), flood volumes, low-flow streamflow volumes or rates, river stages, and reservoir stages. This discussion focuses on streamflow frequency analysis.

In performing a frequency analysis, the objective of the analysis must first be well defined: Is the study to determine storage needs for low-flow augmentation or to estimate the chance of experiencing a critical low-flow value for a particular time duration? The objective will determine the type of data that is required. Important considerations include:

- The appropriate streamflow characteristic needs to be identified. For example, are instantaneous peak discharges of interest (floodplain delineation) or are daily flood flow volumes of interest (storage analysis)?

- The data used in the analysis must represent a measure of the same aspect of each event. For example, mean daily peak discharges cannot be analyzed with instantaneous peak discharges.
- Streamflow data being analyzed should be controlled by a uniform set of hydrologic and operational factors. Natural streamflows of record cannot be analyzed with flows that have been modified by reservoir operations. Likewise, peak discharges from snowmelt cannot be mixed with peak discharges from rainfall.
- If only a few years of streamflow records are available, a *partial-duration series* analysis rather than *annual series* can be used to get a better definition of the frequency curve (Fig. 17.6). However, this does not necessarily make the curve more reliable.

The annual series approach uses only the extreme annual event (for peak discharge analysis, the largest peak flow) from each year's streamflow record, while the partial-duration series uses all independent events above a specified base level in the analysis. For peak discharge analysis, the annual series analysis ignores the second-highest peak discharge in a year (16 m^3/sec in year 1 in Figure 17.6), which can be higher than the highest peak discharge in another year (14 m^3/sec in year 2). Including all major events in a partial-duration series can better define a frequency curve when only a few years of data are available. As the number of years of record increases, a frequency curve developed from either method will yield essentially the same frequency relationship in the region of the curve that defines the rarer events (low probability).

Frequency curves based on the two methods are interpreted differently: for annual series, the exceedance frequency (frequency with which a value is equaled or exceeded) for a given magnitude is the number of *years* that magnitude will be exceeded per 100 yr; for a partial-duration series, the interpretation is the number of *events* that will exceed a certain magnitude per 100 yr.

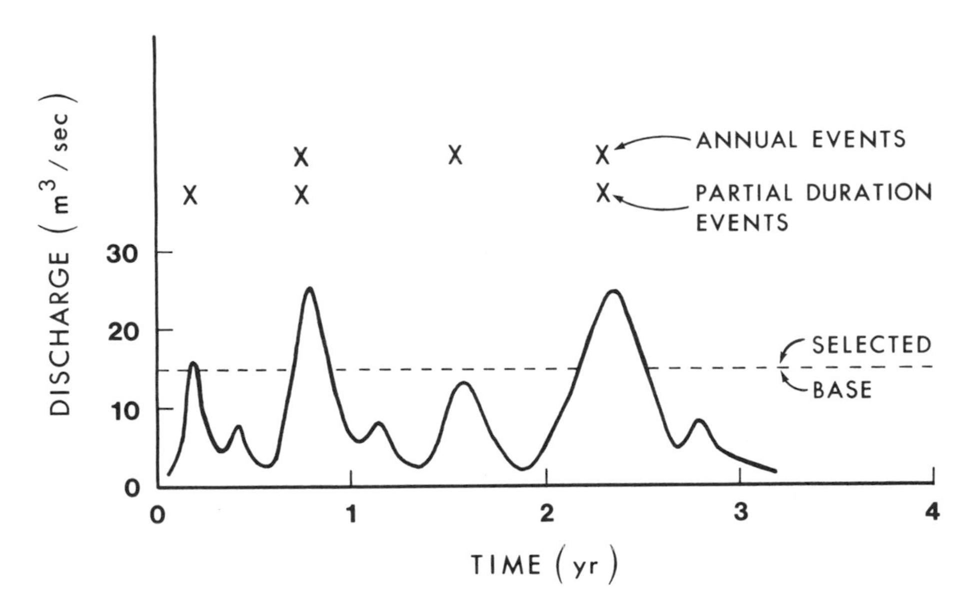

FIGURE 17.6. Annual and partial-duration series approaches for flood peak analysis.

Once the data sets are obtained, either graphical or analytical methods can be used to develop frequency curves.

Graphical Frequency Analysis

The graphical method involves calculating the cumulative probabilities (plotting positions) for ranked events and then drawing the frequency curve through the data points. Several methods can be used to calculate cumulative probabilities. The median formula (Beard 1962) commonly is used for peak discharge frequency curves:

$$P = \frac{m - 0.3}{N + 0.4} \qquad \textbf{(17.12)}$$

where P = the cumulative probability (plotting position) for the event ranked m in N number of years of record.

The cumulative probability for the least severe events (probability < 50%) can be calculated as follows:

$$P = \frac{2m - 1}{2N}$$

The graphical method does not assume a statistical distribution and is quick and easy to apply. Because the frequency curve is hand drawn, one portion of a curve can be weighted more than another. It is usually recommended that the cumulative probabilities, or plotting positions, determined by the graphical method be plotted, even if the analytical method described below is used to define the curve.

Analytical Frequency Analysis

The analytical method requires that sample data follow some theoretical frequency distribution. The U.S. Water Resources Council (1976) recommends the log Pearson type III distribution for peak discharge frequency curves, which requires the following steps (Box 17.8):

1. The data are transformed by taking the logarithms of peak discharges.
2. The mean peak discharge (first moment) is calculated; this corresponds to the 50% probability of exceedance.
3. The standard deviation (second moment) is calculated; this represents the slope of the frequency curve plotted on log-probability graph paper.
4. The skew coefficient (third moment), an index of nonnormality, is then calculated; this represents the curvature in the frequency curve.
5. Adjustments are then made for small numbers of events. The skew coefficient is usually unreliable for short records; a regionally derived skew coefficient is recommended for fewer than 25 yr of records. If 25–100 yr of records are available, a weighted skew should be used. The skew derived from one station should be used only if more than 100 yr of records are available.
6. The frequency curve for the observed annual peaks (Q) is then determined for selected exceedance probabilities (P) by the equation:

   $$\log Q = \bar{x} + k_s s \qquad \textbf{(17.14)}$$

   where $\bar{x}$ = mean of log Q (m³/sec), k_s = factor that is a function of the skew coefficient and a selected exceedance probability, and s = standard deviation of log Q (m³/sec).

7. Confidence limits can be calculated and plotted for the frequency curve as a final step.

The analytical method has the following advantages over the graphical method: the same curve is always calculated from the same data, which makes it more objective and consistent; the reliability of the curve can be calculated with confidence intervals; and the calculation of statistics by regional analysis (described later in this chapter) allows for the development of frequency curves at ungauged sites.

Frequency Analysis for Ungauged Sites

Because of limited stream-gauging sites on most watersheds, methods are needed to estimate streamflow frequency characteristics at ungauged sites. One method is to use hydrologic models. Frequency curves can be estimated for ungauged areas by modeling the streamflow response using the more readily available and longer-record precipitation data. It is necessary that a valid model be developed, one verified with gauged data from the region. A properly tested and verified continuous simulation model can be useful for such studies. The following steps are needed:

- Calibrate the model using gauged data to establish the constants of the mathematical relationships. Hydrologic judgment and a reconstruction of historic streamflow events are essential to this process. Model parameters can be mapped, and a regional-analysis approach (described later) can be used to extend the mathematical relationships to ungauged sites.
- Adapt the model to the ungauged watersheds using measurable data and estimated parameters.
- Enter the historical precipitation data, which are estimated for the ungauged watersheds, in sequence. A standard frequency analysis can then be conducted on the resulting streamflow output from the model.
- The last step should be a critical analysis of the resulting frequency curve. Any wide discrepancies (perhaps based on frequency estimates from gauged streamflow at a nearby station) must be resolved.

Comments on Frequency Analysis for Watersheds in Remote Areas

There are often few gauged stations on watersheds in remote areas and, where gauges are present, the record periods are short. Therefore, the uncertainty associated with frequency curves developed for such regions is substantial. Uncertainty also exists with long-term records because even those records are only a sample of the possible events that have occurred in the past or can occur in the future. The tendency is to guess high to provide conservative answers. However, the consequences of substantial errors in a calculated frequency curve should be the guide for determining the acceptable risk.

Watersheds are also constantly undergoing changes that affect to some degree the hydrologic response. Fires, timber-harvesting operations, livestock grazing, road construction, and urbanization all affect the runoff response of watersheds. Such changes are sometimes abrupt, and the corresponding change in the hydrologic regime is therefore readily identified. The more subtle, long-term changes are perhaps the most common and also the most troublesome for estimating streamflow frequency curves over a period.

Any time watershed modifications are such that a change in streamflow response is expected, care must be exercised in the frequency analysis. One possible solution is to use a watershed model with sufficient sensitivity to simulate both land use change and

Box 17.8

Graphical and Analytical Frequency Curves for a 19-yr Record of Annual Peak Discharges of the Little North Santiam River near Mehama, Oregon

Water Year	Instant Peak Flow (cfs)	Ranking Position of Peak Flow	Cumulative Probability (Graphical Plotting Position)[a] (%)
1932	13,900	7	34.5
1933	10,600	10	50.0
1934	18,900	3	13.9
1935	10,400	11	55.2
1936	12,200	8	39.7
1937	8,200	18	91.2
1938	16,500	4	19.1
1939	8,570	15	75.8
1940	8,200	17	86.1
1941	8,330	16	80.9
1942	9,300	12	60.3
1943	19,400	2	8.8
1944	7,990	19	96.4
1945	11,700	9	44.8
1946	19,900	1	3.6
1947	15,300	5	24.2
1948	14,800	6	29.4
1949	9,120	13	65.5
1950	8,860	14	70.6

[a]Determined from Equation 17.12.

Analytical Frequency Analysis Calculations:

Mean log of peak discharge $\overline{x} = 4.0650$
Standard deviation $s = 0.1399$
Calculated skew $g = -0.6001$

	Initial Plotting				
	1.0	10	50	90	99
k (g = 0)[a]	2.33	1.28	0	–1.28	–2.33
$k_s s$	0.326	0.179	0	–0.179	–0.326
Log $Q = k_s s + \bar{x}$	4.391	4.244	4.065	3.886	3.739
Q (cfs)	24,600	17,500	11,600	7,690	5,484
P_N[b]	1.79	11.4	50.0	88.6	98.2 plotted

[a]g = 0 from Table A below.
[b]From Table B below (N – 1 = 18); the flow values are then plotted at these adjusted exceedence frequency values.

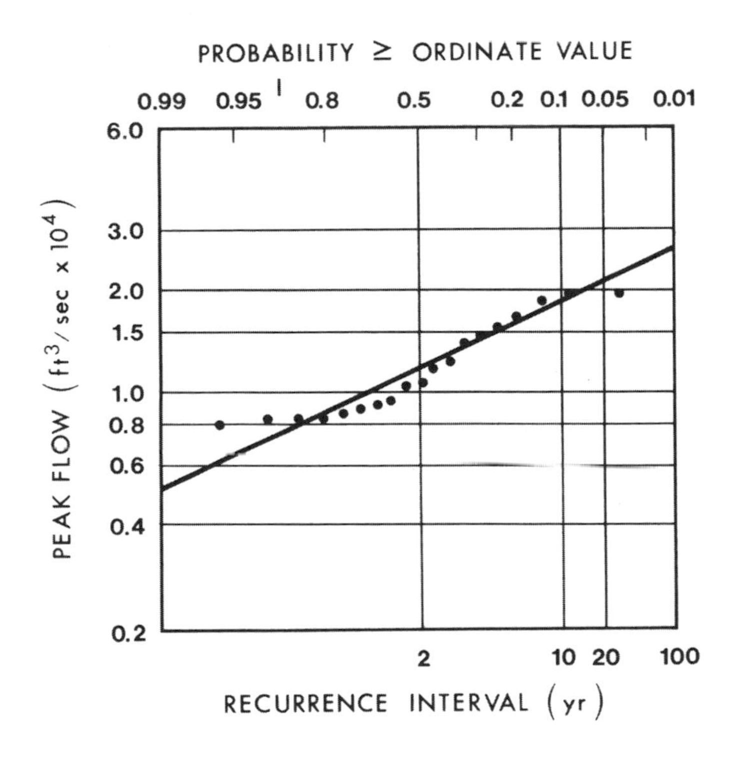

Frequency curve of annual peak discharge for Little North Santiam River near Mehama, Oregon (1932–1950).

(continues)

(continued)

[Box 17.8] Table A. Pearson Type III coordinates

g (Skew Coefficient)	k_s = Magnitude in Standard Deviations from Mean for Exceedence Percentages of								
	1.0	5	10	30	50	70	90	9	99
1.0	3.03	1.87	1.34	0.38	−0.16	−0.61	−1.12	−1.31	−1.59
0.8	2.90	1.83	1.34	0.42	−0.13	−0.60	−1.16	−1.38	−1.74
0.6	2.77	1.79	1.33	0.45	−0.09	−0.58	−1.19	−1.45	−1.88
0.4	2.62	1.74	1.32	0.48	−0.06	−0.57	−1.22	−1.51	−2.03
0.2	2.48	1.69	1.30	0.51	−0.03	−0.55	−1.25	−1.58	−2.18
0.0	2.33	1.64	1.28	0.52	0.00	−0.52	−1.28	−1.64	−2.33
−0.2	2.18	1.58	1.25	0.55	0.03	−0.51	−1.30	−1.69	−2.48
−0.4	2.03	1.51	1.22	0.57	0.06	−0.48	−1.32	−1.74	−2.62
−0.6	1.88	1.45	1.19	0.58	0.09	−0.45	−1.33	−1.79	−2.77
−0.8	1.74	1.38	1.16	0.60	0.13	−0.42	−1.34	−1.83	−2.90
−1.0	1.59	1.31	1.12	0.61	0.16	−0.38	−1.34	−1.87	−3.03

Source: From Beard 1962.

[Box 17.8] Table B. Table of expected probability (P_N) versus initial plotting position from normal populations

	P Less Than						
N–1	50.0	30.0	10.0	5.0	1.0	0.1	0.01
1	50.0	37.2	24.3	20.4	15.4	12.1	10.2
2	50.0	34.7	19.3	14.6	9.0	5.7	4.3
3	50.0	33.6	16.9	11.9	6.4	3.5	2.3
4	50.0	33.0	15.4	10.4	5.0	2.4	1.37
5	50.0	32.5	14.6	9.4	4.2	1.79	.92
6	50.0	32.2	13.8	8.8	3.6	1.38	.66
7	50.0	31.9	13.5	8.3	3.2	1.13	.50
8	50.0	31.7	13.1	7.9	2.9	.94	.39
9	50.0	31.6	12.7	7.6	2.7	.82	.31
10	50.0	31.5	12.5	7.3	2.5	.72	.25
11	50.0	31.4	12.3	7.1	2.3	.64	.21

N–1	50.0	30.0	10.0	5.0	1.0	0.1	0.01
12	50.0	31.3	12.1	6.9	2.2	.58	.18
13	50.0	31.2	11.9	6.8	2.1	.52	.16
14	50.0	31.1	11.8	6.7	2.0	.48	.14
15	50.0	31.1	11.7	6.6	1.96	.45	.13
16	50.0	31.0	11.6	6.5	1.90	.42	.12
17	50.0	31.0	11.5	6.4	1.84	.40	.11
18	50.0	30.9	11.4	6.3	1.79	.38	.10
19	50.0	30.9	11.3	6.2	1.74	.36	.091
20	50.0	30.8	11.3	6.2	1.70	.34	.084
30	50.0	30.6	10.8	5.8	1.45	.24	.046
40	50.0	30.4	10.6	5.6	1.33	.20	.034
60	50.0	30.3	10.4	5.4	1.22	.16	.025
120	50.0	30.2	10.2	5.2	1.11	.13	.017
∞	50.0	30.0	10.0	5.0	1.00	.10	.010

Source: From Beard 1962.
Note: P_N values above are usable approximately with Pearson Type III distributions having small skew coefficients.

the associated runoff response. Frequency analysis can then be performed with the simulated data as previously discussed. Again, the results are largely dependent upon the validity of the model and the adequacy of the input data.

Regional Analysis

A regional analysis is a statistical approach in which generalized equations, graphical relationships, or maps are developed to estimate hydrologic information at ungauged sites (U.S. Army Corps of Engineers 1975a). Runoff factors, unit hydrograph coefficients, and streamflow frequency characteristics can be estimated for ungauged watersheds that are within the same climatic region as the gauged watersheds employed. Any pertinent information within the region should be used to relate watershed characteristics to hydrologic characteristics. For example, a regional analysis can be used to estimate runoff coefficients or peak discharge in cubic meters per second per square kilometer associated with a specified recurrence interval, or to estimate the constants needed to execute a complex hydrologic model.

Regional streamflow frequency analysis is most commonly applied. Equations and maps are developed that allow the derivation of exceedance frequency curves for ungauged areas in the following manner:

1. Select components of interest, such as the mean annual peakflow, 100 yr recurrence interval peakflow, etc.

2. Select explanatory variables (characteristics) of gauged watersheds, such as drainage area, watershed slope, and percentage of area covered by lakes or wetlands.
3. Derive prediction equations with single or multiple linear regression analyses, as previously discussed.
4. Map and explain the residual errors that are the differences between calculated and observed values at gauged sites.
5. Determine frequency characteristics for ungauged locations by applying the regression equation with adjustments as indicated by the mapped residual errors.

The residual errors constitute unexplained variance in the statistical analysis. Since including all variables that influence a hydrologic response is impossible, the mapping of unexplained variances can sometimes indicate other important factors. For example, sometimes mapped residuals will indicate a relationship between the magnitude of the residual error and watershed characteristics such as vegetative type, soils, or land use.

Regional analysis has been applied widely in the United States where streamflow or other hydrologic data are available but are scattered throughout different climatic regions. The method must be used with care, however, since only one or perhaps a few components of the hydrologic system are included in the analysis. The results obtained should be considered a rough estimate of the true hydrologic characteristics in any case.

SUMMARY

To be able to go beyond the purely descriptive treatment of watershed management, it is imperative to understand and be able to apply the methods that quantify the hydrologic response of watersheds. This chapter summarizes methods of quantifying streamflow characteristics that are commonly used in streamflow forecasting for engineering design and other related applications but that have direct application to forest and other wildland watersheds. After completing this chapter and relating the respective methods to other parts of the book, you should be able to:

- Indicate the criteria used to select a hydrologic method for a particular purpose.
- Define a hydrologic model and explain the different types of models, their applications, and their limitations.
- Discuss the use of regression methods and other simple empirical methods for hydrology and the conditions under which it is appropriate to apply them.
- Define a unit hydrograph and explain the conditions under which it can be applied.
- Discuss in general terms the applications of isotopes in hydrology and watershed management.
- Select an appropriate hydrologic method to accomplish specific objectives, such as:
  - determine the size of a culvert for a road,
  - determine the probability that a peak discharge of a certain magnitude will be equaled or exceeded in a specified number of years,
  - determine the peak discharge and stormflow volume for a watershed that has streamflow records,
  - determine how streamflow over the seasons will be affected when vegetation is altered, and
  - examine the effects of different land uses on stormflow peak and volume.

REFERENCES

American Society of Civil Engineers (ASCE). 1969. *Design and construction of sanitary and storm sewers.* Man. Rep. Eng. Pract. 37.

Arnold, J.G., J.R. Williams, A.D. Nicks, and N.B. Sammons. 1990. *SWRRB: A basin scale simulation model for soil and water resources management.* College Station: Texas A & M University Press.

Barnes, B.S. 1939. The structure of discharge recession curves. *Trans. Am. Geophys. Union* 20:721–725

Barten, P.K., and K.N. Brooks. 1988. Modeling streamflow from headwater areas in the northern Lake States. In *Modeling agricultural, forest, and rangeland hydrology,* 347–356. Proc. 1988 International Symp., American Society of Agricultural Engineers.

Beard, L.R. 1962. *Statistical methods in hydrology.* Sacramento, CA: U.S. Army Corps of Engineers, Sacramento District.

Brown, H.E. 1965. Characteristics of recession flows from small watersheds in a semiarid region of Arizona. *Water Resour. Res.* 1:517–522.

Burns, D.A., and J.J. McDonnell. 1998. Effects of a beaver pond on runoff processes: comparison of two headwater catchments. *Jour. Hydro.* 205:248–264.

Dunne, T., and L.B. Leopold. 1978. *Water in environmental planning.* San Francisco: W.H. Freeman and Co.

Fread, D.L. 1993. Flow routing. In *Handbook of Hydrology*, ed. D.R. Maidment, 10.1–10.36. New York: McGraw-Hill.

Goldstein, R.A., J.B. Mankin, and R.J. Luxmore. 1974. *Documentation of PROSPER: A model of atmosphere-soil-plant-water flow.* Oak Ridge, TN: Oak Ridge Natl. Lab. EDFB-IBP-739.

Gray, D.M., ed. 1970/1973. *Handbook on the principles of hydrology.* Reprint. National Research Council, Canada. Port Washington: Water Information Center, Inc.

Guertin, D.P., P.K. Barten, and K.N. Brooks. 1987. The peatland hydrologic impact model: Development and testing. *Nordic Hydrology* 18:79–100.

Haan, C.T., H.P. Johnson, and D.L. Brakensiek, eds. 1982. *Hydrologic modeling of small watersheds.* Am. Soc. Agric. Eng. Mono. 5.

Hall, F.R. 1968. Base-flow recessions: A review. *Water Resour. Res.* 4:973–984.

Hawkins, R.H., and A.V. Khojeini. 2000. Initial abstraction and loss in the curve number method. *Water Resour. in Arizona and the Southwest* 30:29–35.

Hewlett, J.D., and A.R. Hibbert. 1967. Factors affecting response of small watersheds to precipitation in humid areas. In *Forest hydrology,* 275–290. New York: Pergamon Press.

Hewlett, J.D., and J.B. Moore. 1976. *Predicting stormflow and peak discharge in the Redland District using the R-index method.* Georgia For. Res. Pap. 84, Athens.

Holtan, H.N., G.J. Stiltner, W.H. Henson, and N.C. Lopez. 1975. *USDAHL-74 model of watershed hydrology.* USDA Agric. Res. Serv. Tech. Bull. 1518.

Horton, R.E. 1933. The role of infiltration in the hydrologic cycle. *Trans. Am. Geophys. Union* 14:446–460.

Howe, J.W. 1966. *Recession characteristics of Iowa streams.* Iowa City: Iowa State Water Resource Research Institute.

Hydrocomp. 1976. *Hydrocomp simulation programming operations manual.* Palo Alto, CA: Hydrocomp Inc.

Indri, E. 1960. Low water flow curves for some streams in Venetian Alps. *Int. Assoc. Hydrol. Sci. Publ.* 51:124–129.

Kendall, C. and E.A. Caldwell. 1998. Fundamentals of isotope geochemistry. In *Isotope Tracers in Catchment Hydrology*, ed. C. Kendall and J.J. McDonnell, 51–86. Amsterdam: Elsevier Science B.V.

Krabbenhoft D.P., C.J. Bowser, M.P. Anderson, and J.W. Valley. 1990. Estimating ground water exchange with lakes: The stable isotope mass balance method. *Water Resour. Res.* 26:2445–2453.

Larson, C.T. 1973. Hydrologic effects of modifying small watersheds: Is prediction by hydrologic modeling possible? *Trans. Am. Soc. Agric. Eng.* 16:560–564, 568.

Leaf, C.F., and G.E. Brink. 1973. *Hydrologic simulation model of Colorado subalpine forest.* USDA For. Serv. Res. Pap. RM-107.

Linsley, R.K., Jr., M.A. Kohler, and J.L.H. Paulhus. 1982. *Hydrology for engineers.* 3d ed. New York: McGraw-Hill.

Magner, J.A., and C.P. Regan. 1994. *Tools and techniques for the assessment of ground water/surface water interactions in glacial hydrogeologic settings.* Ground Water Management series 18, 685–698. Dublin, OH: National Ground Water Association.

Magner, J.A., Regan, C.P., Trojan, M. 2001. Isotopic characterization of the Buffalo River Watershed and the Buffalo Aquifer near Moorhead Municipal Well One. *Hydrologic Sci. & Tech.* 17:237–245.

Moore, I.D., R.B. Grayson, and A.R. Ladson. 1991. Digital terrain modelling: A review of hydrological, geomorphological, and biological applications. *Hydrologic Processes* 5:3–30.

Singh, V.J. 1992. *Elementary hydrology*. Englewood Cliff, N.J.: Pentice Hall.

Singh, V.J. (ed). 1995. *Computer models of watershed hydrology.* Highlands Ranch, CO: Water Resources Publications.

Toebes, C., and D.D. Strang. 1964. On recession curves. 1 Recession equations. *J. Hydrol.* (N.Z.) 3:215.

U.S. Army Corps of Engineers. 1975a. *Hydrologic frequency analysis.* Hydrologic Engineering Methods for Water Resources Development, vol. 3. Davis, CA: Hydrology Engineering Center.

U.S. Army Corps of Engineers. 1975b. *Program description and user manual for SSARR.* Portland, OR: North Pacific Division, U.S. Army Corps of Engineers.

U.S. Soil Conservation Service. 1972. *National engineering handbook.* Sec. 4, Hydrology.

U.S. Water Resources Council. 1976. *Guidelines for determining flood flow frequency.* Hydrol. Comm. Bull. 17.

Wicht, C.L. 1943. Determination of the effects of watershed management on mountain streams. *Trans. Am. Geophys. Union,* Pt. 2:594–605.

Woolhiser, D.A., and D.L. Brakensiek. 1982. Hydrologic system synthesis. In *Hydrologic modeling of small watersheds,* 3–16. Am. Soc. Agric. Eng., Monogr. 5.

CHAPTER 18 *Tools and Technologies for Watershed Analysis and Research*

Introduction

This chapter provides a general reference on tools and technologies commonly used for watershed analysis and research, including methods used to understand cumulative watershed effects. Watershed analysis and research are often constrained by either gaps in information or investigative methodologies. This chapter, therefore, has been prepared to provide a better understanding of the experimental protocols of watershed analysis and research. The content is more applied than theoretical and covers field studies, statistical methods, computer simulation techniques, and geographical information systems. Most of the tools and technologies presented have applications in both watershed analysis and watershed research.

Field Methods

It is essential that the objectives be clearly understood before instituting field studies so that appropriate measurements are taken and monitoring is properly designed. Field studies are expensive, and therefore, they should be carried out to obtain information that is necessary for a watershed analysis. But in designing field studies, a broad perspective and flexibility in design can enhance the opportunity to gather valuable information that might not have been initially envisioned. For example, many early soil erosion studies related rates of erosion to different soils, vegetative cover conditions, and so forth. However, most of them failed to relate rates of erosion to productivity of the sites; therefore, valuable information was not collected that would have enabled economists to value the "costs" of erosion in meaningful terms.

Plot Studies

Testing a hypothesis in hydrology or watershed management can involve the use of plot studies. In some instances, cause-and-effect information can be obtained through the use of small, homogeneous plots that are isolated under controlled conditions (Sim 1990). Plot studies have been utilized in water balance investigations, precipitation-runoff evaluations, and analyses of the effects of vegetative management practices on water regimes. Results from these studies can indicate a need for larger-scale (e.g., watershed-level) experiments. Process studies on plots are also formulated to obtain a better understanding of causes and effects and as a basis for subsequent interpretations of the results obtained from watershed experiments. Plot studies also help to identify key relationships that can be developed into mathematical algorithms for computer simulation models.

Plot Characteristics

There are many options for plot size and shape for watershed research. The plot size depends largely upon the nature of the hydrologic or watershed management questions being researched or analyzed. Plots can be 1–2 m^2 or several hectares in size. Data obtained from smaller plots are often not as representative of natural systems as those obtained from larger plots, although the use of smaller plots can be sufficient for reconnaissance studies. Larger plots frequently yield better results that are more reliable for extrapolation.

A standard plot size might be specified for a particular type of plot study. In using the universal soil loss equation to measure surface loss, for example, a standard plot of 6×72.6 ft (approximately 2×22 m), which is 0.01 acre (0.004 ha), is specified.

The plot size is ultimately determined by the objectives of the study, the local conditions, access, and ease of construction and maintenance. The expense and types of materials required and available are also factors to consider. In some instances, plot studies are installed on experimental watersheds to allow for multiple analyses.

Experimental Designs

Plot studies allow for a variety of experimental designs. The analyst or researcher must choose one that will provide clear-cut answers to the questions being investigated.

Experiments are always subject to variability. Sources of variation in hydrologic and watershed management research are many. Proper experimental design can help account for the most extraneous sources of variation. Studies should be planned so that known sources of variation are varied deliberately over as wide a range as necessary. The experiments should be designed in such a way that their variability can be eliminated from the estimate of chance variations. Simple and commonly used experimental designs to achieve this purpose include the *completely randomized design*, *randomized block design*, *Latin square design*, and *factorial design*. (For details on these and other experimental designs, see references on statistical methods and inferences at the end of this chapter.)

The data sets required from plot studies should be obtained with the most appropriate instrumentation and equipment available. Collection of data in plot studies has to be consistent and unbiased. The procedure for data collection, including designation of those responsible and the schedule for measurements, must be clearly specified when studies are being planned.

Other Observations

Plot studies are often carried out in small-scale trials in testing a hypothesis. Plot studies are attractive because the study of small, homogeneous systems permits control of the inputs and accurate measures of outputs from the plot.

Box 18.1

Estimating Evapotranspiration Using Plots with Different Vegetative Cover Conditions (from Johnston 1970)

A plot study was conducted in northern Utah to estimate evapotranspiration differences among three cover conditions: bare soil, herbaceous vegetation, and an aspen-herbaceous cover. Soil moisture depletion was measured to a depth of 9 ft in each 1/10 acre plot. Four-year average evapotranspiration losses were 21.0 in., 15.3 in., and 11.3 in. for the aspen-herbaceous, herbaceous, and bare plots, respectively. During the active growing season, soil moisture in the aspen-herbaceous plot was depleted to at least 9 ft; the bare plot exhibited little soil moisture depletion below 2.5 ft (see figure below). Soil moisture in the herbaceous plot was mostly depleted above a depth of 4 ft during the same period. The results of these plot studies helped quantify the differences in evapotranspiration among cover types and indicated the potential for increasing water yield by manipulating vegetative cover in northern Utah.

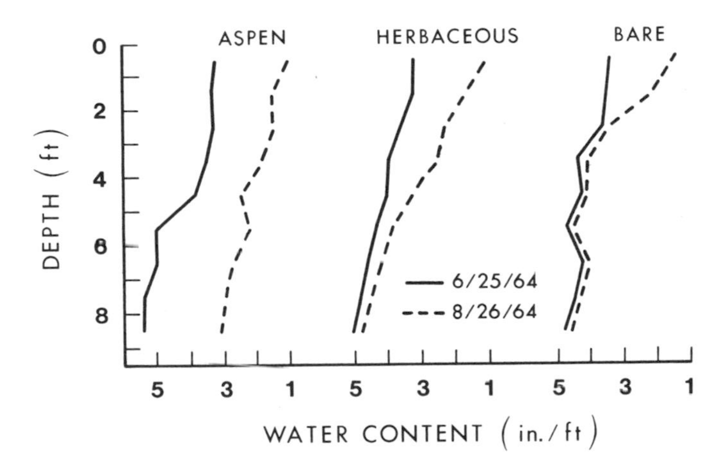

Soil moisture depletion for three plots in northern Utah (from Johnston 1970).

Plot studies have been used to estimate interception losses in forest communities, to estimate soil water depletion under different plant species or communities (Box 18.1), and to quantify infiltration, surface runoff, soil erosion, and other hydrologic properties of soils and land use conditions. Plot studies are easier to establish and less expensive to conduct than experiments involving entire watersheds. For example, artificial barriers can be constructed to constrict flow or prevent leakage, which enables the water budget components to be measured more accurately. Comparisons of several land treatments or different vegetation types can be made with a series of plots. To examine the same number of treatments or systems using watershed-level investigations would be costly.

A major limitation of plot studies, however, is the inability to extrapolate the results obtained to larger systems such as a watershed, which are often more complex in nature. For example, the differences in soil moisture depletion determined from plot studies in Box 18.1 cannot tell us how much water yield would change as a result of vegetation changes over an entire watershed. Plot and watershed studies do not provide the same type of information. Plot studies should be designed to improve our knowledge of certain processes so we can better interpret and understand the integrated response of larger watershed systems; therefore, plot studies should be closely linked to larger experimental watershed studies. The results of plot studies should also provide empirical data for the development and testing of mathematical relationships needed to simulate particular processes as components of computer simulation models. If such objectives are kept in mind in the planning of a plot study, more useful information should result.

The boundary distortions imposed on hydrologic processes by plots can be a problem. Increasing plot size can help to reduce these effects. Another possible solution to this problem is using plots with no artificial boundaries. For example, sections of hillslopes can sometimes be used as plots, with the top of the ridge and toe of the slope defining plot boundaries.

One drawback to plot studies is the cost of replicating the research in those systems in which the results will be applied. Large-scale applications of research results require replications on a wide-scale basis. The inherent meteorological variability of the system in question can require plot studies to be monitored for 5–10 yr to attain the level of precision in the results that will meet the extrapolation standards.

Lysimeters

Lysimeters, large tanks within which a soil-plant system is contained and studied (see Chapter 3), are constructed in such a way as to approximate field conditions. Water inputs and outputs are measured either by weighing or by keeping track of water flow into and out of the lysimeter. Accurate water budgets can be obtained with lysimeters, although the devices have serious limitations:

- Lysimeters generally cannot be used to investigate the hydrologic processes for large shrubs or trees or for plant communities that contain such plants.
- Even for grasses or agricultural crop plants, disturbances of the soil system and the artificial boundary conditions that are created limit the interpretation and subsequent extrapolation of the results.
- Lysimeters are costly and, in most instances, have limited application for watershed management studies.

Watershed-Level Experiments

Watershed-level experiments are generally used to demonstrate or determine the hydrologic response of an experimental watershed to a prescribed management practice; generally, they provide a more realistic response to a management practice than plots. Different types of vegetative cuttings, changes in vegetative overstory from one type to another, and replacing one type of vegetative cover with another are often evaluated on experimental watersheds (Bosch and Hewlett 1982, Whitehead and Robinson 1993, Hornbeck et al. 1993). Watershed-level experiments have been employed worldwide to determine the effects of vegetative management practices on the magnitude, timing, and quality of water regimes. These experiments can also contribute to the understanding of the hydrologic cycle in a region and the effects of watershed management practices on it.

Experimental Watersheds

Experimental watersheds are instrumented for *system-response* studies of the hydrologic cycle and the effects of watershed management practices on the hydrologic cycle. Experimental watersheds are relatively small (e.g., on the order of hundreds of hectares or less) and should be representative of soil, vegetative, and physical characteristics of interest. Experimental watersheds are generally selected for study on the basis of their similarity to other noninstrumented watersheds in a particular area or region. Experimental watersheds can be used to:

- Study the effects of cultural changes on the hydrologic regimes of the watershed. Cultural changes can involve deliberate modifications of one or more prominent characteristics of a basin, such as land use (e.g., deforestation for agriculture or livestock production), land management (e.g., afforestation or terracing), and exploitation of water resources (e.g., diversion of water for irrigation).
- Serve as a part of a network of stations to furnish data sets of the hydrologic processes in the area.
- Provide sites for fundamental hydrologic research. With the installation of instrumentation, experimental watersheds can be sites for the training of personnel on hydrologic processes.
- Test and validate computer simulation models. Three types of experimental watersheds have been used throughout the world, *single*, *paired*, and *nested watersheds*.

Single Experimental Watersheds. Single experimental watersheds are studied to determine the effects of watershed alterations on themselves. However, these experiments have value only in terms of comparisons with their own historical data sets. Periods of calibration for single experimental watersheds, therefore, usually are longer than for multiple experimental watersheds.

In single-watershed experiments, a single watershed is calibrated upon itself by a regression. Streamflow, the dependent variable, is related to those factors that influence it, such as precipitation, groundwater storage, soil moisture storage, evapotranspiration, and so on. The single-watershed approach is less costly than paired- or multiple-watershed approaches because only one watershed is measured. It has also been argued that the analysis of a single-watershed experiment is more informative because it relates streamflow to factors that influence it rather than to streamflow from another (control) watershed (Reigner 1964). Its principal disadvantage is the complexity involved; more tabulation and analysis of data are required than with the control-watershed approach.

The success of single-watershed studies depends largely upon the ability to predict watershed behavior from climatological variables. Because these predictions are difficult in most instances, the paired- (or multiple-) watershed approach is used more often.

Paired Experimental Watersheds. Paired experimental watersheds have been widely used throughout the world. Determining the effects of management changes on the hydrologic regime requires paired experimental watersheds to be calibrated. These watersheds should be as similar as possible in terms of climate, soils, topography, and geomorphology, so that unbiased comparisons can be made. One of the two watersheds is established as a control, and its physical characteristics are kept as constant as possible throughout the pretreatment calibration period and posttreatment evaluation period of the study for comparison with the other watershed, for which some modification or treatment is planned.

The period of pretreatment calibration must be long enough that the control watershed can be used to predict the behavior of the other watershed with some degree of confidence (Box 18.2). Once a satisfactory correlation is achieved, one of the paired experimental watersheds is treated (harvesting of timber, conversion of forest to herbaceous cover, etc.), and the calibration regression is used to detect significant changes in streamflow.

Statistical techniques have been developed to calibrate paired experimental watershed studies (Kovner and Evans 1954, Reinhart 1956, Wilm 1948). These techniques have been used to detect relatively small changes in streamflow resulting from watershed treatments (Bosch and Hewlett 1982, Whitehead and Robinson 1993, Hornbeck et al. 1993).

The use of paired experimental watersheds is assumed to provide statistically more reliable results than single experimental watersheds. However, there are also disadvantages to this approach, including the possibility of a fire or other disturbance that changes the character of the control in the posttreatment evaluation period (Reigner 1964). Another disadvantage can be the costs and time required for installing and maintaining the required instrumentation on the experimental watersheds for the duration of the study (see below) and for obtaining the necessary data sets for analysis. One disadvantage of both the single- and paired-watershed approaches is the uncertainty that correlations at the beginning of the calibration period will be high enough to allow the detection of treatment effects on the modified watershed.

Nested Experimental Watersheds. Nested experimental watersheds are variants of single experimental watersheds. A segment of a larger watershed is demarcated on a river basin. Sometimes, two or more segments of the watershed, termed subbasins, are chosen for study. The subbasins are then deliberately modified to show the effects of changes within the subbasins and within the entire watershed. Nested experimental watersheds have also been used in tandem with a single control watershed. In these cases, the upper region is often demarcated as a subbasin within the watershed, and the effects along the common river system are studied.

Other Observations

While watershed-level experiments have been widely used throughout the world to evaluate hydrologic responses to watershed management practices, there are limitations to this approach. In particular, the use of experimental watersheds has been criticized by many, as summarized by Hewlett et al. (1969) and others:

- Experimental watersheds are often unrepresentative.
- Transferring results obtained from experimental watersheds to other areas is difficult.
- Experimental watersheds are costly to establish.

Box 18.2

Pretreatment Calibration of Paired Experimental Watersheds

A question often asked in paired-watershed experiments is how long the experimental watersheds should be calibrated to detect a significant difference in water yield after one of the watersheds has been treated in some manner. If too short a calibration period is specified, the experiment might lack satisfactory statistical precision; if the period is longer than necessary, obtaining the final results of the experiment are postponed, and as a consequence, the cost of the experiment is likely increased. To provide a solution to this problem, Wilm (1944, 1948) presented a straightforward method for determining the necessary minimum length of a calibration period for a watershed experiment, including both pretreatment calibration and posttreatment evaluation periods. This method has been modified at times to accommodate experiment-specific situations, but its framework remains valuable in watershed analysis and research and is therefore presented here in the form of an example.

Wilm's method of calibration is analogous to solving the statistical formula for estimating *k,* the number of observations required in each of two sets of data to provide a desired standard error of the difference between their averages:

$$k = \frac{2\,s^2}{s_d^2}$$

where s is a pooled estimate of the standard deviation of the populations from which the two sets of data were obtained and s_d is the standard error of the difference obtained by:

$$s_d^2 = \frac{2\,s^2}{k}$$

Wilm made a number of transformations (substituting equalities and making transpositions) to obtain another equation expressing k in terms of the number of observations (annual water yields, stormflows, etc.) likely to be required in the pretreatment and posttreatment periods, and therefore during the entire watershed experiment:

$$k = \frac{S_{y\bullet x}^2 F}{d^2} = \left(2 + \frac{F}{k-1}\right)$$

where $s^2_{y\cdot x}$ is calculated from the available calibration period data set, the F statistic is selected by the investigator to conform to the specifications of the experiment, and d is the smallest difference in water yield that is worthwhile in the opinion of the investigator.

(continues)

(continued)

Values for the term $F\left[2+\frac{F}{(k-1)}\right]$ corresponding to varying values of k are presented in Table A to help the investigator determine the length of a pretreatment calibration period.

[Box 18.2] Table A. Values of $F\left[2+\frac{F}{(k-1)}\right]$ corresponding to values of k

k	$F\left[2+\frac{F}{(k-1)}\right]$
3	71.52
4	27.76
5	19.01
6	15.46
7	13.60
8	12.47
9	11.67
10	11.08
11	10.69
12	10.32
13	10.10
14	9.88
15	9.68
20	9.10

Note: F is taken at the 0.05 level, on $2(k-1)-1$ degrees of freedom. For example, where k is 5, the value of F is associated with $2(5-1)-1$, or 7 degrees of freedom.

Wilm's method of calibration can be illustrated with data obtained from the Beaver Creek watersheds in north central Arizona (Baker 1999) from 1961 to 1971. Paired watersheds in the ponderosa pine forests were considered in this experiment, viewed in retrospect in this example. The experiment was designed to determine the effects of modifying forest overstories on annual water yields. One of the watersheds (Y), an area of 1785 acres, was selected for a silvicultural treatment to optimize the growth of the posttreatment (residual) forest stands, while the other watershed (X), nearly 865 acres in size, served as a control.

Annual water yields for the paired watersheds, expressed in inches, are presented in Table B. Only 10 of the 11 yr of the record were used in the pretreatment calibration analysis because of a failure to obtain unbiased measurements of annual water yields in 1970; it is important that biased measurements be eliminated from any calibration analysis.

[Box 18.2] Table B. Annual water yields, watersheds Y and X, Beaver Creek

	Watershed Y	Watershed X
Water Yield	Annual Water Yields (in.)	
1961	2.44	0.94
1962	9.23	4.68
1963	0.52	0.37
1964	2.60	1.76
1965	13.73	7.78
1966	10.60	6.26
1967	3.23	1.89
1968	8.67	3.90
1969	11.41	5.73
1970	—	—
1971	1.38	0.37
Average	6.38	3.37

The statistics for the regression of annual water yields from watershed Y on those from watershed X are shown in Table C. As indicated by the correlation coefficient (r), the annual water yields for the paired watersheds are correlated well enough that any bias associated with both Y and X is insignificant.

[Box 18.2] Table C. Statistics for the regression of annual water yields

	Sum	Mean	Range
Watershed Y	63.820	6.382	13.210
Watershed X	33.670	3.367	7.410

Watershed Y = 0.372 + 1.785 (Watershed X)

$S_{y \cdot x} = 0.693$

$r^2 = 0.987$

It was determined by the watershed managers responsible for the experiment that the smallest worthwhile difference in annual water yield would be 1 in., or about 15% of the pretreatment annual water yield of watershed Y. Using this value for d, the calculated value for $S_{y \cdot x}$ (see Table C), and the appropriate value of $F\left[2 + \frac{F}{(k-1)}\right]$ from Table A as inputs, a pretreatment calibration period is obtained:

(continues)

(continued)

$$k = \frac{0.693}{1}(11.08) = 7.67 = 8 \text{ yr}$$

Ten years of pretreatment data had been obtained before the calibration analysis, so it was concluded that a sufficient pretreatment calibration had been obtained between the paired watersheds; in fact, again in retrospect, the silvicultural treatment could have been applied to watershed Y earlier in the experiment.

It is often suggested that the posttreatment evaluation period be the same number of years as the pretreatment calibration period. Observed water yields from the treated watershed are compared to water yield values that are predicted from the calibration regression throughout the posttreatment evaluation period to determine whether a significant difference in water yield has occurred (see Box 3.2). However, it is also desirable that one or more preliminary analyses of the posttreatment results be made to ensure that the posttreatment period is not unnecessarily prolonged.

- Experimental watersheds frequently leak.
- Changes in response (e.g., water yield) following treatment are often too small for detection.
- Integrated results obtained from experimental watersheds conceal hydrologic processes.

These criticisms, if valid, question the practical extension of watershed-level experiments to operational situations. However, before a final decision is made on the use of experimental watersheds, these criticisms warrant discussion.

Representativeness. The most serious criticism of experimental watersheds is that they are not representative of river basins and that as a consequence, transferring results to larger areas is difficult. This criticism stems from the frustration of water resource planners who attempt to apply the results of watershed-level experiments to large river basins under the assumption that a small watershed is a microcosm of the larger river basin. However, selecting the location of experimental watersheds is largely based upon the homogeneity of features. Experimental watersheds, therefore, should not necessarily be expected to epitomize larger, heterogeneous river basins.

Transferring Results. Satisfactory results have been achieved when experimental watersheds are representative of the larger river basin. Within major physiographic units, similarities in streamflows from small and large watersheds are often noted as mean values for the entire physiographic unit. These similarities would usually be expected in terms of annual water yields. However, peak discharge and stormflow volumes are greatly affected by the size of the watershed.

A basic problem of representativeness is defining what the experimental watershed is to represent. Certainly, if removing a forest overstory on a 10 ha experimental watershed increases runoff, it should do the same on a 10 km^2 watershed with similar vegetation. However, the magnitude of this increase once the water moves off the experimental watershed and through the channels of a river basin is beyond the scope of experimental watershed research.

Costs. Watershed-level experiments are costly to establish and maintain for the evaluation period. In the past, many of these experiments were hastily conceived and poorly planned, and because of the emphasis on continuity, their termination was delayed far beyond the point of diminishing returns. Nevertheless, such experiments have provided a historical perspective that no longer allows excuses for inadequately planned research efforts. Unfortunately, relatively few benefit-cost analyses of watershed-level experiments have been made, nor have realistic alternatives (in lieu of experimental watersheds) been advanced that would provide quantitative information on the influence of vegetation manipulation on the quantity, timing, and quality of streamflow.

Time is a major cost in watershed-level experiments. Unfortunately, the need for quick results sometimes eliminates a sufficient pretreatment calibration between a control watershed and one or more watersheds to be treated. Calibration periods have often been 5 yr or longer, although some studies have indicated that calibration periods as short as 3 yr can provide useful information when the treatment period is longer. The number of years required to obtain adequate calibration is not fixed; it depends largely upon the objectives of the experiment and the variability of the calibration data sets. If one is interested in annual water yield changes, each year represents only one data point in the calibration equation. However, where stormflow changes are the interest, it is likely that several storms can be analyzed in any given year; in this case, only a few years might then provide adequate calibration data.

Once a watershed has been instrumented, short-term studies can often be superimposed on a watershed-level experiment at considerably less cost. Such studies should provide more information than if conducted independently, for their results are additive to the information collected on the behavior of the watershed as a whole.

Watershed Leakage. Nearly all watersheds leak. Unfortunately, no certain method exists for determining the amount of inflow or outflow across subterranean strata that divide adjoining watersheds. However, where relative differences rather than absolute values are the primary concern, leaks can be less of a problem when experimental watersheds are properly calibrated against each other.

Detection of Changes. The principal advantage of paired or multiple experimental watersheds is the high degree of correlation that is generally obtained between or among the adjacent catchments. Although some problems in refining statistical analyses remain, treatment differences in streamflow of less than 10% have been detected with high confidence levels. Furthermore, one might question whether detecting changes that are less than 10%, for example, are relevant in the first place. Therefore, the criticism that changes in response are often too small for detection is unwarranted in many cases.

Concealing of Hydrologic Processes. From a simplistic viewpoint, where only the difference in streamflow characteristics between experimental watersheds is considered, the criticism that these integrated results conceal hydrologic processes can be valid. However, there is a need to adequately explain experimental watershed responses so that results can be made more broadly applicable to larger river basins; this has led to increased emphasis on studies concerned with the difficult question of why a watershed reacts as it does rather than simply what takes place following a vegetation manipulation.

Experimental watersheds are indispensable to the complete understanding of watershed management effects, as demonstrated by watershed-level experiments worldwide (Bosch and Hewlett 1982, Whitehead and Robinson 1993). But even with a more complete understanding, the circuitous argument of the uniqueness of individual experimental watersheds must still be faced. Therefore, the empiricism of watershed-level experiments must be relied upon, as acknowledged by even the most severe critics of this methodology. Watershed-level experiments are still needed in many situations, but such research efforts must be well planned.

Statistical Methods

In carrying out plot- or watershed-level studies, statistical methods must be used in their analysis to support the results obtained. There are two primary objectives of statistics (Freese 1967): (1) to estimate population parameters and (2) to test hypotheses about these parameters.

An example of the first objective is the application of statistical methods to estimate the mean (average) and variance of a population. A statistician's task is to choose proper sampling techniques for collecting the required source data and then to calculate the desired statistics. An example of the second objective of statistics is to test the hypothesis (theory) that the estimated population mean equals or, perhaps, exceeds a predetermined value. It is the statistician's job to develop appropriate tests of the hypothesis to satisfy this objective.

Sampling Techniques

It may be desirable to have a complete enumeration of a population to support the results of a study. A complete enumeration is rarely possible, however, and usually a sample is taken. The sample size, the size and shape of individual sampling units, and the sampling design must be determined.

Sample Size

A primary objective of sampling is to take enough measurements to obtain a desired level of precision—no more, no less. The size of the sample to be taken depends largely upon two factors: (1) the inherent variability within the population being sampled and (2) the desired level of precision (i.e., the *allowable error*). To calculate the sample size, it is necessary to have an estimate of the variance of the population, which can be obtained from a preliminary sample of the population. Additionally, a t value at the specified level of probability is required, and the level of precision must be specified.

Size and Shape of the Sampling Units

The size and shape of the sampling units will affect the precision of sampling and the costs. In general, small plots exhibit more variability among themselves than large plots. Circular plots are usually easier to establish than rectangular plots. Circular plots also have less edge than a rectangular plot of equal size. Circular plots can leave portions of the sampling frame unsampled, however, unless they are allowed to overlap. Many combinations of plot size and shape are available for use in watershed studies.

The size and shape of individual sampling units must be compatible with the sampling objectives and the inherent variability of the population being sampled.

Sampling Designs

After calculating the sample size and selecting the appropriate size and shape of the individual sampling units, data are then collected to estimate the required population parameters. The problem is deciding upon the most efficient sampling design in terms of the sampling objectives and the characteristics of the population. Many sampling designs and variations of sampling designs exist. Three basic sampling designs that are often employed in watershed measurements are simple random sampling, stratified random sampling, and systematic sampling.

The fundamental idea behind *simple random sampling* is that when allocating a sample of n individual sampling units, every possible combination of the n units has an equal chance of being selected. Furthermore, the selection of any given individual sampling unit is completely independent of the selection of all other units. Often, previous knowledge of the distribution of a population can be used to increase the precision of a sample.

Stratified random sampling takes advantage of certain types of information about a population by grouping homogeneous units of a population together on the basis of some inherent characteristic (vegetative cover, soil parent material, slope-aspect combinations, etc.). Each homogeneous unit, called a stratum, is then sampled by employing a simple random sampling design, and the group estimates are combined to estimate population parameters.

Individual sampling units in *systematic sampling* are not allocated randomly but according to a predetermined pattern. Systematic sampling has been used widely for two reasons: (1) the location of the individual sampling units in the field is often easier and cheaper than is the case with other sampling designs, and (2) there is often a feeling that a sample that is deliberately spread over a population may be more "representative" than a simple random sample.

Statisticians might not argue against the first reason, but they are generally less willing to accept the second. Estimation of sampling errors associated with a systematic sample requires more knowledge about the population being sampled than is usually available.

Basic Statistics

The determination of a "central value of a series of observations" obtained from a sample of a population is required in many situations. The *mean* is the most familiar and commonly used measure of central tendency. Another measure of central tendency is the *median*. The median of a series of random values obtained from a sample of a population, arranged (ranked) in order of size, is the value halfway through the series. Still another measure of central tendency is the *mode*. When a series of random values are arranged by classes and frequencies (a frequency distribution), one class (or a few classes) will generally show the highest frequency of occurrence. The class (or classes) with the highest frequency of occurrence is the mode (or modes).

Measures of dispersion define the extent to which the individual observations in a series vary from the central tendency. The r*ange* is a measure of the total interval between the smallest and the largest values in a series of random values. Often, the range gives preliminary information on the variability of random values in a series.

TABLE 18.1. Formulas for basic statistics

$$\text{Mean} = \bar{x} = \frac{\Sigma X}{n}$$

$$\text{Variance} = s^2 = \frac{\Sigma(X - \bar{x})^2}{n - 1} = \frac{\Sigma X^2 - \frac{(\Sigma X)^2}{n}}{n - 1}$$

$$\text{Standard Deviation} = \sqrt{s^2}$$

$$\text{Coefficient of Variation} = \frac{s}{\bar{x}}$$

$$\text{Standard Error of the Mean} = s_{\bar{x}} = \sqrt{\frac{s^2}{n}} = \frac{s}{\sqrt{n}}$$

$$\text{Confidence Limits} \cdot CI = \bar{x} \pm ts_{\bar{x}}$$

Perhaps the most useful measure of dispersion is the *variance*, which is the variability of individual random values about the estimated population mean. From the variance, one obtains an idea of whether most of the individuals in a series are close to the mean or spread out. The square root of the variance is called the *standard deviation*, which is used in calculating the *coefficient of variation*, among other statistics. The coefficient of variation, the ratio between the standard deviation and the mean, facilitates the comparison of variability about different-sized means.

The *standard error of the mean* is a measure of dispersion among a set of sample means, just as the variance is a measure of dispersion among the individuals in a series of random values. Fortunately, it is not necessary to obtain a set of simple random samples to calculate the standard error of the mean. Instead, a satisfactory estimate can be obtained directly from the source data of one simple random sample.

The reliability of an estimated population parameter is indicated by *confidence limits*. Confidence limits are a function of the standard error of the mean and a t value selected to meet the level of precision specified. Although confidence limits can be structured around many statistics, most commonly they are established about an estimated population mean.

Formulations for some of the more commonly used basic statistics calculated from data sets obtained in simple random sampling are presented in Table 18.1.

Tests of Hypotheses

It was mentioned above that one of the two objectives of statistics is to test a hypothesis about the estimated population parameters. Analysts and researchers are often interested in knowing whether the means of two or more groups of sample data are different. The groups of data can represent types of treatments (e.g., the effects of different types of vegetative cover on water yield or soil loss) that we wish to compare.

Two hypotheses are evaluated in hypothesis tests: the null hypothesis and an alternative hypothesis. The *null hypothesis* states that there is no statistical difference in the means of two or more groups of data, while the *alternative hypothesis* states that there is a statistical difference. The results of hypothesis tests allow statements to be made about the acceptance or rejection of the null hypothesis. If the null hypothesis is accepted, the alternative hypothesis is rejected, and vice versa.

Tests of hypotheses generally involve comparing two or more groups of sample data. To compare only two groups of sample data, t tests are used; analysis of variance is used to compare three or more groups of sample data. Analysis of variance can also be used to compare two groups of sample data, but t tests are normally used for this purpose because of their computational ease. All tests of hypotheses require that the following conditions be met:

- A valid sample has been made.
- The variables measured are normally distributed (see below).
- All groups of sample data have the same population variance.

Tests of hypotheses are outlined in standard references on statistical methods and inferences and therefore will not be detailed in this chapter.

Regression Analysis

An important statistical tool that is frequently used in analyzing watershed measurements is *regression analysis*. Regression analysis requires the selection of appropriate mathematical models to quantify the relationships between a *dependent variable* (e.g., streamflow discharge) and one or more *independent variables* (e.g., rainfall amount, antecedent soil moisture). These mathematical models are approximations based on sample data and therefore are subject to sampling variations.

Simple Regression

A *simple regression* defines a relationship between a dependent variable, Y, and one independent variable, X. A simple *linear regression* is used with a straight-line relationship between the two variables. Linear regression models of the following form have been used extensively in watershed studies:

$$Y = a + bX \tag{18.1}$$

where Y = dependent variable, a and b = regression constants, and X = independent variable.

The dependent variable might be annual runoff in millimeters, and the independent variable might be annual precipitation in millimeters for a given watershed.

Many times, a nonlinear response, such as that represented by Equation 18.1, is found between the two variables, in which case the relationship between the variables is often transformed into a linear form to simplify the analysis. For example, streamflow (Y) versus stage or water surface elevation (X), or suspended-sediment concentrations in parts per million (Y) versus discharge in cubic meters per second (X), are typically nonlinear responses that are represented by log-transformed relationships:

$$y = aX^b \tag{18.2}$$

is transformed to:

$$\log Y = \log a + b(\log X) \tag{18.3}$$

Simple regression analyses can involve other nonlinear relationships (parabolic, exponential, etc.) and, as a consequence, require a transformation of the variables (Box 18.3).

Multiple Regression

The dependent variable can be related to more than one independent variable in many instances. For example, in some regional analyses, peak or low-flow discharge is related to watershed area, annual rainfall, percentage of area in lakes or wetlands, and watershed slope. If this relationship can be estimated by a *multiple-regression analysis*, it can allow a more precise estimation of peak or low-flow discharge (the dependent variable) than is possible with a simple regression. A multiple regression is of the form:

$$Y = a + b_1X_1 + b_2X_2 + \ldots + b_n X_n \tag{18.4}$$

where a = constant and $b_1, b_2, \ldots, b_n$ = slope of the relationships between respective independent variables $X_1, X_2, \ldots, X_n$ and the dependent variable Y.

Frequency Analysis

A *frequency analysis* of data sets is performed for many purposes, for example, to determine the probability of a rainfall event of a specified magnitude or to delineate a watershed in terms of the probabilities of wind speeds of specified magnitudes occurring. A pattern of frequency of occurrence of units in each of a series of equal classes is a *frequency distribution function.*

A frequency distribution function shows the relative frequency of occurrence of different values of a variable (X) for a data set that represents a population of interest. By knowing the frequency distribution function, it is possible to determine what proportions of the individuals in the population are within specific size limits. Each set of data representing a population has its own frequency distribution function. There are certain distribution functions that are frequently used in watershed analyses, however; these include the *normal*, *binomial*, and *Poisson* distributions.

A normal frequency distribution, which is the familiar “bell-shaped” distribution, is used widely in statistical analyses of watershed measurements (Fig. 18.1). Theoretically, a normal frequency distribution exhibits the following properties:

- The mean, median, and mode are identical in value.
- Small variations from the mean occur more frequently than large variations from the mean.
- Positive and negative variations about the mean occur with equal frequency.

A binomial frequency distribution is associated with data where a fixed number of the individuals are observed on each unit. Furthermore, the unit is characterized by the number of individuals having some specific attribute. An asymmetric Poisson fre-

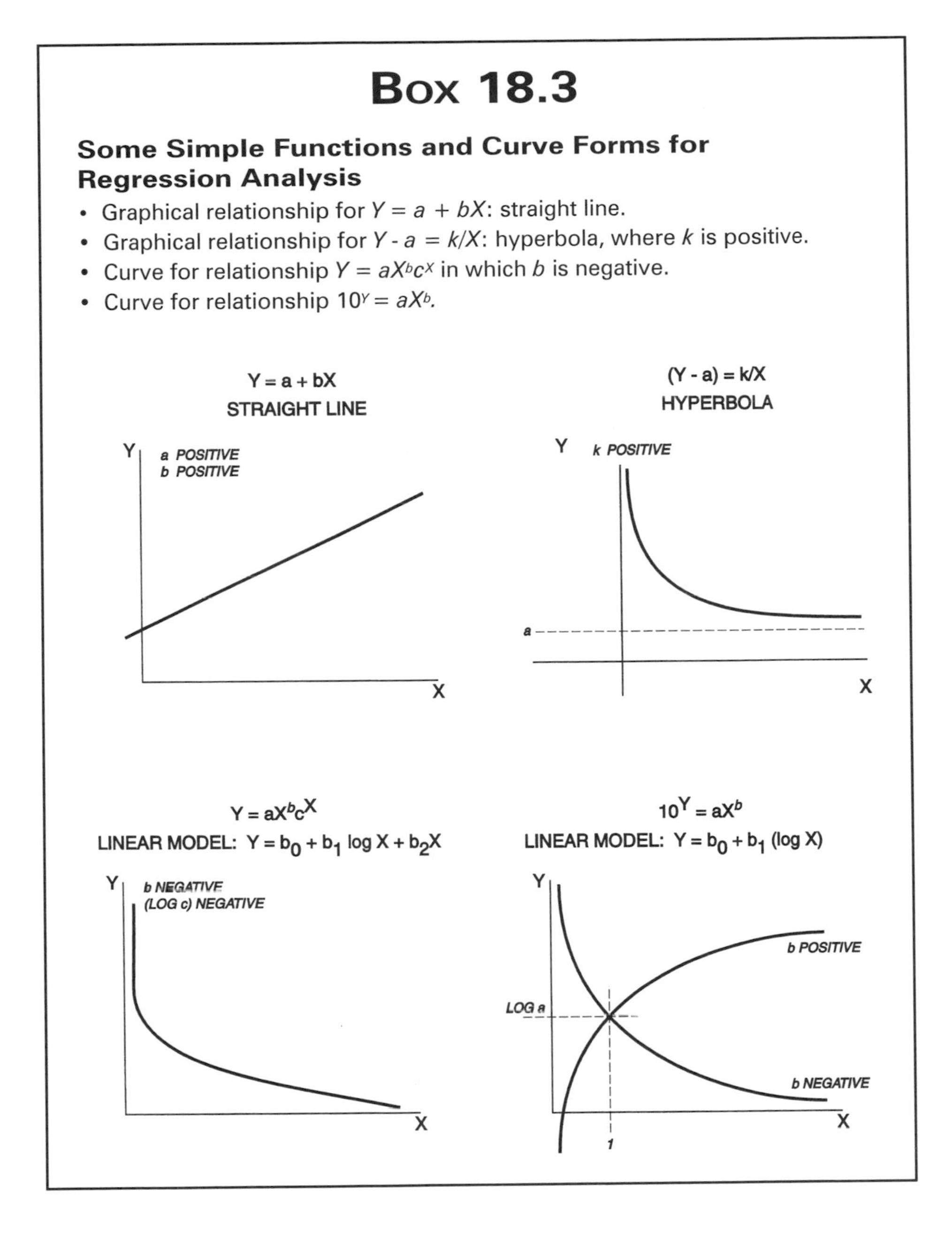

quency distribution can arise where individual units are characterized by a count having no fixed upper limit, especially in situations where zero or low counts dominate.

There are a number of other frequency distribution functions that can be used to describe populations of watershed attributes. One of the more important tasks of a statistician is to identify the most appropriate function for a situation and then apply the proper statistical methods to estimate the population parameters and, when necessary, test hypotheses about these parameters. An example of a frequency distribution commonly used in flood analysis is described in Chapter 17.

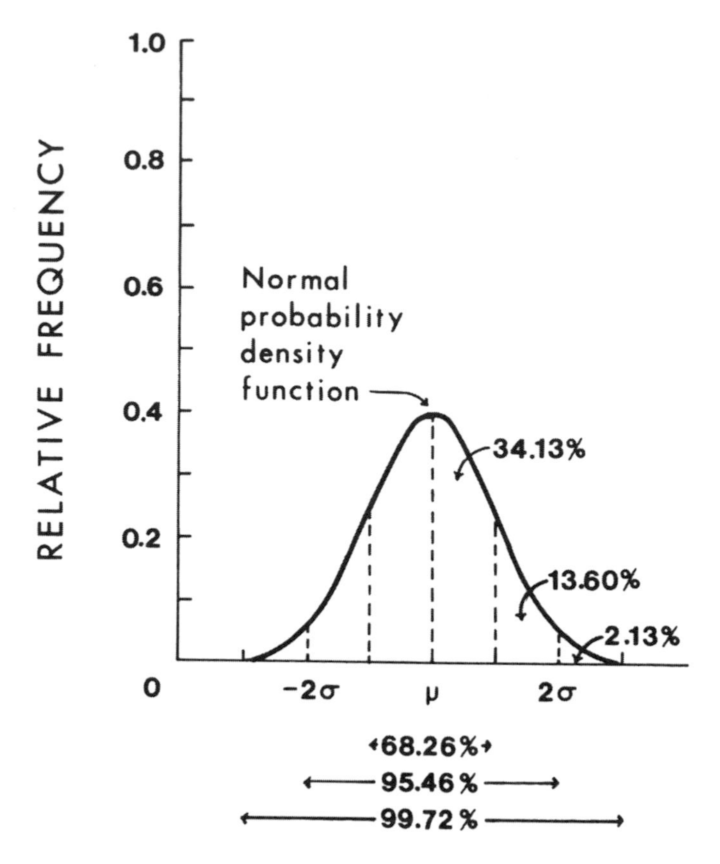

FIGURE 18.1. The bell-shaped curve of the normal frequency distribution function.

In addition to knowing the relative frequency with which values of a variable, X, occur in a population, it might also be necessary to know what proportion of a population lies between specified limits of X. For example, a hydrologist might wish to know what proportion of all precipitation events that occur on a river basin lie within a particular range in precipitation amounts. It is necessary to construct a *cumulative frequency distribution function* to obtain this kind of information. Normally, to develop such a function, values of X are accumulated, starting with the smallest value of X and proceeding to the largest value in the source data (Fig. 18.2). Cumulative frequency distribution functions that are developed by starting with the largest, not the smallest, value of X and proceeding to the smallest value are referred to as *exceedance curves*.

Time Series Analyses

Climatological and hydrologic data sets frequently occur in the form of a time series, resulting in questions such as the following: For a specific climatic measurement (e.g., air temperature), is there evidence that a change in the values is occurring, and, if so, what is the nature and magnitude of the change?

Available statistical methods, such as Student's t test for estimating and testing for a change in the mean values, can often play a role in the analysis of changes in climatological or hydrologic data sets through time. However, a Student's t test is valid only if

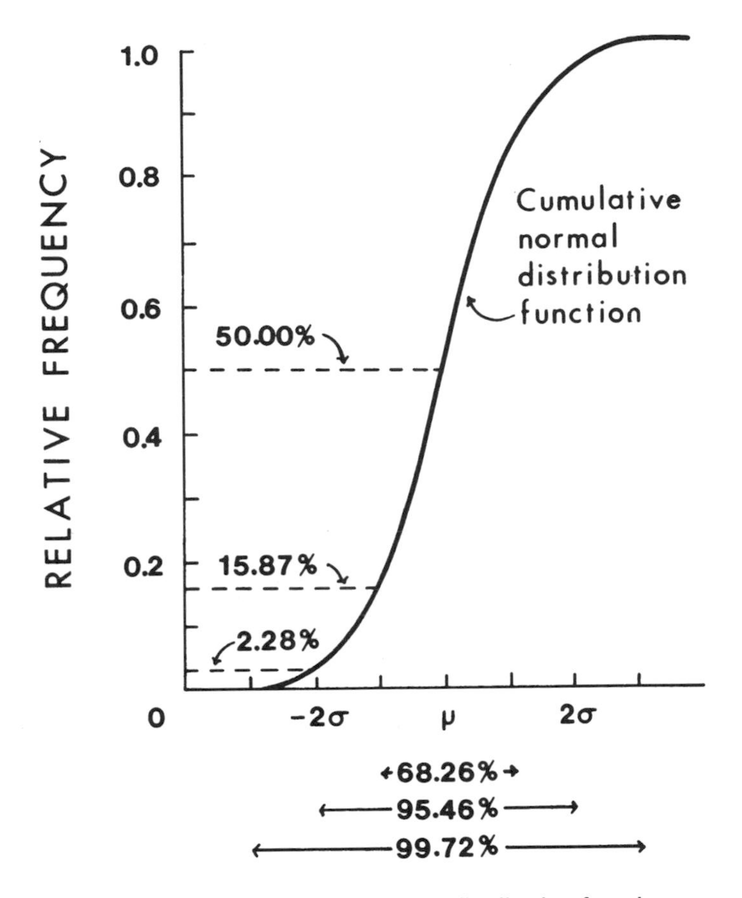

FIGURE 18.2. The cumulative frequency distribution function.

the observations in the data set are independent and vary about the mean values normally with constant variance. A climatological or hydrologic data set is frequently a time series in which successive observations are serially dependent and nonstationary and have a "strong" seasonal effect. Commonly employed parametric and nonparametric statistical procedures, which rely on independence or special symmetry in the distribution function, are not always appropriate in the analysis of changes in climatic measurements through time. An alternative approach is a *time series analysis*.

It is not a purpose of this chapter to describe time series analysis techniques in detail. However, it is important to mention that in applying a time series analysis, or any type of stochastic analysis, to climatic data sets, three steps are recommended:

1. Identify the form of the *mathematical model* that fits the climatic data sets in question.
2. Calculate the model parameters by using the method of *maximum likelihood of occurrence*; this is the *estimation step*.
3. Check the model for possible inadequacies; the appropriate model modifications are made by repeating the first two steps if this check reveals a serious anomaly.

Significant advances in time series analysis techniques have been made in recent years. These advancements, coupled with the availability of computer facilities, allow these techniques to be used with relative ease, assuming that long-term, high-quality data collection has been done.

Computer Simulation Models

Plot studies and watershed-level experiments have provided needed information on the effects of watershed management practices and land use changes on the hydrologic and other natural resources (timber, forage, wildlife, etc.) on watersheds. Information from these investigations is invaluable and should be enlarged upon. However, source data that are obtained from one location can seldom be applied directly to estimate impacts at other locations. Alternative means of estimating these impacts are required in many instances—one means is the application of computer simulation models such as those discussed in Chapter 17.

Computer simulation models are representations of actual systems that allow one to explain and, in many instances, predict the hydrologic response to watershed management practices and, in doing so, gain a better understanding of the impacts of these practices. These models are largely based upon the *systems approach* to simulation (Fig. 18.3) and differ in terms of how and to what extent each component of the hydrologic process is considered.

Computer simulation models are a composite of mathematical relationships, some empirical, some based on theory. As one attempts to explain or predict the impacts of watershed management practices on increasingly complex systems, more detail and complexity are needed in model formulation. However, the reality is that our understanding of hydrology and watershed systems is not sufficient to represent every process mathematically; this can lead to the development of models (or parts of models) that must be calibrated by fitting parameters and relationships to the conditions encountered. This calibration process involves adjusting the parameters and relationships until the computed response approximates the observed response. Once calibrated, models can be used to estimate the hydrologic response of the watershed to new, independent input data.

Development of Computer Simulation Models

Computer simulation modeling is widely used in many fields of science and management, largely because of the savings of time and costs and the flexibility of modeling as an analytical tool. Other reasons include the improved computer literacy of managers and scientists and the wide availability of personal computers and software. The users of computer simulation models should not forget, however, that models are abstractions of actual systems and that the output from models is only an estimate of system response. Moreover, confidence limits of the predicted output values are difficult to ascertain.

Computer simulation models are developed in many ways. One approach is through regression analysis (see above). It is best to use a regression model that expresses a natural relation between the variables in the curve-fitting process. Knowledge of the behavior of variables employed in a relationship allows the selection of one specific regression model over another. This process often leads to the formulation of more detailed plot studies that help to define cause-and-effect relationships.

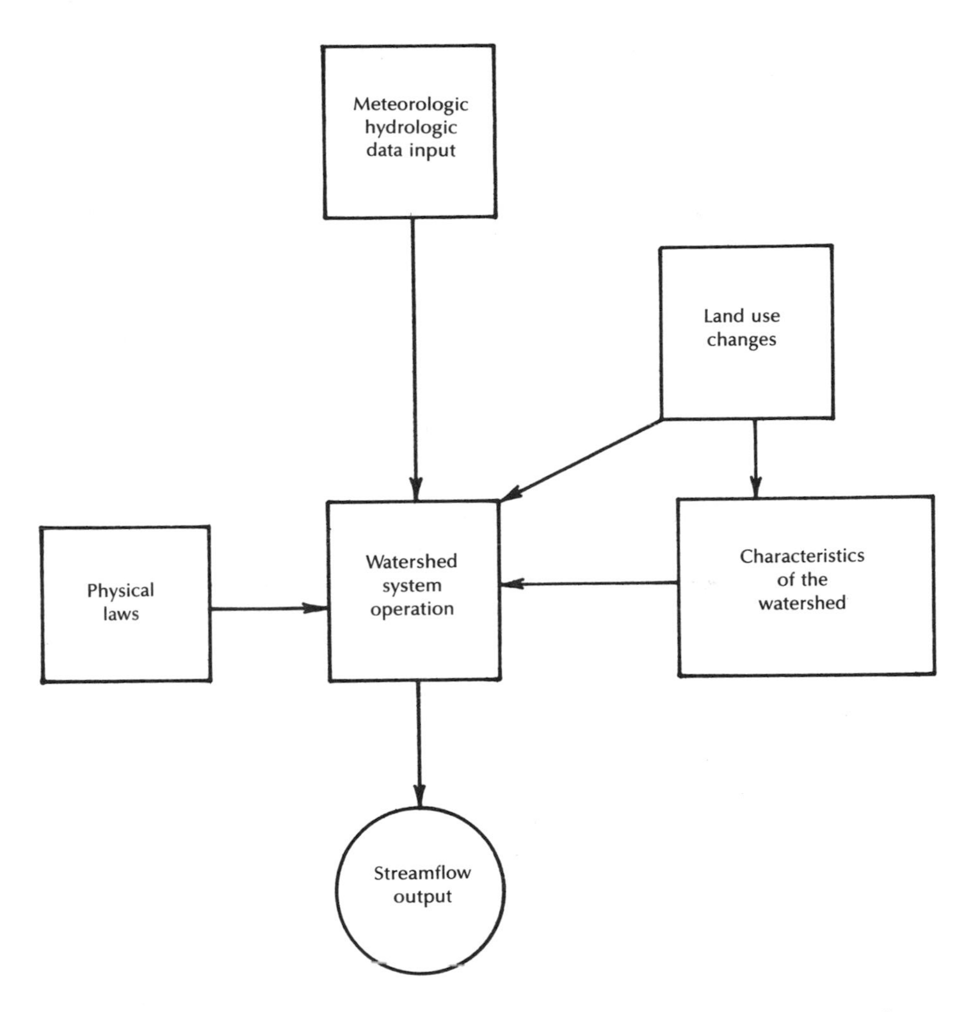

FIGURE 18.3. Systems approach for modeling the hydrologic response of watersheds (from Dooge 1973).

When cause-and-effect relationships cannot be identified, empirical relationships can be derived. Selection of the regression model to represent a particular set of data is somewhat of an art. The choice should be made with an awareness of the statistical properties of various regression models.

The assemblage of one or more appropriate predictive functions, such as those defined by regression analyses, allows a particular watershed system to be simulated. The response of a watershed system to different inputs and levels of inputs can be simulated with such models. Sensitivity of model outputs or response to changes in the value of regression constants (parameters) in the model can be determined, and the effects of modifying the structure of the system on response can also be examined. Furthermore, assessments of subsequent development of a system and requirements for additional support data can be determined.

The mathematical equations representing the predictive functions of a watershed system are assembled in a flow diagram in developing a computer simulation model. These functions and the necessary linkages are then translated into a set of instructions in a computer language. Next, the model components expressed in a computer language are entered into a computer, along with the appropriate descriptive, or input, data. Finally, by executing the model, outputs that predict the response of the system to specified inputs are obtained.

For ease of operation, many simulation models require input data introduced through answers to questions posed to the user by the computer program. These are termed *interactive models*. Assemblages of data records on magnetic disks or tape are input to *batch* (or noninteractive) models.

In developing a computer simulation model to estimate the effects of watershed management practices on hydrologic and other natural resources on watersheds, the following characteristics are generally desired, regardless of the simulation setting:

- The model should be activity oriented. It must be possible to represent the watershed management practices to be simulated and then be able to simulate these effects in the model proposed.
- The model should be capable of simulating time effects. The effects of most watershed management practices can be expected to change over time. It is necessary, therefore, that time-dependent phenomena be represented in the model.
- Spatial effects should be simulated. The spatial distribution of watershed management practices generally has an influence on the effects of these management practices. As a consequence, the model must be able to accept inputs that describe the spatial variability of both the ecosystem landscape and the watershed management practices on the landscape.
- Data availability should be considered. The computer simulation model should not include data requirements that are difficult, costly, or time consuming to collect or acquire for the area on which the model is proposed for use. In many simulation situations, the models must be able to operate on relatively extensive databases that are readily available.

Applications of Computer Simulation Models

Watershed managers are frequently asked what the anticipated hydrologic effects and responses of natural resources to a proposed land use practice are. Applications of computer simulation models, when they have been properly developed for the conditions specified in the simulation exercise, can be useful in formulating the answer to this question in many instances. Hydrologic effects that can often be simulated by computer models, that are generally of concern to a watershed manager, and that are important on a watershed basis include:

- Increases or decreases in streamflow.
- Increases or decreases in peak flows and low flows.
- Increases or decreases in the physical, chemical, and bacteriological quality of surface water.
- Increases or decreases in groundwater supplies.
- Increases or decreases in the quality of groundwater.

Computer simulation models can also be applied to estimate the responses of other natural resources to a watershed management practice. Included among these simulators are those that have been developed to estimate the following responses:

- Growth and yield of forest overstories.
- Composition, production, and utilization of forage plants.
- Development, accumulation, and distribution of organic materials (tree leaves or needles, branches, etc.) on the soil surface, which (in turn) can be inputs to simulators of fuel-loading characteristics for fire management, for erosion and sedimentation rates, etc.
- Livestock carrying capacities.
- Quality of habitats for wildlife species.

Among the criteria that computer models should meet to simulate the responses of hydrologic and other natural resources to a watershed management practice are:

- Accuracy of prediction—It is desirable that computer simulation models be developed whose error statistics are known. Those models with minimum bias and error variance are generally superior.
- Simplicity—Simplicity refers to the number of parameters that must be estimated and the ease with which the model can be explained to potential users.
- Consistency of parameter estimates—Consistency of parameter estimates is an important consideration in the development of models that use parameters estimated by optimization techniques. Computer simulation models are likely to be unreliable if the optimum values of the parameters are sensitive to the period of record used in the simulation exercise or if the values vary widely between similar watersheds.
- Sensitivity of results to changes in parameter values—It is desirable that the models not be sensitive to input variables that are difficult to measure and costly to obtain.

These criteria are also useful in selecting a computer simulation model from the possible alternatives that might be available for a specified computer exercise.

Other Observations

Computer simulation techniques are often available in a variety of scales and spatial resolutions. However, informational needs and data availability are likely to vary from one area to the next, which makes one standard and inflexible computer simulation technique impractical. What is often needed in these situations is a framework in which individual hydrologic processes (interception, infiltration, surface runoff, etc.) and natural resource components (water, timber, forage, etc.) are represented by *modules* that can be linked together to meet specific simulation objectives.

An advantage of this framework is that the modules can be updated or replaced as needed without disrupting other simulators in the modular framework. Users of modular systems can add or delete modules to accommodate particular informational requirements, easily adjusting the mix of modules from one situation to another. Besides being more flexible, a modular system is less expensive to operate, easier to use, and requires fewer data to operate than larger, all-purpose computer simulation models.

Selecting computer simulation models for specific applications involves a compromise between theory (or completeness) and practical considerations. Locally derived models can be based on limited data and be empirical in nature, limiting their application elsewhere. More complete and theoretical simulation models might be better suited for widespread application, but they can require input data or other information for application that are not available.

One can often choose from an array of software that has been developed and that incorporates most standard hydrologic and other simulation methods. In cases where computer simulation models are not available to meet particular needs, the models must be developed. This requires that one have the necessary functional relationships, which are based largely on theory, plot studies, watershed-level experiments, and other previous research. The developmental process, therefore, can be circular, in that computer simulation models are often used when the results of plot studies are lacking, but information from plot studies generally forms a basis for this work in those situations where simulation models are not available and must be constructed.

Geographic Information Systems

There has been an unparalleled evolution of computing technologies in recent years. Geographic information systems (GISs) are examples of this accelerating development and are assuming a more important role in the planning and implementation of management practices on watersheds. These systems are largely designed to satisfy the recurring geographical information needs of watershed analysts and researchers. Therefore, GISs have been optimized to store, retrieve, and update information on watersheds, and they have been programmed to process this information upon demand in a format that meets the users' informational needs.

The Concept of a Geographic Information System

One maintains a set of geographically registered data layers in a GIS for subsequent retrieval and analysis (Fig. 18.4). The stored layers can be updated or otherwise modified as necessary to meet the users' needs (Star and Estes 1990, Bernhardsen 1999). The layers can be stored in either *raster* or *vector form.*

In a raster (or cell-based) system, a layer is represented by an array of rectangular or square cells, each of which has an assigned value. The advantages of a raster system are:

- The geographic location of each cell is implied by its position in the cell matrix. The matrix can be stored in a corresponding array in a computer, provided sufficient storage is available. Each cell, therefore, can be easily addressed in the computer according to its geographic location.
- The geographic coordinates of the cells need not be stored, since the geographic location is implied in the cells' positions.
- Neighboring locations are represented by neighboring cells. Therefore, neighboring relationships can be analyzed conveniently.
- A raster system accommodates discrete and continuous data sets equally well and facilitates the intermixing of the two data types.
- Processing algorithms is simpler and easier to write than is the case in vector systems.

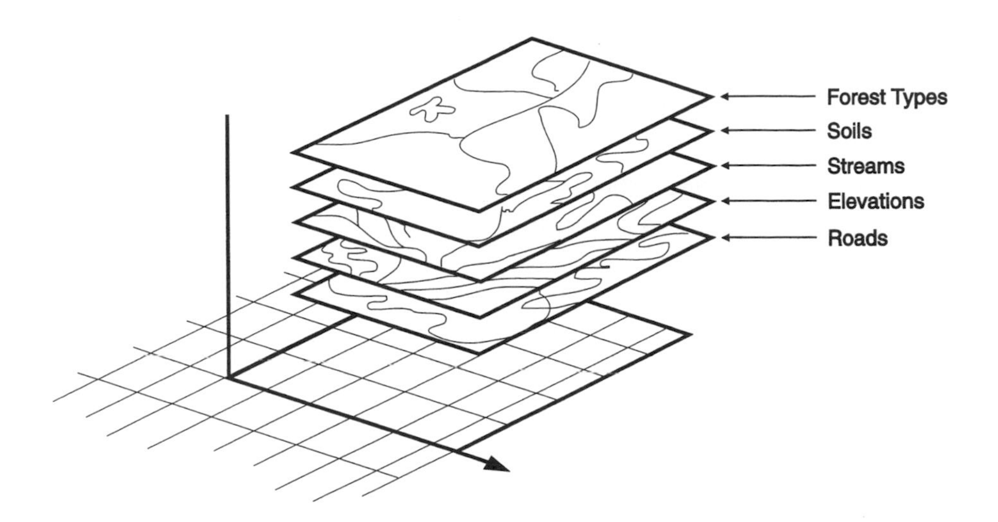

FIGURE 18.4. A GIS conceptualized as a set of geographically registered data layers.

- Map unit boundaries are presented by different values. When these values change, the implied boundaries change.

Disadvantages of a raster system include:

- Storage requirements are larger than those for vector systems.
- The cell size determines the resolution at which the resource is represented. It is difficult to adequately represent linear features.
- Image access is often sequential, meaning that a user might have to process an entire map to change a single cell.
- Processing of descriptive data is more cumbersome than is the case with a vector system.
- Input data are mostly digitized in form. A user, therefore, must execute a vector-to-raster transformation to convert digitized data into a form appropriate for storage.
- It can be difficult to construct output maps from raster data.

In a vector (or line-based) system, the line work is presented by a set of connected points. A line segment between two points is considered a vector. The coordinates of the points are explicitly stored, and the connectedness is implied through the organization of the points in the database. Advantages of a vector system are:

- Much less storage is required than for raster systems.
- Original maps can be represented at their original resolutions.
- Forests, streams, roads, and other resource features can be retrieved and processed individually.

- It is easier to associate a variety of resource data with a specified resource feature.
- Digitized maps need not be converted to a raster form.
- Stored data can be processed into line-type maps without a raster-to-vector conversion.

Disadvantages of a vector system include:

- Locations of the vertex points need to be stored explicitly.
- The relationship of these points must be formalized in a topological structure that can be difficult to understand and manipulate.
- Algorithms for accomplishing functions that are the equivalent of those implemented on a raster system are more complex. Furthermore, the implementation can be unreliable.
- Continuously varying spatial data cannot be represented as vectors. A conversion to raster is required to process these types of data.

The spatial data sets for a GIS are obtained from maps, aerial photographs, satellite imagery, and traditional and global positioning surveys. Each source involves a number of steps and transformations from the original measurements to the final digital coordinates, and at each step, errors can creep into the system. The origin of common errors in a GIS includes (Bolstad and Smith 1992, Choudry and Morad 1998):

- Field measurements—All positional information ultimately relies on field measurements. These measurements can be relatively precise (such as those that define legal property boundaries) or they can be only approximate (such as "eyeball" locations of inventory plots on topographic maps). Global positioning systems (GPSs), consisting of a control segment, a constellation of satellites, and receivers, provide lower costs and higher throughput than traditional surveying in many situations. GPSs are only appropriate for a limited number of data layers, however.
- Maps—Manual or automated map digitization is currently the most common form of spatial data entry and therefore has the greatest impact on spatial accuracy of the data sets digitized. Manual digitizing involves putting a paper or mylar map on a digitizing surface and tracing the features to be entered. Original map accuracy is an important determinant of spatial data accuracy regardless of the digitizing method.
- Imagery—Imagery is a common source of natural resource spatial data, for both initial database development and updates. Most imagery comes from aerial cameras and satellite scanners, although video cameras and airborne scanners are also becoming popular. Aerial photographs have been routinely used in resource mapping for 50 yr or more. Among the factors affecting the accuracy of photo-derived data layers are tilt and terrain distortion and lens or camera distortion. Automated classification of satellite imagery is an established method of land cover mapping. Classification converts multiband reflectance data into a single-layer land cover map, which is registered to a geographic coordinate system. The accuracy of class-boundary location, therefore, is a function of classification accuracy, the geometry of the image, and the quality of the registration.

- Digitization—Positional accuracies during digitization are affected by the equipment and operator skill. Currently used digitizers possess accuracies (and precisions) of better than 0.0025 cm. However, errors of varying magnitudes (e.g., up to 16%) can still be observed for digitized arcs and polygons.
- Coordinate registration—This involves converting from the digitizer coordinates to the coordinate system of the map projection used for printing the source map. Positional errors can occur at any of the steps in the process, including (1) identifying control points in both the geographic and the digitizer space, (2) choosing a mathematical transformation and estimating the coefficients, and (3) applying the transformation to the digitized data in producing the output layer. While large blunders are easily detected, small or random errors are not. Control points must be obtained from ground surveys, or when field measurements are lacking, control is commonly digitized from geographic coordinate points drafted on the source map (e.g., Universal Transverse Mercator [UTM] graticule intersections drafted on 1:24,000-scale base maps).

Applications of Geographic Information Systems

A GIS is essentially a *spatial data management system*. Therefore, any need to estimate the amount and spatial location of parameters represents a potential application of a GIS in watershed analysis and research. One application of a GIS in watershed analysis is predicting the spatial variability of surface erosion (soil loss) from spatial data sets obtained from maps of the vegetative cover, soils, and slope of the area. Solutions of a surface erosion (soil loss) prediction model, for example, the Universal Soil Loss Equation or its modifications (see Chapter 7), combine the spatial data sets, their derivatives, and other information necessary to predict the spatial variability of surface erosion on a watershed. This analysis can determine areas of potentially severe surface erosion, providing an initial step in the appraisal of surface erosion problems.

Another application of a GIS is predicting snowmelt runoff from a watershed, which often includes the use of a statistical model in the form of a regression model. Important parameters in predicting snowmelt runoff are the snow-covered area of the watershed, snow water equivalents, and other spatial data, including topography, soils, vegetation, thermal emittance, and near-infrared reflectance. Some of these data are collected at specific sites on the ground. All of these data are tested by a regression analysis to determine which are significant in a particular time and space. The best model with the highest correlation coefficient is then used to predict snowmelt runoff on the entire watershed.

These two examples demonstrate different modeling approaches using spatial data sets. However, a *final warning* on the application of GISs is warranted. Although a GIS is a powerful tool, there can be a temptation to rely too heavily on the computer output and, in doing so, stop thinking about the problem confronted (Congalton and Green 1992). It is important, therefore, that one clearly define the objectives of a study first and only then decide to what extent a GIS is the appropriate approach for helping to satisfy the objectives (Box 18.4). In some instances, a decision is initially made to use a GIS, and the search then begins for a problem to solve. There can also be a tendency to collect too much data; one more layer of data is not always the answer. If a user is thoughtful and wary, a GIS is a useful tool.

Box 18.4

Applications of Geographic Information Systems (from Franklin 1994)

Geographic information systems (GISs) may well be one of the most important technologies that managers of natural resources have acquired in recent years. GISs, often used with remote sensing, have a number of possible applications for these managers:

- Inventorying and monitoring—GISs and remote sensing can play important roles in storing current inventory information, including quantities of resources available, where they are located, and whether they are growing, shrinking, or holding their own. Of inventorying and monitoring, the latter is probably the major new thrust for watershed and other natural resource managers. Monitoring is crucial for two important reasons:
  - Management agencies often enter into agreements through court-ordered decrees, environmental impact decisions, and purposeful planning that require the monitoring of impacts from specified management practices.
  - Managers should monitor the effectiveness of prescribed management practices to obtain the feedback required to make corrections in the practices when necessary.
- Management planning—GISs can be a valuable tool and technology in management planning within administrative boundaries and, as frequently is the case in watershed management planning, among ownerships because of the importance of larger spatial scales. No longer can managers deal with only one forest stand at a time without placing the management of the stand into a larger context. Management planning in the absence of this larger context can result in unwanted and unexpected outcomes, such as cumulative effects or ecosystem fragmentation.
- Policy setting—With science a focus of policy setting in natural resource issues, GISs offer a means of preparing and displaying information on management alternatives for review by elected and appointed officials charged with policy making. It is possible that policy setting in the early 1990s with respect to protecting late-successional forest ecosystems and their associated species in the Pacific Northwest could have progressed more quickly and efficiently if the alternatives considered by the Scientific Panel and Forest Ecosystem Management Assessment Team (FEMAT) had been better mapped and displayed. Many of the resource maps presented to FEMAT for consideration had to be hand drawn, a tedious, labor-intensive process that, no doubt, introduced errors in transcription.
- Research—Both GISs and remote sensing can be critical to analysis and research. Analysts and researchers experience more problems going into the field and conducting replicated studies at the landscape level than they do at smaller scales. With GISs, analysts and researchers can address the problems involving larger scales by synthesizing resource data, developing concepts, and displaying the findings. Information about patches, edges,

connectivity, cumulative effects, and dispersed and aggregated activities can be included in the process.

- Consensual decision making—GISs make possible interactive collaborations among managers, policy setters, analysts and researchers, and stakeholders in the decision-making process. The time when analysts and researchers prepared and then presented management alternatives to the public without incorporating current science, social perspectives, and economic interests has passed. Today, professionals are participants and, at times, facilitators in the decision-making process, and GISs can play a significant role in expediting that process.

Other Considerations

Other considerations in relation to the use of a GIS as a research tool, regardless of the system, are data organization, database functions, input, query and analysis, user interface, display, reporting, and hardware (Guertin et al. 2000). Data organization relates to the raster versus vector issue. Database functions cover topics such as the operating system used. Input includes digitizing and input from external sources. Overlay is a query and analysis topic. User interface is concerned with menus versus command modes and other methods of control. Display relates to graphics output, and reporting relates to the format, such as tabular. To the system observer, hardware is the most prominent consideration, but it is part of the environment for the experienced user.

Summary

The intent of this chapter has been to provide a general reference for those responsible for the planning and implementation of watershed analysis or research efforts. This chapter, therefore, presents a framework of protocols for watershed analysis and research to help managers, analysts and researchers, and decision makers become aware of the tools and technologies available for these purposes. After completing this chapter, you should be able to:

- Explain how plot studies can be used to obtain cause-and-effect information on hydrologic processes.
- Describe the linkage between plot studies and watershed-level experiments.
- Discuss the differences between single, paired, and nested experimental watersheds.
- Describe how statistical methods are used to answer questions such as:
  - How often is a specified event likely to occur?
  - Is the response observed atypical?
  - Which variable is most influential in producing the result obtained?
- Describe the uses of computer simulation models in predicting the effects of land use change on hydrologic and other responses on watershed lands.
- Discuss the concept, common errors, and possible applications of geographic information systems.

References

Baker, M.B., Jr. 1999. *History of watershed research in the Central Arizona Highlands*. USDA For. Serv. Gen. Tech. Rpt. RMRS-GTR-29.

Bernhardsen, T. 1999. *Geographic information systems: An introduction*. New York: John Wiley and Sons.

Bolstad, P.V., and J.L. Smith. 1992. Errors in GIS. *J. For*. 90(11): 21–29.

Bosch, J.M., and J.D. Hewlett. 1982. A review of catchment experiments to determine the effect of vegetation changes on water yield and evapotranspiration. *J. Hydrol*. 55:3–23.

Choudry, S., and M. Morad. 1998. GIS errors and surface hydrologic modeling: An examination of effects and solutions. *J. Survey. Engineer*. 124(3): 134–143.

Congalton, R.G., and K. Green. 1992. The ABCs of GIS. *J. For*. 90(11): 13–20.

Dooge, J.C.I. 1973. *Linear theory of hydrologic systems*. USDA Agric. Res. Serv. Tech. Bull. 1468.

Feldman, A.D. 1981. Models for water resources system simulation: Theory and experience. *Adv. Hydrol*. 12:297–418.

Franklin, J.F. 1994. Developing information essential to policy, planning, and management decision-making: The promise of GIS. In *Remote sensing and GIS in ecosystem management*, ed. V.A. Sample, 18–24. Covelo, CA: Island Press.

Freese, F. 1967. *Elementary statistical methods for foresters*. USDA For. Serv. Agric. Handb. 317.

Green, R.H. 1979. *Sampling design and statistical methods for environmental biologists*. New York: John Wiley and Sons.

Guertin, D.P., S.N. Miller, and D.C. Goodrich. 2000. Emerging tools and technologies in watershed management. In *Land stewardship in the 21st century: The contributions of watershed management*, tech. coords. P.F. Ffolliott, M.B. Baker, Jr., C.B. Edminster, M.C. Dillon, and K.L. Mora, 194–204. USDA For. Serv. Proceed. RMRS-13.

Haan, C.T. 1977. *Statistical methods in hydrology*. Ames: Iowa State University Press.

Hewlett, J.D., H.W. Lull, and K.G. Reinhart. 1969. In defense of experimental watersheds. *Water Resour. Res*. 5:306–316.

Hornbeck, J.W., M.B. Admas, E.S. Corbett, E.S. Verry, and J.A. Lynch. 1993. Long-term impacts of forest treatments on water yield: A summary for northeastern USA. *J. Hydrol*. 150:323–344.

Johnston, R.S. 1970. Evapotranspiration from bare, herbaceous, and aspen plots: A check on a former study. *Water Resour. Res*. 6:324–327.

Kovner, J.L., and T.C. Evans. 1954. A method for determining the minimum duration of watershed experiments. *Trans. Am. Geophys. Union* 35:608–612.

Longley, P.A., M. Goodchild, D. Maguire, and D.W. Rhind, eds. 1998. *Geographic information systems: Principles, techniques, applications and management*. New York: John Wiley and Sons.

Ramsey, F.L., and D.W. Schafer. 1997. *The statistical sleuth: A course in methods of data analysis*. Belmont, CA: Duxbury Press.

Reigner, I.C. 1964. *Calibrating a watershed by using climatic data*. USDA For. Serv. Res. Pap. NE-15.

Reinhart, K.G. 1956. Calibration of five small forested watersheds. *Trans. Am. Geophys. Union* 39:933–936.

Sim, L.K. 1990. *Manual on watershed research*. Laguna, Philippines: ASEAN-US Watershed Project College.

Singh, V.P., ed. 1995. *Computer models in watershed hydrology*. Highlands Ranch, CO: Water Resources Publications.

Star, J., and J. Estes. 1990. *Geographic information systems: An introduction*. Englewood Cliffs, NJ: Prentice Hall.

Whitehead, P.G., and M. Robinson. 1993. Experimental basin studies—An international and historical perspective of forest impacts. *J. Hydrol.* 145:217–230.

Wilm, H.G. 1944. Statistical control of hydrologic data from experimental watersheds. *Trans. Am. Geophys. Union*, Pt. 2:616–622.

———. 1948. *How long should experimental watersheds be calibrated*? USDA For. Serv., Rocky Mtn. Forest and Range Exp. Stn., Res. Note 2.

Policies need to be established that recognize the importance of sound watershed management: a channelization project that destroyed an important riparian system due to a lack of effective policies.

PART 6

Socioeconomic Considerations in Watershed Management

Much of the earlier part of this text has dealt with the biophysical issues that face managers of watersheds and with the technical opportunities to deal with these issues. However, to be successful, watershed management also requires institutional changes and the inputs of social sciences, so that plans, incentives, and programs are acceptable to the people living in the watershed. Programs will fail if they do not recognize the wants of people affected and the context of the existing laws, property rights, and institutional environments in which people live. Thus, this last part of the book deals with the important themes of how watershed management plans are developed, how policies evolve, and how we go about doing the types of economic analysis needed to support integrated management of watersheds. An understanding of all these themes is necessary to design the market and other incentives to ensure that integrated management leads to outcomes for upstream and downstream dwellers and land and water users that are both effective and efficient.

CHAPTER 19 *Implementing Watershed Management: Policy, Planning, and Economic Evaluation Issues*

INTRODUCTION

Earlier parts of this book deal with the technical aspects of hydrology and the effects of land use on streamflow regimes (Part 1), erosion-sedimentation and channel processes (Part 2), and water quality management and habitat indices (Part 3). The cumulative effects of land use and watershed management practices were highlighted. This chapter focuses on how this technical information can be used to make decisions about future land use. Topics include policies for sustainable development, planning watershed management, and economics of watershed management. The interdisciplinary characteristics of watershed management and the need to take into account both hydrologic and socioeconomic factors when planning and implementing natural resource programs of many kinds are emphasized. Such information should provide the basis for developing sound natural resource policies. Here, we try to point out the relevance of a watershed management perspective in helping achieve sustainable, environmentally sound land and water use.

Watershed management does not "just happen," and we seldom find a "watershed manager" who manages the whole process in a given watershed or river basin. Rather, watershed management is the result of different people and groups doing different things, hopefully in a planned, coordinated, effective, and efficient manner. Watershed management is different and more complex than other types of land management, particularly if we focus on integrated watershed management as defined earlier. These differences and complexities lead to the need for a special understanding in developing successful watershed management programs. The three main differences and needs considered in this chapter are:

1. The policy context for watershed management and the processes by which policies evolve and create the framework within which planning and implementation of watershed management programs and activities take place.

2. The planning and implementation processes for watershed management, which differ in many ways from those employed for most other types of land management activities.
3. The nature of the economic analysis and background information needed for policy formation, planning, and implementation, and the ways in which such are generated and utilized. This chapter focuses on the economic methods and data needs; other chapters of the book focus on the biophysical data needs.

Policy Context and Processes

Watershed management activities are undertaken in a local, state, and national policy context. There is always in place a framework of laws, regulations, institutional mechanisms, cultural and social mores, and market systems that govern and guide the activities. Within this context there are many political, social, and economic challenges facing those trying to improve the effectiveness and efficiency of watershed management programs. Further complicating matters is the fact that polices tend to be evolutionary, changing, and (it is hoped) improving as more is learned, different stakeholder interests come to predominate, and governments change.

Watershed management is not practiced for its own sake—although some people might like to think so! Rather, people devote resources, time, effort, and thought to such activities because they think that people (including themselves) can benefit in the short, medium, and/or longer term. In this chapter, therefore, we come full circle, back to the points emphasized in the first chapter that relate to the human-focused reasons why watershed management is needed and practiced. Whenever there are human interests at stake, there also is a policy context that evolves over time, as governments and citizens try to mediate and settle conflicting interests in land and water use. We hear such terms as "property rights and water use rights," "water allocations," "land use fees," and forest practice laws and regulations. These all refer to the rules that society has set for itself.

The U.S. Policy Context

It is only recently that landowners and users in the United States have come together on a wide scale in a coordinated fashion to plan and achieve mutually agreed-to watershed management objectives. According to Lant (1999), there were few local watershed management associations in the country in the 1980s, although such initiatives had existed for decades in Japan, India, and a few other countries. However, there are over 1800 of these locally led initiatives throughout the United States at the present time. Interactions of the social, political, and economic forces of land stewardship with the technical aspects of watershed management are effectively fostered through the activities of these organizations. What these organizations are actually called is not as important as their general purpose and function (Box 19.1).

Watershed management will become even more important in the United States as the regional and local water scarcities and declines in water quality being faced now and in the future are more widely recognized by elected political officials and citizens. The Congressional Research Service (Copeland 2001) suggests with regard to water quality issues that the key water quality issues that may face the 107th U.S. Congress include actions to implement existing provisions of the Clean Water Act, whether additional steps are necessary to achieve overall goals of the Act, and the appropriate federal role in guiding and paying for clean water activities. Legislative prospects for comprehen-

Box 19.1

Watershed Councils: Examples of Locally Led Watershed Management Initiatives in the United States

Watershed councils are oriented to local issues and function as nonprofit advisory, educational, and/or advocacy organizations that encourage the protection, conservation, and sustainability of watershed resources on a river-basin basis. Some watershed councils act as forums for exchanging ideas, views, concerns, and recommendations, while other watershed councils are actively responsible for the management of river basins. Membership is typically open to public agency personnel and private-sector stakeholders with vested technical and sociocultural interests in a river basin and its tributary watersheds. The proliferation in watershed councils is illustrated by the nearly 55,000 pages found on the Internet when searching for "watershed council" or related terms on July 1, 2001. Watershed councils are found in every region of the United States and reflect the issues of crucial concern to the watershed managers and stakeholders in the river basin and geographic region of interest. For example:

- The *Saugus River Watershed Council* of Massachusetts is concerned with the health and beauty of the Saugus River and its tributary watersheds.
- The *Watershed Agricultural Council* promotes environmentally and economically sound agricultural and forestry practices while protecting the New York City water supply for over nine million consumers.
- The *Connecticut River Watershed Council* is the principal advocacy group for protecting and conserving the Connecticut River and linking the council members to the other organizations and agencies.
- The *Chetco River Watershed Council* manages the Chetco River Watershed, which runs from western Oregon to the Pacific Ocean.

One council—the *Watershed Management Council*—is a nonprofit, nationwide educational organization that is dedicated to the art and science of watershed management. It has largely an informational exchange function. The Watershed Management Council publishes a newsletter on timely themes and events such as conferences on interdisciplinary collaboration, cumulative watershed effects, riparian systems, watershed restoration, and water-quality monitoring.

sively amending the Act have for some time stalled over whether and exactly how to change the law. Many issues that might be addressed involve making difficult tradeoffs between impacts on different sectors of the economy, taking action when there is technical or scientific uncertainty, and allocating governmental responsibilities for implementing the law. Attention will also likely be paid to water infrastructure funding, implementation of current programs for developing total maximum daily loads (TMDLs) to restore pollution-impaired waters, and impacts of agricultural activities on water quality. The Clean Water Act's wetlands permit program, a pivotal and contentious issue in the recent past, also remains on the legislative agenda for some Members.

Water issues remain particularly acute in the western states. The main issues of concern that influence the resources put into watershed management and the nature of the programs designed and implemented (Cody 2001) include:

- The federal role in water resources development and management.
- Riparian and appropriative water rights—or water rights issues in general.
- What to do about competing demands for water, that is, who should get what for what purposes?
- Structural or technical approaches to increase water supplies (i.e., dams, diversion canals, etc.) and watershed management vs. recycling and reusing water and generally increasing the effectiveness with which water is used.
- Water conservation, transfers, and markets.

Similarly, there are several major soil and water conservation issues that face the people of the United States and are directly relevant for watershed management policy formation and planning (Cody and Hughes 2001, Zinn 2001). Most of these issues relate to a recent farm bill (Box 19.2), the Federal Agricultural Improvement and Reform Act of 1996 (P.L. 104–127), and various federal and state water projects, including some that have significant watershed management components. There are many other water and soil conservation and use issues that provide challenges to the existing mosaic of federal, state, and local policies related to land and water use and, therefore, to watershed management. Emerging as a clear issue is water use. In some states, such as California, more water is allocated to environmental uses than to any other use, including agriculture.

Box 19.2

Impact, Implications, and Provisions of the Farm Bill

The farm bill sets out soil and water conservation policy at the national level, to be coordinated with state laws and policies. Provisions in that bill relate to the Conservation Reserve (CRP), Wetland Reserve (WRP), Conservation Compliance, and Swampbuster Programs. The CRP is used to retire highly erodible or environmentally sensitive land (see Chapter 12); Conservation Compliance and Swampbuster Programs reduce incentives to cultivate highly erodible lands and wetlands; and the WRP uses easements to protect agricultural wetlands. New programs created include the Environmental Quality Incentives Program (EQIP) to provide $200 million annually in cost sharing to producers to address conservation problems; a conservation option for producers who receive market transition payments; and farmland and floodplain protection, grazing lands conservation, and wildlife habitat protection programs.

One could continue listing important issues that likely will be considered in public and private legislative battles. The point to be made is that there are many water-related issues that provide the impetus for watershed management programs and the need for the special technical and managerial skills of those who can design and implement watershed management programs. The context is broad and complex, and it needs constantly to be revisited, analyzed, and revised by those responsible for the policy framework that guides water and land use.

The Global Policy Context

Globally, there has been a recent increase in watershed management organization and activity, although some countries in Europe and the Middle East have had various forms and levels of watershed management–related policies and rules of behavior for centuries. In some countries, an authoritarian approach is taken. In others, a grassroots, populist approach is taken. Results have been mixed, as indicated by the reluctance of many landowners to adopt needed soil conservation practices on the steep slopes of mountains and hills in Taiwan, even though government expenditures in soil conservation have been increasing yearly (Cheng et al. 2000). High production costs and low profitability in "slopeland agriculture" as a result of labor shortages in rural areas, a rapid appreciation of the country's currency, and "strong" competition from imported agricultural products after a relaxation of restrictions are some of the reasons for this reluctance.

The Bedouin of eastern Jordan, while few in number and with only a limited dependence on material goods and technologies, have banded together in effectively working with government technicians in developing the water and other natural resources of this marginal arid land (Dutton et al. 1998). An *integrated watershed management approach* has been chosen as the basis for jointly planning and implementing three interrelated types of development: beneficial economic change such as that gained through improving livestock productivity; conservation of the environment, achieved largely through a reduction in "mining" the scarce but depended upon groundwater resources; and improved delivery of health and educational services. The smallest development unit is considered as a water basin in forming policies, planning, and managing this development and use in a participatory manner.

Managing water and, more comprehensively, watershed resources is recognized as being paramount to meeting the challenge of restoring the highly eroded watersheds of India, the seventh largest and second most populous country of the world, to a more productive state (Chandra and Bhatia 2000). A key to formulating programs to effectively meet this expanding challenge has been an increased concern for and awareness of the importance of "local human resources" in all aspects of managing and conserving water and other natural resources, at the highest governmental levels. Capacity building at the "individual level" is a centerpiece to this activity. To this end, incentives, policies, and programs to encourage and support efforts to alter people's attitudes and perceptions, increase technical and managerial knowledge, and provide participatory skills at the local level are achieved through linkages involving government representatives, personnel of foreign-assisted watershed management projects, and local stakeholders.

Policy Processes

Those dealing with watershed management need to understand the policy environment and the decision-making processes affecting land and water resource use if they are to implement successful management practices. We need to develop answers to a number of questions including:

- How do decision makers choose which issues to deal with and which policies to establish in response to issues?
- How can the different stakeholders coordinate their actions when there are often conflicts among stakeholder groups, and overlaps and interactions among decisions and policies?
- Once decisions are reconciled into a coordinated set of policies, how can they be implemented on the ground, recognizing that policies represent decisions, not actions?
- How can various policies be enforced when voluntary action is required by most of the stakeholders affected?

In other words, we need to determine what is involved in the policy design and implementation process. The policy development process is dynamic; it is not a one-time process, but rather it is an interactive process of successive approximations as society moves toward long-range goals on which it can reach a consensus. Therefore, policy changes are continuously taking place, often incrementally, with a few major changes defining new directions occasionally.

The policy development process for watershed management is the same as that in any other sector or field (Brooks et al. 1994, Gregersen et al. 1994, Quinn et al. 1995). Figure 19.1 outlines the policy development process in general terms. Each stage of the process is discussed briefly below.

Identifying and Assessing Issues

There are many issues (problems and opportunities) related to watershed management that may warrant a policy response. Many of these issues deal with policy objectives concerning:

- Land use stability and soil conservation.
- Flood damage reduction and the timely flow of water.
- Maintaining or improving water quality.
- Fair and equitable water allocations.
- Sustainable economic growth and development.

Issues to be addressed by a policy statement arise when one or more of these objectives are not being satisfied, or when conditions related to any of the objectives are worsening. Similarly, policies can be developed to respond to opportunities to improve people's welfare, for those living both on the watershed and off the watershed, but who are affected by its health and management. Four points to keep in mind about policy issues related to watershed management objectives are:

1. Issues arise largely because of *different perceptions* by *different stakeholders* of the best uses for natural resources. Specific uses of natural resources are valued

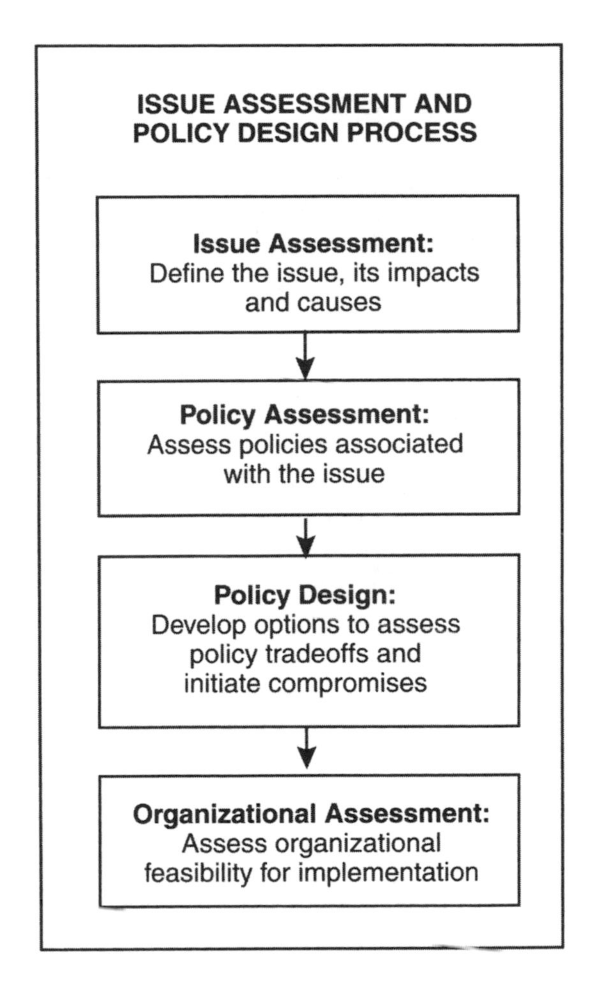

FIGURE 19.1. The process of policy development.

differently by the various stakeholders, depending especially upon whether the stakeholders live on or off a watershed and, if they live on the watershed, whether they live upstream or downstream from potential or actual problem areas.

2. Policy issues related to watershed management generally involve *disagreement* or *controversy* among competing interests over the current and planned uses of natural resources and the associated societal goals to be achieved by such uses.
3. Policy issues differ from management (technical) issues in terms of how they should be addressed. In the case of management issues, technical experts can analyze the situation and make recommendations for resolution of the issue, using technical criteria established to meet the goals specified in existing policy. In the case of policy issues, however, it is often disagreement over the goals themselves that leads to the emergence of the issue. Because different value frameworks exist, different sets of criteria for choice among options to resolve an issue also exist. These differences need to be addressed as a first stage in the policy development process.

4. Policy issues evolve over time as people's views on the uses of natural resources change. For example, conversion of forests to agricultural or livestock grazing lands was not much of a policy issue in the early 1900s in most countries. However, this conversion has become a major issue in subsequent years as incomes have grown, technology has advanced, relative political power has shifted, and changes have taken place in the values and in the influence of environmental groups.

If everyone agreed on the management objectives for a watershed, the process of policy formation and implementation would be relatively simple and straightforward. As discussed above, however, all stakeholders in a watershed generally do not agree on the management objectives; and even if they have common goals, they often do not agree on the best means of achieving them. Therefore, opinions on the best approaches to resolving the identified issues, such as deteriorating water quality, increased flooding, or reductions in the availability of water at certain times of the year, tend to differ.

Since differences in watershed management goals and objectives relate directly to differences in values, it is important to understand the different value perspectives and valuation frameworks involved. We need to identify and analyze the various perspectives and the expectations of the different persons or groups involved in the issue. The last major section of this chapter examines valuation in detail. Here, we will focus on the following questions:

- From whose point of view is the identified issue a problem?
- Who benefits and who gains?
- What are the various opportunities and threats that the issue poses?

We have to obtain answers to these questions to ensure a constructive debate to resolve the issues confronted. Those stakeholders involved in the debate need to identify the underlying reasons why controversial activities occur that lead to disagreements. In the case of water rights and allocations, for example, it is well known that some of the most bitter battles occurred in the western United States (Box. 19.3). Similarly, there are major battles waged internationally over water rights and pollution issues. As indicated in Table 19.1, a number of countries depend largely upon water supplies that originate outside their boundaries. Consequently, a number of countries are dependent on international agreements and policies to secure their water supplies.

Barring wars between the countries, all of the disagreements over water rights have to be resolved through the policy process. Once disagreements have been defined, regardless of how the definition was achieved, a process of discussion and debate generally takes place in trying to develop a consensus on the general policy directions and approaches that will at least be considered acceptable by the stakeholders. Resolutions to these discussions and debates often include market solutions, regulatory solutions, or combinations of both. In others words, the trade-offs for all of the parties involved are defined, debated, and negotiated.

Defining Workable Solutions and Proposing Polices

Once general agreement is reached—once the disagreeing parties come together constructively in terms of objectives—a more detailed analysis takes place to define in operational terms the trade-offs involved to reach the objectives, the compensation needed to make the trade-offs acceptable to all parties, and the policies that will best fit

Box 19.3

The Colorado River Controversy

It is almost impossible to discuss the policies of water in the western United States without mention of the Colorado River controversy—especially of the political and legal issues involved (Mann 1963). Solutions to the problems of how to obtain and best use the Colorado River water has been one of the dominant political issues for many years, making or breaking a number of politicians. It has been said, for example, that during the 1920s and 1930s, one could not be in favor of the Colorado River Compact, which specifies the allocation of water between the Upper Basin states (Colorado, New Mexico, Utah, and Wyoming) and the Lower Basin states (Arizona, California, and Nevada) and hope to win election. Many earlier issues of contention on the use of Colorado River water (such as water transfers to satisfy rural and urban needs, settling obligations in reference to Indian water rights, and the role of groundwater recharge in helping to assure water supplies) continue to fester to the present time (Eden and Wallace 1992).

the context and views of the stakeholders involved. The result of the debate, analysis, and negotiation is identification of a new policy or set of policies.

In the policy development process, it is essential that the existing policy context, and its impacts on the various stakeholders and their actions, be assessed before designing and proposing new policy options. In most situations, successful proposals will be those that recommend incremental change—evolution rather than revolution.

Of particular interest are two main types of weakness in existing policies. First, there are *ineffective policies*, that is, when the policy is meant to accomplish a given objective but ends up not accomplishing that objective. Consider, for example, a policy with an objective of reducing the pressures of timber harvesting to reduce the destruction of natural forests. Such a policy can encourage the planting of trees through subsidies and technical assistance. However, implementation of the policy could result in the establishment of tree plantations that produce wood that is not in demand. The deforestation continues as before. In this case, the policy is ineffective.

Second, there are *conflicting policies*, where a policy in one sector has unintended and unanticipated negative effects in another sector, or a policy intended to have a particular positive impact also has a negative impact. For example, the policy mentioned in the previous paragraph might improve the wood supply situation, but also result in the establishment of tree plantations in areas where the added uptake of water by the trees and the consequent increased water loss through evapotranspiration creates serious water problems downstream. In this case, the policy conflicts with policies and goals related to water supply. Another example of a conflicting policy is one that promotes agricultural exports through the provision of price supports. This policy in turn encourages deforestation of watershed lands critical to watershed protection that otherwise would have been marginal for agricultural production.

TABLE 19.1. Dependence on imported surface water, selected countries

Country	Percentage of Total Flow Originating Outside of Border	Ratio of External Water Supply to Internal Supply[a]
Egypt	97	32.3
Hungary	95	17.9
Mauritania	95	17.5
Botswana	94	16.9
Bulgaria	91	10.4
Netherlands	89	7.9
Gambia	86	6.4
Cambodia	82	4.6
Romania	82	4.6
Luxemburg	80	4.0
Syria	79	3.7
Congo	77	3.4
Sudan	77	3.3
Paraguay	70	2.3
Czechoslovakia	69	2.2
Niger	68	2.1
Iraq	66	1.9
Albania	53	1.1
Uruguay	52	1.1
Germany	51	1.0
Portugal	48	0.9
Yugoslavia	43	0.8
Bangladesh	42	0.7
Thailand	39	0.6
Austria	38	0.6
Pakistan	36	0.6
Jordan	36	0.6
Venezuela	35	0.5
Senegal	34	0.5
Belgium	33	0.5
Israel[b]	21	0.3

Source: Gleick 1993.

[a]Using national average annual flows. "External" represents river runoff originating outside national borders; "internal" includes average flows of rivers and aquifers from precipitation within the country.

[b]Although only 21% of Israel's water comes from outside current borders, a significant fraction of Israel's fresh water supply comes from disputed lands, complicating the calculation of the origin of surface water supplies. This percentage would be affected by a political settlement of the Middle East conflict.

After analyzing the policies associated with the watershed issues at hand, the next task is to identify the key policies that require change. This task represents a transition to the third stage of the policy development process.

Formalizing Policies and Their Implementation Mechanisms

In the case of public action on policies, proposals generally move to the legislature for laws or to the administrators of agencies if it is merely a matter of creating regulations for implementation of policies already covered by legislation. At this stage, there will be a more explicit discussion of the policy instruments needed to most effectively implement policies. These instruments generally fall into three categories:

1. Regulatory mechanisms—restrictions on land uses, land rights and water quotas, and controls on riparian vegetation management, and so on.
2. Fiscal and financial mechanisms—higher taxes on excessive water use, subsidies for soil conservation practices, fines for improper timber-harvesting practices, and so on.
3. Direct public investment and management—increased or decreased levels of public management of infrastructure and resources, public soil conservation education, research activities, and so on.

Often, a combination of the three types of policy instruments will be most effective and is chosen for implementation (Box 19.4).

The implementation of policies depends largely upon the nature of the policies established and the ways in which the public and private sectors are involved in the implementation process. In many situations, the key policies affecting watershed management are established by government agencies. However, direct or at least indirect conflicts between the various agencies occur, with one pursuing one approach and others pursuing other approaches. Consistency and coordination, therefore, become essential.

Policies have to be implemented and enforced by both public and private groups to be effective. At this stage, it is important to assess in detail the institutional context and the different stakeholders who will be involved in implementation. In some cases, institutional strengthening will be necessary to provide for effective implementation. It needs to be remembered, however, that in the United States most of the ongoing watershed management activities in place are undertaken voluntarily by landowners and users, largely in response to public and private incentives and guidelines for best management practices.

Monitoring and Evaluating Policy Implementation

Monitoring and evaluating policy implementation is also an integral part of the policy process (Fig. 19.1). Things seldom remain stationary and stagnant. We need to know how conditions and contexts are changing so policies can be adjusted to conform to the new conditions.

Monitoring and evaluating policies involve addressing the questions that are relevant to a given situation and set of decision makers. Such questions can differ markedly from one situation to another. However, the following dimensions of people's welfare associated with watershed management are relevant in general to decision makers regardless of the situation:

Box 19.4

Water Resource Problems and Policy Actions

Policy Instruments	Examples of Alternative Policy Actions for Specific Problems		
	Variable Water Supplies	Water Shortages	Declining Water Quality
Fiscal incentives	Use peak load pricing Legalize water markets	Use opportunity cost pricing Legalize water markets	Make polluter pay for damages Use tradeable pollution permits
Regulatory mechanisms and institutions	Use flood plain zoning Establish priorities for water use in droughts	Establish water user associations Establish river basin entities Impose restrictions on use	Set water quality standards Establish land use zoning around streams and in watersheds
Direct public investments	Public groundwater development Expand reservoir storage capacity	Transfer water from surplus regions Install water meters Install water-saving toilets	Install waste treatment plants Install aerators on polluted rivers

Source: Easter et al. 1995.

One can think of a number of objectives related to watershed management, such as those listed in Box 12.2, which also provides examples of policy instruments that fit with each objective.

- The *distributional dimension* refers to who gains and who loses and how to increase benefits and local participation in decisions. Included are impacts on different stakeholders classified in various ways, for example, by income classes, by regions or location, at different points in time, or by gender, age, or occupation type.
- The *sustainability* or *livelihood security dimension* concerns whether the positive changes in welfare are sustained over time. The Brundtland Commission's Advisory Panel on Food Security, Agriculture, Forestry, and Environment used livelihood security as an integrating concept and defined it as a combination of adequate stocks to meet basic needs, security of access to productive resources, and maintenance of resource productivity on a long-term basis (Brundtland Commission 1987).
- The *economic efficiency dimension* concerns such issues as changes in total benefits and costs to the region or nation due to the implementation of water-

shed management, measured in terms of people's willingness to pay for things (which might not be equal to what they actually have to pay for things); and the relation between incremental benefits and costs (economic efficiency measures such as rates of return, net present worth, and benefit-cost ratios).

Within each of these welfare dimensions, there are questions of concern about monitoring and evaluating the progress in watershed management and, therefore, the usefulness and effectiveness of policies that guide the specified watershed management practices. In assessments of the distributional dimension of watershed management, the following questions might be relevant:

- How are the benefits and costs involved distributed among different groups?
- Who loses as a result of watershed management and how?
- In what way does watershed management encourage the equitable distribution of benefits and costs among the local and regional populations?
- How does watershed management affect local empowerment and participation in resource development?
- In cases where the success of watershed management depends on groups being involved financially in watershed management, is that involvement financially acceptable (profitable) to them?
- What are the budgetary impacts for different agencies and/or groups involved?
- How does watershed management impact other sectors and the stability in the region or country?

Questions commonly addressed in assessments of the sustainability and livelihood security dimension of watershed management include the following:

- What are the impacts on livelihood security, focusing primarily on resource control and income generation?
- Are the benefits sustainable over time?
- What are the budget and financial sustainability (recurrent cost) implications?

The third dimension of welfare, economic efficiency, deals with one basic question: What is the overall relationship between the economic costs and benefits associated with the watershed management practices used to implement policies? This topic is discussed in the final section of this chapter.

A Watershed Management Framework and Its Policy Implications

As stated elsewhere in this book, watershed management practices are undertaken to achieve one or more of the following:

- Maintain or increase land and resource productivity.
- Ensure that adequate quantities of usable water are available to users.
- Ensure adequate water quality.
- Reduce flooding and the damage from flooding.
- Reduce the incidence of mass soil movement and other forms of soil loss.

Institutional mechanisms—regulations, market and nonmarket incentives, public investments in research, education, and extension—ease the application of the watershed management practices in many instances (Quinn et al. 1995). To be successful, however, these institutional mechanisms require an appropriate policy environment that:

- Recognizes the existence of watersheds and the processes that operate on a watershed scale.
- Allows for the explicit accounting of the environmental benefits and costs associated with forestry, water resource, and other development programs.
- Encourages the formation of institutions that help to achieve the goals of watershed management.

Properly planned and implemented watershed management practices have potentials to benefit both upland and downstream land and water users. Such benefits might be realized immediately in the form of increased levels of production, conservation of soil and water resources upon which current and future production depends, and protection and/or enhancement of environmental quality. Unfortunately, there are often a number of reasons why achieving successful watershed management through policy intervention is difficult, including:

- Watershed boundaries and political boundaries rarely coincide. Biological and physical processes, on the one hand, and social, political, and economic activities, on the other, do not normally take place on the same scale or within the same boundaries.
- Watershed management responsibilities, when they do exist, are often shared by a multitude of organizations, including those responsible for forestry, agricultural extension, water management, rural development, and transportation. Frequently, no single organization has coordination responsibility or authority.
- Watershed management goals and benefits are often not considered, not adequately understood, or given low priority in formalizing the policies that deal with the management of natural resources (timber, water, forage, wildlife) on watersheds. Consequently, these policies can hinder the practice of watershed management, although this generally is not intentional.
- Policies that encourage one or another land use activity on upland watersheds are often determined by a desire for increased production of commodities. It is not uncommon that little consideration is given to the sustainability of the land use activity and of the desired level of production or to the negative externalities that can result.

Developing an appropriate policy environment to encourage watershed management on a sustainable basis generally requires that landowners, private companies, government agencies, and other interested parties become involved in:

- Making responsible decision makers at all levels and the general public aware of the importance of sustainable land and water use and, especially, the linkages upon which watershed management attempts to build.
- Identifying all the major stakeholder groups involved with watershed use, as well as their perceptions and attitudes with respect to these issues.
- Clarifying the jurisdiction over watershed management and the coordinating roles of the organizations involved in the management of watershed lands.
- Internalizing externalities by more equitably distributing the benefits and costs associated with the use of natural resources on upland watersheds.
- Assessing the impacts of the policies for watershed management as they evolve.

Once again, it is essential that there be some sort of monitoring and evaluation to assess the status of the natural resources on watershed lands and the impacts of policy on these resources, if watershed management policy is to be responsive to changing conditions (Quinn et al. 1995). Also critical is the presence of institutional response mechanisms by which the lessons learned from monitoring and evaluation can be incorporated into future policy design and application.

Planning and Implementation Processes

Activities in one area can impact activities in other areas, and they can have impacts—both positive and negative—on many different groups of people. Because of these interactions, it is critical that land management activities on a watershed be coordinated through a practical, integrative planning process. Planning can help to avoid the negative impacts and encourage and take advantage of the positive ones.

The Planning Process: An Overview

Planning for watershed management involves an integration of three major sets of elements. First, there are the *objectives* established on the basis of problem analyses and directives from government and other stakeholders. Second, there are the various *constraints*: budget, biophysical limitations, and social, cultural, and political conditions associated with the specific situation. Third, there are the *techniques* or methods for carrying out watershed management activities. Watershed management planning involves the integration of objectives, existing constraints, and available techniques to improve the effectiveness and efficiency of decision making and implementation.

Many constraints involved in watershed management are physical, such as climate, soil conditions, and landforms. But, just as many constraints are social, economic, or institutional in nature, for example, cultural restrictions, available budget, and political acceptability. The challenge for the watershed planner and manager is to bring all these factors into the planning process, using accurate and detailed data where available and using approximations and order-of-magnitude estimates in other cases.

In only a few instances will there be enough data and information to make risk-free decisions concerning actions and reactions. Planning, therefore, requires many judgment calls and a great deal of flexibility. If we have learned one thing from the past, it is that projects and programs seldom develop as planned. The manager who is ready to adjust to changes in basic conditions is further ahead than one who rigidly adheres to an original plan despite changes in conditions. A major purpose of this chapter, and indeed the whole book, is to provide background and practical support for managers and planners so that they can be better prepared to anticipate these needed adjustments, identify critical information and concepts, and work more easily with the inevitable changes in constraints and opportunities that will take place as planning and implementation proceed.

In practice, the administrator of a watershed management program is a synthesizer and an integrator of many disciplines and fields. Therefore, the need for synthesis, for blending social, physical, and biological sciences, and for blending science with art are emphasized, since in the final analysis planning and management are as much art as science.

Each watershed-related problem has its own unique set of technical elements and characteristics. Therefore, in terms of physical planning (such as the identification and design of needed physical structures, vegetative manipulation, etc.), each project requires a somewhat different technical approach. The same does not hold for the planning process itself. Regardless of the type of watershed program, the same planning process can be used. It is only the relative emphasis on each step in the process that will differ. This chapter lays out the basic steps in the process and what they involve. The fine-tuning of the planning process will be discussed in terms of adjustments in emphasis and content needed to deal with particular types of problems.

The Watershed Management Planning Context

In a watershed management planning framework, such as that indicated in Figure 19.2, land use units on which production takes place are the central elements. Within a watershed, the land use units interrelate as indicated in the figure. Activities undertaken on upstream land use units have "off-site" effects on land use units lower down the watershed. These units in turn have off-site impacts still farther down the watershed. Such off-site impacts can come about through, for example, erosion, pollution from chemicals used on the upstream lands, diversion of water, or changes in water flow during the year. These impacts from upstream are part of the "inputs" or costs in the production process for any given land use unit. In addition, other inputs—labor, human-produced capital, and institutional and social capital—are applied on site in the production process. These inputs, combined with the on-site soil, biological, and water resources, result in the four categories of outputs shown in Figure 19.2.

The first category of outputs comprises those produced on site—agricultural, forestry, livestock, and other goods and services, including environmental services. The other three categories of outputs shown in the figure are off-site ones that have various impacts downstream. They include impacts of soil moving off the land use unit, impacts of changes in streamflow pattern and volume caused by on-site activities, and impacts

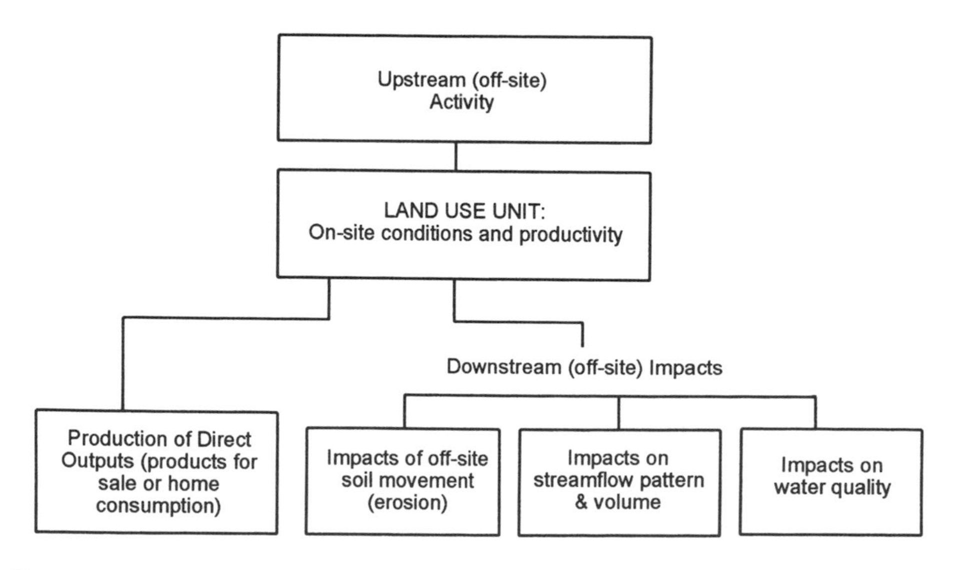

FIGURE 19.2. On-site and downstream physical effects of land use in the context of watershed management (from CGIAR 1996).

downstream of changes in water quality due to the activities on the land use unit being assessed. Watershed management practices can modify the relationships between the inputs and outputs to increase the positive social and economic contributions of the land and water resources both by improving sustainable productivity on site and by reducing downstream damages from soil erosion, flooding and drought, and pollution.

A key point to keep in mind is that, in the final analysis, watershed management activities are undertaken to benefit humans. We do not undertake activities to reduce erosion as an end objective. Rather, we undertake them to prevent erosion that causes losses that have direct impact on human values, such as reduced food production or loss of reservoir storage, which leads to reduced hydropower production and energy availability.

The relationships between watershed management activities (inputs) and their physical effects (outputs) are discussed throughout the book. Valuation of inputs and outputs is discussed in the following major section of this chapter.

A basic principle of analysis in assessing the impacts (both positive and negative) of a watershed management project (such as those impacts indicated in Figure 19.2) is application of what is called the *with-and-without concept*. In other words, we want to assess the changes or differences that occur with and without the watershed management project or activities being assessed and planned. For example, when we talk about the objective of increasing productivity or reducing sedimentation, we are referring to the differences in productivity or sedimentation with and without the project or activity being planned.

A second and related principle is that losses prevented due to the project have to be treated in the same way as actual gains due to the project when applying the with-and-without concept. For example, productivity can still be declining with the project or activity, but at a slower rate than without the project. Therefore, there is a higher level of productivity (an increase or gain in productivity) with the project than without the project (Fig. 19.3). This gain represents a benefit due to the project. This is an important principle, since so many of the benefits of soil conservation and watershed management activities are losses prevented rather than net gains due to, for example, rehabilitation of degraded lands.

Both principles have to be applied with a time dimension in mind. An example illustrates this point. Assume that the benefits of a watershed project are illustrated in Figure 19.4A. In this instance, which is common for severely degraded watersheds, the initial phase of a project causes productivity to fall below the level of the without-project condition (AG), because critically disturbed areas are removed from agricultural cultivation and production is temporarily reduced on areas that are undergoing structural and revegetation measures. After some time, productivity increases and is maintained at a higher level (BC). The area GBCE in Figure 19.4A represents the benefits of the project in upland productivity. Benefits are sometimes mistakenly identified as only FBCD if the with-and-without approach is not applied. In the example of sedimentation of a reservoir (Fig. 19.4B), the watershed management activity results in a slower rate of reservoir capacity reduction; the downstream benefits of the project are area ABC. Note again that the project initially has little effect on the rate of loss of reservoir capacity. However, as structural and vegetative measures become effective, the rate of loss declines and eventually reaches a relatively stable condition. The reservoir capacity still declines with the project, but at a lower rate.

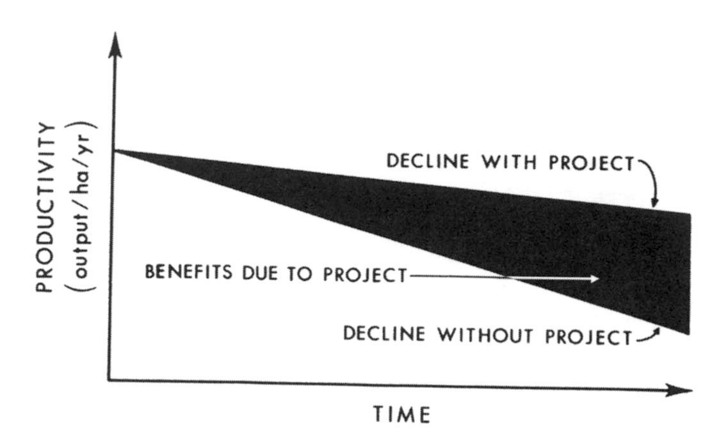

FIGURE 19.3. Productivity differences with and without a management project.

Steps in the Planning Process

The planning process involves the following steps:

1. Monitor and evaluate past activity and identify problems and opportunities.
2. Identify main characteristics of the problems and/or opportunities and define constraints and objectives to overcome problems and/or take advantage of opportunities; develop strategies for action.
3. Identify alternative ways to implement strategies, given the constraints.
4. Appraise and evaluate the impacts of alternatives, including environmental, social, and economic effects; assess uncertainty associated with results.
5. Rank or otherwise prioritize alternatives and recommend action when recommendations are requested.

In other words, we deal with the questions: How do we decide that something needs to be done? What is it we want to do? What alternatives do we have available to do it? Which alternative(s) is (are) best, given the circumstances?

Planning as an Iterative Process

Project planning is an iterative process of successive approximations, where we learn from experience and then incorporate that learning into the ongoing planning process. Watershed management planning, therefore, involves incremental learning. In practice, planners seldom go into a situation and immediately get down to the business of detailed and often expensive project design. Rather, planners go through a series of iterations, from a quick, low-cost assessment of the situation through progressively more sophisticated and detailed design and appraisal stages until they have all the information needed, wanted, or affordable to make a decision.

This incremental process of successive approximations makes good sense, since a major purpose of planning is to help decision makers and resource managers reject unsuitable options before too much time and too many resources have been spent study-

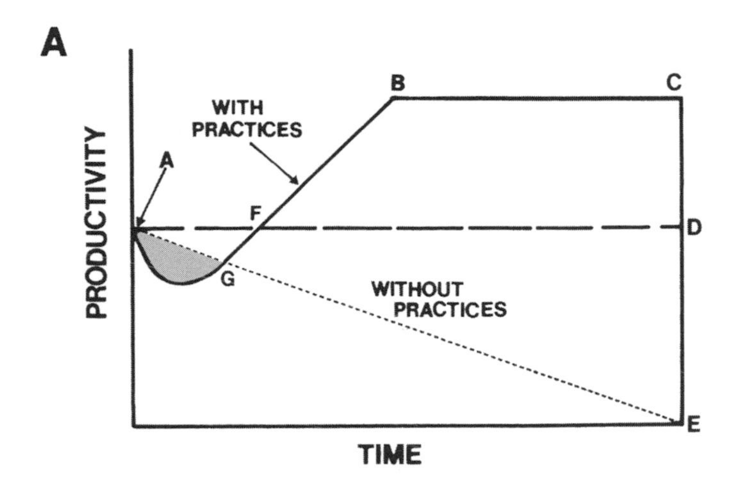

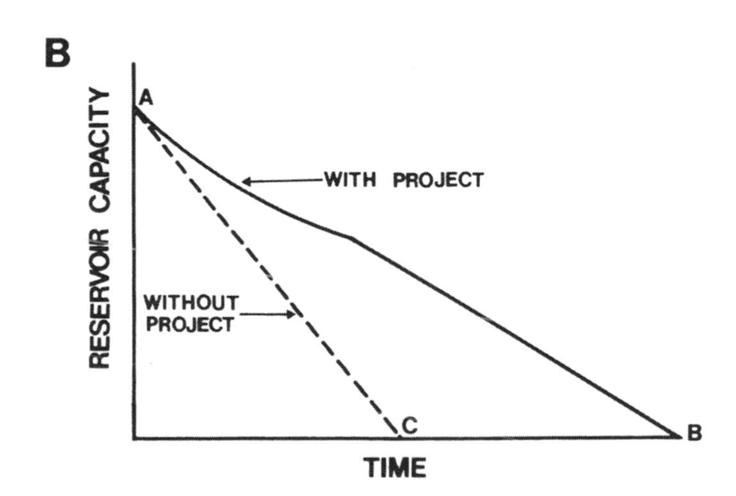

FIGURE 19.4. Relationships of upland productivity (A) and downstream sedimentation of a reservoir (B) under conditions "with" and "without" watershed management practices (from Gregersen et al. 1987, " FAO, by permission).

ing and developing them. An experienced planner can focus almost immediately on an appropriate strategy and a few viable and suitable options for dealing with a given situation. The objective is to discard unsuitable alternatives with a minimum expenditure of time and effort and to spend more time on the few alternatives that experience has taught are the best in a given situation. Figure 19.5 provides a view of (1) how the design and appraisal stages develop together, with appraisals being made at each iteration to decide whether or not to continue; (2) which alternative solutions to a problem should be rejected; and (3) which should be looked at in more detail. There is nothing fixed about the number of iterations involved; the four shown in Figure 19.5 are illustrative only.

Many persons will be quick to point out that by following rules of thumb and experience, we risk missing more efficient or effective solutions that could be identified if more time and resources were devoted to the effort. This contention is essentially correct. However, the optimum solution is constrained by the availability of time and resources.

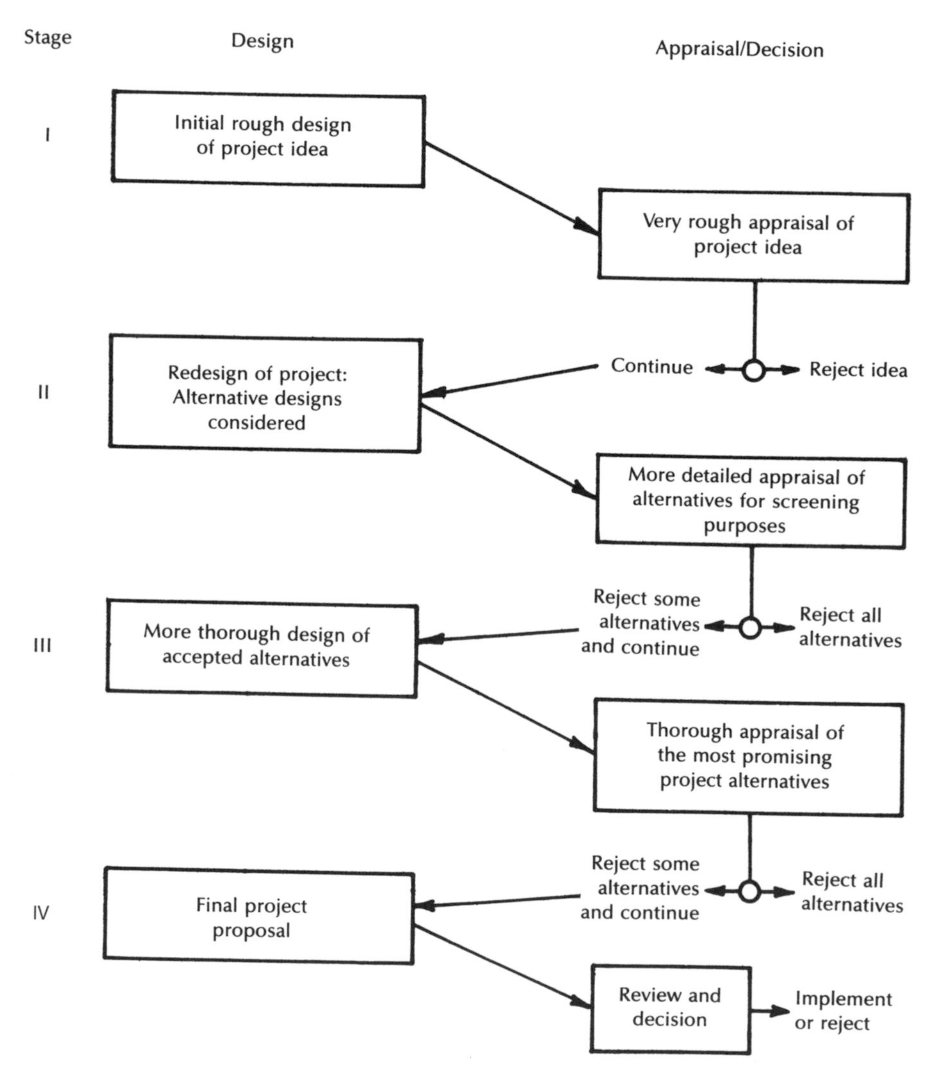

FIGURE 19.5. Stages in the project-planning process (OECD 1986, by permission). The number of iterative stages will vary depending on the project situation.

Each step in the planning process is discussed in more detail in the following sections. In keeping with our objective and orientation, we dwell on the practical solutions and avoid the impractical ones, even though the latter may be theoretically more appealing than the former.

Evaluating Past Activity and Identifying Problems

The process of planning has no beginning and no end in an ongoing or operating watershed management situation. However, for the purpose of exposition and explanation, the logical starting point is before a problem or opportunity is identified by monitoring.

In almost every part of the world, someone is collecting data about land and water uses and abuses, for example, the changes that are taking place in the biological and physical environment. This information can be used to determine whether a watershed management problem exists and what can be done to resolve it.

The types of information needed to identify problems are many and diverse. Key information needed to describe the environment is found in the categories of hydrology, climatology, soils, and vegetation. These data are collected on both a long-term, continuing basis to define major problems and a short-term, sometimes intermittent basis to provide information to solve particular crises. Ideally, long-term monitoring allows observation and analysis of resource responses over time to help identify problems. For example, precipitation and streamflow data need to be collected continuously for many years before we can characterize flooding, drought, and erosion conditions.

In addition to long-term continuous data acquisition and monitoring, data are often collected on a one-time basis to analyze a specific problem. This type of analysis might be performed to quantify the severity of a particular problem or to determine if corrective actions being taken on the watershed actually are solving the problem.

How data are generated by hydrologists, biologists, soil scientists, and so forth to meet these informational needs is discussed throughout this book and in the literature. Different types and levels of detail and accuracy are needed in different situations, depending upon the intensity of land use, the rate at which problems appear to be developing, or the severity of an already identified problem. Therefore, other things being equal, collection of more detailed information on sediment loads and siltation can be better justified for a watershed that drains into a multipurpose reservoir than for a relatively unpopulated watershed that drains directly into the ocean. Similarly, the monitoring of water quality in a river passing through large population centers should normally be much more intense than similar efforts for an isolated river that flows through wildlands.

Often, no formal measuring or monitoring system is used to produce the information that leads to identification of watershed management concerns and eventual action. Rather, problems are observed directly after they have occurred, as when a reservoir is silting up rapidly, when the scars of erosion begin to appear on the landscape, or when floods become more frequent and serious.

Regardless of how problems and opportunities are noticed, their definition becomes one of the first steps in watershed management planning. For the purpose of this book, we have broken down the types of problems commonly encountered into the categories discussed in Chapter 1 (see Table 1.1). In many instances, more than one solution to a problem can be possible. For example, water supplies might be enhanced for livestock production on the range by developing reservoir projects and water-harvesting schemes. In other cases, some solutions are mutually exclusive: increased agricultural cultivation to enhance food production may require that grazing by livestock be discontinued entirely.

Not explicitly discussed previously are other, non-production–oriented objectives, such as sustaining or enhancing wildlife habitat or providing scenic beauty. In many parts of the world, societies set aside watershed lands that have particular beauty, historical significance, or other amenities; these can be valid objectives and therefore must be considered in a planning effort together with objectives for adjacent or nearby lands managed for quite different purposes.

The types of problems that are important in watershed management are many and varied. In some cases, planning involves actions to prevent the problem from occurring; in other cases, the problem already exists and the planning involves development of a program to reduce the problem or (at least) ameliorate the conditions that caused it. While the specific actions taken in each case may differ, the planning process will be the same.

Identifying Constraints, Setting Objectives, and Developing Strategies

The next stage in the planning process involves setting objectives and developing strategies to solve the problems or respond to the opportunities. The objectives generally flow directly from the problem analysis. In their most general form, statements of objectives merely indicate that there is a need to develop a cost-effective means of overcoming or preventing the problem or to ameliorate the conditions causing it, eventually leading to benefits such as those identified in the bottom row of Figure 19.2. If flooding is the problem, the objective is to reduce flood damage. If poor-quality drinking water is the problem, the objective is to improve water quality and reduce health problems. With greater levels of specification, objectives can be translated into targets constrained, for example, in terms of the riskiness of approaches taken, the level of cost applied, and the level of achievement of other objectives.

Single versus Multiple Objectives

Single objectives, even when constrained in many ways, are not too difficult to deal with in the planning process. In most cases where one is dealing with single objectives, clear decision criteria can be developed for determining the extent to which alternative project designs (sets of activities) are acceptable and how they rank relative to each other.

Difficulties arise when more than one objective exists. Decision-making models have been developed to deal with *multiple objectives*. In practice, however, the use of such models has not always been successful. Focusing on one main objective, with other objectives expressed as constraints on the main one, is easier for decision-making purposes. For example, a major project objective can be reduction of sedimentation in a given reservoir. Additional objectives, such as better water quality or enhanced river transportation, then can be expressed as constraints on the main objective. With a given budget level, the objective can be to maximize reduction of sedimentation in the reservoir subject to an associated maximum permissible level of water pollution, a maximum allowable reduction in livestock herds in upland areas, and a minimum reduction in water yield.

Some planners avoid the problem of dealing quantitatively with multiple objectives by developing an array of effects of alternative project designs for the various objectives. The decision maker then has to provide subjective weightings to compare different alternative combinations of outputs related to the various objectives.

Developing a Strategy to Achieve Objectives

Once objectives have been established and agreed upon, a general strategy for action needs to be developed. The distinction between a strategy and a plan is subtle and probably more a matter of degree. Here, the term *strategy* is used to describe the general *direction* taken to achieve the objectives. A plan further includes the *magnitudes* or *targets* to be achieved and the timing of the actions to be taken to achieve them.

A strategy to reduce erosion that is causing reduced agricultural productivity and downstream sedimentation might consist of the following:

- Identify and define cost-effective land use practices that can decrease erosion.
- Develop an extension system that will give people the information they need to choose the appropriate practices.
- Develop incentive mechanisms to get local populations interested in using these land use practices.
- Provide credit facilities to give people the ability to adopt the new methods.

The important thing is not the strategy statement itself, but the process by which it was developed. If we just look at the problem statement, we could think of a number of alternative strategies to reduce erosion. For example, in an autocratic situation, we could suggest regulation against certain land uses with appropriate enforcement. In other cases, the best strategy might be to leave the situation alone because the independent nature of the local people would preclude any reasonable chance for success, and therefore, spending our scarce resources and time elsewhere would be better. Many other scenarios could be developed and transformed into strategy statements. So how do we decide on one specific strategy?

Considering Constraints

We arrive at a logical strategy by looking systematically at information on the constraints and conditions that surround the problem. Constraints that might have been considered in the above example include:

- Mass communication facilities do not exist in the region; currently no extension organization is in operation.
- Landforms and soils are such that only certain types of changes of land use practices will have a good chance of success in the area.
- The political structure is such that effective regulation and enforcement would be difficult; an approach using incentives is needed.
- There do not appear to be rigid cultural constraints that stand in the way of changing land use practices, but tradition is important to the local people; therefore, incentive mechanisms have to be included, at least initially.
- Currently, the financing mechanisms available for long-term agricultural crop and livestock management are inadequate; since some tree crops will probably have to be established on steep slopes, the need for improved credit mechanisms, alternative incomes, and other ways to support tree crop establishment and other long-term activities are needed.

Many preliminary questions would be asked during this strategy formation stage. The idea is not to formulate and quantify exact plans of action—that is, how many hectares will have to be subjected to a certain treatment; how much of an incentive payment and what kind will be needed; and so forth. Instead, we go rather quickly through the whole context of the project to (1) eliminate some approaches that obviously would not work, given the existing constraints; (2) determine which constraints need to be removed to have a reasonable chance for success; and (3) identify any obvious strong points that should be taken advantage of in project design and implementation.

Identifying Alternative Ways to Implement the Strategy

After an acceptable strategy has been developed, we get down to the details of designing alternatives to implement the strategy. The task becomes much more specific and technical. The need is to identify the various actions that could be used to implement the strategy and produce desired results. This is where technical specialists, social scientists, politicians, and others dealing with economic and cultural elements, come into the picture.

The types of actions needed in different instances include (1) physical/engineering actions (terraces, dams, gabions), (2) biological actions (planting, cutting vegetation), (3) regulatory measures to discourage or require certain actions (regulating grazing, timber harvest), (4) actions to create incentives (provision of free goods and services, subsidies, credit, outright cost sharing), and (5) educational activities (information materials, demonstration). The task at this stage is to identify the possibilities and the array of options that are available, given the constraints and circumstances surrounding the project.

Appraising Alternatives

While the alternatives are being developed, they are also being appraised or evaluated (Fig. 19.6). In most cases, an appraisal is an ongoing activity. In its broadest meaning, *appraisal* refers to the process of identifying, defining, and quantifying the likely or expected impacts of an action (a practice) or closely related set of actions (a project). Some of these impacts will be positive and some will be negative.

Nature of Impacts Being Appraised

A list of areas that can be affected by watershed management activities is presented in Table 19.2. The separation into economic and financial, environmental, and social effects relates to the different types of impact a change in watershed management can cause.

For example, assume that a watershed management project results in 100 more people being employed as road workers, tree planters, and guards to prevent illegitimate woodcutting or livestock grazing on certain watershed areas; this increase in people employed is an actual physical change due to the project—it is a fact. The economist looks at allocation of resources to these newly employed people, the redistribution of income that takes place, implications for public budgets, and the implications in terms of regional and national levels of production over time. The watershed manager may look at the implications of increased forest protection for soil and water quality, ecological stability, water yield, and other physical-biological impacts. The social scientist

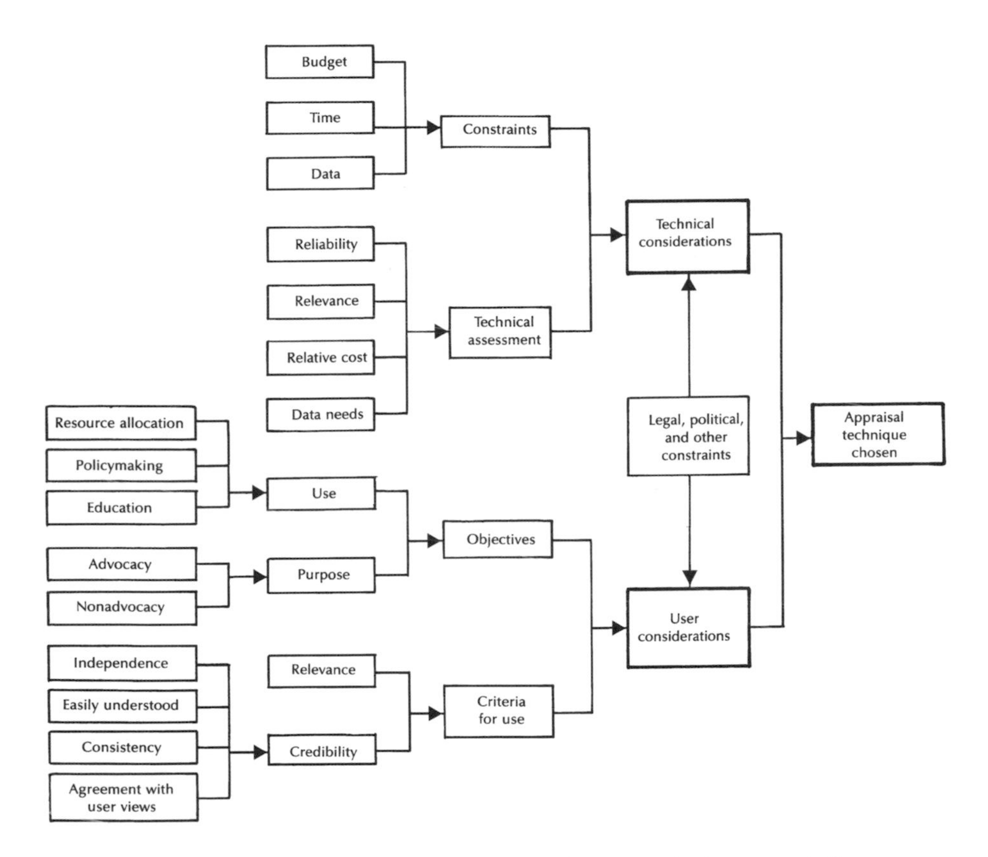

FIGURE 19.6. Factors affecting the choice of an appraisal approach (from Gregersen and Lundgren 1986).

may look at regional employment changes in terms of effects on cultural systems, social cohesion, and conflicts with previous institutional traditions (e.g., informal tenure traditions). The point is that the employment of the 100 additional persons due to the project has more than just environmental effects.

Some analysts worry about double counting of effects when (for instance) we say that the increase in employment increases income, ecological stability, community cohesion, and feelings of self-worth. These are merely different measures of concern related to the same physical change. They are complementary measures that relate to different decision criteria and describe different dimensions of the same physical change. All the impacts may be of interest to decision makers.

Making Appraisals Useful and Relevant

Project appraisals are useful only if they provide timely information of relevance to decision makers. A clear distinction needs to be made between the technical analyst's considerations in choosing a good appraisal approach and the user's view of what characterizes a good, acceptable, usable appraisal. Figure 19.6 outlines these two points of view and the characteristics or criteria of relevance. The task of a good planner is to bring these two sets of criteria or considerations together into the final appraisal.

TABLE 19.2. Scope of effects of watershed management activities

Economic/financial effects on:
Regional and national level of production
Allocation of resources
Regional and national income
National balance of payment
Stability of income over time
Distribution of income (both interpersonal and intertemporal)
Public budgets
Environmental effects on:
Ecological diversity
Ecological stability
Wildlife protection
Soil protection
Landscape aesthetics
Water yield and timing
Water quality
National patrimony
Social effects on:
Regional employment
Working conditions
Public participation
Migration flows
Cultural traditions
National vulnerability
Political stability

Source: Organization for Economic Cooperation and Development 1986, by permission.
Note: For convenience in exposition, we have divided the effects into three categories: economic/financial, environmental, and social. Other categories could equally as well have been chosen.

Appraisals should use the minimum amount of resources necessary to reach an acceptable decision on how best to achieve an objective. In some cases, this point is reached after a "quick-and-dirty" appraisal. The evidence and political agreement concerning the best alternative are so clear that one need go no further to reach a clear decision. A more detailed and thorough second-stage appraisal is needed when the evidence is not quite clear enough to make a judgment after the first-stage appraisal. Finally, in cases involving major commitments of resources, a formal feasibility study is needed to arrive at the point where a comfortable decision can be made. Sometimes, a formal appraisal (a feasibility study) is also required by the institutions involved. Table 19.3 indicates considerations for each of three stages of appraisal.

It is important to pursue appraisal of projects at sequential levels because resources for appraisal are limited in most cases and it encourages initial consideration of a number of alternatives for achieving objectives. Starting out with only two alternatives—do nothing and option A—is unduly restrictive. The preferable approach is to start with a number of alternatives and then narrow them down systematically in stages (Fig. 19.5). This approach also encourages the introduction of economics into planning and design rather than tacking it on at the very end of project planning through a feasibility study. Such involvement at early stages is desirable.

A fundamental question in any evaluation is: What criteria are we going to use? The most common economic criteria used for watershed management decision making are discussed later in this chapter.

Table 19.3. Stages in the evaluation process

Stage I. Rough appraisal of the project idea.

Make tentative calculation of the economic effects of the "most obvious" project alternative and the "without" alternative.
Make quick assessment of financial, administrative, and political feasibility.
Attempt to detect adverse environmental effects (i.e., long-term and system effects), social effects, and effects on different groups concerned (i.e., distributional effects).
Consider means to mitigate negative effects.

Outcome: Recommendation on whether to continue with project idea.
Performed by: Project initiator, using existing available information.

Stage II. More detailed appraisal for screening purposes, using the results of Stage I.

Design several project alternatives that seem relevant in light of existing objectives and of the major problems arising in the environmental and social fields, as identified in Stage I.
Acquire economic, financial, environmental, and social expertise for the appraisal.
Make calculations of the economic and financial effects of the alternatives, possibly improving upon existing forecasts and shadow prices.
Identify and describe the major social effects, possibly with the help of a representative discussion group.
Identify and describe the major environmental effects, particularly indirect and long-term effects.
Sample public opinion on the project alternatives.
Exclude alternatives that are not feasible for administrative or political reasons.
Establish rankings of the remaining alternatives (as seen by various groups), possibly using the representative discussion group.

Outcome: Identification of several promising project alternatives, elimination of alternatives with obvious flaws, decision on whether to continue the project.
Performed by: Appraisal team in collaboration with external expertise and possibly a representative discussion group.

Stage III. Thorough appraisal of the most promising project alternatives, given the results of Stage II.

Redesign the project alternatives in light of results obtained in Stages I and II.
Complete the detailed analysis of economic/financial, environmental, and social effects, collecting new data where necessary and utilizing the insights of a representative discussion group (level of detail depends on the time and budget available, on the purpose of the appraisal, and on the nature of the project).
Complete an appraisal report on the most promising alternatives, in the form of a scenario of likely developments over time for each alternative, including the "without" alternative.
Rank the most promising alternatives as seen by various groups, in collaboration with the representative discussion group, and possibly with input from local hearings.
Prepare summary presentations of the scenarios and the ranking for the decision makers, public groups, financial institutions, and other authorities.

Outcome: The necessary basis for choosing between project alternatives or terminating the project.
Performed by: Appraisal team, in collaboration with the discussion group and whatever expertise is available given budgetary and time constraints.

Source: Organization for Economic Cooperation and Development 1986, by permission.

Dealing with Risk and Uncertainty in Appraisals

One faces a situation of uncertainty rather than risk in most watershed projects. The distinction is simply that in the case of *risk* one can apply probabilities to various outcomes, whereas in the case of *uncertainty*, no such quantitative measures of probability of occurrence can be generated. In a situation of uncertainty, one can always develop some subjective probability estimates for different aspects of a project that are of interest. However, such estimates often do more harm than good, since subjectivity in the planning process should not be hidden. We suggest using a standard, straightforward sensitivity analysis, which is an analysis of how the measures of project worth or desirability would change under different assumptions concerning the values of key parameters (Gregersen et al. 1987).

Recommending Action

In some instances, the planner's task stops when the alternatives and the implications of risk and uncertainty for the different options have been evaluated. In other cases, however, the planner is asked for recommendations regarding which alternative should be undertaken and the timing and approach of implementation.

The appraisal results can be presented in different ways depending on the nature of the planning situation. Presenting a ranked set of alternatives is often preferable, or perhaps several rankings utilizing different appraisal criteria. However, ultimately, only the responsible decision maker can decide which alternative is chosen.

Planning as a Continuous Process

The planning process is an ongoing and a continuous process. It is an iterative process, with information concerning results of actions and emerging problems constantly being fed back into it. This information is then used to suggest incremental changes in the ongoing program or project. More formally, the process of collecting and disseminating information on ongoing operations is part of the monitoring and evaluation task referred to earlier. There is a constant feedback of information, which is synthesized and analyzed and then applied to the development of alternative strategies and suggestions for action. This type of continuous process leads to a healthy interaction between planners, technical personnel, and managers of watershed activities.

Economic Assessment of Watershed Practices, Projects, and Programs

Although technical personnel in watershed management work deal predominantly with physical and biological processes, they should also have some understanding of the economic and financial implications of what they are doing. This understanding is important for those charged with developing budgets and allocating funds to projects or programs. Such people want to know the economic value of the proposed activities, how much they will cost, and what the cash flow (including recurrent costs) will be over time. Technical personnel need to understand economics so they can provide the information for decisions or, in some cases, choose the most cost-effective (that is, most economically efficient) alternative for achieving a given purpose themselves. In this section, we explore the economic aspects of watershed management, interpreting economic impacts in the broadest sense to include not only the traditional economic-efficiency impacts but also the broader distributional and sustainability impacts associated with programs. An overview of the ways in which economists value and analyze the benefits and costs associated with watershed management is also presented.

Traditionally, the main economic impact of interest in watershed management is that related to the *economic-efficiency dimension*, that is, the relation between total benefits and costs to the nation or region due to the activity being assessed. More specifically, the relationship between incremental benefits and costs with and without the activity being assessed can be expressed in terms of a number of measures, such as the net present worth, rate of return, and benefit-cost ratio (discussed later).

In addition to efficiency measures, economic assessments of watershed management attempt to provide pertinent and timely information on two other dimensions of economic impact identified above:

1. The *distributional dimension*, that is, who gains and who loses.
2. The sustainability or *livelihood security dimension*, which is concerned with the question of whether the positive benefits or changes in welfare can be sustained over time.

The Distinction Between Economic and Financial Efficiency

One point that often confuses people is the difference between *economic efficiency* (the relation between total benefits and costs) and *financial efficiency* (profits, or the difference between market-priced returns and costs) for the private participants (individuals, companies, etc.) involved in a project. Both are important for their own purpose, although we focus on economic efficiency in this chapter because it more clearly reflects society's broader interests in watershed management beyond the income-generation concerns of private individuals and firms.

The two significant differences between economic and financial efficiency measures relate to:

1. The benefits and costs (or positive and negative impacts) included in the assessment—Only direct market-traded returns and costs are considered in the financial analysis, while as many as possible of the nonmarket benefits and costs (positive and negative impacts) are also included in the economic-efficiency analysis.
2. How those costs and benefits (impacts) are valued—A financial analysis always uses market prices, while the best estimates of people's willingness to pay for goods and services are used in an economic-efficiency analysis. In economic analysis, market prices are often adjusted to more accurately reflect social or economic values. These prices are referred to as *accounting* or *shadow* prices.

A financial analysis is not just of interest to private individuals and companies. The public sector also needs to consider financial aspects of projects in looking at the distributional impacts of project activities, that is, who actually pays and who gains from such activities. Some of the financial analysis questions that arise in the public sector include:

- What is the budget impact likely to be for the management agencies involved?
- Will the project increase economic and financial stability in the affected regions?
- Will it have balance-of-payments impacts?
- Will the project be attractive to the various private entities (such as those upstream) who will have to put resources into the project to make it work? What will the income redistribution impacts of the project be?

A financial appraisal deals strictly with market-traded goods and services, actual money inflows and outflows, and who gets paid and who pays. Market-traded goods and services refer to those that are openly bought and sold and therefore have identified market prices attached to them. An economic appraisal also considers market-traded goods and services, but it attempts to value them in terms of society's true willingness to pay for them. Sometimes, these values differ from market prices, as discussed later. An economic analysis also includes the social benefits and costs of goods and services that are not traded in the marketplace; for example, it attempts to consider the values of such things as health effects, flood prevention, aesthetic benefits, and wildlife habitat preservation. The major economic and financial components of an appraisal are summarized in Table 19.4.

Time Value of Money, Discount Rate, and Present Value

Economists most often look at economic efficiency in terms of what they call *present values*. This merely means that all estimated future benefits and costs are brought back to a common point in time, generally the present—therefore, the present value idea. This task is accomplished by discounting the benefits and costs, using an acceptable *discount rate*. If the present value of benefits is greater than the present value of costs for a watershed management project and there is no cheaper way of achieving the objective being sought, the project is considered an economically efficient use of resources.

How is the discount rate used to adjust, or normalize, benefits and costs that occur at different times so they all reflect value at the same point in time? Discounting assumes that society values a dollar of present benefit or cost more highly than a dollar of benefit received or cost paid sometime in the future. For example, when we lend money as a business, we expect to get interest as a payment for forgoing use of the loaned money for consumption today. If we lend \$100 at 8% interest, we expect 10 yr from now to get $\$100 \times (1.08)^{10}$, or \$215, that is, our original \$100 plus \$115 of interest. We could get this value using the simple compound interest formula. Thus, future value is determined by:

$$V_f = V_p (1 + r)^t \qquad \textbf{(19.1)}$$

where V_f = future value; V_p = present value; r = annual compounding rate, or interest rate; and t = time, from year 1 through year n.

Similarly, if we borrow \$100 at 8% interest, we expect to have to pay back \$215, 10 yr from now. We could discount \$215, 10 yr at 8% interest, by using the discount formula and arrive back at \$100 of present value. Present value is determined by:

$$V_p = \frac{V_f}{(1 + r)^t} \qquad \textbf{(19.2)}$$

For the above example, $\$215/(1.08)^{10} = \$100$.

A more detailed discussion of discounting is found in Gregersen and Contreras (1992) and other references on economic and financial appraisal.

Steps in the Economic Appraisal Process

Economic appraisals of watershed management should include at least the following steps, once alternatives for achieving an objective have been identified in the planning process. For each alternative identified:

TABLE 19.4. A comparison of financial and economic analyses

	Financial Analysis	Economic Analysis
Focus	Net returns to equity capital or to the private group or individual.	Net returns to society.
Purpose	Indication of incentive to adopt or implement.	Determine if government investment is justified on economic efficiency basis.
Prices	Prices received or paid either from the market or administered.	May require shadow prices, e.g., monopoly in markets, external effects, unemployed or under-employed factors, overvalued currency.
Taxes	Cost of production.	Transfer payment and not an economic cost.
Subsidies	Source of revenue.	Transfer payment and not an economic cost.
Interest and loan repayment	A financial cost; decreases capital resources available.	Transfer payment and not an economic cost.[a]
Discount rate	Marginal cost of money; market borrowing rate; opportunities cost of funds to individual or firms.	Opportunity cost of capital; social time preference rate.
Income distribution	Can be measured re: net returns to individual factors of production such as land, labor, and capital but not included in financial analysis.	Is not considered in economic efficiency analysis. Can be done as separate analysis or weighted efficiency analysis with multiple objectives.

Source: From Gregersen et al. 1987, as adapted in part from F.J. Hitzhusen 1982.
[a]Unless external loan.

- Define and quantify the physical inputs and outputs involved; create tables that show inputs and outputs as they occur over time.
- Determine unit values (both actual financial-market prices and economic values) for inputs and outputs, and estimate likely changes in such values over time, for example, growth in wages or fuel costs.
- Compare costs and benefits by calculating relevant measures of project worth and other indices and measures needed to answer relevant questions raised by decision makers; consider the implications of risk and uncertainty, for example, through a *sensitivity analysis*, which indicates how measures of project worth might change with changes in assumptions concerning input and/or output values.

These steps are interrelated as shown in Figure 19.7, which outlines both the financial and the economic considerations involved in watershed management appraisals designed to obtain the information indicated in the bottom row of the figure. The term *project* refers to both self-contained projects and watershed practices incorporated in other projects.

As indicated earlier, the appraisal process is generally iterative; one passes through increasingly detailed and complex stages of evaluation. This process also holds for economic and financial appraisals, which should be adapted to the stage in the planning process and the needs of the decision makers. We will explore each step in an economic analysis of a watershed management project and then apply the process to a case example.

Identifying and Quantifying Physical Inputs and Outputs

Developing information on the relationships between inputs and outputs is one of the major tasks facing the watershed manager; it is also a major subject of this book. Such information, put in a specific project context, provides the basis for estimating and quantifying the inputs and outputs associated with the project being analyzed. It is in this first step that the main interaction between an economist and a technical expert takes place.

Information Needed

In the process of identifying and defining the alternatives to be evaluated, the technical watershed management experts have to identify and quantify most of the inputs and associated outputs in physical terms. For an economic analysis, information is needed on (1) the units in which the inputs and outputs are measured, (2) the source of the

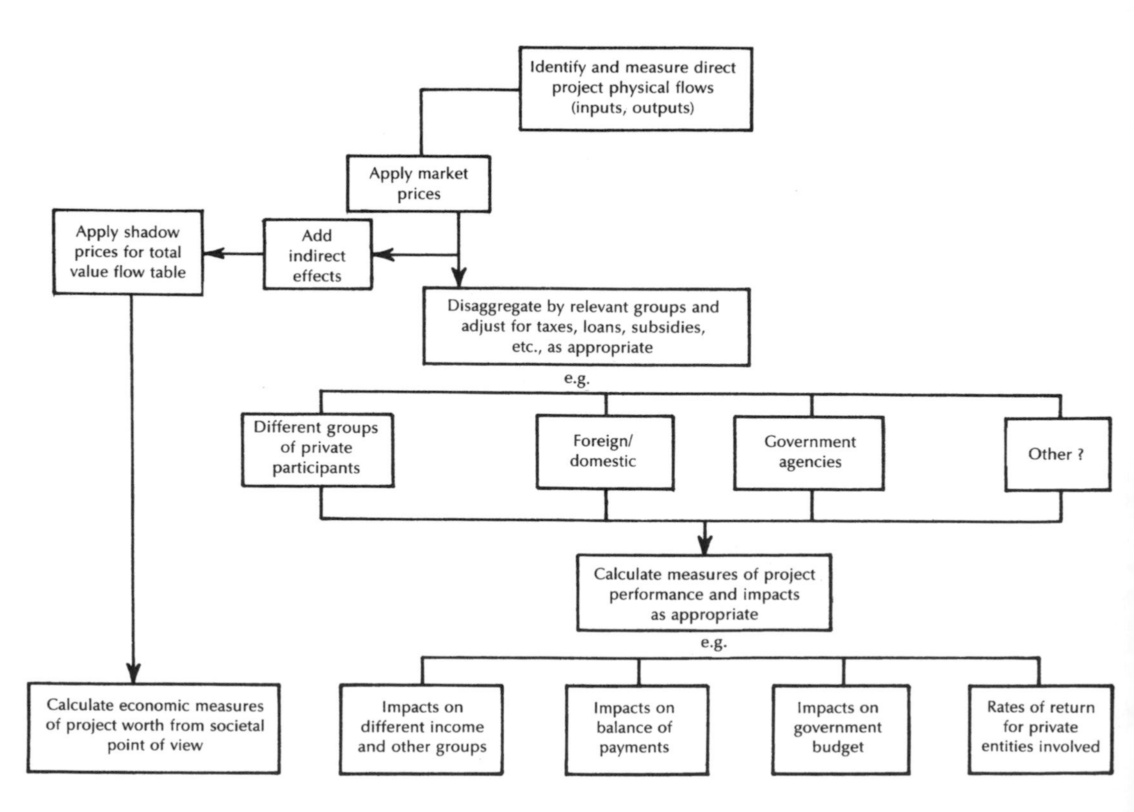

FIGURE 19.7. Overview of economic and financial appraisal processes (from OECD 1986, by permission).

inputs (e.g., whether they are to be purchased or provided by project participants on a work-share or other basis), and (3) when inputs will be needed and when outputs will occur. This information is provided in what is generally called a *physical-flow table*, which shows the flow of physical inputs and outputs over time.

The technical experts can generally meet these informational needs if they are aware of them early in the planning process. Common categories of inputs used in watershed management projects are shown in Table 19.5.

With-and-Without Principle

A point that bears reemphasis is that the input and output quantities included in a physical-flow table need to reflect the differences with and without the project. As an example, assume that two forest guards have been and will continue to be stationed on a given watershed to protect fragile areas from encroachment and deforestation—the "without project" situation. With the project, the total number of guards will be increased to four. Only the additional two guards, not the total number of guards, should be included in the "with project" analysis since two would be there with or without the project.

Many outputs associated with watershed management are expressed in the form of *losses prevented*, for example, forestry productivity losses prevented and flood or drought losses prevented. Even though these outputs are often difficult to quantify using the with-and-without principle, they have to be included because they are as real and important in human-value terms as increases in production.

TABLE 19.5. Examples of inputs for watershed management activities

Category of Inputs	Examples—Description
Work force	Resource managers—forest, range, watershed managers and planners. Engineers and hydrologists—design of erosion control structures, floodplain analyses, water yield estimate, etc. Skilled labor—construction. Unskilled labor. Training/extension specialists to facilitate adoption of project.
Equipment	Detailed listing of equipment needed for project construction and maintenance. Schedule of needs and equipment maintenance, i.e., timing.
Land	Land classified according to suitability for various uses. Designate sensitive areas to be protected (benefits foregone). Areas to receive treatments followed by management.
Raw materials	Utilities (energy, fuels, etc.). Wood (construction, fence posts, etc.). Other construction materials (concrete). Water.
Structures and civil works	Housing, roads, other facilities needed for project that are not part of project itself; if part of project, they are included in work force, materials, etc., listed above.

Source: From Gregersen et al. 1987, by permission.

Relating Inputs and Outputs to Human Use and Value

Another point that needs to be stressed is that an economic analysis of a watershed management project looks at inputs and outputs in relation to *human value*. Therefore, tons of soil loss prevented is not an adequate output measure. People normally do not put a value on soil loss as such; rather, such losses prevented have to be related to food losses prevented or other losses prevented that can be linked directly to human values. From the economist's point of view, a discussion of input and output relationships is complete when it has established the relationship between inputs used and goods and services consumed and valued by *society*.

Dealing with Nonmarket, Nonquantifiable Outputs

Some beneficial effects (or outputs) of watershed management, such as health and aesthetic benefits, are not easily measured. If these effects cannot be quantified and valued, they should still be mentioned explicitly and described to the extent possible in a final appraisal. Many nonquantifiable benefits relate to fundamental issues associated with the sustainability of human activity, including such benefits as ecosystem preservation and protection of biodiversity. These relate to fundamental human needs that are thought to be important but which cannot be quantified and valued other than in an anecdotal or descriptive fashion.

Valuing Inputs and Outputs

To answer the budget and other financial questions raised earlier, the valuation approach is straightforward: market prices are used for those inputs and outputs traded in the market. Nonmarket benefits and costs are not considered in financial analyses. Since market prices are generally determined straightforwardly, we will concentrate on the valuation of benefits and costs for the economic analysis. A number of detailed treatments of valuation issues and methods are available.

Measures of Economic Value and Shadow Pricing

What is economic value in practical terms? A basic measure of value used is *willingness to pay* (WTP). WTP is a measure that reflects society's willingness to pay for goods and services at the margin, that is, if another unit of the good or service were made available. WTP is a reflection of scarcity value in the sense that the more that is available, the less any individual generally is willing to pay for the good or service at the margin.

Another common measure used is *opportunity cost* (OC), which is a measure of value of the opportunity foregone when a resource is used for one thing rather than another. A typical example would be the OC of setting aside an area as a watershed protection area. The OC applied to the land set aside would be the value of the timber, minerals, livestock grazing, and other benefits foregone. As another example, the OC of using labor to produce X is the value of the production foregone by not using that same labor in its next-best use, producing Y. For example, assume that a worker is producing one cord of firewood per day and selling it for $30. That worker then finds a job near home that will pay $40 a day. The wage is $40, but the OC is the value given up when the worker has to stop producing fuelwood, namely, $30 in this case.

There is a direct relationship between WTP and OC. The OC values are those used in measuring the WTP for the goods and services foregone. In the above fuelwood example, the OC was the willingness to pay for one cord of wood, assumed to be $30.

In a competitive economy with no constraints on the movement of prices, one can assume that market prices adequately reflect WTP at the margin. It is for this reason that market prices are widely used in economic as well as financial analyses. However, WTP and OC can diverge from market prices when regulations are placed on the prices, as in the establishment of price ceilings or minimum prices, or when subsidies or taxes affect the prices (Gregersen and Contreras 1992).

When divergence between market prices and the true WTP occurs, existing market prices are inadequate measures of economic value and are therefore adjusted to reflect true scarcity in the economy. These adjusted prices are called *shadow prices*. These adjustments are frequently based on observed market prices but increase or decrease the price of the good or service to reflect true scarcity value. For example, the $30 OC from the fuelwood-cutter example might be used as a shadow price for labor instead of the government-set minimum daily wage of $40.

A second instance where shadow pricing is required is when the goods and services do not have observable market prices. Many environmental services are of this type. In this case, the economic analyst attempts to derive shadow prices that reflect society's WTP for the good.

Developing Shadow Prices

A general classification of approaches to valuing inputs and outputs or deriving measures of WTP or shadow prices is shown in Figure 19.8. Most watershed management benefits and costs can be handled with one of the three approaches indicated.

Using Market Prices. In the case where a market price is considered to adequately reflect WTP for a good or service, the market price itself can be used in the economic analysis as a reflection of economic value at the margin (Gregersen et al. 1995, Gittinger 1982, Hufschmidt et al. 1983, Dixon and Hufschmidt 1986). A guideline could be to use the market price for a good or service unless there is a reason to believe that it is significantly distorted. There are many reasons for this suggestion:

- Market prices are often accepted more readily by decision makers than are artificial values derived by the analyst.
- Market prices are generally easy to observe, both at a single point in time and over time.
- Market prices reflect the decisions of many buyers, not just the judgment of one analyst or administrator, which is the case with subsidized prices.
- The procedures for calculating shadow prices are imperfect, and estimates can therefore (in certain cases) introduce larger discrepancies than the simple use of even imperfect market prices.

Using Surrogate Market Prices. In the case of benefits and costs not themselves valued in the market but for which clear substitutes exist in the market, one can use the market prices of the substitutes to develop surrogate or proxy values for the benefits or costs being valued. For example, as mentioned earlier, there is no market for

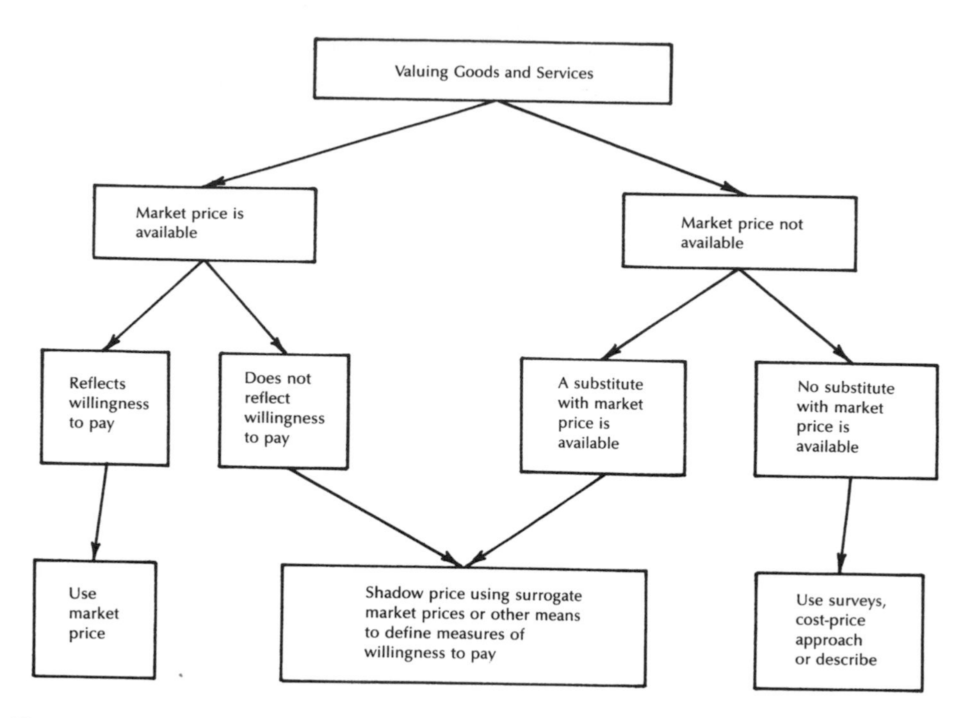

FIGURE 19.8. Valuation conditions and approaches (from Gregersen et al. 1987, © FAO, by permission).

the soil eroded from uplands and the sediment deposited on lowlands. However, values can be placed on these effects by several different means. One approach examines the market prices of eroded uplands or silted lowlands and then compares them to market prices of other, comparable land unaffected by erosion. In this with-and-without analysis, the difference in land values acts as a surrogate or proxy market price for the damage caused by erosion. In another approach, the market value of crop losses due to erosion is used to value the cost of erosion, and the value of changes in production on the fields on which sediment is deposited is used to measure the value of the soil moved through the erosion process.

Surrogate market approaches can also be used to develop appropriate shadow prices when market prices are felt to be distorted. A common shadow-pricing problem is that of labor for a development project in an area of high unemployment. If the project generates new employment and if there is a government-mandated minimum wage in the economy, then that minimum wage is not likely to adequately reflect opportunity cost for the use of the previously unemployed labor. The minimum wage will be higher than the true opportunity cost of employing additional workers. In such cases, one can make an estimate of what the unemployed were producing—such as home repairs, growing food for the family, taking odd jobs as they arose—while out of a full-time job. The value of this production is then taken as the shadow price or measure of economic value for the labor, as in the woodcutter example above (see Winpenny 1991, Dixon et al. 1994).

Using Hypothetical Valuation Approaches. Where there are no possibilities to derive acceptable market price measures of value, one might derive some value information through surveys or expert judgment. Further, it is possible to derive minimum values for some benefits through analysis of the cost of producing or deriving them. This approach is commonly called a *cost-price analysis*, since it uses costs to derive some information by which to estimate the minimum value of benefits that would be required to break even with costs. It should be borne in mind, however, that such values represent only a lower boundary on the value of the good or service in question. In other words, the resulting value is the minimum value the decision maker would have to place on the output to ensure a break-even relationship between economic costs and benefits.

Because neither surrogate nor hypothetical market prices rely on actual market prices, the results have to be interpreted carefully. These approaches should be used in a project analysis only when market-based approaches are impractical or impossible to apply. Of course, a hypothetical valuation approach is the ultimate basis for many political decisions. The decision maker makes the judgment that the value of certain watershed management outputs is greater or less than the resources needed to produce them. This type of decision is made every day without any quantitative analysis of monetary values. Frequently, this form of valuation is implicit and is never described explicitly.

The above three approaches provide a means of generating at least some information about the monetary values of the benefits from watershed management projects. Gregersen et al. (1987) discuss specific valuation methods and techniques appropriate for each category of watershed management benefits and costs. Other references of interest include Dixon et al. 1994.

Comparing Costs and Benefits: Measures of Project Worth

Once the physical-flow tables and the unit-value assumptions are formulated, it is then generally a straightforward task to bring the two together to develop two basic tables used in economic and financial analyses: the *economic value–flow table* for economic analyses and the *cash-flow table* for financial analyses. The economic value–flow table shows the flows of economic benefits and costs (e.g., shadow-priced market and nonmarket benefits and costs) over the life of the project. The cash-flow table shows the flows of actual money expenditures and receipts over the life of the project, generally by 1 yr intervals.

The reason for developing the economic value–flow and cash-flow tables is to organize information so it can be used to evaluate and compare project alternatives. At least two evaluation questions are always of interest to decision makers: Is the proposed project worth doing? If so, is the project better than other alternative uses of available scarce resources?

Questions of Interest

Six questions concerning costs and benefits are relevant in an appraisal of a watershed management project, five of which are related to the economic and financial assessment processes shown in Figure 19.7. The sixth question relates to economic stability and growth considerations. These questions, which should be asked for each project alternative being evaluated, are:

1. What is the budget impact likely to be for the agencies and private entities involved?
2. Will the project be attractive to all the private entities (including upstream landowners) who will have to put resources into the project to make it work?
3. What are the income distribution impacts of the watershed management activities proposed for the project?
4. Will the project have balance-of-payment impacts?
5. Are economic benefits greater than costs; that is, is the project an economically efficient use of resources?
6. Will the project increase economic stability of the affected region?

Budget Implications for the Public Sector

A public treasury normally requires budget analyses using market prices. It will require an analysis of a proposed watershed management project in terms of what is taken into the treasury and what goes out as public expenditure. Since most watershed management projects mainly involve costs with few monetary returns from a government's perspective, a budget analysis of watershed management will usually show a negative cash flow. In some cases, mitigation or preventive costs avoided because of the project (such as channel or canal dredging, water purification, or reconstruction of flood-damaged infrastructure) will be counted as benefits.

A budget analysis using market prices is useful for indicating the financial resources required from the treasury to carry out the proposed project. Since it is not an economic analysis, it does not provide enough information on the benefit or value of a watershed management project for a decision maker to make an informed decision about the merit of the project. This decision requires an economic-efficiency analysis (described below).

Financial Implications for Private Entities

A market-based, *discounted* cash-flow analysis provides enough information to private entities (e.g., a farmer or a firm) to enable them to make decisions. Since private entities are concerned with their actual costs and returns, they should use market-based prices. Since private entities are also concerned with when costs and benefits occur, they should use discounting in their cash-flow analyses so they can better compare alternatives in present-value or comparable terms.

The results of the financial analyses are crucial for three main reasons: (1) a planner or analyst has to know whether the project or activities within the project being proposed will be attractive to those private entities that will have to put resources into the project to make it work; (2) the information from the cash-flow analysis can be used to develop information needed to design appropriate incentive or tax packages to help ensure project implementation; and (3) the same information is ultimately needed for budgeting purposes.

In developing appropriate incentive packages for a watershed management project, institutional issues also need to be considered, including those related to land tenure and the distribution of costs among regions. In general, the upstream land users incur the costs of watershed management while the downstream population benefits; as a consequence, some transfer mechanism may be needed. The absorptive capacity of government and local institutions to handle increases in investment also needs to be considered.

Income Redistribution Implications

Most watershed management projects have income redistribution effects: hiring labor, buying inputs and land, and paying administrative expenses. In addition, if the project has important productivity effects, these effects are also translated into income when goods or services are bought, sold, or consumed.

Decision makers are frequently concerned with the income redistribution effects of a watershed management project, especially if the people living in the project area are politically sensitive or poor. The analyst can estimate the amount of new money going into a region because of the project; it is also possible to determine the likely primary recipients of these new resources. Note that this analysis examines the benefits and costs of the project in market-price terms. These prices include subsidies (a return) and taxes (a cost). As an example, laborers get wages in money, not the shadow price of labor that might be used in the economic analysis. Since laborers also spend money, not shadow money, using market prices in analyzing income redistribution implications is appropriate.

As such, market-based analyses describe the actual flows of money and market-traded goods and services in the economy and in the project area. However, the results of these analyses do not necessarily indicate whether a watershed management project is economically efficient, equitable, or desirable, nor do the results say anything about the external effects of the actions of decision makers. That is the appropriate role for economic analyses and political considerations.

Balance-of-Payments Effects

A single watershed management project would not normally be expected to have major balance-of-payments (BOP) effects. Other than imported inputs and their valuation, the most likely BOP effects of a single project will be expressed through associated production activities. Taken together, however, a set of watershed management projects contained in a watershed management program can have major BOP effects. For example, increased exports of agricultural, forestry, and fish products resulting from increased site productivity due to improved watershed management can affect the BOP either through exports or substitution of domestic production for imports.

The two main concerns in valuing BOP effects are (1) a correct determination of the actual effect of the project, using a with-and-without analysis in an OC framework, and (2) the use of the correct shadow price for foreign exchange. For example, because of improved watershed management, land use might shift from production of cotton to rice. Both commodities might presently be exported. The BOP effect, therefore, is the value of increased rice exports less the value of decreased cotton exports, both valued using the correct shadow price for foreign exchange.

Economic-Efficiency Analysis

Some questions of interest to decision makers take on a social context that requires going beyond the consideration of market prices and market-traded goods and services. These questions relate to the basic economic efficiency of alternatives, for example, the relationship between the total costs to society and the total benefits it receives from investment in a given project or set of activities. They also relate to economic stability.

Once the two value-flow tables have been set up, as discussed earlier, the streams of benefits and costs can be evaluated to compare alternative watershed management projects. An economic-efficiency analysis is a systematic way to do this comparison. In practice, there are three principal value measures used in an economic-efficiency analysis: (1) *net present worth*, (2) *economic rate of return*, and (3) *benefit-cost ratio*. All three measures are calculated using the same benefit and cost data and assumptions.

Net Present Worth. Net present worth (NPW), also known as net present value, is based on the desire to determine the present value of net benefits from a project. If the goal of the analysis is to determine the total *net contribution* (net benefits) of a project to society, the use of the NPW criterion will provide a systematic ranking of alternatives. The formula for the NPW calculation is:

$$\text{NPW} = \sum_{t=1}^{n} \left[\frac{B_t - C_t}{(1 + r)^t} \right] \qquad \textbf{(19.3)}$$

where B_t and C_t = benefit or cost in year t; and r = social discount rate.

Economic Rate of Return. The economic rate of return (ERR) is frequently used to evaluate projects. Unlike the NPW or a benefit-cost ratio, the ERR does not use a predetermined discount rate. Rather, the ERR is the discount rate that sets the present value of benefits equal to the present value of costs. That is, the ERR is the discount rate, r, such that:

$$\sum_{t=1}^{n} \left[\frac{B_t}{(1 + r)^t} \right] = \sum_{t=1}^{n} \left[\frac{C_t}{(1 + r)^t} \right] \qquad \textbf{(19.4)}$$

or

$$\sum_{t=1}^{n} \left[\frac{B_t - C_t}{(1 + r)^t} \right] = 0 \qquad \textbf{(19.5)}$$

Although a discount rate is determined as a result of the calculation of the ERR, not prescribed, this does not eliminate the use of a reference discount rate. The calculated ERR is compared with some reference discount rate to decide whether the project is economically efficient. For example, if the ERR calculated is 15% and the OC of project funds is 10%, a watershed management project would be economically attractive. Conversely, if project funds cost 18%, the project would be financially unattractive.

Benefit-Cost Ratio. A benefit-cost (*B/C*) ratio simply compares the present value of benefits to the present value of costs:

$$B/C \text{ ratio} = \frac{\sum_{t=1}^{n} \left[\frac{B_t}{(1 + r)^t} \right]}{\sum_{t=1}^{n} \left[\frac{C_t}{(1 + r)^t} \right]} \qquad \textbf{(19.6)}$$

If the *B/C* ratio is greater than 1, the present value of benefits is greater than the present value of costs and the project is an economically efficient use of resources, assuming there is no lower-cost means for achieving the same benefits.

Relationships between Measures of Product Worth. The above three measures of project worth are related; this is not surprising, since they use the same data in their calculations. These relationships are shown as follows (Gittinger 1982):

$$\text{NPW} = (\text{present value of benefits}) - (\text{present value of costs})$$

The ERR is equal to the discount rate that results in:

(present value of benefits) = (present valu of costs)

$$B/C \text{ ratio} = \frac{\text{present value of benefits}}{\text{present value of costs}}$$

This symmetry naturally extends to the results of the calculations. The values of these three measures have the following relationships:

If NPW > 0, then ERR > r and $B/C > 1$
If NPW < 0, then ERR < r and $B/C < 1$
If NPW = 0, then ERR = r and $B/C = 1$

Even though all three measures use the same data and assumptions and have a symmetry in their results, it is possible that when a set of alternative projects is examined, the use of different evaluation criteria will give different project rankings. This raises the question of which criterion (NPW, ERR, or *B/C* ratio) to use.

Deciding Which Criterion to Use. Maximum total NPW is the economic objective (the objective function) we seek for investment of available scarce resources. Therefore, the NPW measure must always be a part of any choice criterion (or ranking scheme) for accepting or rejecting watershed management practices, projects, or programs (Dixon and Hufschmidt 1986).

The ERR and *B/C* ratio are measures of benefits per unit of cost. Thus, they give no indication of the total magnitude of the net benefits or NPW. Since this is what we want to maximize for a given investment budget, reliance on just the ERR or *B/C* ratio could lead to selection of projects that provide total net benefits that are smaller than those resulting when projects are selected using the NPW criterion.

In cases where projects are not mutually exclusive and there are no constraints on costs, all projects that result in a positive NPW can be accepted. In cases where not all projects can be selected because of a cost constraint, the goal is to select that set of projects or activities that yields the greatest total NPW.

Gittinger (1982) made a comparative analysis of these three measures of present value and presented the results in tabular form. In Table 19.6, an adaptation of the Gittinger table, we distinguish between project selection or ranking under three conditions: (1) independent projects with no constraints on costs, (2) independent projects with an overall constraint on costs, and (3) mutually exclusive projects. For independent projects in the absence of cost constraints (a highly unrealistic case), each of the three measures can be used to select or reject projects, because each distinguishes between efficient and inefficient use of resources.

When there is a constraint on the costs for independent projects such that not all economically justifiable projects can be selected, only the *B/C* measure can give correct rankings for project selection. Assume there are five (A–E) independent projects available, and assume a cost constraint of $300,000 (Table 19.7). Ranking by NPW, one would choose project A, which has the highest NPW, and just use up the available budgeted cost. However, ranking based on *B/C*, one would choose B and C, then E, and then D. Note that these four projects add up to a total cost of $300,000, which is the budget constraint. However, the four projects have a total NPW of $21,000, as opposed to $10,000 for project A. The reason NPW does not work for ranking projects when costs

TABLE 19.6. Comparison of the three measures of present value

	NPW	ERR	*B/C* Ratio
Selection or Ranking Rule for			
Independent projects			
No constraint on costs	Select all projects with NPW > 0; project ranking not required.	Select all projects with ERR greater than cutoff rate of return; project ranking not required.	Select all projects with *B/C* ratio > 1; project ranking not required.
Constraint on costs	Not suitable for ranking projects.	Ranking all projects by ERR may give incorrect solution.	Ranking all projects by *B/C* ratio where C is defined as constrained cost, will always give correct ranking.
Mutually exclusive projects (within a given budget)	Select alternative with largest NPW.	Selection of alternative with highest ERR may give incorrect result.	Selection of alternative with highest *B/C* ratio may give incorrect result.
Discount Rate	Appropriate discount rate must be adopted.	No discount rate required, but reference rate of return must be adopted.	Appropriate discount rate must be adopted.

Source: Adapted from Gittinger 1982 and Dixon and Hufschmidt 1986.

TABLE 19.7. Comparing use of NPW and B/C for ranking projects

	Cost	Benefit	NPW	B/C
A	300,000	310,000	10,000	1.03
B	100,000	108,000	8,000	1.08
C	100,000	108,000	8,000	1.08
D	50,000	52,000	2,000	1.04
E	50,000	53,000	3,000	1.06

are constrained is that NPW says nothing about returns per unit of scarce factor, in this case cost or budget. For the *B/C* ranking to be correct, it must be formulated so that the costs constrained appear in the denominator of the ratio.

For mutually exclusive projects (such as two or more projects that would use the same site), the NPW measure is the only one that will always lead to the correct selection. Again, assume that alternatives A–E (Table 19.7) are mutually exclusive alternatives and are the only ones available for a given area; the budget constraint is still $300,000, which has to be spent in the area. It is evident in this case that choice A maximizes NPW on the area. If the budget were reduced to $100,000, then either B or C would be the logical choice; A is no longer an alternative, given the cost constraint.

Suppose that all the NPWs in the table were calculated at a discount rate of 10%, and assume that you could borrow funds up to $300,000 at 8%. Project A would then be chosen for this particular area. In all of these cases, the correct selection or ranking is defined as the one that yields the largest NPW when one uses up the constrained factor, in this case budget, land, and so forth.

Discount Rate for Public-Sector Evaluations. Whereas in financial analysis an interest rate that reflects market rates for investment and working capital is usually used and is therefore sensitive to inflation rates, the discount rate used by governments in an economic analysis is usually not readily observable in the economy. Economists have developed many approaches for determining and justifying a discount rate for economic analysis. These include the OC of capital and the social rate of time preference (Baumol 1968, Gittinger 1982, Hufschmidt et al. 1983, Dixon and Hufschmidt 1986).

The actual discount rate used in an economic analysis will likely be country specific, and it should be established as a matter of government policy. Important factors governing the choice of rate will be the OC of capital (or the cost of money to the government) and the government's current view of the consumption and investment mix of the private sector in relation to its concerns for future generations. However, whatever rate is selected, it should be used to evaluate all projects. A comparison of projects evaluated with different discount rates is logically untenable.

Regional Impacts of Watershed Management Projects

One of the questions of relevance to decision makers concerns the effect of watershed management on regional economic stability. This concern, while real enough, does not fit directly into a benefit-cost analysis. It is best answered separately and then presented along with the economic-efficiency analysis.

Concerns about the regional economic impacts of projects are related to the income distribution effects described earlier. Whenever there is a change because of a project (new jobs, additional agricultural crop areas, increased forest production), the analyst must determine the *incremental* net benefit of the project, not the gross benefit. For example, a hillside stabilization program might create new jobs. The appropriate measure of economic impact is not the total wages paid to laborers, but the total *minus* the amount they would have earned if the new jobs had not been created (their OC). Similar care must be taken with the use of land or capital resources.

The types of potential regional impacts associated with watershed management projects can include:

- Job creation in construction and maintenance of management structures and facilities.
- Job creation through new or expanded production in agriculture, forestry, fisheries, or transportation.

- Increased productivity of existing cultivated agricultural areas, forests or woodlands, and fishery resources.
- New production on previously unused fields or water resources.
- Forestry- or fuelwood-related job creation and increased production.
- Secondary impacts of the previously listed changes, which have to be handled conservatively and carefully to avoid double counting (see discussion above).

Although the actual elements in a regional-impact analysis vary from case to case, such an analysis will be useful to decision makers when alternative watershed management projects are considered.

On a larger scale, a regional analysis should address the question of economic stability and likely changes because of the project. One would expect that successful watershed management would help to stabilize production (agricultural, forestry, fishery), and therefore incomes, in the region. There can also be project-induced migration within the region and between regions. These effects are addressed both quantitatively and qualitatively. Schuster (1980) discusses techniques for determining magnitudes of regional distributional impacts in the case of forestry-related projects.

Assessment of Nonmonetary Benefits and Costs

In spite of all the advances made in economic valuation of nonmarketed goods and services, there are always some effects of watershed management projects that are impossible to either quantify or value. For example, the construction of a capital infrastructure, such as a dam, a flume, or a power transmission line, can have a negative aesthetic impact on an upland area; that is, the view is not as natural or pleasing as it previously was. This kind of aesthetic effect is almost impossible to quantify, largely because there are no accepted units of measurement or expressions of value for scenic beauty. Similarly, a watershed management project that would require a change in the lifestyle of a traditional community would have a *cultural* impact. These impacts are also difficult to quantify.

In these cases, effects that cannot be quantified or converted into dollar amounts should still be recognized and described, kept within the analysis, included qualitatively, and presented to the decision maker. In this way, the effects will not be ignored, even though they cannot be entered directly into the economic analysis.

Summary

This chapter has dealt with watershed management policy and planning issues and what economists do to analyze the costs and benefits associated with watershed management. Such information is needed in considering the trade-offs involved, both between alternative watershed management approaches and between the best watershed approach and other productive uses of public and private funds. When you are finished with the chapter, you should understand:

- That an appropriate policy environment is a prerequisite for being able to implement watershed management practices that satisfy the objectives specified by the interested parties.
- The general stages in the policy development process.
- Why achieving successful watershed management through policy intervention has been difficult.

- Since land use activities on watersheds have interrelated impacts, both spatially and temporally, planning of such activities and their coordination is a necessary undertaking.
- What is involved in planning for watershed management, what steps need to be taken in the planning process, and how one appraises watershed management plans and projects.
- The general nature of an economic analysis, and the economic and social welfare impacts (both positive and negative) of watershed management practices, projects, and programs.
- The specific ways in which an economic analysis is applied to watershed management practices, projects, and programs.

References

Baumol, W. 1968. On the social rate of discount. *Am. Econ. Rev.* 58:788–802.

Brooks, K.N., P.F. Ffolliott, H.M. Gregersen, and K.W. Easter. 1994. *Policies for sustainable development: The role of watershed management.* Environ. Nat. Resour. Pol. Train. Proj., Pol. Brief 6. Washington, DC: U.S. Agency for International Development.

Brooks, K.N., H.M. Gregersen, P.F. Ffolliott, and K.G. Tejwani. 1992. Watershed management: A key to sustainability. In *Managing the world's forests: Looking for balance between conservation and development*, ed. N.P. Sharma, 455–487. Dubuque, IA: Kendall/Hunt, for the World Bank.

Brundtland Commission. 1987. *Food 2000: Global policies for sustainable agriculture.* Report of the Advisory Panel on Food Security, Agriculture, Forestry, and Environment to the World Commission on Environment and Development. London: Zed Books.

CGIAR (Consultative Group on International Agricultural Research). 1996. *Priorities and strategies for soil and water aspects of natural resources management research in the CGIAR.* Document SDR/TAC:IAR/96/2.1 for the Mid-term Meeting, May 20–24, 1996, Jakarta, Indonesia. Rome: FAO of the United Nations, TAC.

Chandra, A., and K.K.S. Bhatia. 2000. Water and watershed management in India: Policy issues and priority areas for future research. In *Land stewardship in the 21st century: The contributions of watershed management*, tech. coords. P.F. Ffolliott, M.B. Baker, Jr., C. B. Edminster, M.C. Dillon, and K.L. Mora, 158–165. USDA For. Serv. Proceed. RMRS-P-13.

Cheng, J.D., H.K. Hsu, W. Jane Ho, and T.C. Chen. 2000. Watershed management for disaster mitigation and sustainable development in Taiwan. In *Land stewardship in the 21st century: The contributions of watershed management*, tech. coords. P.F. Ffolliott, M.B. Baker, Jr., C. B. Edminster, M.C. Dillon, and K.L. Mora, 138–148. USDA For. Serv. Proceed. RMRS-P-13.

Copeland, C. 2001. *Clean Water Act issues in the 107th Congress.* Congressional Research Service Issue Brief 10069 for Congress. (Updated May 23, 2001) Washington, DC: Congressional Research Service.

Cody, B.A. 2001. *Western water resource issues.* Congressional Research Service Issue Brief 10019 for Congress. (Updated May 3, 2001). Washington, DC: National Council for Science and the Environment. (http://www.cnie.org/nle/h2o-31.html)

Cody, B., and H.S. Hughes. 2001. *Water resource issues in the 107th Congress.* Congressional Research Service Issue Brief 20569 for Congress. (Updated January 16, 2001). Washington, DC: National Council for Science and the Environment (http://www.cnie.org/nle/h2o-28.html)

Dixon, J., and M. Hufschmidt, eds. 1986. *Economic valuation techniques for the environment: A case study workbook.* Baltimore: Johns Hopkins Univ. Press.

Dixon, J.A., L. Fallon Scura, R.A. Carpenter, and P.B. Sherman. 1994. *Economic analysis of environmental impacts*. London: Earthscan Publications.

Dutton, R.W., J.I. Clarke, and A.M. Battikhi. 1998. *Arid land resources and their management: Jordan's desert margin*. London: Kegan Paul International.

Easter, K.W., H.J. Cortner, K. Seasholes, and G. Woodard. 1995. *Water resources policy issues: Selecting appropriate options*. EPAT/MUCIA Project Draft Policy Brief. St. Paul: Dept. of Forest Resources, Univ. of Minnesota.

Eden, S., and M.G. Wallace. 1992. *Arizona water: Information and issues*. Issue Pap. 11. Tucson: Water Resources Research Center, Univ. of Arizona.

Gittinger, J.P. 1982. *Economic analysis of agricultural projects*. 2nd ed. Baltimore: Johns Hopkins Univ. Press.

Gleick, P.H. 1993. *Water in crisis*. New York: Oxford Univ. Press.

Gregersen, H.M., and Contreras, A. H. 1992. Economic assessments of forestry project impacts. FAO For. Pap. 106. Rome.

Gregersen, H.M., and A. Lundgren. 1986. An evaluation framework. In *Alternative approaches to forestry research evaluation and assessment*, 26. USDA For. Serv. Gen. Tech. Rep. NC-110.

Gregersen, H.M., K.N. Brooks, J.A. Dixon, and L.S. Hamilton. 1987. *Guidelines for economic appraisal of watershed management projects*. FAO Conserv. Guide 16. Rome.

Gregersen, H.M., K.N. Brooks, P. Ffolliott, A. Lundgren, B. Belcher, K. Eckmand, R. Quinn, D. Ward, T. White, S. Josiah, Z. Xu, and D. Robinson. 1994. *Assessing natural resources policies*. EPAT/MUCIA Project Draft Pol. Brief. St. Paul: Univ. of Minnesota.

Gregersen, H.M., J.E.M. Arnold, A. Lundgren, and A. Contreras H. 1995. *Valuing forests: Context, issues, and guidelines*. FAO For. Pap. 127. Rome.

Hitzhusen, F.J. 1982. *The economics of biomass for energy: Toward clarification for non-economists*. Mimeogr. Ohio State Univ., Columbus.

Hufschmidt, M.M., D.E. James, A.D. Meister, B.T. Bower, and J.A. Dixon. 1983. *Environment, natural systems, and development: An economic valuations guide*. Baltimore: Johns Hopkins Univ. Press.

Lant, C.L. 1999. Introduction: Human dimensions of watershed management. *J. Amer. Water Resour. Assoc.* 35:483–486.

Mann, D.E. 1963. *The politics of water in Arizona*. Tucson: Univ. of Arizona Press.

Organization for Economic Cooperation and Development (OECD). 1986. *The public management of forestry projects*. Paris.

Quinn, R.M., K.N. Brooks, P.F. Ffolliott, H.M. Gregersen, and A.L. Lundgren. 1995. *Reducing resource degradation: Designing policy for effective watershed management*. EPAT Working Paper 22. Washington, DC.

Schuster, E. 1980. Economic impact analysis of forestry projects: A guide to evaluation of distributional consequences. In *Economic analysis of forestry projects*, 63–132. FAO For. Pap. 17, Suppl. 2. Rome.

Winpenny, J.T. 1991. *Values for the environment: A guide to economic appraisal*. London: Her Majesty's Stationery Office.

Zinn, J.A. 2001. *Soil and water conservation issues*. Congressional Research Service Issues Brief 96030. (Revised April 18, 2001). Washington, DC: National Council for Science and the Environment. (http://www.cnie.org/nle/ag-18.html)

Appendix

Standard Conversion Factors from Metric to English Units

Quantity	Metric Unit	English Unit	To Convert Metric to English Multiply by:
Length	centimeters (cm)	inches (in.)	0.394
	millimeters (mm)	inches (in.)	0.0394
	meters (m)	feet (ft)	3.28
	meters (m)	yards (yd)	1.09
Area	square millimeters (mm^2)	square inches ($in.^2$)	0.00155
	square meters (m^2)	square feet (ft^2)	10.76
	square meters (m^2)	square yards (yd^2)	1.196
	square meters (m^2)	acres	0.000247
	hectares (ha)	acres	2.47
	square kilometers (km^2)	square miles (mi^2)	0.386
Volume	cubic centimeters (cm^3)	cubic inches ($in.^3$)	0.0610
	liters (l)	cubic feet (ft^3)	0.035315
	cubic meters (m^3)	cubic feet (ft^3)	35.3
	cubic meters (m^3)	cubic yards (yd^3)	1.31
	cubic meters (m^3)	acre-feet	0.000811
	liters (l)	pints	2.113376
	liters (l)	quarts	1.056688
	liters (l)	gallons	0.264174
Velocity	kilometers/hour (km/hr)	miles/hour (mi/hr)	0.621
	meters/second (m/sec)	feet/second (ft/sec)	3.28
Acceleration	meters/second^2 (m/sec^2)	feet/second^2 (ft/sec^2)	3.280839
Flow	cubic meters/second (m^3/sec)	cubic feet/second (ft^3/sec)	35.3
	liters/second (l/sec)	gallons/minute (gpm)	15.850322
Rates and yields	kilograms/hectare (kg/ha)	pounds/acre (lb/acre)	0.892183

(continues)

(Continued)

Quantity	Metric Unit	English Unit	To Convert Metric to English Multiply by:
	metric tons/hectare (t/ha)	short tons/acre	0.446091
	millimeters/hour (mm/hr)	inches/hour (in./hr)	0.03937
	centimeters/day (cm/day)	inches/day (in./day)	0.393701
Mass	grams (g)	ounces [avdp] (oz)	0.0353
	kilograms (kg)	pounds [avdp] (lb)	2.20
	metric tons (t)	short tons (ton)	1.10
Density	grams/cubic centimeter (g/cm^3)	pounds/cubic foot (lb/ft^3)	62.4
	kilograms/cubic meter (kg/m^3)	pounds/cubic foot (lb/ft^3)	0.0625
Force	newtons (N)	pounds force (lbf)	0.00986
Pressure or stress	kilopascals (kPa)	atmosphere (standard) (atm)	0.00987
	kilopascals (kPa)	inches of mercury @ 60°F	0.296134
	kilopascals (kPa)	millibars (mb)	10.0
	kilopascals (kPa)	feet of water @ 30.2°F	0.33456
	kilopascals (kPa)	inches of water @ 60°F	4.018655
	kilopascals (kPa)	pounds/square foot (lb/ft^2)	0.145038
	kilopascals (kPa)	pounds/square inch ($lb/in.^2$)	20.885459
Temperature	degrees Celsius (°C)	degrees Fahrenheit (°F)	°F = (1.8) (°C) + 32
Energy	joules (J)	British thermal units (mean) (Btu)	0.00095
	joules (J)	calorie (cal)	0.239
	joules (J)	watt-hours	0.00028
Power	watts	foot-pounds/second (lbf/sec)	0.73756

Source: Adapted from American Society for Testing and Materials (ASTM), 1976, *Standard for Metric Practice* (Philadelphia: ASTM).

INDEX